# Environmental Monitoring and Control

# Environmental Monitoring and Control

Edited by Kane Harlow

SYRAWOOD
PUBLISHING HOUSE
New York

Published by Syrawood Publishing House,
750 Third Avenue, 9th Floor,
New York, NY 10017, USA
www.syrawoodpublishinghouse.com

**Environmental Monitoring and Control**
Edited by Kane Harlow



International Standard Book Number: 978-1-68286-472-2 (Hardback)



**Cataloging-in-Publication Data**

Environmental monitoring and control / edited by Kane Harlow.
p. cm.
Includes bibliographical references and index.
ISBN 978-1-68286-472-2
1. Environmental monitoring. 2. Pollution--Measurement. 3. Environmental impact analysis.
4. Environmental engineering. 5. Environmental protection. I. Harlow, Kane.
QH541.15.M64 E58 2017
628--dc23

Printed in China.

# TABLE OF CONTENTS

# PREFACE

This book has been an outcome of determined endeavour from a group of educationists in the field. The primary objective was to involve a broad spectrum of professionals from diverse cultural background involved in the field for developing new researches. The book not only targets students but also scholars pursuing higher research for further enhancement of the theoretical and practical applications of the subject.

Environmental Monitoring evaluates the quality of the environment along with the impact of human activities on it. The increasing environmental degradation has fueled the research in this field. It includes the monitoring of air, water, soil, etc. This book attempts to understand the multiple branches that fall under the discipline of environmental monitoring and how such concepts have practical applications. It elucidates the concepts and innovative models around prospective developments with respect to this field. This book will serve as a reference to a broad spectrum of readers including environmentalists, ecologists, researchers, professionals and students associated with this field at various levels.

It was an honour to edit such a profound book and also a challenging task to compile and examine all the relevant data for accuracy and originality. I wish to acknowledge the efforts of the contributors for submitting such brilliant and diverse chapters in the field and for endlessly working for the completion of the book. Last, but not the least; I thank my family for being a constant source of support in all my research endeavours.

**Editor**

# Removal of water hardness causing constituents using alkali modified sugarcane bagasse and coffee husk at Jigjiga city, Ethiopia

**Adhena Ayaliew Werkneh[1, *], Angaw Kelemework Abay[1], Anbisa Muleta Senbeta[2]**

[1]Department of Chemistry, College of Natural Science, Jigjiga University, PO. Box: 1020, Jigjiga, Ethiopia
[2]Department of Food Science and Nutrition, College of Dryland Agriculture, Jigjiga University, PO. Box: 1020, Jigjiga, Ethiopia

**Email address:**
adhena1988@gmail.com (A. A. Werkneh)

**Abstract:** Alkaline modified sugarcane bagasse and coffee husk were used for the adsorption of water hardness causing constituents ($Ca^{+2}$ and $Mg^{+2}$). The water hardness sample was collected using polyethylene bottle from Jigjiga city drinking water supply, Ethiopia. The adsorbents were characterized using FTIR and BET surface area techniques. The concentration of the constituents were determined using AAS Spectroscopy. It was found that, using the ABC and ACHC as an adsorbent, the maximum sorption capacity obtained for Ca and Mg hardness adsorption are 46.8 and 37.35, and 52.9 and 41.23 mg $g^{-1}$ for ACHC and ABC respectively. Activated carbon filtration also depends on various parameters such as pH, contact time, adsorbent dose, temperature and initial Ca and Mg ion concentrations. The maximum recovery of the adsorbed calcium and magnesium was achieved in less than 200 minutes leading to 78% and 73% respectively. After treating synthetic water solution simulating an actual water stream with the alkali-modified bagasse and coffee husk, total hardness of the treated sample meets the required standard for drinking water, below 60 mg/L of $CaCO_3$. Therefore, ABC is more suitable for the removal of hardness ions than ACHC from drinking water; and are considered as effective low cost adsorbents.

**Keywords:** Water Hardness, Activated Carbon, Bagasse, Coffee Husk

## 1. Introduction

Quality of water is one of the most important natural resources of the world. It plays a vital role in the development of communities; hence a reliable supply of water is essential. It needs to be maintained all the time for human and industrial use. As for human consumption, quantity and quality of drinking water have been recognized as increasingly critical issues. Addressing the deterioration of water quality in developing countries, where an estimated one billion people lack access to potable quality water, is a primary motivating factor for many community development efforts and is a key component of the Millennium Development Goals [1]. The provision of safe water to the people is an urgent development priority of any country in the world [1], [2].

Most of the water resources should be treated for purification before consumption. In some countries, groundwater is the main safe drinking water resource [1]. In some cases, the resource does not satisfy to the desirable levels regarding their chemical properties, such as hardness, nitrate contamination, heavy metals, soluble iron, etc. [2]. Among them, water hardness can appear problematic in some cases; it can also be considered as an important aesthetic parameter. However, because public acceptance of hardness differs remarkably according to local conditions, a maximum acceptable level has not been defined. In general, water supplies with total hardness higher than 200 mg/L can be tolerated by consumers but are considered as poor resources; while values higher than 500 mg/L are not acceptable for most of the domestic consumptions [3], [4].

Recently, various methods including electro deionization process, electro membrane processes, capacitive deionization, membrane and fluidized pellet reactor, ion exchange process and adsorption have been studied for the removal of a wide variety of ionic and molecular species from various water streams, including those responsible for hardness (e.g. $Ca^{+2}$ and $Mg^{+2}$ cations). Amongst the developed processes, adsorption has been widely studied for the uptake of various ionic and molecular species from water [6]-[8].

In an effort to reduce the proportion of people without sustainable access to safe drinking water, there is a need to optimize the production of activated carbon from locally available wastes and apply it for water treatment in diverse communities. This has led to the development of alternative low-cost technologies such as activated sugar cane bagasse for the treatment of drinking water in the developing world. Research has also been focused on the indigenous production of water treatment chemicals using locally available raw materials [1]. Carbon adsorption offers significant advantages including low cost, availability, profitability, ease of operation and efficiency in comparison with conventional methods especially from economically and environmentally points of view [4],[5].

In countries with poor economic base, the high cost of importing the water treatment chemicals prevents consistently good drinking water quality being achieved in many cases. The activated carbon has been widely used worldwide as an effective filtration or adsorption material for removing chemical contaminants from drinking water. In most developing countries, the activated carbon is imported at high cost, limiting the quantities of safe drinking water available to the people. The high cost of importing the activated carbon puts a significant burden on the water treatment budget since foreign currency is scarce [6],[7]. The main aim of the study was to apply activated carbon prepared from coffee husk and bagasse wastes in drinking water treatment and assess the efficiency of the carbons in the purification of hardness of water as a function of operating parameters [3].

# 2. Materials and Methods

## *2.1. Description of the Study Area*

Jigjiga is a city in eastern Ethiopia and the capital of the Somali Region of the country. The city is located in the Jigjiga Zone approximately 80 km (50 mi) east of Harar and 60 km (37 mi) west of the border with Somalia. The city has an elevation of 1,609 meters above sea level and are found with coordinates of 9°21′N42°48′E. The climate of Jigjiga is a subtropical highland climate (Köppen climate classification), with the influence of mountain climate, with hot and dry summers and cold winters. The temperature range of the city was between 25 and 29 °c. As of 2008, Jigjiga has about 34.1% of the total population has access to drinking water from underground water

***Figure 1.1.*** *Map of the study area.*

***Figure 1.2.*** *Shows (A) Raw sugarcane bagasse, (B) Alkali activated carbon, (C) Raw Coffee Husk.*

### *2.2. Chemicals and Solutions*

Apparatus and instruments: pH meter (MP 220, METTLER TOLEDO), FTIR spectrometer, AAS spectrophotometer (BUCK SCIENTIFIC MODEL VGP210, USA), Rotary Shaker (VRN - 480, GEMMY Orbit Shaker, Taiwan), Balance (OHAUS, E11140, Switzerland), Desiccators, Electrical mill (IKA-WERKE, M20 GMBH & CO.KG, GERMANY), Filter Paper (Whatman 542, 90 mm diameter), Sieve no of different size (IMPAOT, UK), Deionizer, hot air Oven (OV150CGENL ABWIDNES, England).

#### *2.2.1. Reagents and Chemicals*

Analytical grade $CaCl_2$ and $MgSO_4$. $7H_2O$ purchased from Avishkar LAB TECH CHEMICALS, LOT which were used as a model compound to stimulate the total hard water, Sulphuric acid ($H_2SO_4$) purchased from Reagent chemical service limited Company, Runcorn Cheshire used to activate bagasse and coffee husk, Sodium hydroxide (NaOH) from Avishkar LAB TECH CHEMICALS, LOT used to adjust the pH, HCl from Reagent chemical service limited Company, Runcorn Cheshire used for titration, Sodium Chloride (NaCl) purchazed from TITAN BIOTECH LIMITED, BHIWADI which was used for titration in Sear's surface area analysis. All experiments were conducted according to the standard methods for the examination of water and wastewater [21].

#### *2.2.2. Sampling and Sample Collections*

Tap ground water samples were collected in clean 1000 ml plastic bottles. The containers were first washed with de-ionized water, and then several times with the sample water before collection in order to avoid any contamination. The samples were then carried in ice-packed coolers to the laboratory for analysis within 24 hours.

### *2.3. Collection and Preparation of Adsorbent*

Raw Sugarcane Bagasse and coffee husk were collected from Wonji sugar factory and coffee refinery S. C in Ethiopia, which are collected as a waste. The samples were soaked for 24 hours and washed with distilled water before use in order to remove any impurities. Raw sugar cane bagasse sample was boiled for 30 min to remove remain soluble sugars. They were kept in drying oven maintained at 105 °C for a period of 24 hours. The dried materials were grounded with electrical grinder to get the desired particle size of 500 $\mu$m. Then, they were treated with concentrated sodium hydroxide, 2 M NaOH for 24 hours at room temperature with the ratio of 1:1 (base to carbon ratio) then stirred for 30 min and left for overnight. The purposes of treating carbon using 2 M NaOH were to create a suitable environment for its ring opening which increase the number of adsorption sites. Finally, the treated carbons were washed with distilled water to remove excess bases and any other soluble substances before the sample were dried in Furnace and adjusted its pH to about 6.5. These materials are referred as activated Sugarcane bagasse and coffee husk. All experiments were conducted according to the standard methods for the examination of water and wastewater.

### *2.4. Batch Filtration Experiments*

All experiments were conducted in batch mode in 250 mL conical flasks. Several operating parameters including pH (2-10), temperature (10-50 °C), adsorbent mass (2-10 g/L), initial calcium and magnesium concentrations (40-120 mg/L) and contact time (40-120 minutes) were investigated. Optimized adsorption times for modified adsorbents were first examined by varying the contact time at room temperature, pH= 6.0 and for an adsorbent mass of 2 g per liter of solution. For this purpose, 2 g of adsorbent were added to 1 L of solution in a conical flask containing $Ca^{+2}$ or $Mg^{+2}$ cations at a concentration in the range of 40 to 120 mg/L. The mixture was then shaken at 200 rpm.

The removal efficiency (%R) and sorption capacity (Qe) was determined as follows (Eq. 1 and 2):

$$R(\%)=[(Ci-Ce)/Ci]\times 100 \quad (1)$$

$$Q_e\ (mg/g) = [(Ci-Ce)/m]V \quad (2)$$

Where, $C_o$ and $C_f$ are the initial and final concentration of metal in solution (mg $l^{-1}$), V is the volume of solution (l) and m is the mass of sorbent (g).

After investigation of the effect of the contact time and the initial ion concentration, the effects of pH and adsorbent mass were examined. The effect of temperature was then investigated in the optimal conditions for pH, adsorbent mass and contact time, for three initial concentrations of $Ca^{+2}$ or $Mg^{+2}$ cations (60, 120 and 180 mg/L). All the experiments had performed in duplicate and the mean values were reported.

***Analysis of the Adsorbate Solution:***

Spectrophotometric measurements were carried out using Atomic Absorption spectrophotometer using calcium and magnesium hollow-cathode lamp at respective wavelengths and an air/acetylene flame with strict adherence to standard calibration guidelines.

***Characterization of Adsorbents:*** In order to understand the mechanism of the sorption, large quantities of work was done to investigate the influence of the sorption process using different kinds of techniques. Fourier Transform Infrared (FTIR) Spectroscopy analysis was conducted at Ethiopia Pharmaceutical Factory, Addis Ababa, Ethiopia which is used to asses functional groups of the adsorbents.

***Determination of BET surface area:*** The specific surface areas of the adsorbents were determined using the Sear's method (1956). For this 1.5 g of modified adsorbents were acidified with 0.1 M HCl to pH value of 3-3.5. The volume in the beaker was made to 150 ml with distilled water after addition of 10.0 g of NaCl. Titration was then carried out with 0.1 M of NaOH to pH value of 4.0 and then to pH value of 9.0. The volume V (ml), required to raise the pH from 4.0 to 9.0 was noted and the specific surface area was computed from the following equation [20].

$$S(m^2/g) = 32V - 25 \quad (3)$$

***Determination of the zero point charge:*** The zero point charge was determined using 0.01 M solution of NaCl as an electrolyte and by adding 0.1 M solutions of HCl. For this purpose, the pH of eight beakers containing 50 mL of electrolyte was set to the desirable values in the range of 2 to 12. Then 2 gram of adsorbent was added into each beaker and shaken for 48 hours. After completion of the reaction, the adsorbent was filtered and the final pH of each beaker was measured. By plotting the initial pH versus the pH after 48 hours of agitation, the zero point charges of the adsorbents were determined, which were found to be 6.5 for modified bagasse and coffee husk.

### *2.5. Regeneration of the Spent Adsorbents*

Regeneration tests for saturated modified adsorbents were carried out by adding 2 g/L of spent adsorbent in 2 M solution of NaOH. For adsorbent saturation, 2 g/L of adsorbent were let in contact with 250 mL solution containing 100 mg/L calcium or magnesium and stirred at 200 rpm until equilibrium time was reached (120 minutes). The spent of the adsorbents were filtered, washed and dried at 55 °C for 24 h. The dried spent adsorbents were let in contact of 2 N NaOH for 2 hours; then filtered, washed several times with deionized water and dried at 55 $^0$C for 24 h. The regenerated adsorbents were then tested for the adsorption of calcium and magnesium and the regeneration percentage were calculated based on the comparison of the removal efficiencies of fresh and regenerated adsorbents.

## 3. Result and Discussion

### *3.1. Adsorbent Characteristics*

The various physical and chemical characteristics of the AC for both coffee husk and bagasse are represented in table 3.1.

***Table 3.1.*** *Physico-Chemical properties.*

| Parameters | ABC | ACHC |
|---|---|---|
| pHzpc | 7.58 | 103 |
| BET ($m^2/g$) | 546.6 | 410 |
| Mesh size | 500$\mu$m | 500$\mu$m |

The modified adsorbents were characterized by means of instrumental techniques called Transform Infrared spectroscopy (FTIR) and BET surface area.

The specific surface area of the adsorbents was measured using the BET technique. It showed a significantly higher specific surface area for the ABC and CHC, 546.60 and 410 $m^2/g$ respectively. The remarkable improvement of the surface area can most likely be attributed to the removal of components occupying the pores of the AC resulting in more accessible pores and consequently larger surface area.

#### *3.1.1. Infrared Spectral Analysis*

The adsorption of Ca and Mg ions on bagasse and coffee husk were also affected by other interactions between functional groups of Ca and Mg and bagasse, coffee husk in addition to electrostatic interaction. FTIR analyses were conducted in order to identify possible locations for these interactions. The result is presented in figure 3.1 (a and b). As it can be seen, the FTIR spectra of Ca and Mg ions, bagasse and coffee husk and after adsorption are discussed below.

#### *3.1.2. FTIR Spectra of Bagasse*

FTIR spectra of bagasse has a broad band centered between 3154.63 $cm^{-1}$ and 3334.98 $cm^{-1}$ ( hydrogen bonded OH), the band at 2853.73 to 2954.03 $cm^{-1}$( $-CH_2$ and $-CH_3$ asymmetric and symmetric stretching), the peak at 1774.54 $cm^{-1}$ (associated with C-O carbonyl), the peaks at 1590.34 $cm^{-1}$ and 1462.07 $cm^{-1}$ (associated with the aromatic ring of lignin) and the large peak at 1022 to 1249.89 $cm^{-1}$ (associated with the C-O bond bending of cellulose). FTIR spectra of SBC show peaks at 3568.37 $cm^{-1}$. This could be due to (N-H stretching), 1590.34 $cm^{-1}$(N=N stretching), 1462.07 $cm^{-1}$ (aromatic C-C stretching), and 1377.20 $cm^{-1}$ (S-O bending).

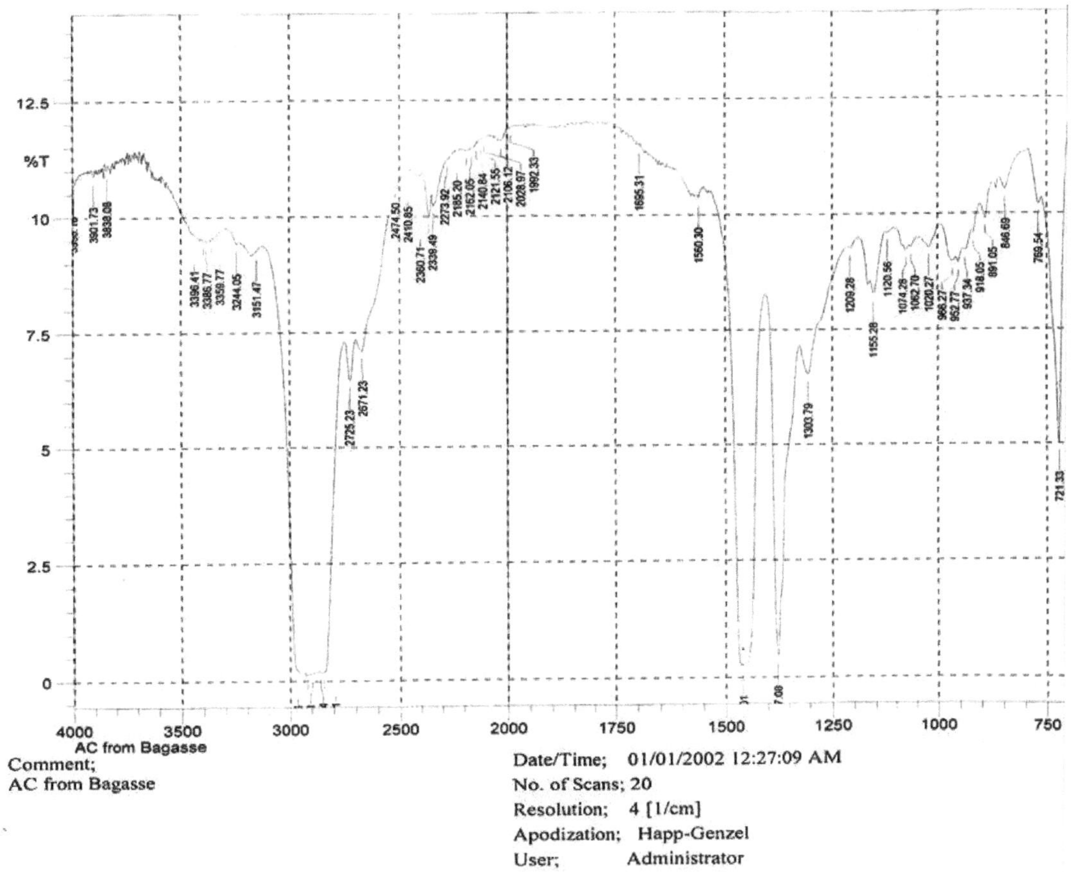


***Figure 3.1. (a)*** *FTIR spectral analysis of AC from Sugar cane bagasse.*

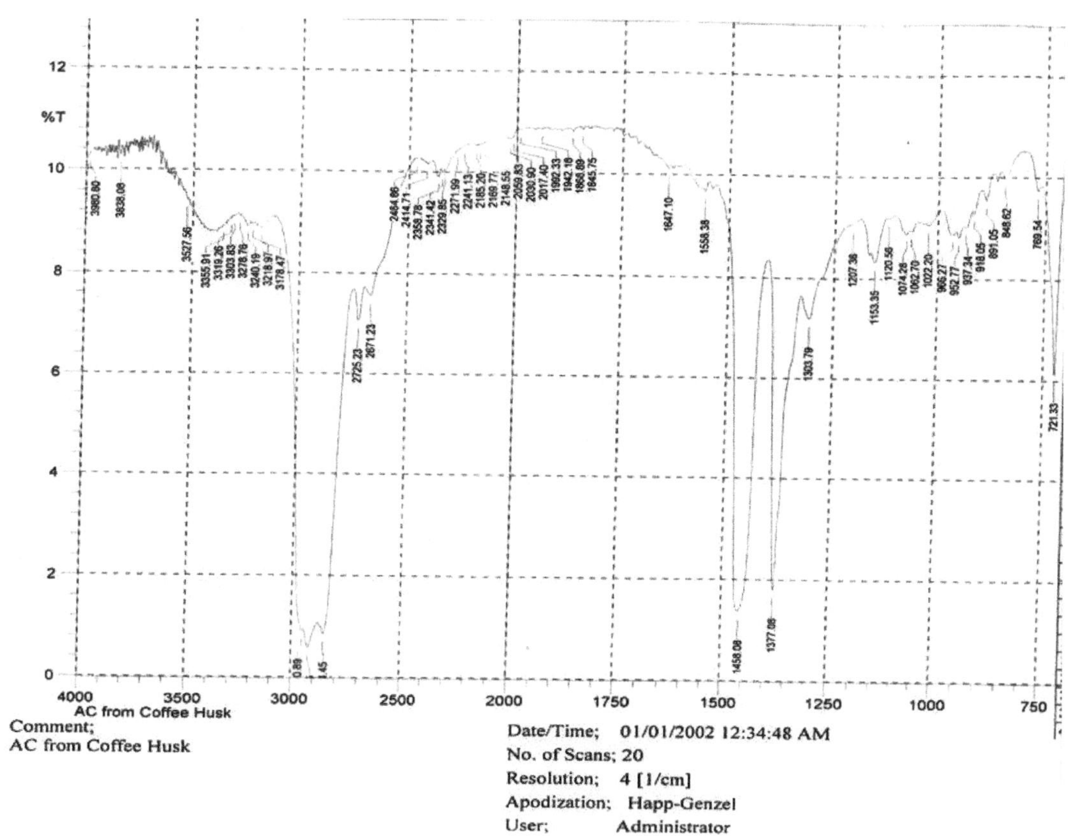


***Figure 3.1. (b)*** *FTIR spectral analysis of AC from coffee husk.*

### 3.1.3. FTIR Spectral Analysis of Coffee Husk

The FTIR spectral analysis shows that several functional groups are available on the surface of coffee husk for binding hardness causing agents hardness causing divalent ions. The result was shown in fig 3.2 (b), Wavelengths corresponding to their respective functional groups. Therefore, the wavelengths appeared in the graph i.e. 3431, 2925, 1726, 1652 and 1450 indicates the applicability of -OH, -CH, C=O, C=C, -COO respectively. The FTIR spectra obtained for the activated coffee husk samples are shown in Fig. 1. The broad band at about 3400 $cm^{-1}$ included many vibration modes mainly attributed to -OH groups with a minor contribution of -NH functional groups. The presence of methyl and methylene groups is confirmed by the two sharp peaks at 2925 $cm^{-1}$ and 2855 $cm^{-1}$ attributed to asymmetric and symmetric stretching of C- H bonds in aliphatic chains.

The peaks located at 1737 and 1633 $cm^{-1}$ are characteristics of carbonyl group stretching from cellulose and ketones. The presence of -OH group, along with carbonyl group, confirms the presence of carboxylic acid groups in the biosorbent. The peaks at 1508 $cm^{-1}$ are associated with the stretching in aromatic rings. The peaks observed at 1071 and 1024 $cm^{-1}$ are due to C-H and C-O bonds. The -OH, -NH, carbonyl and carboxylic groups are important sorption sites.

### 3.2. pH Point of Zero Charge

The point of zero charge of the adsorbent ABC and ACHC were assessed from the graph of final pH versus initial pH for 0.5 g of the adsorbents. The results are presented in figure 3.2 (a and b). As shown from the graphs, the values of pHpzc of an adsorbent are determined from the points where the initial pH equals the final pH. The pHpzc values are 7.50 and 8.103 for ABC and ACHC respectively. As presented in the graphs it seems that, the adsorbents were negatively charged at pH greater than the pHpzc and below pHpzc there was a charge reversal.

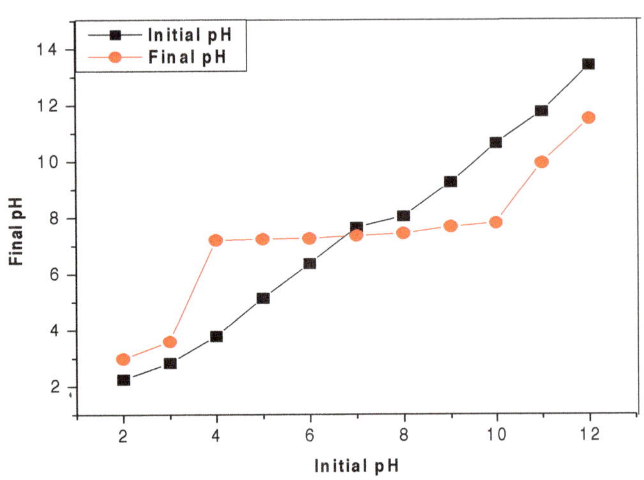


*Figure 3.2. (a) pH point zero charge (pHpzc) for ABC.*

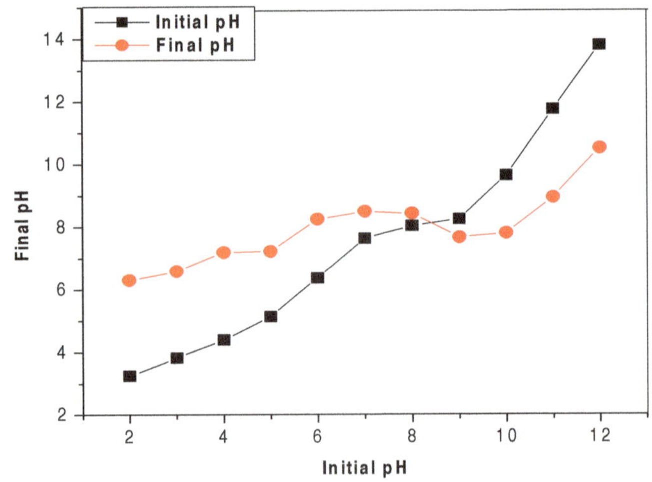


*Figure 3.2. (b) pH point zero charge (pHpzc) for ACHC.*

It has been reported by earlier researchers that, the pHpzc of an adsorbent increases with increase in basic groups on the surface of the adsorbents [19]. This is due to that, the adsorbents have basic surfaces since the pHpzc values is greater than 7.

From the results, it can be concluded that alkali modification of the adsorbent gave a negative (basic) surface charge for the adsorbent. The relationship between pHpzc and adsorption capacity is that cations adsorption on any adsorbent will be expected to increase at pH value higher than the pHpzc while anions adsorption will be favorable at pH values lower than the pHpzc [21].

### 3.3. Batch Activated Carbon Filtration

In the present study, alkali treated bagasse and coffee husk are used as an adsorbent for water hardness causing constituents from aqueous solutions and polluted water. Based on the results obtained, the effects of these parameters are discussed in the subsequent sections.

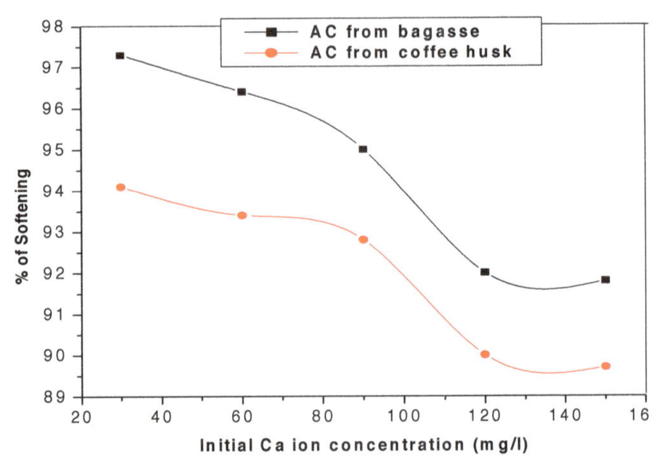


*Figure 3.3. Effect of initial Ca hardness concentrations on softening efficiency (2 g adsorbent, pH 6.5, room temperature, agitation speed 200 rpm.*

***Effect of the contact time and initial hardness agent concentration:*** The effects of the initial calcium and magnesium concentrations and the contact time on the softening efficiency of ABC and ACHC are illustrated in figure 3.3. As it can be seen, adsorption efficiency was improved by increasing the contact time and decreased by increasing the initial ions concentration.

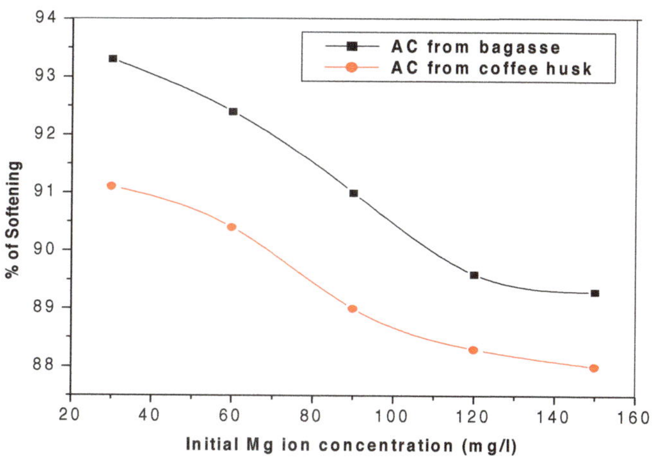


***Figure 3.4. (a)*** *Effect of initial Mg ion concentrations on softening efficiency (2 g adsorbent, pH 6.5, room temperature, agitation speed 200 rpm.*

Accordingly, for an initial metal concentration of 120 mg/L, 97% and 94.1% of calcium and 90.8% and 93.3% of magnesium were adsorbed by ABC and ACHC respectively, showing for both adsorbents a higher affinity for calcium over magnesium. Furthermore, alkali modification of both adsorbents improved the adsorption capacity toward both tested cations.

The effect of contact time on Ca and Mg ions was investigated by varying the contact time (30-150 min), while other parameters were kept constant. The result is presented in figure 3.4 and 3.5. As it can be seen in the figure, softening efficiency increased with an increasing in contact time before equilibrium is reached and after equilibrium removal efficiency would be constant.

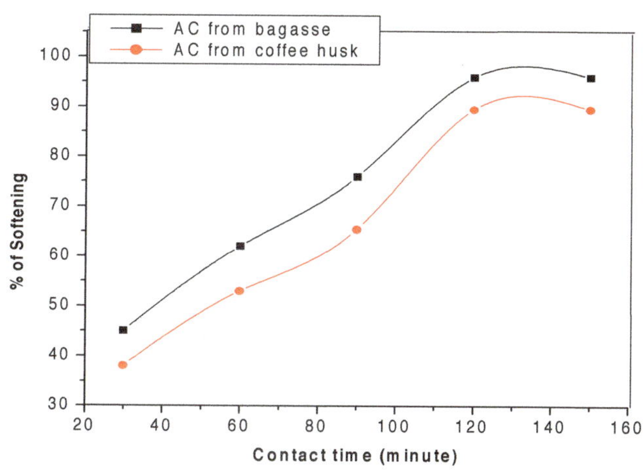


***Figure 3.4. (b)*** *Effect of contact time on softening efficiency of calcium hardness (2 g/L adsorbent, pH 6.5, room temperature, agitation speed 200 rpm and at 120 mg/l initial concentration*

The optimum time observed for removal of 96.1% and 93.4 % of Ca and Mg hardness were 120 min. There was no appreciable increase in percentage removal of total hardness after these optimum times. As shown in Figure 3.4(b) and 3.5, the adsorption process took place in two stages. The first stage was rapid. This may be due to that, at the start large number of vacant surface site may be available for adsorption process. The second stage represented a slower progressive adsorption. The reason is that, the remaining vacant surface sites may be exhausted due to repulsive forces between the solute molecules of solid and bulk phase [24],[26]. With the progressive occupation of these sites, the process becomes slower in the second stage. Moreover the initially deposited metal ions penetrate to the interior of the sorbent through intra-particle diffusion which was slower process. This was similar with the observations of other studies [21], [24,].

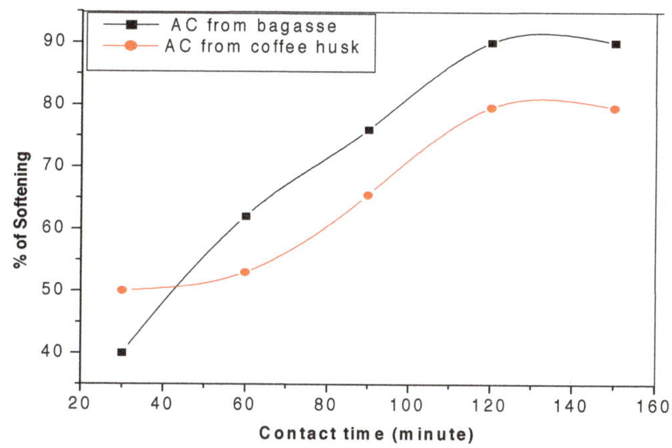


***Figure 3.5.*** *Effect contact time on softening efficiency of magnesium hardness (2 g/L adsorbent, pH 6.5, room temperature, agitation speed 200 rpm, 120 mg/l.*

Surface adsorption and ion exchange can be considered as the driving forces of ion removal. While bonding of metal ions to the surface can be considered as the main mechanism responsible for metal uptake by natural pumice, in addition to surface adsorption, ion exchange can also be involved in the case of the alkali-modified pumice leading to a remarkable enhancement of the adsorption capacity [19]. Furthermore, alkali modification can contribute to the removal of impurities, which can unblock some pores improving accessibility to the active adsorption and ion exchange sites of the modified sample. Removal efficiency also showed an increase with the initial metal concentration, which can be attributed to an increase of the concentration gradient increasing the driving force [24].

***Effect of the pH and the adsorbent mass:*** The pH of the solution should be considered as an important factor affecting metal adsorption process due to its impact on the degree of ionization of metal specie and the surface charge of the adsorbent. The effect of pH of the reaction mixture on the adsorption efficiency at various adsorbent doses was examined in order to optimize the adsorbent dosage and the pH. According to the results summarized in fig 3.6 and 3.7, the highest adsorption capacity towards $Ca^{+2}$ and $Mg^{+2}$ ions was achieved at pH = 6.5 for ABC and ACHC and for all tested dosages. This optimal pH was in accordance with the zero point charge values (6.50 and 8.310 for ABC and ACHC respectively).

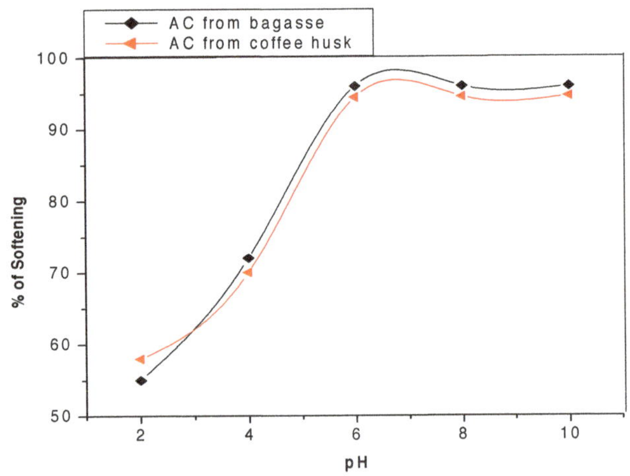


*Figure 3.6. Effect of pH on softening of calcium hardness (contact time 120 min, ions concentration 120 mg/L, room temperature, 200 rpm agitation).*

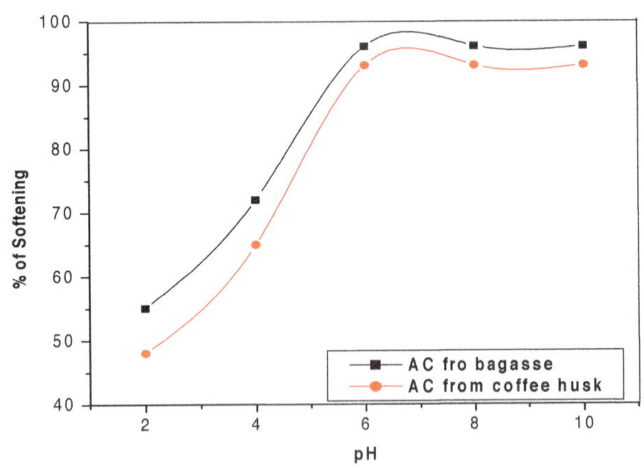


*Figure 3.7. Effect of pH on softening of magnesium hardness (contact time 120 min, ions concentration 120 mg/L, room temperature, 200 rpm agitation).*

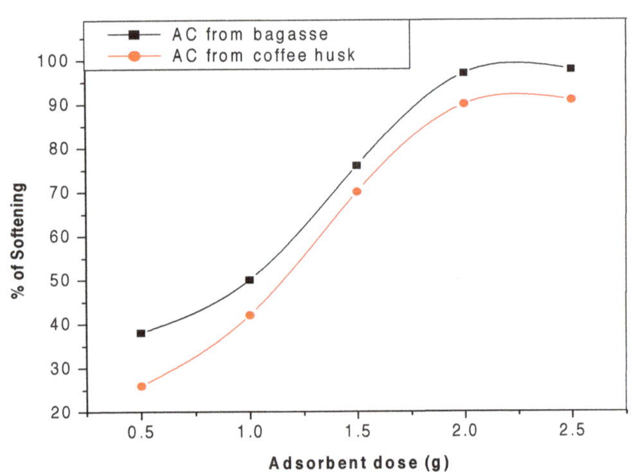


*Figure 3.8. Effect of adsorbents dose on softening of calcium hardness (contact time 120 min, ions concentration 120 mg/L, room temperature, 200 rpm agitation).*

The removals of water hardness causing constituents by ABC and ACHC at different dose (0.5 - 2.5 g) for the constant Ca and Mg concentration of 120 mg/l are investigated. Results are presented in figure 3.8 and 3.9. The percentage removal of Ca and Mg increases from 37.0 to 96.2% with an increase in the activated bagasse and coffee husk carbon amount from 0.5 to 2.5 g respectively. This is due to the increasing of the adsorption sites available for adsorption.

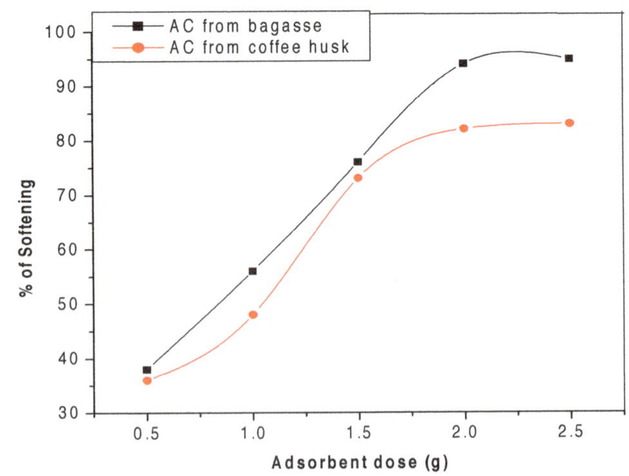


*Figure 3.9. Effect of adsorbent dose on softening of magnesium hardness (contact time 120 min, ions concentration 120 mg/L, room temperature, 200 rpm agitation).*

As it shows in fig 3.6 and 3.7 of the pH, 79 and 96% of calcium and 51 and 93% of magnesium were removed by 2 g/L of the ABC and ACHC adsorbents, respectively. The lower removal efficiencies observed in acidic medium (i.e. pH in the range 2-5) can be attributed to the protonation of adsorbent functional groups or competition of $H^+$ with metal ions to bind and occupy the active sites of the adsorbents. On the other hand, in alkaline environment (i.e. pH in the range 7-10), the formation of metal hydroxide can be considered as the main reason for decreasing metal uptake. Similar results were reported [26]. The highest removal capacity at pH=7.5, namely close to the zero point charges of both adsorbents, confirms that the studied adsorbents can be considered as promising low cost adsorbents suitable for the softening of hardness ions from drinking water.

The linear increase of the adsorption capacity for increasing adsorbent dosage indicated the accessibility of a larger number of sorption sites at higher dosage to adsorb calcium and magnesium ions. Contrarily, in the removal of calcium using raw and modified sugar cane bagasse, adsorption capacity remained constant above 100 mg adsorbent.

***Effect of the temperature:*** The temperature effect in the range of 10 °C to 60 °C was studied and thermodynamic parameters were calculated. As shown in Fig 3.10, while the effect of temperature on magnesium adsorption seemed to be negligible, a low maximum for the adsorption of calcium was absorbed at 20 °c.

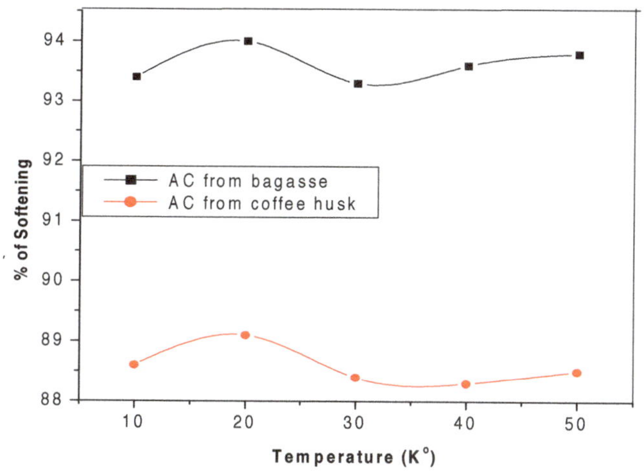


*Figure 3.10. Effect of temperature on softening of Calcium hardness (contact time 120 min, ions concentration 120 mg/L, 2 g adsorbents, 200 rpm agitation).*

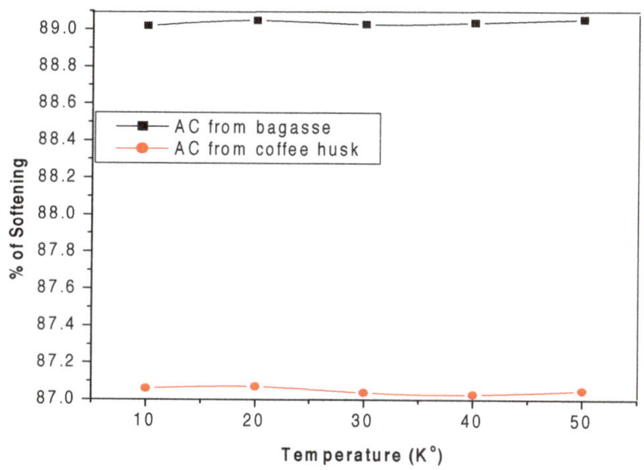


*Figure 3.11. Effect of temperature on softening of magnesium hardness.*

### 3.4. Regeneration of the Saturated Adsorbents

Regeneration experiments were conducted to study the reusability of the spent adsorbents, which is a very important parameter in terms of economic feasibility of the developed process. Regeneration using sulphuric acid solution was carried out on the spent bagasse and coffee husk samples. As it were shown in the figure below, maximum recovery of the adsorbed calcium and magnesium was achieved in less than 200 minutes leading to 78 and 73% desorption of the adsorbed calcium and magnesium at 97 and 93% of the adsorbed calcium and magnesium from ABC and ACHC respectively. Furthermore, maximum cation desorption for saturated carbon were observed after 300 min of regeneration, while only 200 min of regeneration was needed to achieve maximum cation desorption of the saturated modified adsorbent. It should be noticed that even if the modified sugar cane bagasse showed higher sorption capacity for calcium and magnesium rather than modified coffee husk, its regeneration potential was lower than that of the activated carbon from bagasse.

Generally, Ca and Mg recovery and adsorbent regeneration under 8 cycles of adsorption and desorption (n) are presented in fig 3.12 and 3.13. As the n increased, the percentage of Ca and Mg recovered and adsorbent regeneration and also Ca and Mg removal, until 4 of these cycles slightly decreased and after that, until 8, sharply decreased. It is shown that ABC and ACHC can be reused for softening/regeneration processes; but after that, it loses its softening/regeneration ability and its performance drops down.

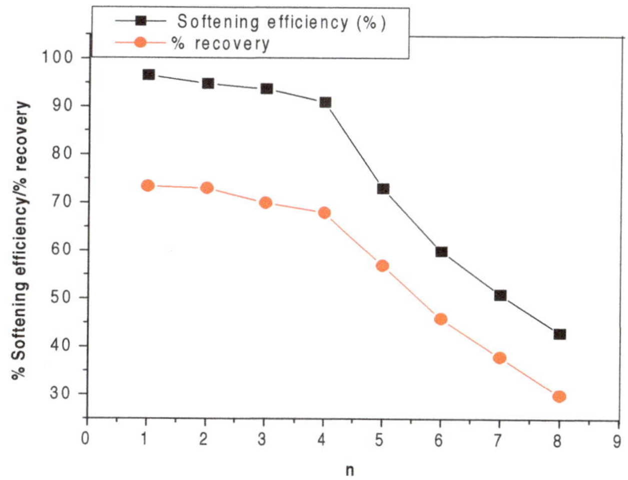


*Figure 3.12. Successive softening/regeneration efficiency of Ca hardness.*

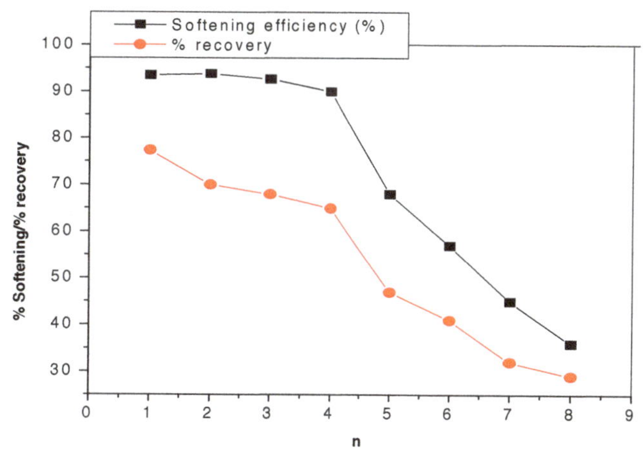


*Figure 3.13. successive softening/regeneration efficiency of Mg hardness.*

## 4. Conclusion

Softening of hard water by removing $Ca^{2+}$ and $Mg^{2+}$ cations was studied using alkali-modified sugar cane bagasse and coffee husk as adsorbents. Increasing the mass of adsorbent, the contact time and decreasing the initial ions concentration leads to an increase of cations removal. The studied adsorbents showed a higher selectivity for calcium adsorption if compared to magnesium in ABC than ACHC. Both adsorbents are efficient to adsorb water hardness causing constituents; but, ABC is more critical than ACHC. After treating synthetic water solution simulating an actual water stream with the alkali modified activated bagasse and activated coffee husk, total hardness of the treated sample met the required standard for drinking water.

## Acknowledgement

This paper was sponsored by Jigjiga University under Research and Community Service Directorate as part of the 2014 funded proposal. I am grateful to Haramaya University and Ethiopian pharmaceutical factory for their cooperation in AAS and FTIR analysis and characterization respectively.


## References


[1] AWWA, Standard Methods for the Examination of Water and Wastewater, 20th ed., Washington, DC, 2005.

[2] O.K. Jinior, L.V.A. Gurgel, L.F. Gil, Removal of Ca(II) and Mg(II) from aqueous single metal solutions by mercerized cellulose and mercerized sugarcane bagasse grafted with EDTA dianhydride (EDTAD), Carbohydr. Polym. 79 (2010) 184–191.

[3] A. Zuorro, R. Lavecchia, S. Natali, Magnetically modified agro-industrial wastes as efficient and easily recoverable adsorbents for water treatment; chemical engineering transactions Vol. 38, 2014.

[4] Ami Cobb, Low-Tech Coconut Shell Activated Charcoal Production, International Journal for Service Learning in Engineering Vol. 7, No. 1, pp 93-104, (2012).

[5] D. Bruggen, C. Vandecasteele, Removal of pollutants from surface water and groundwater by nano filtration: Overview of possible applications in the drinking water industry, Environ. Pollut. 122 (2003) 435–445.

[6] M.R. Teixeira, M.J. Rosa, The impact of the water background inorganic matrix on the natural organic matter removal by nanofiltration, J. Membr. Sci. 279 (2006) 513–520.

[7] M. Yan, D. Wang, J. Ni, J. Qu, Y. Yan, C.W.K. Chow, Effect of polyaluminum chloride on enhanced softening for the typical organic-polluted high hardness North-China surface waters, Sep. Purif. Technol. 62 (2008) 401–406.

[8] K. Suzuki, Y. Tanaka, T. Osada, M. Waki, Removal of phosphate, magnesium and calcium from swine wastewater through crystallization enhanced by aeration, Water Res. 36 (2002) pp 2991–2998.

[9] A. Dimirkou, M.K. Doula, Use of clinoptilolite and an Fe-overexchanged clinoptilolite in $Zn^{2+}$and $Mn^{2+}$ removal from drinking water, Desalination. 224 (2008) 280–292.

[10] R. Sheikholeslami, Composite scale formation and assessment by the theoretical Scaling Potential Index (SPI) proposed previously for a single salt, Desalination 278 (2011) 259–267.

[11] L. Fu, J. Wang, Y. Su, Removal of low concentrations of hardness ions from aqueous solutions using electrodeionization process, Sep. Purif. Technol. 68 (2009) 390–396.

[12] J.S. Park, J.H. Song, K.H. Yeon, S.H. Moon, Removal of hardness ions from tap water using electromembrane processes, Desalination 202 (2007) 1–8.

[13] S.J. Seo, H. Jeon, J.K. Lee, G.Y. Kim, D. Park, H. Nojima, J. Lee, S.H. Moon, Investigation on removal of hardness ions by capacitive deionization (CDI) for water softening applications, Water Res. 44 ( 2010 ) 2267–2275.

[14] C.W. Li, J.C. Liao, Y.C. Lin, Integrating a membrane and a fluidized pellet reactor for removing hardness: effects of NOM and phosphate, Desalination 175 (2005) 279-288.

[15] J.N. Apell, T.H. Boyer, Combined ion exchange treatment for removal of dissolved organic matter and hardness, Water Res. 44 (2010) 2419–2430.

[16] H. Faghihian, M.G. Maragheh, H. Kazemian, The use of clinoptilolite and it's sodium form for removal of radioactive caesium, and strontium from nuclear wastewater and $Pb^{++}$, $Ni^{++}$, $Cd^{++}$, $Ba^{++}$ from municipal wastewater, Appl. Radiat. Isot. 4 (1999) pp 655-661.

[17] H. Kazemian, H. Modarres, H.G. Mobtaker, Evaluating the performance of an Iranian natural clinoptilolite and its synthetic zeolite P for removal of Cerium and Thorium from nuclear wastewaters, J. Radioanal. Nucl. Chem. 258 (2003) 551-556

[18] Seifi *et* al. Adsorption of BTEX on surfactant modified granulated natural zeolite nanoparticles: parameters optimizing by applying Taguchi experimental design method, Clean-Soil, Air, Water, 39 (2011) pp 939–948.

[19] Sepehr *et*, al. (2013) Removal of hardness agents, calcium and magnesium, by natural and alkaline modified pumice stones in single and binary systems, Applied Surface Science, pp 295-305, Vol. 03 (42)

[20] N. Feng, X. Guo, S. Liang, Y. Zhu, J. Liu, Biosorption of heavy metals from aqueous solutions by chemically modified orange peel, J. Hazard. Mater. 185 (2011) 49–54.

[21] B. Ersoy, A. Sariisik, S. Dikmen, G. Sariisik, Characterization of acidic pumice and determination of its electrokinetic properties in water, Powder Technol. 197 (2010) 129–135.

[22] B. Ozturk, Y. Yildirim, Investigation of sorption capacity of pumice for $_{SO2}$ capture, Process Saf. Environ. Prot. 86 (2008) 31– 36.

[23] M.R. Panuccio, A. Sorgona, M. Rizzo, G. Cacco, Cadmium adsorption on vermiculite, zeolite and pumice: Batch experimental studies, J. Environ. Manage. 90 (2009) 364-374.

[24] F. Akbal, Sorption of phenol and 4-chlorophenol onto pumice treated with cationic surfactant, J. Environ. Manage. 74 (2005) 239–244.

[25] H. Kazemian, M.H .Mallah, Elimination of $Cd^{2+}$ and $Mn^{2+}$ from Wastewaters Using Natural Clinoptilolite and Synthetic Zeolite-P, Iran. J. Chem. Chem. Eng. 25 (2006) 91-94.

[26] F. Gode, E. Moral, Column study on the adsorption of Cr(III) and Cr(VI) using Pumice, Yarıkkaya brown coal, Chelex-100 and Lewatit MP 62, Bioresour. Technol. 99 (2008) 1981–1991.

# 2

# Operating Conditions of A Locally Made Fixed-Bed Incinerator, a Case Study of Bagamoyo – Tanzania

**Arthur Mngoma Omari[2, *], Geoffrey Reuben John[1], Karoli Nicholas Njau[3], Peter Lucas Mtui[1]**

[1]College of engineering and Technology, University of Dar es Salaam, Dar es Salaam, Tanzania
[2]College of engineering and Technology, Mbeya University of Science and Technology, Mbeya Tanzania
[3]School of Materials, Energy, Water and Environmental Sciences, Nelson Mandela African Institution of Science and Technology, Arusha Tanzania

**Email address:**

mngoma2003@hotmail.com (A. M. Omari), geofrey.john@nm-aist.ac.tz (G. R. John), karoli.njau@nm-aist.ac.tz (K. N. Njau), plmtui@yahoo.com (P. L. Mtui)

**Abstract:** To minimize the pollution from municipal solid waste incinerators, the study of operating conditions is imperative. The local design incinerators can be used for high performance combustion and minimize pollutions. The incinerator located at Bagamoyo hospital in Tanzania is used as a pilot for this experiment. The emission and operating conditions shows that the performance of incinerator is at maximum peak of about 70% when the oxygen at exit is about 6.9% and when secondary temperature is between 1073 and 1173K and primary temperature maintained at 673K. The experiment shows that when the primary chamber temperature increases beyond 673K, the secondary temperature decreases this is due to complete combustion at primary chamber which will cause insufficient incomplete gases and therefore limited combustible gases to burn.

**Keywords:** Waste incineration, emission, fixed bed, solid waste

## 1. Introduction

Waste is material or object which the holder may dispose-off or plan to dispose-off [1]. Municipal solid waste generation has been increased as the results of human development and population growth. This is a big challenge of waste disposal in most countries [2]. The rate of waste generation of these countries including Tanzania; has been increasing yearly with unchanging methods of waste disposal. This increase of municipal solid waste generated necessitates municipals to improve the method of waste disposal and handling [3]. Waste disposal method with improper safety conditions and regulations can cause environmental problems [4, 5]. Common method of waste disposal in most of developing countries, especially those in Sub Sahara Africa is dumping , open burning and landfill [6]. The possibilities of recycling waste materials and energy recovery from waste are better methods of waste management system currently used. The utilization of municipal solid waste to energy will assist to mitigate fossil energy use [4].

The municipal solid waste production rate is varying from place to place due to different levels of inhabitant economic and the seasonal or climatic conditions of the year and therefore the net energy content of municipal solid waste of different place also differs from each other [7]. The waste materials are dynamic and different in sizes and shape [8, 9]. Neither it's physical nor chemical properties of municipal solid waste are the same; the properties are varying seasonal and regional. When compared the properties of fossil fuels or biomass, the properties of municipal solid waste are dynamic. In this way, different type of waste disposal system designs needed to be developed. There are three basic method of waste disposal; these include physical, biological and thermal methods [10]. The physical method of waste disposal, is the method of waste disposal by converting waste to physical forms such as pellets, briquettes or chips for further use in thermal process to acquire energy from waste [11]. Biological methods of waste disposal are the process of waste disposal in which the organic waste recovering by composting or digestion. The results from compost materials can be used as compost manure [12]. The emitted gas during the processes such as methane can be tapped and used as fuel for electricity generation, for driving internal combustion engines and heat generator in combine heat and power systems [7, 13]. Thermal waste treatment is the process of waste disposal that use heat to treat waste materials [1]. The waste management disposal by using thermal method has many challenges [14, 15]. The incineration, gasification and pyrolysis are amongst the known techniques [16]. The waste disposal by incineration may reduce waste volume by 90% and weight by 75-80%. [17, 18].

The incineration is the common method of waste disposal in some hospitals in Tanzania, most of these incinerators are locally brick made with poor performance [19]. The utilization of the energy in combine heat and power can be used in boilers for the electricity generation [14]. The interactions that take place within the process including pyrolysis, combustion, heat transfer, mass transfer and gas flow [20].

In order to identify the proper operating procedures of incinerator that should be practiced on municipal solid waste incinerators, the study of flue gas emission associated with incinerator operating conditions is necessary. The incinerator can become a big polluter if not properly operated [9]. The flue gas from incinerator contain particles and gases which are harmful and toxic to human and environment [21]. If the mixing level of the burnt wastes are better operated, the amount of pollutants to the environment will be reduced [22]. The emission gases from incinerator indicate the combustion behavior of the incinerator, design, operation and the mixing of waste materials in the combustion chamber [23, 24].

Pollution gases such as $SO_X$, $NO_X$ and $CO_2$ which can be formed during the combustion of municipal solid wastes can dissolve in the atmosphere and form an acid rain [25, 26]. The $NO_X$ can be produced in large amount when the temperature of combustion exceeds 1100°C. The concentration of $SO_X$ courses the formation of $H_2SO_4$, which increases the concentration of H+ ion and therefore makes the rain water drops to be harmful to human and environment [27]. When rain water is too acidic, it can cause problem such as killing fish in water, damage crops [28, 29], erode buildings and roads [27, 30].

The aim of this study is therefore, to analyze the flue gas emission from an existing incinerator for the purpose of studying the operating conditions of the existing design in order to improve incinerator design and operating condition.

## 2. Materials and Methods

### *2.1. Materials*

#### *2.1.1. Waste Sample Composition*

The municipal solid wastes were collected from Bagamoyo municipality dump located at Sanzale. The waste sample composition is studied and taken from the average composition ratio of the municipal solid waste studied from previous works reported by [11, 31-34]. This composition will be taken as a standard for the current work. The wastes were sorted to separate the combustible and non-combustible wastes to conform to standardize ASTM D5231-92 [35]. The waste mass of about 150 kg from dump were packed in polythine bags of about 50 kg each and then transported to the incinerator ready for incineration. Diesel is used as auxiliary fuel to run the burners.

#### *2.1.2. Waste Sample Formula for Reactions*

The samples formula of municipal solid waste used for this experiment were adopted from ultimate and proximate analysis results found in Arusha, Tanzania [36].

#### *2.1.3. Prototype Incinerator Design*

The pilot incinerator for this study was the fixed bed incinerator located at Bagamoyo hospital, Figure 1, 2 and 3

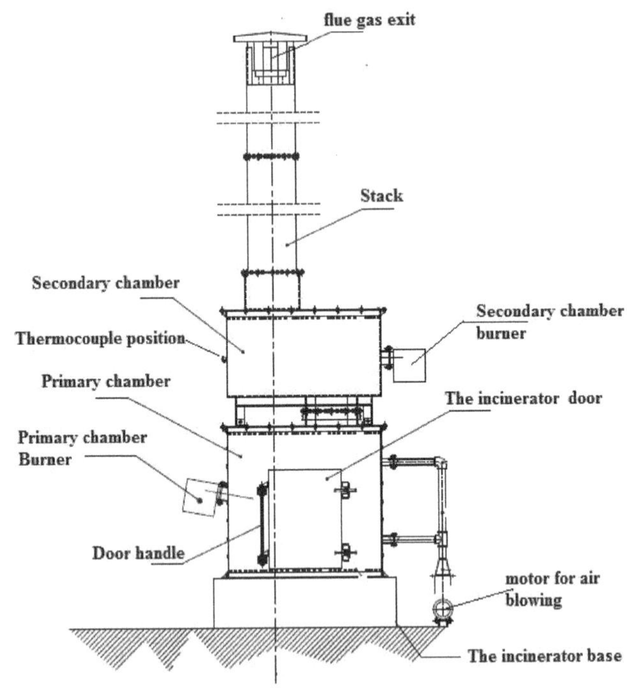


***Figure 1.*** *The incinerator layout at Bagamoyo.*

### *2.2. Working Principles of the Incinerator*

#### *2.2.1. Background*

This is the starved air incinerator; it composes of two combustion chambers with batch type feeding as shown in Figure 1. The two chambers have two different functions. The primary (ignition) chamber is to ignite the waste and secondary (combustion) chamber is for combustion. The operating temperature for primary chamber can be varied between 400 and 500$^{O}$C by adjusting the fuel supply to the burner while the secondary chamber temperature can vary from 800 to 1000$^{O}$C. The primary chamber maintains minimum temperature sufficient to sustain combustion and killing microorganism in the waste. The minimum temperature need to prevent refractory damage and minimum generation of volatiles to the secondary chamber. This is because high temperatures at primary chamber causes vigorous burning and also cause solid metal to evaporate and enter the secondary chamber as vapor which will increase the particulate matters to the flue gases. The high temperature at primary chamber can affect the combustion gas volume and rapid increase in flue gas volume at secondary chamber.

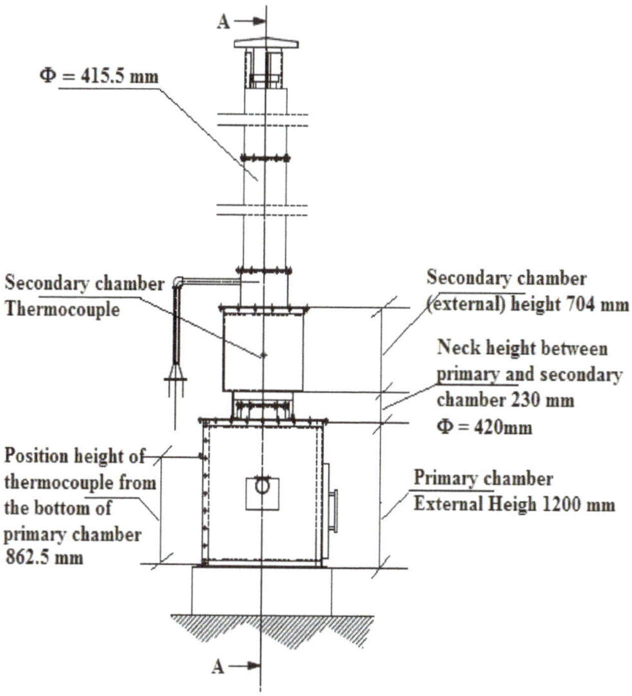


***Figure 2.*** *The incinerator layout side view.*

This incinerator is not included an air control device because it is believed that with proper combustion control through proper and precise operation, the system can meet the emission regulations. The Chinese standards of emission regulations for pollution for power plants shown in Table 1

***Table 1.*** *Emission standards for waste to energy power plant.*

| Standard | Item | Threshold limit value | Unit |
|---|---|---|---|
| GB 13233-2003 | $SO_2$ | 400 | ppm |
| | $NO_X$ | 450 | ppm |
| GB 13233-2011 | $SO_2$ | 100 | ppm |
| | $NO_X$ | 100 | ppm |

Source: [37]

#### *2.2.2. Operating Procedure*

The first thing is to remove the ash which was left from the last cycle. The ash has been cooled down for more than 8 hours overnight. Second is charging the waste to the incinerator. This is done manually. At this stage the thing to consider is the heating value of waste materials and moisture contents.

These affect the performance of an incinerator. The high moisture contents will provide insufficient thermal input and need excessive fuel consumption while the higher heating value waste may exceed thermal capacity of the incinerator.

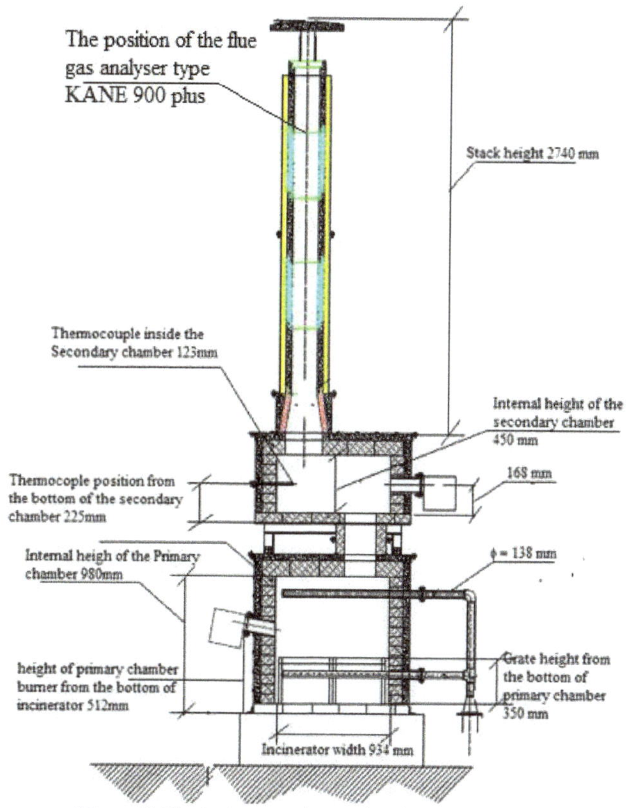


***Figure 3.*** *The incinerator layout A – A cross section view.*

Prior ignition of waste, the secondary chamber is pre heated to a predetermined temperature of $400^{0}C$ by the secondary burner. The combustion process in the primary chamber is controlled to obtain starved air supply. The starved air supply causes the incomplete combustion products. The incomplete combustion products formed are the pyrolysis gases which will burn again in the secondary chamber to complete the combustion. The air blowers are installed on the side of the incinerator for supply air to secondary and primary chambers as shown in Figure 1. The two thermocouples type K are installed in the combustion chamber near the exit point in order to read the representative temperature of each chamber shown in Figure 2

The outlet of secondary chamber is supplied with excess air to ensure the complete combustion (Figure 3) while the primary chamber is supplied by starved air to pyrolysed the waste. The burners are installed within the combustion chambers to ignite the waste and to maintain the combustion temperature shown in Figure 3.

The flue gas measurements are taken from the stack using Hand-held combustion analyzer type Kane 900 plus placed at 300 mm below the exit of the stack and 150 mm inside the stack. The flue gas analyzed including CO, $CO_2$, $NO_X$ and $SO_2$. After the complete incineration process the incineration is switched off for cooling.

### *2.3. Methods*

#### *2.3.1. The Operating Incinerator under Normal Conditions.*

The running of incinerator was done parallel with the

emission measurements of flue gas from the incinerator at the stack shown in Figure 3.Emission gases and temperature results from the incinerator under no load and with load were tabulated and recorded. The operation of the incinerator was set with load of 50kgs with variation temperatures of primary and secondary chambers. The temperatures were in the range of 400 $^{o}$C against 800, 900 and 1000$^{o}$C and 500$^{o}$C against 800, 900, and 1000$^{o}$C for primary and secondary chambers respectively. The flue gas emission released during the incineration process and the ash left out after the completing the incineration were recorded.

#### *2.3.2. The flue Gas Analysis with Variation of Primary Chamber Temperature*

The flue gas emissions were recorded by varying the primary chamber temperature for analyzing the performance of the incinerator. The primary chamber temperature is set to 400$^{o}$C and 500 $^{o}$C The flue gas emission readings were recorded. The combustion products gases $O_2$, $CO_2$, CO and $NO_X$ are recorded and plotted.

The efficiency of incinerator was also analyzed by using $CO_2$ an indicator, in this way a ratio in equation (1) was utilized

$$\frac{CO_2}{CO_2+CO} x\ 100\% \qquad (1)$$

Where:
$CO_2$ – concentration of Carbon dioxide
CO – concentration of Carbon monoxide
The results were tabulated in Table 4 and 5 and plotted in Figure 6 and 7.

## 3. Results and Discussion.

### *3.1. The Waste Sample Composition.*

The average value of waste samples composition from [11, 31-34] were tabulated in Table 2(a). These were used as the standard composition for this experiment.

***Table 2(a).*** *Waste sample composition of Municipal solid waste.*

| Country | Composition | | | | | Source |
|---|---|---|---|---|---|---|
| | Combustible | | | | Non combustibles | |
| | Plastics | Papers | organic waste | Textiles | | |
| Tanzania | 7 | 11 | 67 | 2 | 13 | [11] |
| Kenya | 13.8 | 11.3 | 58.8 | 7.8 | 8.3 | [31] |
| Nigeria | 10 | 6 | 47 | 7 | 30 | [32] |
| Ghana | 3.3 | 6.6 | 73 | 2.2 | 14.9 | [33] |
| Zambia | 5 | 5 | 50 | - | 40 | [34] |
| Uganda | 1 | 3 | 78 | - | 18 | [34] |
| Ethiopia | 2 | 4 | 88 | - | 6 | [34] |
| Average values (%) | 6.0 | 6.7 | 66.0 | 4.8 | 18.6 | |

***Table 2(b).*** *The combustibles fraction of Municipal solid waste*

| Combustible wastes | Plastics | Papers | Food and organic | Textiles | Total |
|---|---|---|---|---|---|
| (%) | 7 | 8 | 82 | 3 | 100 |

The sample obtained shows that the composition of municipal solid waste has about 81.4% of combustible waste and 18.6 % of non-combustible waste as shown in Table 2(a). These values are also found by in [38] in his research in Mirpurkhas city in Pakistan, where he revealed that the 60 – 70% of total waste generated from municipal solid waste is combustible.

The combustible waste fraction has the average of 82% of food and organic waste, 8 % of paper waste, 7% of plastics and 3% of textiles. This gives the fraction of combustible waste 93% biodegradable and 7 % non-biodegradable wastes as shown in Table 2(b).

This suggest that the energy from municipal solid waste can be recovered by either using biogenic method or by using thermal treatment method. Textiles and paper wastes are easily combusted because it composed of lignocelulosic biomass. The food and organic waste contains more than 60% of lignocelulosic biomass [39]. Lignocelulosic biomass are poor and slow in hydrolysis rates and low cellulose digestibility [40]. Compared to non cellulosic biomass, cellulosic materials are not efficiently disrupt orderly hydrogen bonds among them and therefore slow down hydrolysis rates [40].The cellulose in biomass consists of unorganized crystaline structure, amorphous structure bundled together to form cellulose fibrils. Owing to the location of these cellulose within the cell walls, the accessibility of enzymes are restricted and interfered [39]. The presence of lignin from the food and organic waste fraction of municipal solid waste, cause the waste impermeable and resistance against microbial, thus make degradation of waste by microbial very tough [41]. The incineration is better option and faster energy recovery method than biogenic methods [42].

### *3.2. Empirical Formula for Waste Sample Formulation*

For development of empirical formula for municipal solid waste, the ultimate analysis results were used. The values are tabulated in table 3.

***Table 3.*** *Proximate and ultimate analysis of Arusha municipal solid waste.*

| | | | | | | | |
|---|---|---|---|---|---|---|---|
| | **Proximate analysis** | | | | | | |
| **Location** | **MC as received (wt.%)** | | **VM (wt.%)DB** | **Ash (wt.%)DB** | **FC (wt.%) DB** | | **HHV (MJ/kg)** |
| Kaloleni | 59.67 | | 74.43 | 8.16 | 17.41 | | 11.90 |
| Sakina | 63.99 | | 84.00 | 10.00 | 6.00 | | 11.37 |
| Central market | 55.70 | | 78.30 | 13.48 | 8.22 | | 12.76 |
| Average | 59.79 | | 78.91 | 10.55 | 10.54 | | 12.01 |
| | **Ultimate analysis** | | | | | | |
| Location | C | H | O | N | S | Cl | P |
| Kaloleni | 55.57 | 5.34 | 34.88 | 2.09 | 0.31 | 0.04 | 0.10 |
| Sakina | 55.70 | 5.29 | 34.27 | 2.13 | 0.22 | 0.07 | 0.13 |
| Central market | 53.20 | 5.24 | 34.71 | 2.86 | 0.37 | 0.04 | 0.11 |
| Average | 54.82 | 5.29 | 34.62 | 2.36 | 0.30 | 0.05 | 0.11 |
| Normalize to 100% | 56.20 | 5.42 | 35.49 | 2.42 | 0.31 | 0.05 | 0.11 |
| Calculate number of moles | $\frac{x_C}{m_C}$ | $\frac{x_H}{m_H}$ | $\frac{x_o}{m_o}$ | $\frac{x_N}{m_N}$ | $\frac{x_s}{m_s}$ | $\frac{x_{cl}}{m_{cl}}$ | $\frac{x_p}{m_p}$ |
| Number of moles in 100kgs | $\frac{56.20}{12}$ | $\frac{5.42}{1}$ | $\frac{35.49}{16}$ | $\frac{2.42}{14}$ | $\frac{0.31}{32}$ | $\frac{0.05}{32}$ | $\frac{0.11}{28}$ |
| | 4.683 | 5.423 | 2.218 | 0.173 | 0.010 | 0.001 | 0.004 |

Source:[43]

Where: MC - Percentage of moisture in municipal solid waste

VM - Volatiles matter

FC - Fixed Carbon

$x_C$, $x_H$, $x_O$, $x_N$, $x_S$, $x_{Cl}$, $x_P$ are the fraction composition of Carbon, Hydrogen, Oxygen, Nitrogen, Sulphur, Chlorine and phosphorus respectively.

$m_C$, $m_H$, $m_O$, $m_N$, $m_S$, $m_{Cl}$, $m_P$ are the Atomic number of the Carbon, Hydrogen, Oxygen, Nitrogen, Sulphur, Chlorine and phosphorus respectively.

The chemical formula for municipal solid waste for our case will be

$$C_{4.683}H_{5.423}O_{2.218}N_{0.173}S_{0.010}Cl_{0.001}P_{0.004} \quad (2)$$

- The empirical formula of a chemical compound is a simple expression of the relative number of each type of atom in it. It shows the proper ratio of atom in the compound.
- It can help in predicting chemical reactions which will assist in predicting the output result of reaction.

### *3.3. The Operations of an Incinerator under Normal Conditions*

The incinerator operates as normal, the variation of temperature in primary and secondary chambers with the flue gas emission were tabulated in Table 4 and plot in the Figure 4, Figure 5 and Figure 6.

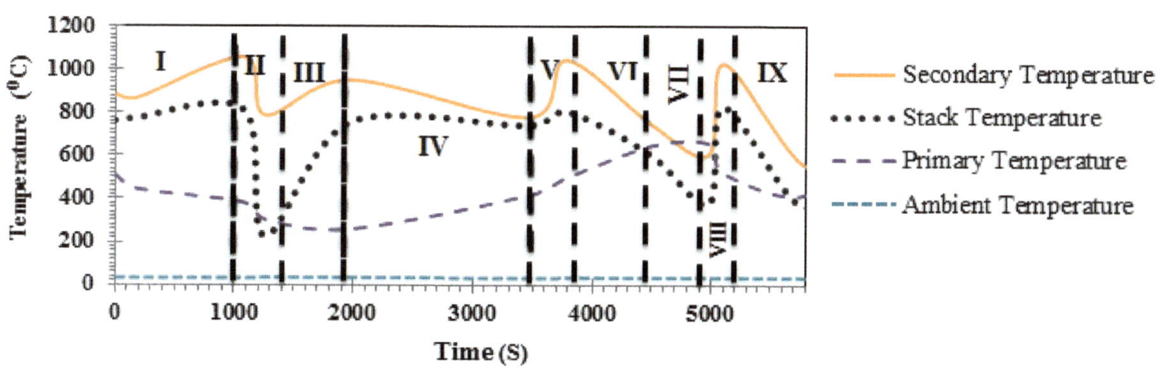


***Figure 4.*** *Typical temperature distribution in an incinerator for one complete cycle.*

***Table 4.*** *The operation for one cycle with emission from incinerator.*

| **Time sec** | **Emissions** | | | | | | **Temperature** | | **The comments on the operating conditions** |
|---|---|---|---|---|---|---|---|---|---|
| | **$O_2$** | **CO** | **$CO_2$** | **$T_S$** | **$T_{Amb}$** | **$NO_x$** | **$T_{Pri}$** | **$T_{Sec}$** | |
| | **%** | **ppm** | **%** | **°C** | **°C** | **ppm** | **°C** | **°C** | |
| 0 | 5.5 | 101 | 11.4 | 760.3 | 30.9 | 71.3 | 509.3 | 883.3 | Without Load |
| 180 | 3.9 | 5832 | 12.6 | 774 | 31.5 | 2374 | 445 | 865.7 | At start after feeding |
| 1080 | 8.3 | 12 | 9.3 | 811 | 31.7 | 165 | 382 | 1054.3 | Mid of incineration (Sec burner On) |
| 1260 | 19.6 | 317 | 1 | 225.7 | 32.3 | 199 | 303.3 | 786.7 | Sec. Burner OFF |
| 1980 | 8.9 | 0 | 8.9 | 755.3 | 32.3 | 92 | 259 | 949.7 | Sec. Burner ON, Near end |
| 3480 | 7.4 | 100.7 | 9.8 | 739 | 32.1 | 136 | 418.3 | 779.3 | 400/800 |

| Time | Emissions | | | | | | Temperature | | The comments on the operating conditions |
|---|---|---|---|---|---|---|---|---|---|
| sec | $O_2$ | CO | $CO_2$ | $T_S$ | $T_{Amb}$ | $NO_x$ | $T_{Pri}$ | $T_{Sec}$ | |
| | % | ppm | % | $^oC$ | $^oC$ | ppm | $^oC$ | $^oC$ | |
| 3780 | 7.8 | 0 | 9.63 | 807.7 | 32.4 | 67 | 496.3 | 1044 | 500/1000 |
| 4500 | 6.9 | 1159 | 10.4 | 600.7 | 32.9 | 582 | 646 | 745.7 | Door open slightly and burner off |
| 4980 | 16 | 716.3 | 3.6 | 377.3 | 33.1 | 302.3 | 655.7 | 606.7 | Burners off |
| 5100 | 8.2 | 0 | 9.4 | 828 | 33.3 | 87 | 520.7 | 1027.7 | Sec Burner on/Initial stage |
| 5760 | 17.6 | 162 | 2.4 | 371.7 | 33.7 | 99 | 413.3 | 569 | End of incineration (Sec burner off) |
| 6180 | 11.7 | 776.7 | 6.8 | 528.3 | 33.7 | 371.7 | 617.7 | 572 | Burners off /Door slightly open |

Where:

$O_2$ Mass of oxygen to the total mass of the flue gas in percentage (%)

CO Mass of carbon monoxide to the total mass of the flue gas in parts per millions (ppm)

$CO_2$ Mass of carbon dioxide to the total mass of the flue gas in percentage (%)

$NO_x$ Mass of Nitrogen oxides to the total mass of the flue gas in parts per millions (ppm)

$T_s$ Temperature of the stack (°C)

$T_{pri}$. Primary chamber temperature of an incinerator (°C)

$T_{sec}$ Secondary chamber temperature of an incinerator (°C)

$T_{amb}$ Ambient temperature (°C)

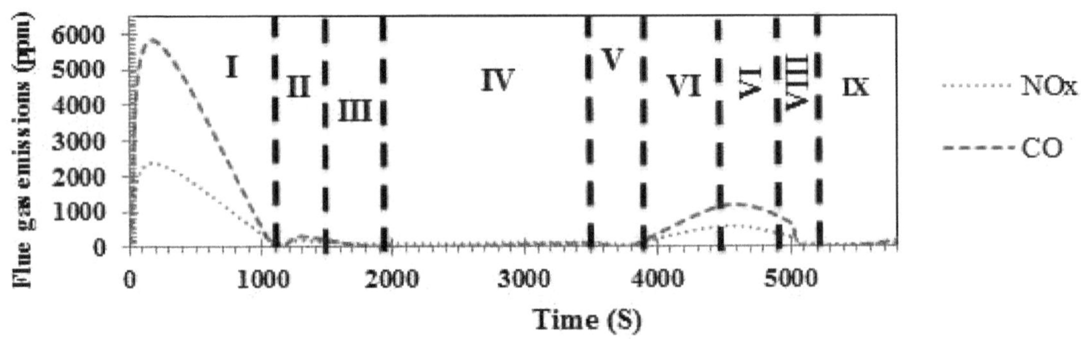


***Figure 5.*** *$NO_X$ and CO emissions from incinerator for one complete cycle*

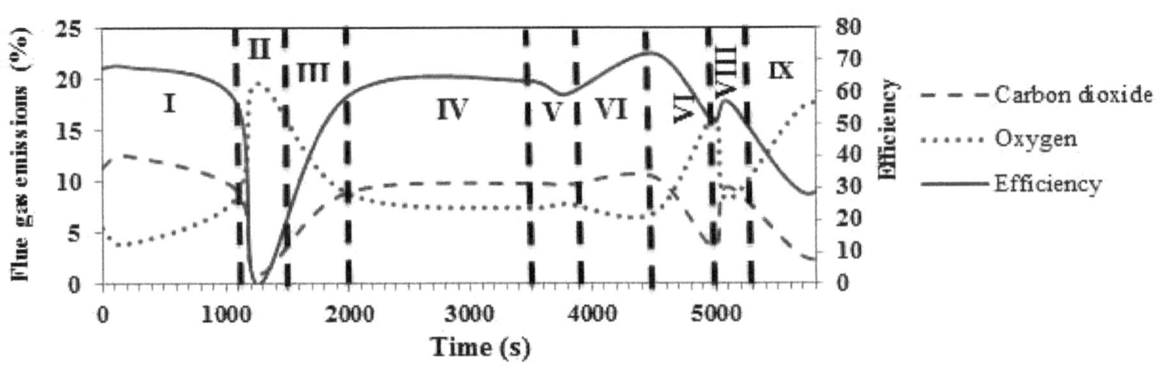


***Figure 6.*** *Efficiency, $CO_2$ and $O_2$ emission from incinerator.*

The measurement and results observed from an incinerator are divided into regions. There are 9 regions division. Region I show that when the incinerator working under no load, the maximum combustion temperature attained for the stack was 1073K as shown in Figure 4. The emission of $NO_X$ and CO seems to be much higher at the beginning and then suddenly decreases. This is due to the incomplete combustion of the fuel at starting shown in Figure 5. The $NO_X$ formed in this case is prompt $NO_X$ which normally coming from the combustion of Nitrogen bearing fuels. The fuel released nitrogen bound in them as free radical which are oxidized to $NO_X$ [44].

At the second region the temperature of the incinerator falls. The stack temperature decreases to 520K. The primary chamber temperature decreases to 576K while secondary temperature falls to 556K, this is because at this time the incinerator burners are switched off and the feeding stock is added to the incinerator. The incinerator efficiency also falls to zero since there is no combustion during the feedstock and all the burners are switched off.

At region III, all the burners are switched on. The temperature of the incinerator changes, the secondary and stack temperature rises while the primary temperature falls. The cause of primary chamber temperature to fall is due to the feedstock temperature which is at room temperature from the dump. The emissions show that the $CO_2$ concentration increases while the $O_2$ concentration decreases as shown in Figure 6. The concentration of carbon monoxide decreases while $NO_X$ emissions become constant at 92 ppm, Figure 5.

At region IV the temperature of the primary chamber increases while that of secondary slightly decreases, this showing that the feed stock is already acquired the uniform temperature with the combustion chamber. The emission of $NO_X$ and CO is at constant of about 136 and 100.7 ppm Figure 5 while the $O_2$ and $CO_2$ emission are about 7.4 and 9.8% respectively. The efficiency of the incinerator combustion fluctuates between 63 and 59% as shown in Figure 6. The decrease in secondary chamber temperature is caused by reduction of incomplete combustion gases which are expected from primary chamber to the secondary chamber such as CO which decreases from 136 to 100.7 ppm shown Figure 5 [45].

At region V, the incinerator temperature of all chambers increases Figure 4. The combustion efficiency is slightly decreases Figure 6. The concentration of $NO_X$ and CO is slightly decreases. The $NO_X$ concentration decreases from 136 ppm to 67 ppm while concentration of CO decreases from 100.7 to 0 ppm (Figure 5). The decrease of $NO_X$ and CO and the efficiency of the combustion are caused by switching off the second burner for trying to maintain the desired temperature. The $NO_X$ concentration decreased because the amount of diesel of combustion decrease as the burner is switched off. The only $NO_X$ formed at this temperature is a prompt $NO_X$ [44] which is normally comes from the fuel used.

At region VI, the burner is switch on again to maintain the desired temperature. The temperature of secondary chamber and stack decreases while the temperature of the primary chamber increases. The emission shows that there is increase in CO and $NO_X$ concentration. The CO increases from 0 to 1159.3 ppm and that of $NO_X$ increase from 67 to 582 ppm (Figure 5). The efficiency of combustion is also increased from 59% to 71.7% (Figure 6). This is the highest efficiency of this incinerator in which the concentration of $O_2$ is 6.9% and that of $CO_2$ is 10.4%. These value approaches the recommended value of $O_2$ at the exit point which was studied in simulation of fixed bed thermal oxidizer for solid waste disposal. The value obtained and recommended for oxygen at exit point is 6% [46].

At region VII, the incinerator door is slightly opened to increase the excess air supply and the primary chamber burner is switched off. The secondary and stack temperature are decreasing while the primary temperature is still increasing (Figure 4). This may be caused by complete combustion in the primary chamber by supplying excess air by opening the door and therefore decrease the incomplete combustion gases in the secondary chamber. The $NO_X$ and CO is decreases from 582 to 302.3 ppm and from 1159.3 to 716.3 ppm respectively (Figure 5). The efficiency also is decreased from 71.7 to 50.7% (Figure 6).

At the region VIII, the secondary burner is switched on for about 100s while the primary chamber burner is still off. This is done in order to maintain the required temperature. The secondary and stack temperature rise rapidly (Figure 4). The efficiency is slightly increased to 56% and then raised and fluctuate to 50% (Figure 6)

At region IX, all the temperature decreases. The efficiency fall to 27% and $CO_2$ concentration decreased to 6% while $O_2$ increases to 20.1% (Figure 6). The cause of temperature rise from secondary chamber is due to the high concentration of incomplete combustion gases at the chamber. These gases were combusted by the burner and suddenly they diminished.

### *3.4. Flue Gas Analysis with Variation of Primary and Secondary Chambers Temperatures*

#### *3.4.1. Analysis of $O_2$, $CO_2$, CO and $NO_x$ at Primary Chamber Temperature Fixed at 400°C*

***Table 5.*** *Flue gas emission at primary temperature 400°C.*

| Primary Temp °C | Secondary Temp °C | $O_2$ (%) | CO (ppm) | $CO_2$ (%) | $NO_X$ (ppm) |
|---|---|---|---|---|---|
| 400 | 808.00 | 8.76 | 209.20 | 9.16 | 204.00 |
| | 902.00 | 11.96 | 47.20 | 5.66 | 92.00 |
| | 1005.00 | 11.70 | 3.20 | 4.20 | 51.00 |
| | 1069.00 | 12.20 | 0.60 | 4.26 | 57.60 |

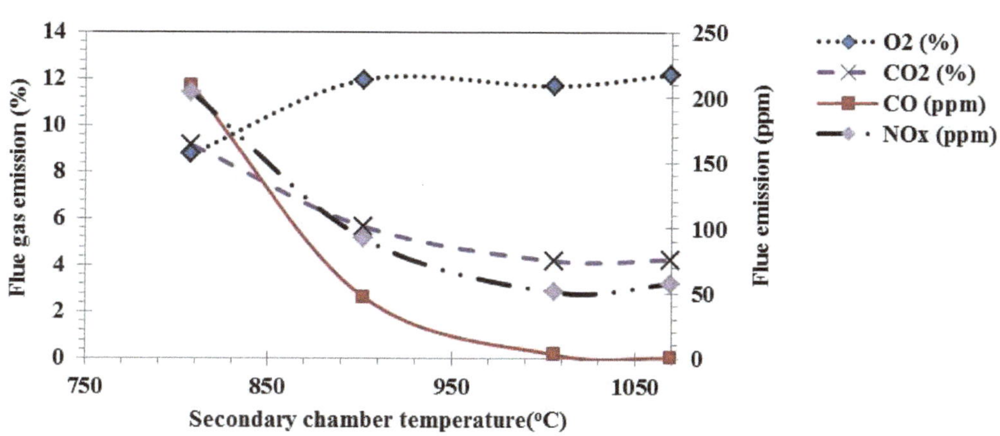


***Figure 7:*** *Flue gas emission at primary chamber temperature 400°C*

### *3.4.2. Analysis of $O_2$, $CO_2$, CO and $NO_X$ at Primary Chamber Temperature Fixed at 500°C*

***Table 6.*** *Flue gas emission at primary temperature 500°C.*

| Primary Temp. °C | Sec. Temp. °C | $O_2$ (%) | CO (ppm) | $CO_2$ (%) | NOx (ppm) |
|---|---|---|---|---|---|
| 500 | 810.30 | 17.24 | 301.80 | 1.84 | 184.20 |
| | 913.70 | 17.32 | 250.40 | 1.80 | 156.60 |
| | 1055.30 | 8.64 | 64.40 | 6.06 | 351.80 |
| | 1088.00 | 12.26 | 0.00 | 4.06 | 47.00 |
| | 1093.70 | 10.64 | 0.80 | 4.96 | 168.20 |
| | 1102.70 | 12.26 | 0.00 | 3.70 | 155.80 |

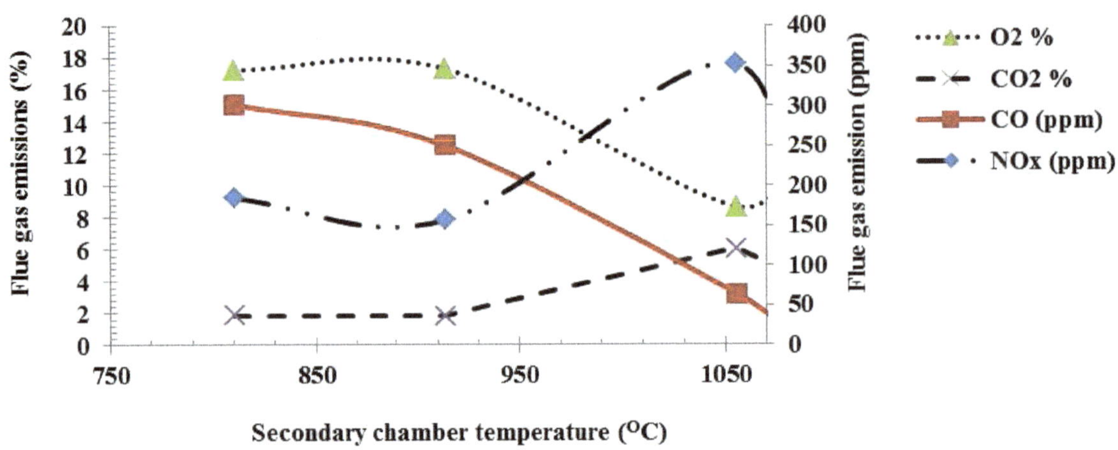


***Figure 8.*** *Flue gas emission at primary chamber temperature fixed 500°C.*

The flue gas shows that the relation between $O_2$ and $CO_2$ that, as the amount of $CO_2$ is increased the amount of $O_2$ decreases. This indicates that at this point the efficiency of combustion is better

### *3.4.3. Analysis of Efficiency, Excess Air and $O_2$ in Flue Gas when Primary Chamber Temperature Fixed at 500°C.*

The variation of stack temperature, excess air and the oxygen in the flue gas when changing the primary chamber temperature were analyzed. The results were recorded and tabulated in table 6 and plotted in Figure 7

***Table 7.*** *The stack temperature against excess air and efficiency of incinerator with primary chamber temperature 400°C.*

| $T_{Pri/sec.}$ °C | $T_{Sec}$ °C | $O_2$ (%) | η (%) | $CO_2$ (%) | $T_{Stack}$ °C | $T_{Amb.}$ °C | $X_{Air}$ (%) |
|---|---|---|---|---|---|---|---|
| 400/(800,900, 1000,1100) | 808.00 | 8.76 | 57.54 | 9.16 | 749.00 | 31.58 | 72.58 |
| | 902.00 | 11.96 | 44.30 | 5.66 | 705.20 | 31.80 | 276.18 |
| | 1005.00 | 11.70 | 39.80 | 4.20 | 679.40 | 32.00 | 151.96 |
| | 1069.00 | 12.20 | 36.08 | 4.26 | 800.80 | 31.90 | 147.56 |

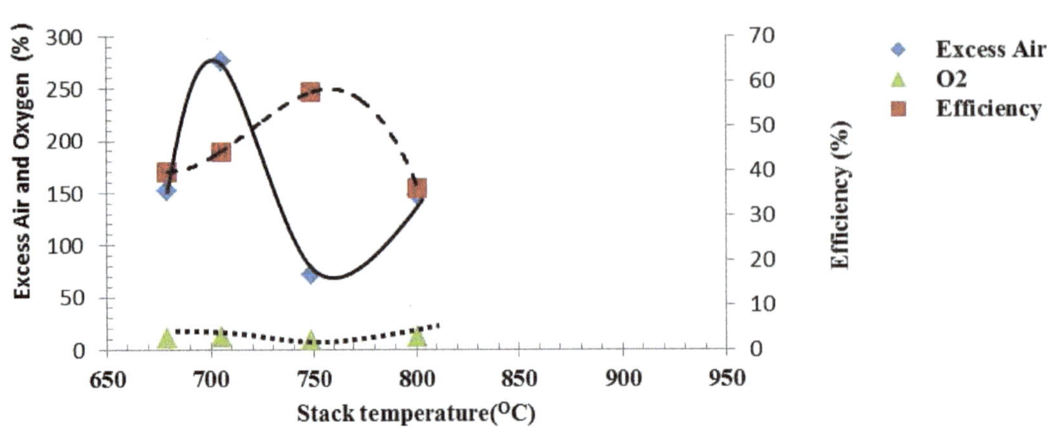


***Figure 9.*** *Variation of efficiency against Excess air and oxygen with primary chamber temperature 400°C.*

The curves shows that when the $O_2$ is about 6.9% the efficiency is at the highest and when the oxygen is closer to 20% the efficiency is lowest. This indicates that the combustion is poor when the flue gas composition has much $O_2$ concentration. The excess air also indicates the variation of temperature. As the excess air increases the temperature is decreases and the efficiency is decreases. The efficiency of incinerator seems to increase with increase in temperature.

#### 3.4.4. Analysis of Efficiency, Excess Air and O2 in Flue Gas when Primary Chamber Temperature Fixed at 500°C.

***Table 8.*** *The stack temperature against excess air and efficiency of incinerator with primary chamber temperature 500°C.*

| Prim/ Sec Temp. °C | $T_{sec}$ °C | $O_2$ (%) | η (%) | $T_{stack}$ °C | $T_{Amb}$ °C | $X_{Air}$ (%) |
|---|---|---|---|---|---|---|
| 500/(800, 900, 1000,1100) | 810.30 | 17.24 | 2.78 | 522.60 | 32.40 | 469.60 |
| | 913.70 | 17.32 | 0.00 | 557.40 | 32.30 | 480.50 |
| | 1055.30 | 8.64 | 47.68 | 897.80 | 32.30 | 70.46 |
| | 1088.00 | 12.26 | 26.24 | 893.20 | 33.80 | 214.24 |
| | 1093.70 | 10.64 | 36.96 | 816.00 | 33.80 | 107.40 |
| | 1102.70 | 12.26 | 21.36 | 888.60 | 33.80 | 268.34 |

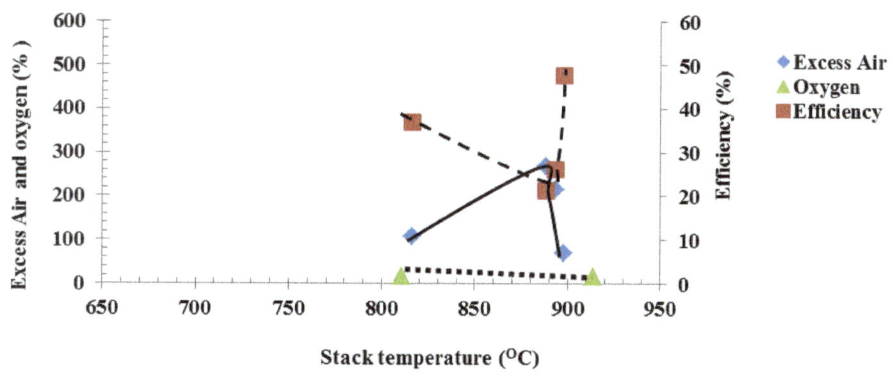


***Figure 10.*** *Variation of efficiency against Excess air and oxygen with primary chamber temperature 500°C.*

With primary chamber at 500°C, the excess air shows to increase at the flue gas composition, as the stack temperature increases, and then it reaches a point when the excess air start to decrease. At this point the $O_2$ concentration at flue gas decreases to about 8%, the efficiency start to increase, the efficiency reach its maximum value when the $O_2$ concentration reaches 6.9%. Showing there is a better combustion at this stage.

## 4. Conclusion and Recommendations

- The waste composition analysis shows that the waste can be used as energy source of as it contains more composition of biodegradable and combustible waste.
- The excess air concentration in the flue gas varies with temperature; it needs to reach a certain point so as to attain the maximum efficiency of the incinerator.
- The incinerator performance is better when the primary chamber set to about 400°C and the secondary chamber ranges between 800 °C and 900 °C

## References

[1] P. T. Williams, *Waste treatment and disposal*: John Wiley & Sons, 2005.

[2] M. Hizami, M. Yusoff, and R. Zakaria, "Combustion of Municipal Solid Waste in Fixed Bed Combustor for Energy Recovery," *Asian Network for Scientific Information,* vol. 12, pp. 1176-1180, 2012.

[3] R. K. Henry, Z. Yongsheng, and D. Jun, "Municipal solid waste management challenges in developing countries–Kenyan case study," *Waste management,* vol. 26, pp. 92-100, 2006.

[4] M. Rodionov and T. Nakata, "Design of an optimal waste utilization system: a case study in St. Petersburg, Russia," *Sustainability,* vol. 3, pp. 1486-1509, 2011.

[5] M. Sharholy, K. Ahmad, G. Mahmood, and R. Trivedi, "Municipal solid waste management in Indian cities–A review," *Waste management,* vol. 28, pp. 459-467, 2008.

[6] S. Rothenberger, C. Zurbrugg, I. Enayetullah, and M. Sinha, *Decentralised composting for cities of low and middle income countries: a userïs manual*: EAWAG, 2006.

[7] H. Cheng and Y. Hu, "Municipal solid waste (MSW) as a renewable source of energy: Current and future practices in China," *Bioresource technology,* vol. 101, pp. 3816-3824, 2010.

[8] V. Nasserzadeh, J. Swithenbank, C. Schofield, D. Scott, A. Loader, and A. Leonard, "Design optimization of Coventry municipal solid waste incinerator," *Journal of Environmental Engineering,* vol. 120, pp. 1615-1629, 1994.

[9] K. A. Arshad, T. Ahmad, and S. Baharun, "Mathematical Modeling Of A Clinical Waste Incineration Process," 2003.

[10] T. Kaosol, "Sustainable solutions for municipal solid waste management in Thailand," *Journal of Civil Engineering and Architecture,* vol. 4, pp. 57-65, 2010.

[11] A. Omari, B. Kichonge, G. John, K. Njau, and P. Mtui, "Potential of Municipal Solid Waste, as renewable energy source - a case study of Arusha, Tanzania.," *International Journal of Renewable Energy Technology Research,* vol. 3, pp. 1-9, 2014.

[12] P. Bundela, S. Gautam, A. Pandey, M. Awasthi, and S. Sarsaiya, "Municipal solid waste management in Indian cities–A review," *International Journal of Environmental Sciences,* vol. 1, pp. 591-606, 2010.

[13] O. Eriksson, M. Carlsson Reich, B. Frostell, A. Björklund, G. Assefa, J.-O. Sundqvist, *et al.*, "Municipal solid waste management from a systems perspective," *Journal of Cleaner Production,* vol. 13, pp. 241-252, 2005.

[14] C. Ryu and D. Shin, "Combined heat and power from municipal solid waste: Current status and issues in South Korea," *Energies,* vol. 6, pp. 45-57, 2012.

[15] D. Shin, S. Choi, J.-E. Oh, and Y.-S. Chang, "Evaluation of polychlorinated dibenzo-p-dioxin/dibenzofuran (PCDD/F) emission in municipal solid waste incinerators," *Environmental science & technology,* vol. 33, pp. 2657-2666, 1999.

[16] EA, "Advanced Thermal Treatment of Municipal Solid Waste," E. F. R. Affairs, Ed., ed. London: Defra, 2013.

[17] C. H. Lam, A. W. Ip, J. P. Barford, and G. McKay, "Use of incineration MSW ash: a review," *Sustainability,* vol. 2, pp. 1943-1968, 2010.

[18] E. Akkaya and A. Demir, "Predicting the heating value of municipal solid waste-based materials: An artificial neural network model," *Energy Sources, Part A: Recovery, Utilization, and Environmental Effects,* vol. 32, pp. 1777-1783, 2010.

[19] S. Manyele and H. Anicetus, "Management of medical waste in Tanzania hospitals," *Tanzania Journal of Health Research,* vol. 8, 2006.

[20] M. Annunziato, I. Bertini, A. Pannicelli, and S. Pizzuti, "Evolutionary Control and On-Line Optimization of an MSWC Energy Process," *Journal of Systemics, Cybernetics and Informatics,* vol. 4, 2006.

[21] M.-Y. Wey, W.-Y. Ou, Z.-S. Liu, H.-H. Tseng, W.-Y. Yang, and B.-C. Chiang, "Pollutants in incineration flue gas," *Journal of hazardous materials,* vol. 82, pp. 247-262, 2001.

[22] Y. Yang, M. J. Pijnenborg, M. A. Reuter, and J. Verwoerd, "Modeling the combustion behavior of hazardous waste in a rotary kiln incinerator," *Journal of Environmental Science and Health,* vol. 40, pp. 1823-1842, 2005.

[23] P. Glarborg, "Hidden interactions—Trace species governing combustion and emissions," *Proceedings of the combustion institute,* vol. 31, pp. 77-98, 2007.

[24] S. V. Manyele and I. S. Kagonji, "Analysis of medical waste incinerator performance based on fuel consumption and cycle times," 2012.

[25] D. Pal, "A numerical model of acidification of raindrops by oxidation of SO2," *International journal of environmental studies,* vol. 57, pp. 167-177, 2000.

[26] EPA. (1998, Nitrogen Oxides (NOx), Why and How They Are Controlled.

[27] D. W. Schindler, "Effects of acid rain on freshwater ecosystems," *Science(Washington),* vol. 239, pp. 149-157, 1988.

[28] T. A. Haines, "Acidic precipitation and its consequences for aquatic ecosystems: a review," *Transactions of the American Fisheries Society,* vol. 110, pp. 669-707, 1981.

[29] C. T. Driscoll, K. M. Driscoll, M. J. Mitchell, and D. J. Raynal, "Effects of acidic deposition on forest and aquatic ecosystems in New York State," *Environmental Pollution,* vol. 123, pp. 327-336, 2003.

[30] A. A. Mustabeen and I. Khan, "Phytotoxic effect of simulated acid rain on black gram," *Archives of Phytopathology and Plant Protection,* vol. 44, pp. 1343-1346, 2011.

[31] E. M. Khamala and A. A. Alex, "Municipal Solid waste composition and characteristics relevant to the waste to energy disposal method for Nairobi City," *Global Journal of Engineering, Design and Technology,* vol. 2, pp. 1-6, 2013.

[32] T. Ogwueleka, "Municipal solid waste characteristics and management in Nigeria," *Iranian Journal of Environmental Health Science & Engineering,* vol. 6, pp. 173-180, 2009.

[33] K. O. Boadi and M. Kuitunen, "Environmental and health impacts of household solid waste handling and disposal practices in third world cities: the case of the Accra Metropolitan Area, Ghana," *Journal of environmental health,* vol. 68, pp. 32-36, 2005.

[34] D. Hoornweg and P. Bhada-Tata, "What a waste: A Global review of solid waste Management," USA2012.

[35] ASTM, "ASTM D5231-92, Standard Test Method for the Determination of the Composition of Unprocessed Municipal Solid Waste," ed. West Conshohocken, PA: ASTM International, 2003.

[36] A. Omari, M. Said, K. Njau, G. John, and P. Mtui, "Energy recovery routes from Municipal Solid Waste, A case study of Arusha-Tanzania," *Journal of Energy Technologies and Policy,* vol. 4, pp. 1-7, 2014.

[37] Y. Zhao, W. Xing, W. Lu, X. Zhang, and T. H. Christensen, "Environmental impact assessment of the incineration of municipal solid waste with auxiliary coal in China," *Waste management,* vol. 32, pp. 1989-1998, 2012.

[38] K. C. Mukwana, S. R. Samo, M. M. Tunio, A. Q. Jakhrani, and M. R. Luhur, "Study of Energy potential from municipal solid waste of Mirpurkhas City," *Journal of engineering, Science & Technology,* vol. 13, pp. 26-28, 2014.

[39] L. Laureano-Perez, F. Teymouri, H. Alizadeh, and B. E. Dale, "Understanding factors that limit enzymatic hydrolysis of biomass," *Applied Biochemistry and Biotechnology,* vol. 124, pp. 1081-1099, 2005.

[40] Y. H. P. Zhang, S. Y. Ding, J. R. Mielenz, J. B. Cui, R. T. Elander, M. Laser, *et al.*, "Fractionating recalcitrant lignocellulose at modest reaction conditions," *Biotechnology and Bioengineering,* vol. 97, pp. 214-223, 2007.

[41] D. Fengel and G. Wegener, *Wood: chemistry, ultrastructure, reactions*: Walter de Gruyter, 1983.

[42] T. Atabarut and E. Ekinci, "Thermal treatment of landfill leachate and the emission control," *Journal of Environmental Science and Health Part A,* vol. 41, pp. 1931-1942, 2006.

[43] A. Omari, M. Said, K. Njau, G. John, and P. Mtui, "Energy recovery routes from Municipal Solid Waste, A case study of Arusha-Tanzania.," *Journal of Energy Technology and Policy,* vol. 4, pp. 1-7, 2014.

[44] G. Stubenberger, R. Scharler, S. Zahirović, and I. Obernberger, "Experimental investigation of nitrogen species release from different solid biomass fuels as a basis for release models," *Fuel,* vol. 87, pp. 793-806, 2008.

[45] W.-H. Chen and J.-C. Chen, "Combustion characteristics and energy recovery of a small mass burn incinerator," *International communications in heat and mass transfer,* vol. 28, pp. 299-310, 2001.

[46] P. Mtui, "Computational Analysis of a Fixed Bed Thermal Oxidizer for Solid Wastes Disposal," *Advanced Materials Research,* vol. 699, pp. 326-334, 2013.

3

# Hydrogeochemical Characteristics and Quality Assessment of Groundwater in University of Science and Technology Port Harcourt

**Tubonimi Joseph Kio Ideriah, Udonna Joseph Ikoro**

Institute of Pollution studies, Rivers State University of Science and Technology Nkpolu Oroworukwo Port Harcourt, Rivers State, Nigeria

**Email address:**
tjkideriah@ymail.com (T. J. K. Ideriah)

**Abstract:** The quality and suitability of boreholes water quality in the Rivers State University of Science and Technology were assessed for potability and irrigation purposes by analyzing the water for physico-chemical parameters, microbial contents and irrigation indices using standard methods. The results obtained were compared with permissible limits for drinking water provided by World Health Organization and Standard Organization of Nigeria The results showed pH ranged from 4.09 - 6.77 at 26.4 - 30.3°C, Turbidity <0.01 NTU in all the samples, Electrical Conductivity 20 - 407 µS/cm, Salinity <0.01 - 0.20‰,TDS 12 - 274 mg/l, Chloride <1.0 - 12.3 mg/l, Sulphate <1.0 - 15.5 mg/l, Phosphate <0.05 - 1.9 mg/l, Nitrate 0.30 - 6.20 mg/l, Total Alkalinity (as $CaCO_3$) 2 - 8 mg/l, Total Hardness <0.1-34.6 mg/l, Calcium <0.08 - 9.2 mg/l, Magnesium <0.05 - 2.8 mg/l, Sodium <0.01 - 44.62 mg/l and Potassium <0.01 - 11.88 mg/l. The results of microbial analyses showed Total Heterotrophic Bacteria population ranged from Nil - 3000 cfu/ml, Total Coliform Bacteria 0 - 210 MPN/100ml while Faecal Coliform Bacteria were not present in all the samples. The Groundwater within the University is fresh, soft and has low pH. The water in some parts had high microbial count and therefore not suitable for drinking. The ground water in the area should be regularly monitored and treated to avoid serious pollution problems. The irrigation indices showed the water is suitable for irrigation and other purposes.

**Keywords:** Groundwater Quality, Physicochemical Parameters, Irrigation Indices, Microbial Population, University

## 1. Introduction

The Niger Delta region is one of the largest wetlands in the world. Some rural inhabitants take what they can from the creeks, ponds and rivers. The Federal Ministry of Water Resources says efforts over the past century to develop National water resources have not yielded much. [1] quoted Central Bank of Nigeria statement that "the population of Nigerians with access to potable water rose from 30% in 1999 to 65% in 2005".

Water is an essential requirement for human and industrial developments. It is also used directly and indirectly by many people for several purposes. Water in general plays a critical part in the maintenance of plant and animal life. Owing to the presence of water in cells and body fluids, such as blood, human beings are approximately 60 -75% water [2].

The major sources of water include springs, ponds, streams, rivers, oceans, rain and ground water. The variety of water sources brings in water with different degrees of impurities. The presence of impurities therefore reduces the use to which the water may be deployed. In this study on Rivers State University of Science and Technology (RSUST), it is limited to bore holes.

For potable water, it must be safe for human consumption while the one for irrigation and some industrial processes may not be as pure as that for human consumption. So water must therefore be analyzed to determine its acceptability for the intended purpose. The levels of parameters obtained may be the cause for rejection or acceptance of the sample.

Sources of water for various uses include atmospheric (i.e. rainwater), surface water (i.e. streams, rivers, ponds, lakes and dams) and groundwater (i.e. springs, wells and mono pumps / boreholes). As a result of prevailing conditions connected with some of these sub-water sources; most industrial, governmental and private sectors in urban areas resort to boreholes as water source for their potable and various needs [3].

Sources of groundwater contamination or pollution include leachate from landfill / refuse dumpsite, industrial liquid effluent, domestic waste, agricultural waste, salt water intrusion, oil pollution and geological formations [4]. Many wells (i.e. groundwater sources) in the United States of America have been closed because of contamination by various toxic substances [5].

In the Niger Delta, water is mostly abstracted from the aquiferous Benin Formation [6]. In urban and industrial areas; hydrochemistry, geochemistry and processes at solid / liquid interfaces are among the important issues in environmental risk assessment studies in water resources policies [7]. The chemical composition of groundwater is dependent on the geology and the geochemical processes within the aquifer [8], [9].

Thus, water quality analyses are focal and imperative in groundwater investigation by monitoring both the water level (where possible) and trends of the water quality parameters that are influenced by the geological formations and the anthropogenic activities in a given area [10].

The contamination of shallow groundwater sources leading to incidents of water borne diseases like abdominal disorders, typhoid fever, dysentery and urinary track infectious has been reported in some communities [11],[12]. The main objective is to determine, establish and document the current status of boreholes water quality on the University campus

Port Harcourt, with population well over 1 million lies within latitudes $4^{\circ} 43' 07''$ and $4^{\circ} 54' 32''$ N and longitudes $6^{\circ} 56' 04''$ and $7^{\circ} 03' 20''$E with a mean annual rainfall of over 2000mm and mean annual temperature of $29^{\circ}$C [13]. The study area Rivers State University of Science and Technology (Fig. 1.1) is located within Port Harcourt metropolis in the Niger Delta sedimentary basin of Nigeria. Port Harcourt is covered on the surface by the Benin Formation which is otherwise called the Coastal Plain Sands. The Benin Formation is Miocene to Recent in age [14]. The domination of loose sands in the Benin Formation makes the ground in Port Harcourt porous and permeable to wastes on the soil surface. This is because during the rainy season, rainwater will cause leachates from the wastes to percolate downwards and pollute the groundwater over time.

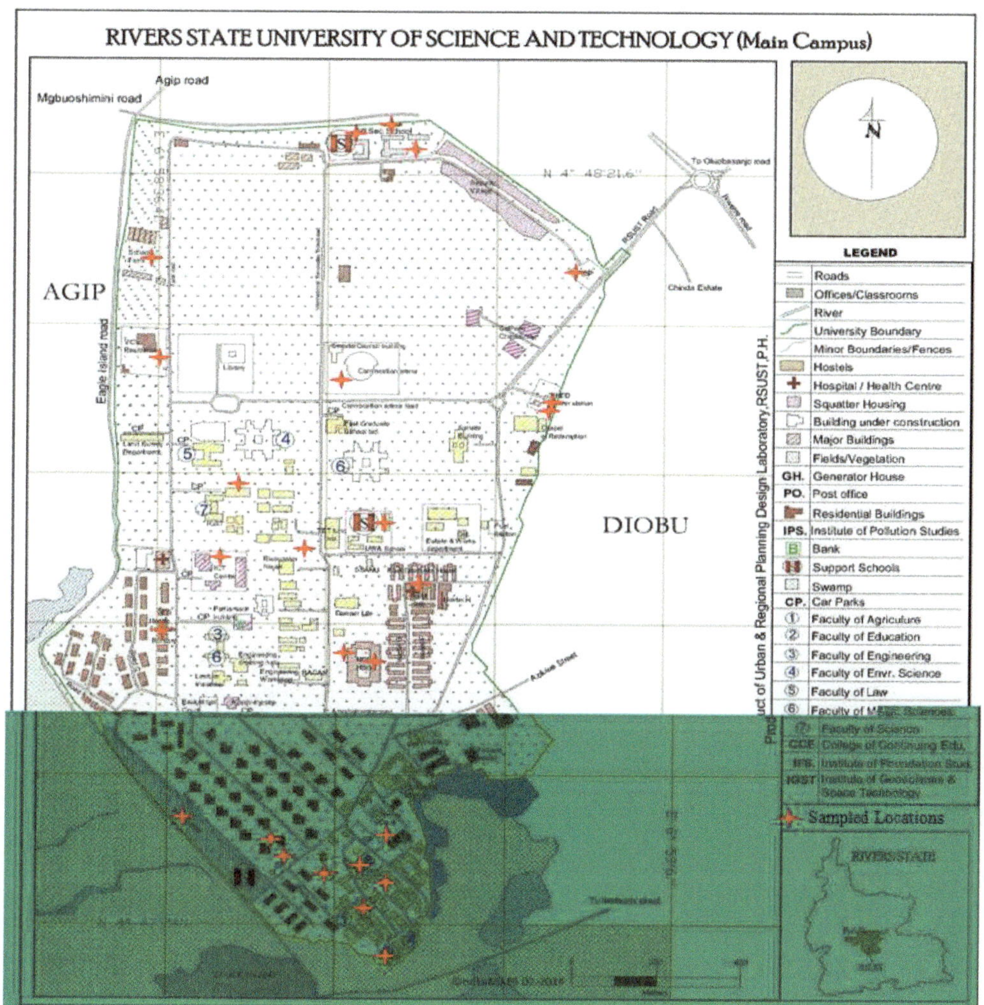


*Source*: Urban & Regional Planning Dept. RSUST, P.H.

***Figure 1.*** *Map of RSUST showing the sampling stations.*

The ground water level, measured from newly drilled boreholes show that water levels are close to the surface, commonly between 3.05 – 9.14metres. For example, the water level at East - West (Nkpolu, Rumuigbo) is 3.05metres but about 6.10metres at Eneka, etc. [15].

# 2. Materials and Methods

## 2.1. Sample Collection

Samples for physico-chemical analysis were collected in 1.5-litre plastic bottles while those for microbiology were collected in sterile McCartney bottle. All the samples were preserved in iced cool box and transported to the laboratory for analyses. Table 1 shows the sampling stations and GPS Co-ordinates taken with a hand held GPS by Garmin.

***Table 1.*** *Sampling Locations and their GPS Co-ordinates.*

| No. | Station | GPS (Latitude) N | GPS (Longitude) E |
|---|---|---|---|
| 1 | RSUST Main BH | 4° 43' 17.70" | 7° 12' 11.88" |
| 2 | Road B Bungalow 1 | 4° 40' 08.16" | 7° 13' 17.94" |
| 3 | Sick Bay | 4° 46' 42.84" | 7° 09' 21.66" |
| 4 | Biological Sciences | 4° 44' 47.52" | 7° 04' 51.96" |
| 5 | IPS Lab | 4° 50' 42.66" | 7° 03' 53.88" |
| 6 | Hostel E | 4° 51' 25.38" | 7° 03' 25.98" |
| 7 | VC's Office | 4° 44' 46.26" | 7° 02' 43.38" |
| 8 | DVC's Office | 4° 43' 00.72" | 7° 15' 09.06" |
| 9 | Fisheries Aq | 4° 39' 54.48" | 7° 22' 24.48" |
| 10 | Inst. of Education | 4° 41' 02.88" | 7° 18' 16.50" |
| 11 | Staff School | 4° 44' 03.12" | 6° 46' 19.98" |
| 12 | NDDC Hostel 1 | 4° 44' 45.18" | 6° 45' 58.62" |
| 13 | NDDC Hostel 2 | 4° 44' 17.16" | 6° 51' 43.02" |
| 14 | Health Services | 4° 28' 53.64" | 7° 25' 22.74" |
| 15 | ICT | 4° 26' 42.06" | 7° 10' 01.68" |
| 16 | Med Lab Sciences | 4° 53' 31.26" | 7° 07' 42.42" |
| 17 | UWA Day Care | 4° 57' 29.70" | 6° 35' 33.24" |
| 18 | Chapel 1 | 4° 56' 33.42" | 6° 35' 14.82" |
| 19 | Chapel 2 | 4° 30' 53.40" | 7° 32' 29.16" |
| 20 | Main Gate Park | 5° 06' 07.47" | 7° 07' 48.32" |
| 21 | Council Unit | 5° 05' 47.90" | 6° 49' 10.88" |
| 22 | ISS 1 | 5° 10' 56.04" | 6° 39' 57.21" |
| 23 | ISS 2 | 5° 09' 12.30" | 6° 38' 03.86" |
| 24 | ISS VP | 5° 04' 22.06" | 6° 39' 06.98" |
| 25 | VC's Lodge | 5° 04' 25.93" | 6° 33' 46.62" |
| 26 | Agric. Farm | 5° 04' 00.16" | 6° 26' 29.61" |

## 2.2. Field Measurements

Some *in situ* measurements were taken in the field for pH, Temperature, Conductivity, Salinity and Total Dissolved Solids using Extech DO700 meter calibrated with buffer pH 4.0, 7.0 and 10.0 as well as 1413 μS/cm conductivity solution.

## 2.3. Laboratory Analyses

Except otherwise stated, the laboratory methodologies used were from *Standard Methods for the Examination of Water and Wastewater* by American Public Health Association [16] and American Society for Testing & Material [17].

Total Alkalinity was determined by titration with 0.02N $H_2SO_4$ – using methyl orange indicator [16]. Chloride was determined titrimetrically by the Argentometric method in the presence of potassium chromate as the indicator. Phosphate was determined by the Ammonium Molybdate $(NH_4)_2MoO_4$) method. Nitrate was determined using the Brucine Method [16]. Sulphate was determined by Turbidimetric method [16].

100mls of water sample was taken. 2mls of buffer solution was added along with a pinch of Eriochrome black T indicator and titrated with 0.0IM EDTA until a blue colour was observed.

Sodium and Aluminium were determined by direct aspiration into Flame Atomic Emission Photometer.

## 2.4. Microbiology

The ten-fold serial dilution was used to obtain appropriate dilutions of the samples. Aliquots of the required dilutions were plated in duplicates onto the surface of dried sterile nutrient agar plates for total heterotrophic bacteria. In the case of total/faecal coliform bacteria, the most probable number (MPN) technique was employed for estimation of their numbers in water. Appropriate volumes of undiluted water samples were inoculated into test tubes of MacConkey broth containing Durham tubes. All inoculated media were incubated at 37°C for 24 hours except for faecal coliform bacteria which was incubated at 44°C [16].

# 3. Results

The range, mean and standard deviation values of the physico-chemical as well as microbiological properties of groundwater samples within the Rivers State University of Science and Technology for July and September 2013 are shown in Table 2. Also contained in the Table 2 are Standard Organization of Nigeria (SON) limits for drinking water and World Health Organization (WHO) standards [18],[19].

## 3.1. Physico-Chemical Parameters pH

The ground water pH ranged from 4.09 - 6.77 with a mean of 4.93±0.58. In July all the samples (100%) were in the range 4.30-5.40 while in September 12.5% of the samples were within 6.73-6.77 range whereas 87.5% of the samples had pH ranging from 4.09-6.39. The lowest pH 4.30 came from ISS2 in July while in September pH 4.09 was obtained at the Main Gate Motor Park (Fig. 2).

*Table 2. Ranges and Means of RSUST Groundwater in July and September 2013.*

| No. | Parameters | Min | Max | Mean | ±SD | SON (2007) | WHO (2006) |
|---|---|---|---|---|---|---|---|
| 1 | pH | 4.09 | 6.77 | 4.93 | 0. 58 | 6.5 - 8.5 | 6.5 - 8.5 |
| 2 | Temperature (°C) | 26.4 | 30.3 | 28.3 | 1.2 | NS | NS |
| 3 | Conductivity (µS/cm) | 20 | 407 | 103 | 102 | 1000 | 1200 |
| 4 | Salinity (‰) | <0.01 | 0.2 | 0.03 | 0.06 | NS | NS |
| 5 | TDS (mg/l) | 12 | 274 | 67 | 67 | 500 | 600 |
| 6 | Turbidity (NTU) | <0.05 | <0.05 | <0.05 | <0.05 | 5 | 5 |
| 7 | Chloride, $Cl^-$ (mg/l) | <1.0 | 12.3 | 2.9 | 2.8 | 250 | 250 |
| 8 | Sulphate, $SO_4^{-2}$ (mg/l) | <1.0 | 15.5 | 2.4 | 3.0 | 100 | 250 |
| 9 | Phosphate, $PO_4^{-3}$ (mg/l) | <0.05 | 1.90 | 0.05 | 0.29 | NS | NS |
| 10 | Nitrate, $NO_3^-$ (mg/l) | 0.30 | 6.20 | 2.65 | 1.32 | 50 | 50 |
| 11 | Alkalinity ((mg/l as $CaCO_3$) | 2 | 8 | 3.8 | 1.8 | 100 | 100 |
| 12 | Carbonate, $CO_3^{-2}$ (mg/l) | 0 | 0 | 0 | 0 | NS | NS |
| 13 | Bicarbonate, $HCO_3^-$ (mg/l) | 4 | 16 | 7.5 | 3.7 | NS | NS |
| 14 | Hardness (mg/l as $CaCO_3$) | <0.1 | 34.6 | 8.0 | 7.8 | 150 | 200 |
| 15 | Calcium, $Ca^{+2}$ (mg/l) | <0.08 | 9.2 | 1.7 | 1.9 | NS | NS |
| 16 | Magnesium, $Mg^{+2}$ (mg/l) | <0.05 | 2.8 | 0.9 | 0.8 | 0.2 | NS |
| 17 | Sodium, $Na^+$ (mg/l) | <0.01 | 44.62 | 11.21 | 13.02 | 200 | 200 |
| 18 | Potassium, $K^+$ (mg/l) | <0.01 | 11.88 | 1.50 | 2.64 | NS | NS |
| 19 | THB (cfu/ml) | Nil | 3000 | 109 | 570 | NS | <100 |
| 20 | TCB (MPN/100ml) | Nil | 210 | 8.6 | 43.6 | 10 | 0-2 |
| 21 | FCB (MPN/100ml) | Nil | Nil | Nil | Nil | 0 | 0 |

Note: <0.05 = Less than detection limit; NS = Not Specified.

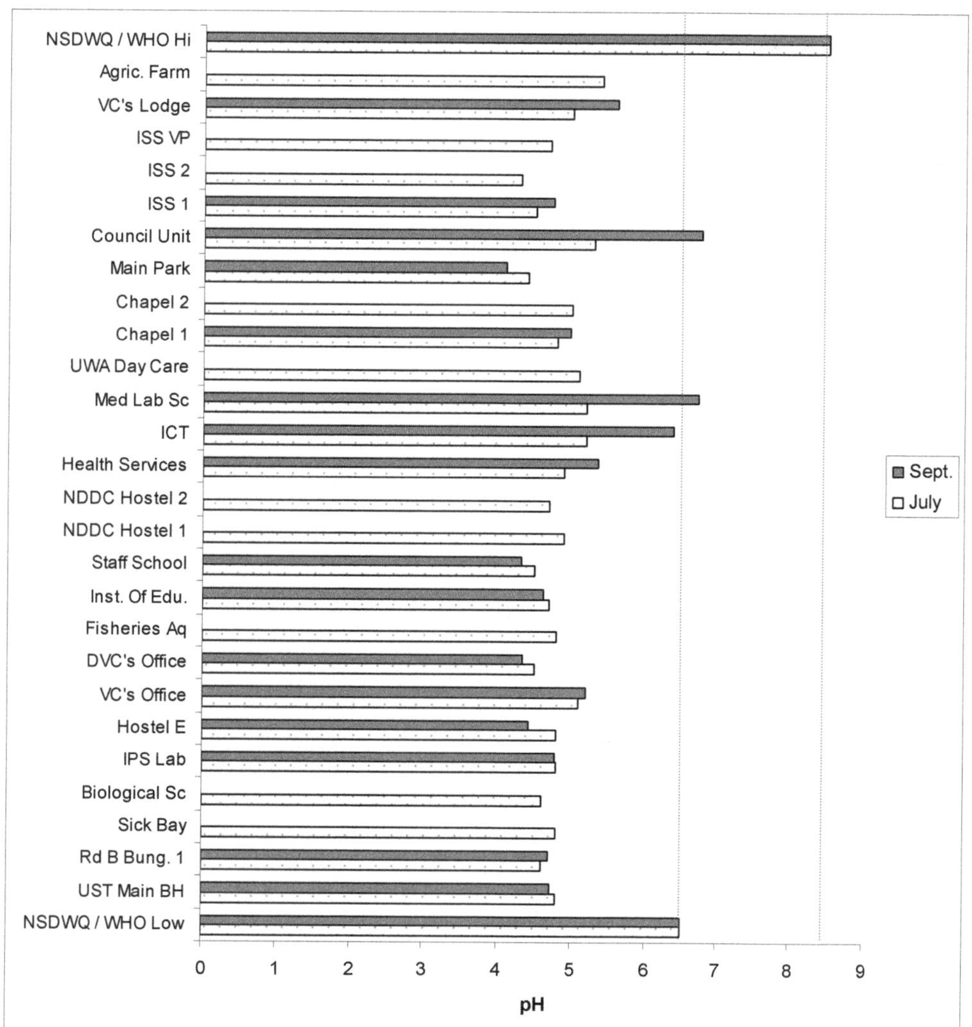


*Figure 2. The pH of groundwater in RSUST in July and September 2013.*

### 3.2. Temperature and Turbidity

Groundwater Temperature ranged from 26.4 - 30.3°C with a mean of 28.3±1.2°C. Turbidity was <0.05 NTU in all the samples.

### 3.3. Conductivity, Salinity and TDS

Conductivity values ranged from 20 - 407 µS/cm with a mean of 103±102 µS/cm. Salinity values ranged from <0.01 - 0.20‰ with a mean of 0.03±0.06‰. Similarly, TDS values varied from 12 - 274 mg/l with a mean of 67±67 mg/l (Fig. 3).

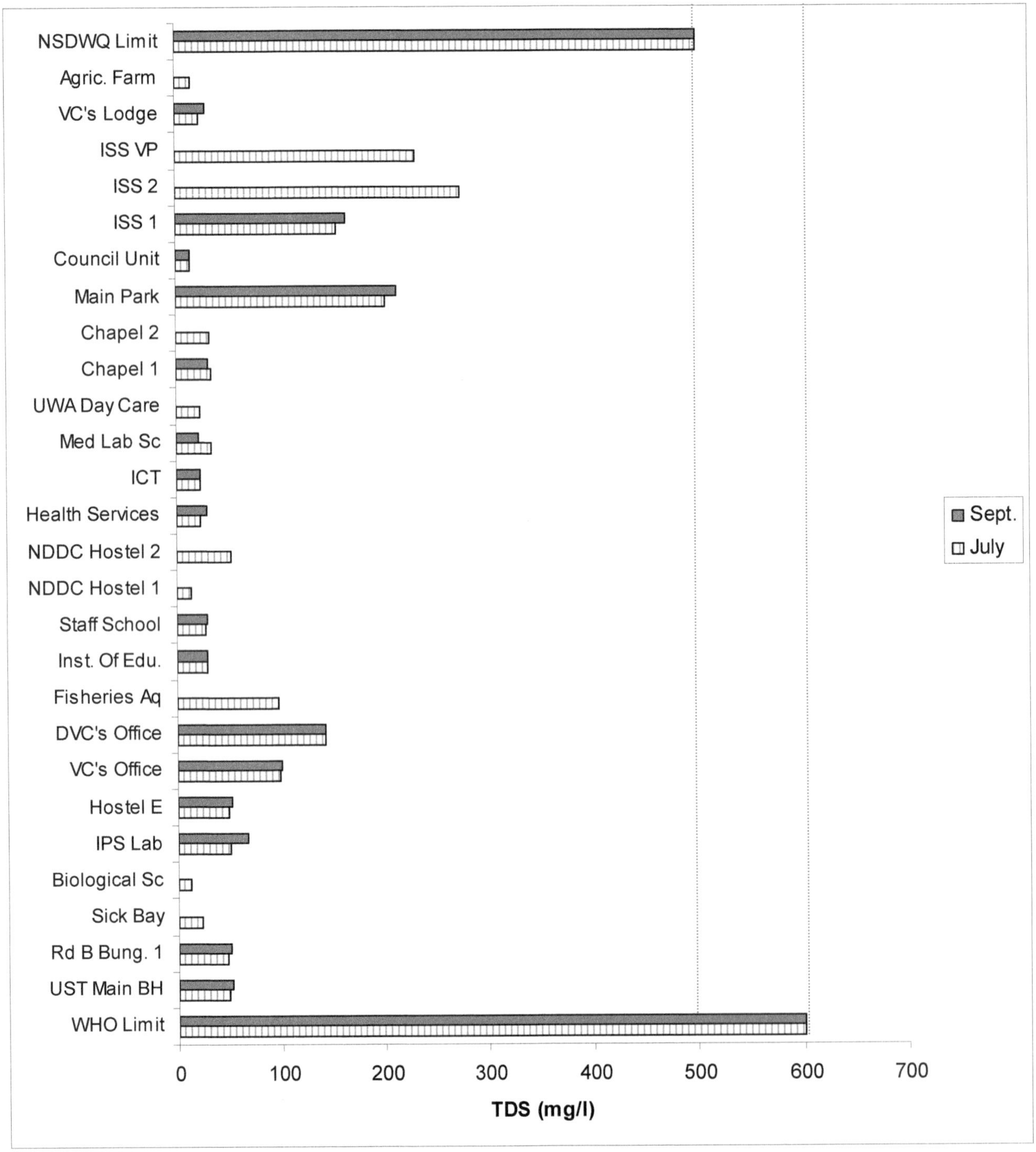


*Figure 3. TDS levels of RSUST groundwater in July and September 2013.*

The correlation existing among the parameters measured are shown in Table 3. TDS correlated strongly with the Conductivity, Salinity, Chloride, Hardness, Calcium, Magnesium, Sodium and Potassium.

### *3.4. Chloride, Sulphate, Phosphate and Nitrate*

Chloride concentrations ranged from <1.0 to 12.3 mg/l with a mean of 2.9±2.6 mg/l. Similarly, Sulphate concentrations varied from <1.0 - 15.5 mg/l with a mean of 2.4±3.0 mg/l. Phosphate concentrations ranged from <0.05 - 1.9 mg/l with a mean of 0.05±0.29 mg/l while the nitrate concentrations varied from 0.30 - 6.20 mg/l with a mean of 2.65±1.32 mg/l.

### *3.5. Total Alkalinity and Hardness*

Total Alkalinity values (as $CaCO_3$) were from 2 - 8 mg/l with a mean of 3.8±1.8 mg/l whereas the Hardness values varied from <0.1-34.6 mg/l with a mean of 8.0±7.8 mg/l.

### *3.6. Calcium and Magnesium*

The Calcium concentrations ranged from <0.08 - 9.2 mg/l with a mean of 1.7±1.9 mg/l whereas Magnesium values varied from <0.05 - 2.8 mg/l with a mean of 0.9±0.8 mg/l.

### *3.7. Sodium and Potassium*

Sodium concentrations ranged from <0.01 - 44.62 mg/l with a mean of 11.21±13.2 mg/l whereas Potassium concentrations ranged from <0.01 - 11.88 mg/l with a mean of 1.50±2.64 mg/l.

***Table 3.*** *Correlation matrix of parameters measured in RSUST groundwater.*

| | pH | Temp | Cond | Sal | TDS | Turb | Cl | $SO_4$ | $PO_4$ | $NO_3$ | Alk. | $CO_3$ | $HCO_3$ | Hard | Ca | Mg | Na | K | THB | TCB | FCB |
|---|---|---|---|---|---|---|---|---|---|---|---|---|---|---|---|---|---|---|---|---|---|
| pH | 1 | | | | | | | | | | | | | | | | | | | | |
| Temp | 0.098 | 1 | | | | | | | | | | | | | | | | | | | |
| Cond. | -0.462 | 0.041 | 1 | | | | | | | | | | | | | | | | | | |
| Sal | -0.373 | 0.042 | 0.958 | 1 | | | | | | | | | | | | | | | | | |
| TDS | -0.458 | 0.038 | 0.998 | 0.962 | 1 | | | | | | | | | | | | | | | | |
| Turb. | DIV/0! | DIV/0! | DIV/0! | DIV/0! | DIV/0! | 1 | | | | | | | | | | | | | | | |
| Cl | -0.402 | 0.272 | 0.780 | 0.769 | 0.766 | DIV/0! | 1 | | | | | | | | | | | | | | |
| $SO_4$ | -0.261 | -0.134 | 0.448 | 0.477 | 0.454 | DIV/0! | 0.208 | 1 | | | | | | | | | | | | | |
| $PO_4$ | -0.009 | -0.176 | -0.107 | -0.089 | -0.105 | DIV/0! | -0.096 | -0.024 | 1 | | | | | | | | | | | | |
| $NO_3$ | -0.251 | 0.357 | 0.437 | 0.366 | 0.413 | DIV/0! | 0.583 | 0.007 | -0.113 | 1 | | | | | | | | | | | |
| Alk. | 0.364 | 0.395 | -0.162 | -0.081 | -0.177 | DIV/0! | 0.001 | -0.055 | 0.021 | 0.323 | 1 | | | | | | | | | | |
| $CO_3$ | DIV/0! | DIV/0! | DIV/0! | DIV/0! | DIV/0! | DIV/0! | DIV/0! | DIV/0! | DIV/0! | DIV/0! | DIV/0! | 1 | | | | | | | | | |
| $HCO_3$ | 0.364 | 0.395 | -0.162 | -0.081 | -0.177 | DIV/0! | 0.001 | -0.055 | 0.021 | 0.323 | 1 | DIV/0! | 1 | | | | | | | | |
| Hard | -0.380 | 0.135 | 0.662 | 0.625 | 0.662 | DIV/0! | 0.669 | 0.662 | -0.005 | 0.301 | -0.154 | DIV/0! | -0.154 | 1 | | | | | | | |
| Ca | -0.341 | 0.178 | 0.578 | 0.553 | 0.578 | DIV/0! | 0.644 | 0.646 | -0.014 | 0.310 | -0.091 | DIV/0! | -0.091 | 0.971 | 1 | | | | | | |
| Mg | -0.401 | 0.068 | 0.733 | 0.682 | 0.734 | DIV/0! | 0.654 | 0.612 | 0.000 | 0.278 | -0.215 | DIV/0! | -0.215 | 0.937 | 0.827 | 1 | | | | | |
| Na | -0.459 | 0.014 | 0.953 | 0.897 | 0.953 | DIV/0! | 0.722 | 0.321 | -0.098 | 0.415 | -0.230 | DIV/0! | -0.230 | 0.559 | 0.487 | 0.627 | 1 | | | | |
| K | -0.371 | 0.151 | 0.761 | 0.714 | 0.762 | DIV/0! | 0.736 | 0.213 | -0.065 | 0.510 | -0.229 | DIV/0! | -0.229 | 0.589 | 0.556 | 0.595 | 0.798 | 1 | | | |
| THB | -0.121 | 0.227 | -0.113 | -0.103 | -0.102 | DIV/0! | -0.139 | -0.158 | -0.053 | 0.057 | 0.034 | DIV/0! | 0.034 | -0.078 | -0.071 | -0.061 | -0.118 | -0.146 | 1 | | |
| TCB | -0.121 | 0.374 | -0.137 | -0.130 | -0.132 | DIV/0! | -0.146 | -0.121 | -0.043 | -0.016 | -0.089 | DIV/0! | -0.089 | -0.055 | -0.053 | -0.048 | -0.140 | -0.141 | 0.733 | 1 | |
| FCB | DIV/0! | DIV/0! | DIV/0! | DIV/0! | DIV/0! | DIV/0! | DIV/0! | DIV/0! | DIV/0! | DIV/0! | DIV/0! | DIV/0! | DIV/0! | DIV/0! | DIV/0! | DIV/0! | DIV/0! | DIV/0! | DIV/0! | DIV/0! | 1 |

Table 4 contains the July and September means of the parameters required in the characterization of RSUST groundwater. A plot of these hydro-geochemical data on the Piper diagram [20] is shown in Fig. 4.

***Table 4.*** *Mean values of the hydro-geochemical data of RSUST groundwater.*

| | | $Ca^{+2}$ | $Mg^{+2}$ | $Na^{+}$ | $K^{+}$ | $CO_3^{-2}$ | $HCO_3^{-}$ | $Cl^{-}$ | $SO_4^{-2}$ | TDS |
|---|---|---|---|---|---|---|---|---|---|---|
| S/No | Station | mg/l | mg/l | mg/l | mg/l | mg/l | mg/l | mg/l | mg/l | mg/l |
| 1 | UST Main BH | 1.5 | 1.4 | 13.49 | 2.33 | 0 | 6 | 2.8 | 0.9 | 49.5 |
| 2 | Rd B Bung. 1 | 1.5 | 0.9 | 8.73 | 0.71 | 0 | 6 | 2.6 | 1.5 | 48.7 |
| 3 | Sick Bay | 1.5 | 0.9 | 3.36 | 0.41 | 0 | 8 | 2.4 | 2.1 | 22 |
| 4 | Biological Sc | 0.0 | 0.0 | 1.58 | 0.21 | 0 | 8 | 0.0 | 1.9 | 12 |
| 5 | IPS Lab | 1.15 | 0.7 | 13.50 | 2.12 | 0 | 10 | 4.6 | 0.9 | 74.5 |
| 6 | Hostel E | 3.8 | 1.85 | 13.60 | 1.32 | 0 | 12 | 4.9 | 5.9 | 95 |
| 7 | VC's Office | 1.15 | 0.7 | 5.95 | 0.41 | 0 | 8 | 3.7 | 1.4 | 63.5 |
| 8 | DVC's Office | 4.6 | 1.4 | 11.11 | 0.91 | 0 | 8 | 3.8 | 7.75 | 85.5 |
| 9 | Fisheries Aq | 1.5 | 0.9 | 18.26 | 1.62 | 0 | 8 | 4.4 | 4.6 | 97 |
| 10 | Inst. of Edu. | 1.15 | 0.7 | 2.57 | 0.01 | 0 | 6 | 1.4 | 0.9 | 25 |
| 11 | Staff School | 0 | 0 | 2.57 | 0.21 | 0 | 10 | 1.2 | 0.7 | 24 |
| 12 | NDDC Hostel 1 | 0.0 | 0.0 | 0.78 | 0.21 | 0 | 4 | 0.0 | 2.2 | 13 |
| 13 | NDDC Hostel 2 | 1.5 | 0.9 | 6.34 | 3.03 | 0 | 4 | 1.0 | 2.8 | 51 |
| 14 | Health Serv. | 1.5 | 0.9 | 1.68 | 0.21 | 0 | 8 | 1.1 | 1.55 | 18 |
| 15 | ICT | 1.15 | 0.7 | 20.32 | 1.32 | 0 | 6 | 4.4 | 1.35 | 92.5 |
| 16 | Med Lab Sc | 1.55 | 0.45 | 3.56 | 0.11 | 0 | 8 | 1.5 | 1.85 | 30.5 |
| 17 | UWA Day Care | 0.0 | 0.0 | 3.56 | 0.61 | 0 | 8 | 1.4 | 2.1 | 22 |
| 18 | Chapel 1 | 0.75 | 0.45 | 2.38 | 0 | 0 | 2 | 0.85 | 1.05 | 16.5 |
| 19 | Chapel 2 | 1.5 | 0.9 | 4.75 | 0.00 | 0 | 4 | 1.2 | 2.6 | 32 |
| 20 | Main Park | 3.85 | 0.95 | 22.31 | 5.94 | 0 | 2 | 5.15 | 1.2 | 101 |
| 21 | Council Unit | 0.75 | 0.45 | 0 | 0 | 0 | 4 | 0.5 | 1.35 | 7 |
| 22 | ISS 1 | 0.75 | 0.45 | 18.84 | 1.31 | 0 | 2 | 0.7 | 1.2 | 76.5 |
| 23 | ISS 2 | 3.1 | 2.8 | 43.63 | 3.03 | 0 | 4 | 8.1 | 2.7 | 274 |
| 24 | ISS VP | 1.5 | 1.9 | 39.66 | 6.45 | 0 | 4 | 1.0 | 11.3 | 230 |
| 25 | VC's Lodge | 0 | 0 | 2.28 | 0.11 | 0 | 4 | 1.35 | 1.25 | 11 |
| 26 | Agric. Farm | 1.5 | 0.9 | 1.77 | 0.21 | 0 | 8 | 1.2 | 2.0 | 15 |

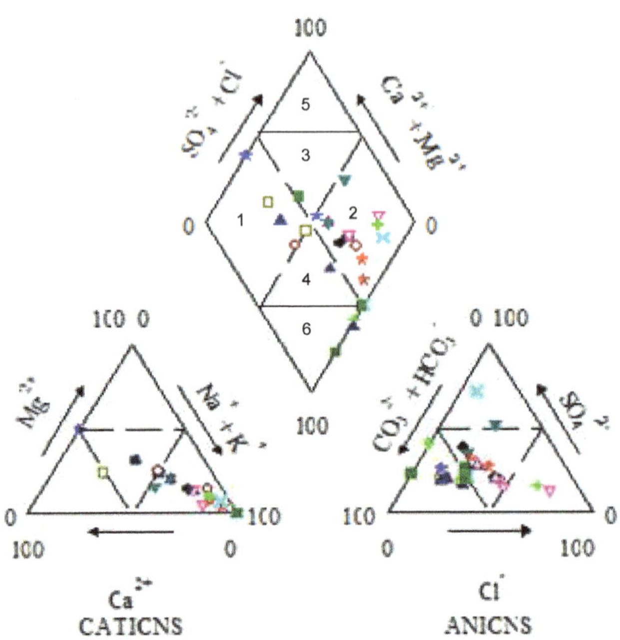


***Figure 4.*** *Piper tri-linear diagram showing groundwater classification.*

Graphical presentation showing order of occurrence of the major anions ($HCO_3^-$, $Cl^-$, $SO_4^{2-}$ and $CO_3^{2-}$) and the corresponding major cations (Na, Ca, K and Mg) are shown in the Pie chart (Fig. 5).

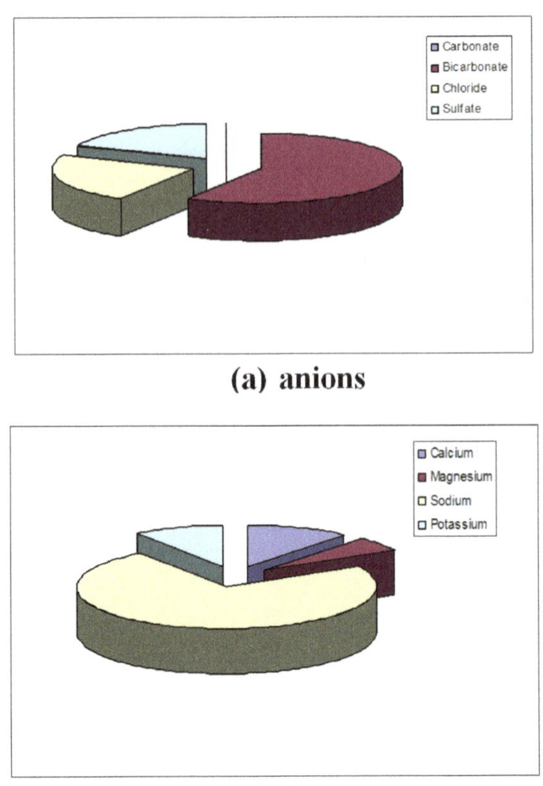


***Figure 5.*** *Pie chart showing major ions in groundwater.*

### 3.8. Microbiological Analysis

The results of the microbial analyses showed Total heterotrophic bacteria (THB) population ranging from Nil - 3000 cfu/ml with mean of 109±570 cfu/ml; while total coliform bacteria (TCB) values varied from Nil - 210 MPN/100ml with a mean of 8.6±43.6 MPN/100ml. Faecal coliform bacteria (FCB) were not present in all the samples. The highest THB (3000 cfu/ml) and TCB (210 MPN/100ml) occurred at the Institute of Education (Figs. 6 and 7).

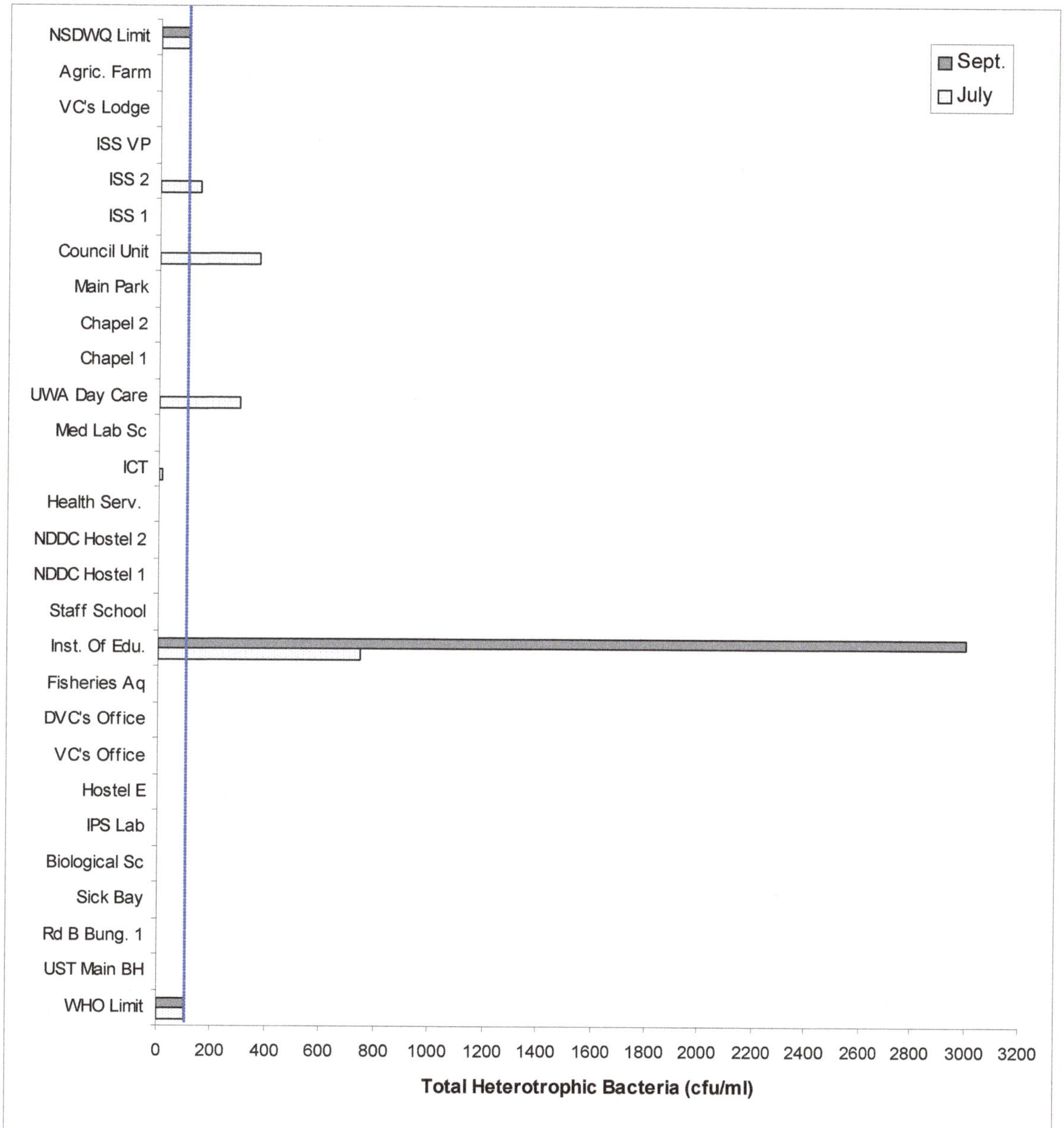


*Figure 6. Total Heterotrophic Bacteria levels in RSUST groundwater July/Sept. 2013.*

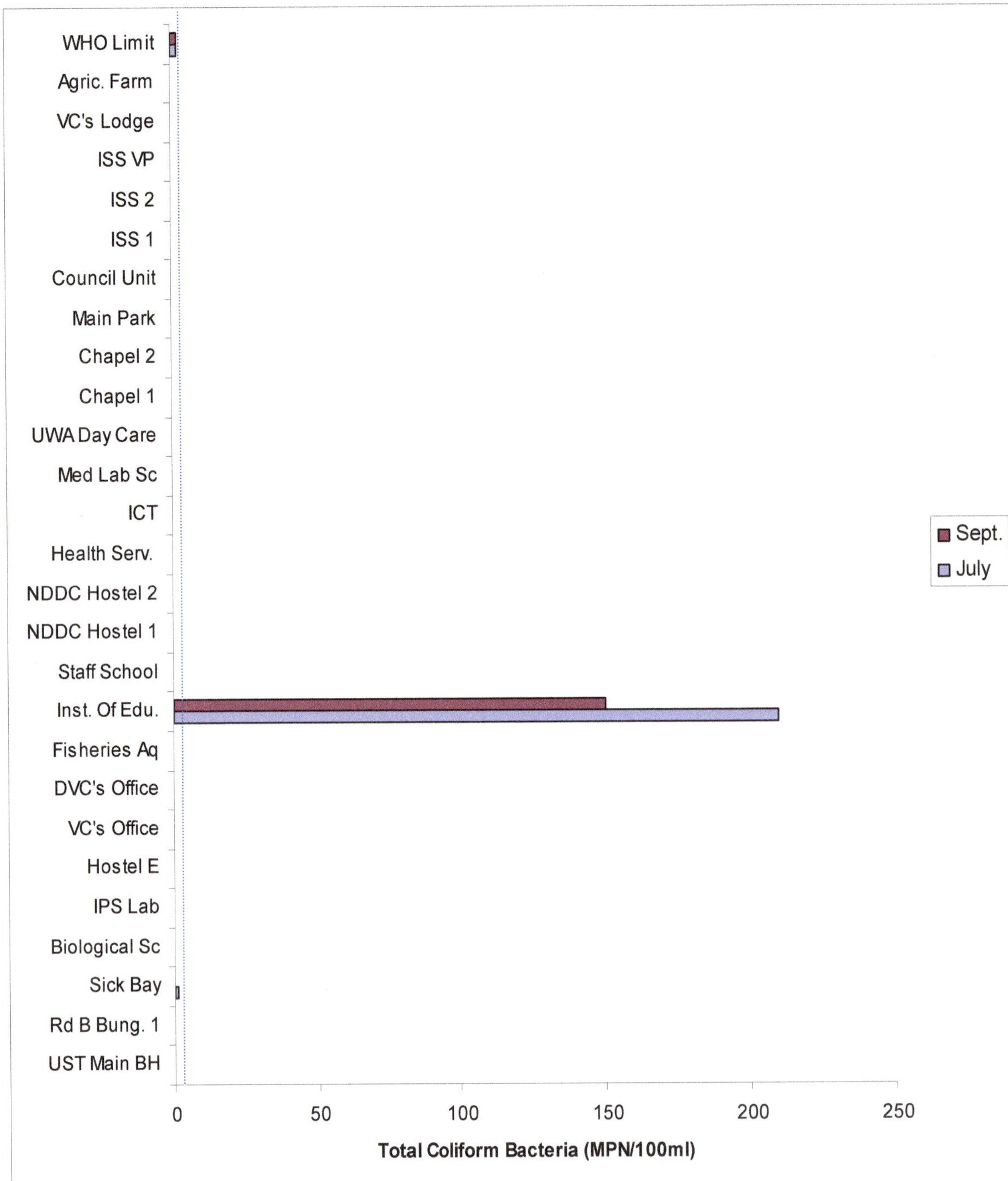


***Figure 7.*** *Total Coliform Bacteria levels in RSUST groundwater July/Sept. 2013.*

# 4. Discussions

### *4.1. Physico-Chemical Parameters pH*

The groundwater pH, a measure of the hydrogen ion concentration depicted variation between acidic (pH 4.09) and near neutral (pH 6.77). This implies that the groundwater in the area is acidic. The low pH values (4.0-4.9) indicates that the aquifer may be associated with waters containing free acids derived from oxidizing Sulphide minerals of the parent bedrock materials. Those that have moderately low pH values (5.0-5.9) might be associated with small amount of mineral acids from Sulphide sources or with organic acids from decaying vegetation. The moderate pH values (6.39-6.77) which are slightly acidic in reaction occurred around ICT, Medical Lab. Sc. and Council Unit. This could be attributed to aquifer having high bicarbonate content [21].

It is observed in Fig 2 that all (100%) of the water samples were below both national and international (SON and WHO) drinking water standards in July while only two stations (Medical Laboratory Sciences and Council Unit with pH of 6.73 and 6.77 respectively) representing 12.5% of the samples had pH values within acceptable limits in September. The remaining 87.5% of the samples were below the acceptable limit (pH 6.5-8.5).

At low pH, dissolution of metals / absorption of toxic substances increase especially in high Carbonate and low Silicate soils. It promotes colour, affects alkalinity, TDS, $CO_2$ absorption and Total Coliform presence [22], [23] - [24].

Prolonged intake of acidic water may predispose one to the dangers of acidosis, which according to Health Experts may lead to cancer or cardiovascular damage including the constriction of blood vessels and reduction in Oxygen supply even at mild levels [25]. It could also cause leaching of valuable minerals such as Calcium, Magnesium, Sodium and Calcium from the body.

Closely related to the pH is the alkalinity, which is a measure of the buffering capacity of the system. The recorded values were low (2 - 8 mg/l as $CaCO_3$) and they are mostly due to bicarbonate contents.

Water hardness is another quality parameter that determines the use of water for drinking / domestic and industrial purposes. Simply put, the hardness is the capacity of water to lather on the application of soap and these increases with the softness of the water. The hardness level of <0.1 - 34.6 mg/l as $CaCO_3$ is within the 0 - 60 mg/l as $CaCO_3$ classification of soft water. These hardness values are within the national and international limits. However, continued intake of soft water has been linked to cardiovascular diseases incidents [26], [27].

The TDS, which is a measure of all the dissolved substances in the water, was from 12 to 274 mg/l. These values are within WHO (600 mg/l) and SON (500 mg/l) limits for drinking water. TDS correlated positively with conductivity, salinity, chloride, sulfate and nitrate. The levels of these parameters in the groundwater were within the stipulated standards (Table 1). All the conductivity values are within acceptable limit (1000 µS/cm, SON and 1200 µS/cm WHO). The TDS values ranged from 12 mg/l to 274 mg/l with mean of 67±67 mg/l. These values are within both [18] and [19] potable water limits (Fig. 3).

The values obtained in the study for Conductivity, Salinity and TDS indicate that the ground water is fresh [21].

### 4.2. Temperature and Turbidity

The groundwater temperatures (26.6-30.3°C) in July and September showed no significant difference indicating similarity in chemical behavior in the water characteristics within the study area. Turbidity values in both periods were <0.01 NTU indicating clear water devoid of suspended solids caused by clay, silt and other substances that enter boreholes from the aquifer or from the soil surface.

### 4.3. Chloride, Sulphate, Phosphate and Nitrate

The low Chloride concentrations (<1.0 - 10.3 mg/l) indicate that the aquifer recharge is high due to high rainfalls, it is not overdrawn and there are no contact with water of marine origin or leaching from the upper soil layers. RSUST is situated close to the mangrove swamp and overdrawn aquifer may give rise to saltwater intrusion.

Sulphate contents are attributable to the sedimentary basin of the Niger Delta region. The low Sulphate levels (<1.0 - 15.5 mg/l) could be related to the removal of Sulphate by Sulphur bacteria in the sub-surface water [28].

Phosphate concentrations are mostly less than 0.05 mg/l; with the exception of water at Health Services station which had 1.90 mg/l. This is indicative of absence of Phosphorus - containing mineral apatite in the area. There is no potable water standard for Phosphorus by SON and WHO.

Nitrate levels in July (0.30 - 4.30 mg/l) are lower than those of September (1.0 - 6.20 mg/l). These Nitrate concentrations are within both [18] and [19] limits of 50 mg/l. The higher levels in September could be due to leaching by percolating water that reached the groundwater fast because nitrate compounds are highly soluble [28].

### 4.4. Microbial Analysis

Total heterotrophic bacteria (THB) population in some boreholes exceeded WHO limit (100 cfu/ml) [18]. The boreholes (15.4%) that exceeded stipulated WHO limit for THB were ISS2, UWA Day Care, Council Unit and Institute of Education. This implies that 15.4% of the boreholes studied had unacceptable THB values.

Total coliform bacteria (TCB) in the borehole at Institute of Education exceeded SON limit of 0-2 MPN/100ml [19]. The non-detection of faecal coliform bacteria in all the samples indicates no pollution with faecal matters.

The presence of microbes could be attributed to myriads of activities of microorganisms in the subsurface, shallow depth and water pressures not being high enough to deter microbial activity as many bacteria can survive under high osmotic pressures [29]. Also indigenous bacterial activity and active micro-flora exist in deep formations due to contamination by surface water seepage to an aquifer unprotected by relatively fine textured soil [30]. Droppings from birds into open water tanks also contributed to microbial contamination of the water. This could be the case of the Institute of Education borehole water which is exposed, located near the mangrove swamp forest and may be vulnerable to migration of contaminants from the creek water if the subsoil is coarse-textured.

### 4.5. Water Characterization

The groundwater in Rivers State University of Science and Technology has been classified based on the hydro-geochemical characteristics obtained in the Piper's diagram (Fig. 4).

There are five water types characterizing the aquifer and they include: Ca - $HCO_3^-$, Na - Cl, Ca - Mg - $Cl^-$, Ca - Na - $HCO_3^-$ and Na - $HCO_3^-$ types. The order of occurrence were: Na - $Cl^-$ type (57.7%), Na - Ca - $HCO_3^-$ type (17.3%), $HCO_3^-$ type (13.5%), Na - $HCO_3^-$ type (7.7%) and Ca - Mg - $Cl^-$ type (3.8%). The Ca - Cl type was not observed in this study.

### 4.6. Suitability for Irrigation

Sodium gets to the aquifer from rainwater in coastal areas

and / or dissolution of rock as rainwater percolates and the groundwater flows through the aquifer. As a result of effects of sodium on soil and plants; it is considered one major factor that governs the use of groundwater in irrigation [31], [32].

The suitability of groundwater for agricultural purposes (such as irrigation) is based on its Sodium Adsorption Ratio (SAR). The SAR was calculated using the formula (Richards, 1954):

$$SAR = \frac{(Na^{+})}{\sqrt{\{ ½[(Ca^{2+}) + (Mg^{2+})]\}}}$$

Where, $Ca^{2+}$, $Mg^{2+}$ and $Na^{+}$ are in mili-eqivalent per litre (meq/l) concentration of the metals in the groundwater.

There are four sodium hazard classifications [33]: low/S1 (SAR <10), medium/S2 (SAR 10 - 18), high/S3 (SAR 18 - 26) and very high/S4 (SAR >26).

Also, the soluble sodium percent (SSP) is another parameter used to indicate water that is suitable for irrigation. It was calculated from the formula:

$$SSP = [Na^{+} / (Ca^{+2} + Mg^{+2} + Na^{+})] \times 100$$

Where, $Ca^{2+}$, $Mg^{2+}$ and $Na^{+}$ are concentrations in meq/l. SSP values less or equal to 50 indicates good quality water while values greater than 50 are contrary and unsuitable for irrigation. The SAR and SSP values obtained for RSUST water samples are in Table 5.

***Table 5.*** *SAR and SSP values and status of RSUST water samples.*

| S/No. | Station / Location | SAR | SAR Status | SSP | SSP Status |
|---|---|---|---|---|---|
| 1. | UST Main BH | 1.903 | Excellent | 60.616 | Unsuitable |
| 2. | Road B Bungalow 1 | 1.392 | Excellent | 49.734 | Good |
| 3. | Sick Bay | 0.536 | Excellent | 49.734 | Good |
| 4. | Biological Sc | 1.079 | Excellent | 50.755 | Unsuitable |
| 5. | IPS Lab | 2.448 | Excellent | 50.094 | Fair |
| 6. | Hostel E | 1.430 | Excellent | 44.531 | Good |
| 7. | VC's Office | 1.078 | Excellent | 50.094 | Fair |
| 8. | DVC's Office | 1.164 | Excellent | 33.417 | Good |
| 9. | Fisheries Aq | 2.911 | Excellent | 49.734 | Good |
| 10. | Inst. Of Education | 0.466 | Excellent | 50.094 | Fair |
| 11. | Staff School | 1.759 | Excellent | 50.755 | Unsuitable |
| 12. | NDDC Hostel 1 | 0.533 | Excellent | 50.755 | Unsuitable |
| 13. | NDDC Hostel 2 | 1.011 | Excellent | 49.734 | Good |
| 14. | Health Services | 0.268 | Excellent | 49.734 | Good |
| 15. | ICT | 3.686 | Excellent | 50.094 | Fair |
| 16. | Med Lab Sc | 0.648 | Excellent | 32.376 | Good |
| 17. | UWA Day Care | 2.432 | Excellent | 50.755 | Unsuitable |
| 18. | Chapel 1 | 0.535 | Excellent | 49.734 | Good |
| 19. | Chapel 2 | 0.757 | Excellent | 49.734 | Good |
| 20. | Main Park | 2.640 | Excellent | 28.922 | Good |
| 21. | Council Unit | 0.002 | Excellent | 49.734 | Good |
| 22. | ISS 1 | 4.246 | Excellent | 49.734 | Good |
| 23. | ISS 2 | 4.325 | Excellent | 59.831 | Unsuitable |
| 24. | ISS VP | 5.074 | Excellent | 67.625 | Unsuitable |
| 25. | VC's Lodge | 1.558 | Excellent | 50.755 | Unsuitable |
| 26. | Agric. Farm | 0.282 | Excellent | 49.734 | Good |

Based on the SAR all or 100% of the water samples are excellent for irrigation since the SAR ranged from 0.002 at Council Unit to 5.074 at ISS VP office. These values are below 10 and thus contain low sodium levels. From SSP standpoint; 53.8% of the samples are suitable for irrigation whose values ranged from 28.922 to 49.734. Four that had SSP of 50.094 representing 15.4% are deemed as fair. The remaining 30.8% having SSP values that ranged from 50.755 to 67.625 are unsuitable for irrigation.

# 5. Conclusions and Recommendations

Based on the findings of this study the tap water within RSUST is fresh and soft with low to moderate dissolved solids. All the borehole water quality on Campus are not potable due to low pH; in addition, borehole water at Council Unit, ISS, UWA and Institute of Education had high microbial count and therefore not suitable for drinking. The use of the water in its present state for aqua-culture might be detrimental to fishes. The water is suitable for irrigation and other purposes except drinking.

There is need to urgently commence treatment of water supplied to the University community and create awareness to educate people on the need to boil and/or filter the water prior to consumption.

# Acknowledgement

The authors sincerely thank Professor B. B. Fakae, Vice Chancellor of University of Science and Technology, Port Harcourt for vision, encouragement and support; and the support and understanding of their colleagues.

# References

[1] Uzochukwu, A. (2009). Sanitation Sector Status and Gap Analysis: Nigeria. In Global Sanitation Fund / Water Supply & Sanitation Collaborative Council. p 15.

[2] EFSA, (2010). European Food Safety Authority. EFSA Panel on Dietetic Products, Nutrition, and Allergies (NDA); Scientific opinion of dietary reference values for water. EFSA Journal. 8(3);1459.

[3] Udoessien, E. I. (2003). Basic Principles of Environmental Science Etiliew Int. Publishers, Uyo, Akwa Ibom State, Nigeria. pp 27 - 63.

[4] Majolagbe, A. O. Kasali, A A. and Ghaniyu, L O. (2011). Quality assessment of groundwater in the vicinity of dumpsites in Ifo and Lagos, Southern Nigeria. Advances in Applied Science Research, 2 (1), pp 289-299.

[5] Redmond, W. A. (2009) [DVD] Microsoft Corporation, USA. *www.scribd.com/doc/52782458 /waterpollution-and-the-law*.

[6] Onyeagocha, A.C. (1980). Petrography and depositional environment of the Benin Formation. Jour. Min. Geol., Vol. 17. No. 2, pp. 147-151.

[7] Ganoulis, J. (1994). Risk Analysis of Water Pollution: Probabilities and Fuzzy Sets. VCH, Weinheim, Oxford, NY, 306 pp.

[8] Back, W (1966). Hydrochemical facies and ground-water flow patterns in northern part of Atlantic Coastal Plain. Hydrology of aquifer systems: U.S. Geol. Survey Prof. Paper 498-A, 42.

[9] Nwankwoala, H. O. and Udom, G. J. (2011). Hydrogeochemical Evaluation of Groundwater in Parts of Eastern Niger Delta, Nigeria. Journal of Academic and Applied Studies, Vol. 1, No. 2, pp 33 - 58.

[10] Subramani, T., I. Elango and S.R. Damodarasamy (2005). Groundwater quality and its suitability for drinking and agricultural use in chithar river basin, Tamil nadu, India. Environ. Geol. 47, 1099-1110.

[11] Nyanganji, J. K., Abdullahi, J. and Noma, I. U. S. (2011) Groundwater quality and related water borne diseases in Dass Town, Bauchi State, Nigeria. Journal of Environmental Issues and Agriculture in Developing Countries, Vol. 3, No. 2, pp 133-148.

[12] Sandhyarami, N. (2009). Groundwater Pollution. buzzle.com.*http://www.buzzle.com/articles/ groundwaterpollution.*

[13] N.M.S. (1998). Nigerian Meteorological Services, pp. 1-10.

[14] Simpson, A. (1954). The Nigerian coal fields, the geology of parts of Onitsha, Owerri and

[15] Ideriah, T.J.K. (2003). Levels of heavy metals and physicochemical parameters in soils, leachates, water and sediments around selected municipal waste dumpsites in Port Harcourt and Environs, Nigeria. Ph.D degree thesis in Rivers State University of Science and Technology Port Harcourt, pp. 40-43.

[16] APHA (1995). Standard Methods for the Examination of Water and Wastewater, 18th ed. APHA-AWWA-WPCF, Washington DC.

[17] ASTM (1999). American Society for Testing & Material (International), Bar Harbor Drive, P O Box C700, West Conshohocken, PA 19428-2959, USA.

[18] WHO (2006) World Health Organization Guidelines for Drinking-water Quality. First Addendum to Third Edition, Vol. 1, Recommendations. Geneva.

[19] SON (2007). Nigerian Standard for Drinking Water Quality. Lagos: Standard Organization of Nigeria.

[20] Piper, A M (1944) A graphic procedure in geochemical interpretation of water analyses. American Geophysical Union Transactions 25:914-923.

[21] Davis, S. N. and Dewest, R. J. M. (1966). Hydrology. John Wiley and Sons, Inc. London.

[22] Gerard. K. (1998). Environmental Engineering, Irwin/ McGraw-Hill Book Company, Singapore pp.456.

[23] Ming-HO, YU (2001). Environmental Toxicology, Impact of Environmental Toxicants on Living Systems. Lewis Publishers, London. pp.49, 151.

[24] Lecherallein, M.W., Welch, N.J and Smith, D.B (1996). Full-scale study of Factors related to Coliform regrowth in Drinking Water. Applied and Environmental Microbiological Journal 1 (62): p.2201.

[25] Ogundipe, S and Obinna, C (2008). "Safety of Table Water goes beyond the bottle" In: Good Health Weekly, Vanguard Newspapers Tuesday, May 20, 2008 p.42. River Basin. J. Hydrological Sciences, London.

[26] Miroslav, R. and Vladimir, N.B. (1999). Practical Environmental Analysis, Royal Society of Chemistry, U.K pp. 152, 175.

[27] Encyclopedia of Chemical Technology (1970). Second Edition Vol. 21, p 693.

[28] McNeely, R. N. Nemimanis, V.P. Dwyer, (1979). Water Quality Source book. A Guide to water Quality Parameter. Inland waters Directorate, water Quality Branch, Ottawa, Canada pp.

[29] Bower, H. (1978). Groundwater Hydrology, International Student Edition. McGraw-Hill Kogakusha, Ltd. Tokyo

[30] Romero, J. C. (1970). The movement of bacteria and viruses through porous media. Ground Water 8 (2): 37-49.

[31] USDA, (1954). U.S. Department of Agriculture. Diagnosis and Improvement of Saline and Alkali Soils Handbook 60, Washington D.C.

[32] Offodili, M E (2002) Groundwater study and development in Nigeria. Mecon Geology Eng. Services Ltd. Jos; 453 pages.

[33] Richards, L. A. (1954) Diagnosis and Improvement of Saline and Alkali Soils Agric Handbook 60, USDA and IBH Publishing Company Ltd., New Delhi, India, pp 98-99.

# 4

# Problems of Solid Waste Management in Small and Medium Enterprises

**Alie Wube Dametew**

Department of Mechanical Engineering, and Research and technology Transfer Head, KIOT Wollo University, Dessie, Ethiopia

**Email address:**
wubealie@gmail.com

**Abstract:** This study describes the problems of solid waste generation in small and medium scale enterprise. The analysis was conducted in Meserte Gabi metal and machine work enterprises using purposive sampling method with 16 sample sizes of this enterprise employer. Filed observation, unstructured interview, and questioners were instruments that use pre and post assessment tools for problems of solid waste management. The study explores the existing situation of the enterprises about solid waste management. The result shows that Solid waste issues were the major problems of Meserte Gabi metal and machine work enterprise. Findings show that waste prevention, recycling; collecting, and awareness, training and awareness are the major problems of the enterprises. After actions were done (conducting training, implementation of kaizen and capacity building) result predicts that, the impact of solid waste problem were changed and proper solid waste management systems were developed. The study highlights that it is essential that the government starts to recognize the mechanisms of supporting of enterprises in order to improve the overall solid waste management system in Dessise-Ethiopia. From the action obtained that conducting training, capacity building, resource utilization, and implementation of kaizen were mechanisms that developed solid management systems in the enterprises.

**Keywords:** Solid Waste, Reuse, Prevention, Kaizen, Waste Management

## 1. Background and Introduction

### 1.1. Background

In both developed and developing countries, the government is turning to small and medium scale industries and entrepreneurs, as a means of economic development and a veritable means of solving problems. Presently in Ethiopia, small and medium scale entrepreneurs assist in promoting the growth of the country's economy. Hence all the levels of government at different times have to promote policies which the growth and sustenance of small and medium scale entrepreneurs. But several wastes are observed in most small and medium scale entrepreneurs. These wastes are reducing the growth and maturity of enterprises and have impacts of sustainable developments. The purpose of waste management is used to have concentration for productivity, environmental problems and understand human responsibilities and role for environment, to care for attitude of participating in the solid waste management, and to develop mechanism and to solve the problem of enterprises. This project is prepared as a stimulating outcome of developing the method of solid waste management small and medium scale enterprises.

### 1.2. Introduction

Wastes are all forms of materials that are dangerous, toxic, reactive, explosive, flammable, organic or bacterial nature, are hazardous to the humankind and the environment as determined as waste. Waste represents the loss of both material and energy resources. Because excessive waste generation is a symptom of inefficient production processes, low durability of goods and unsustainable consumption patterns, waste quantities can be considered as an indicator of how efficiently society uses raw materials. Waste management is closely related to a number of issues such as urban lifestyle, resource consumption patterns, jobs and income levels, and other socio-economic and cultural factors.

Solid waste management is a mixture of human activities which tend to increase with rapid development, improved living standard and changing consumption of products. If solid waste is properly used, it can be a valuable resource, but if it is not effectively managed, it can result in serious adverse impacts on the growth and development of the enterprises. But Solid waste Management is the major challenge in small and medium scale enterprises in Dessise-Ethiopia. Solid waste management is therefore a critical component within the maturity and growth of the company. Hence this research aimed to develop the systems of solid waste management in small and medium scale enterprises.

### *1.3. Statement of the Problem*

In Small and Medium scale metal and machine manufacturing enterprises are produce many products with different problems. One of the main problems is due to improper Waste management and lack of sustainable development. In this research the investigation was conducted to develop problems of solid waste management in Small and Medium scale metal and machine manufacturing enterprises. Also find out factors of waste formation and to develop waste minimization technique.

### *1.4. Objectives*

#### *1.4.1. General Objectives*

The main objective of this study was to identify the major problems of Solid Waste and to develop Solid Waste Management systems in Small and Medium scale metal and machine manufacturing enterprises.

#### *1.4.2. The Specific Objective of This Study was*

- To identify the cause and main type of solid waste
- To assess the existing solid waste management systems of the enterprise.
- To apply the method of waste management for other enterprises in the Town.
- To develop solid Waste Management techniques
- Improve solid waste management for the enterprises

### *1.5. Scope of Research*

The study is limited to problems of solid waste management of small and medium scale metal and machine work enterprises.

### *1.6. Significance of the Study*

- The study contributes control the solid waste pollution effect of the enterprise in the town.
- indirectly to the implementation of change the economic development of the enterprise
- Contribute the building of environmental friendliness of the enterprise
- Contribute to resource utilization by re-use the waste.

## 2. Literature Review

### *2.1. Introduction for Related Works*

A great deal of research has been conducted in order to reduce solid waste problems, which would optimize productivity of small enterprises by reducing the factors that improve the formation of solid waste in the enterprises. This research has played a major role in promoting waste reduction mechanisms in the small enterprises. This chapter tries to summarize some key background references and their contributions to the problem of solid waste. Dr Kalpana Sankar (2004) shows Key environmental problems Caused by poor sanitation, poor waste disposal and , these may lead to serious risks of environmental pollution and public health impacts

Christian Zurbrügg (2003) was develop successful innovative ideas and approaches have been implemented on different levels of the solid waste management system Seadon (2006) results revealed that solid waste was identified as the second most important problem of urban populations. Di Vita (2004). Shows that general awareness of site based waste management issues and the study recognizes the need for holistic approaches to be adopted across all sectors to improve solid waste problems. Jones (2007) to analyze the waste management practice, in particular of used in an industrial and to find out its critical features. Investigated possible links between the environmental management system and the improving and more correct administration of waste. Significant statistical associations improved administration of waste under multiple aspects was found. In the year 2007, China had the largest population in the world. The population of China is expected to grow, and due to the strong correlation between increases in population and increases in waste, waste in China is also expected to increase. Waste can be a by-product of a productive process or it can be an object that has reached the end of its life cycle and that no longer posses any value to the current user (Ferguson 1995). According to the International Solid Waste Association (ISWA) part of sustainability entails viewing waste as a resource (ISWA & UNEP 2004).

If waste is viewed as a resource, resource management theory can be used to enlighten the underlying waste management theory. Zimmerman proposed: "Resources are not, they become; they are not static but expand and contract in response to human wants and actions… they evolve out of the triune interaction of nature, man, and culture, in which nature sets outer limits, but man and culture are largely responsible for the portion of physical totality that is made available for human use." (Zimmermann 1951). Applying principles to waste management, three assumptions can be inferred that contribute to how one can view the relationship between waste and resources. Waste is a subjective matter relative to humans, and fundamentally, it is anthropocentric. A plastic bag is not a resource or waste in itself: it depends on a human's ascription of value to it.

Secondly, the relationship between humans and waste is shaped by the interaction between humans, nature, and culture: that is, economics, process and technology. Processes such as recycling change the availability of plastics, metals, and paper in absolute terms. Finally, the definition of waste and resource expands and contracts in response to human want and action. Thus, if there were shortages in virgin products, then the value of the recycled products would likely increase and there would be more demand for recycled products .In many cases, value can be added and some utility can be restored to the waste, either through process or through a new type of usage (McDougall et al. 1995.The industrial ecology paradigm uses the practice of loop closing and industrial symbiosis to increase resource productivity and reduce waste (Ehrenfeld and Gertler 1997). According to Germany (Emscher Park) (Gibbs, 2003), and most recently China (*Guitang Group* sugar industrial cluster, and *Nan hai National Eco-Industrial Park* in Guangdong) ,These parks attempt to stimulate the symbiotic relationship between industries, where the waste produced by one industry is used as productive inputs into other industries.

(Maclaren 2005) synthesizes the literature on waste management systems in developing countries, she presents two approaches which describe preferred objectives of waste management systems The first approach she proposes is similar to applying the waste management hierarchy of developed countries. The developing country waste management hierarchy begins with waste prevention, toxicity reduction, reuse, recycling and composting, incineration, and disposal of any remaining wastes, ideally with energy recovery systems. The unique feature is toxicity reduction, which may be a paramount concern in developing countries. The second approach recognizes the pragmatic reality of the situation of waste management in many developing countries that 30-60% of waste is not collected. According to a Survey done by Nepal's central Bureau of Statistics (CBS), most Urban residents consider solid waste management as the most important environmental problem in urban areas of Nepal.

According to the data provided by the Municipality and the field survey conducted in May 2008, the average per capita household waste generation rate is 0.30kg/capita/day and the total daily waste generation is 25 tons. Vivian W. Y. Tam (2009) conducted that necessary to recycle concrete waste to preserve landfill areas. To reduce the amount of construction waste generated from sites, coordination among all those involved in the design and construction processes is essential. (Zurbrugg 2006) point out that there are both short-term and long-term objectives of waste management's. The short-term objectives are waste collection and disposal; while the longer-term objectives are waste reduction and recycling.

Gloria Elizabeth Ichim (2007) identified seven necessary key factors that if not given sufficient attention could potentially expose the successful implementation and operation of composting: governmental support must be present, funds must be made available since operating cost of the waste management system will increase, best practices for composting must be established, training for waste workers must be provided, a market (or end use) for the final product must be established.

### 2.2. The Relationship Between Climate Change and Solid Waste

Prof. R. Rameezdeen (June 2010) The Earth's atmosphere contains many types of gases, including those known as "greenhouse gases," which hold in the sun's warmth. Scientists call this naturally occurring phenomenon the "greenhouse effect." Greenhouse gases help regulate global temperatures. Certain human activities such as burning fossil fuels and dumping solid waste, however, produce additional greenhouse gases and upset the natural balance by raising global temperatures. Even before a material or product becomes solid waste, it goes through a long cycle that involves removing and processing raw materials, manufacturing the product, transporting the materials and products to markets, and using energy to operate the product.

The cause of a particular problem is due to improper managed hazardous waste disposal sites. Ethiopian government to develop solid waste management systems to reduce their problems. Despite the fact that controlling the effects of solid waste is under problems thus to solve such problem identification of the problem is developed using research is the beast method. Hence, the key research question addressed in this study is:

- Are problems of solid waste properly and effectively manage in Small and Medium scale metal and machine manufacturing enterprises?
- Is there any change of solid waste management from the past to the existing?
- What are the problems that delay waste management of Small and Medium scale metal and machine manufacturing enterprises?

## 3. Methodology of the Study

### 3.1. Introduction

This research the method of solid waste management systems was developed in Small and medium scale enterprise. Pre and post assessment was used to determine the change of improvements. The study was done through seven month with duration purposive was s done 16workers of In Small and medium scale enterprise. The researcher used descriptive method of research to study the specified problem.

### 3.2. Sample Size and Sampling Procedure

Since the researcher's intention was to identify a single problem and its solution in In Small and medium scale enterprise with 1manager and 15 enterprise employees, purposive sampling method was used. And the researcher also used only one enterprise considering the time and monetary problems to include more enterprises.

### 3.3. Source of Data

The main sources of data for this research were 1manager and 15 employees of In Small and medium scale enterprise. There were five female and ten meal employers a total of about 16 participants, in order to gather the necessary data required to conduct this study.

### 3.4. Instruments and Data Collection Method

In this study data was collected by different methods like

field observations, questionnaires and interviews are presented.

### *3.5. Data Analysis Method*

The collected data were analyzed and presented through Pareto charts, tables and Fish bone diagrams. Finally, the analyzed data are interpreted, discussed and the findings were put in summary form.

### *3.6. Procedure*

***Table 3.1.*** *Action plane*

| Duration period | Activity |
|---|---|
| Observation (before oct16) | During the cooperative training, it was observed that the enterprises are filling with wastes and the wastes are not properly controlled. Whenever some questions asked to the employer about the problem, they have different ideas and they have no common understanding. |
| Oct 16/2013 | Communicate with enterprise managers and discus with solid waste problems and proper management systems |
| Oct 21-24/2013 | Pre assessment questioner to the enterprise |
| Oct 27-30/2013 | Introduction to action research |
| Nov 7&8/2013 | Orientation In Small and medium scale enterprise employer regarding the problem associated with solid waste wastes and its' proper management.<br>-promoting of training and awareness to the employer<br>-Demonstration on Solid waste problem and proper management systems.-solid managing by reuse, prevention, training, collecting, kaizen and recycling |
| Nov11-19/2013 | Feedback and corrective action developed |
| Nov 23-Dec7/2013 | Implementation of kaizen |
| Dec11/2013 | Selecting of waste collecting area<br>-preparation of solid waste dust bin by the employer |
| Jun16/2013 | Discussing with employer about the improvement/change |
| Jun24-Feb12/2013/14 | Evaluation of the progress(Mid term assessment) |
| Feb18-march3/2014 | Post assessment questioner and<br>-post observation of the enterprise |
| March3-19/2014 | Analyzing data |

### *3.7. Data Presentation*

People said that if you want to change the world, begin from yourself. My intention is, I have develop/crate solid waste management system in In Small and medium scale enterprise. I have started the action in Small and medium scale enterprise and am very hopeful to expand the territory of this work in other enterprises and convey the message of solid waste management in the entrepreneurs.

I organized my data chronologically as I collected it in my file throughout the study. For seven month study can take gate proper solid waste management mechanism within observations and questioner response from the enterprise employer, I feel that more time needs to be spent investigating this solid waste management in order to gate better management systems.

## 4. Result and Discussion of the Study

### *4.1. Observation Result*

As the researcher observed and assess at the first period, in the enterprise solid waste management does not know by the employer and wastes are not properly controlled .From observation obtained that awareness and training is a great problem of the enterprise worker for problems of solid waste management systems.

After conducting training for awareness creation, solid waste management problems are slightly changed. Starting of kaizen implementation and capacity building solid waste management is good. For mid and post assessment period the employers use waste collecting facilities, there is know how about waste, use waste collecting facility, isolating of wastes by type and size, some waste materials use as reusing purpose.

### *4.2. Interview*

The interviews conducted in the first period of field work revealed employers does not know solid waste management, even they cannot define the term solid waste. In my, interview, this employer described to me, how even though they felt they were not good at using the waste management system; in using technology during production, it consumes our time and many. They told that because of dust bin preparation we spent/lose many.

After crating good awareness, training and kaizen implementation developed in the enterprise. From mid evaluation time obtained that, solid waste management systems are changes in some cases like awareness, reusing, shop cleaning and waste collecting system. On the post assessment time most factors are controlled and the waste management systems is crated .the respondent says that this changes are obtained because of awareness, training, implementation and using of kaizen, using waste collecting facility and capacity building of the enterprise. But some factors like recycling are still problems.

### *4.3. From the Questionnaire*

For Meserte Gabi metal and machine work enterprises 10 questionnaires distributed to the enterprise 16 workers. At different time the employer were responded to responds and the response of questionnaires were considered as essential sources for the study.

#### *4.3.1. First Assessment Time*

This (first) assessment is conduction before training, implementation of kaizen and the result is predicts the problems of solid waste management in the enterprises.

***Table 4.1.*** *solid waste management awareness among the employer*

| Activity/task/action | Valid | Frequency | Percent | Valid Percent | Cumulative Percent |
|---|---|---|---|---|---|
| Do you know waste management? | no | 21 | 100 | 100 | 100 |

From the above table it is seen that 100% of employer don't know Solid waste management practice they have been practicing is sustainable or not. So there is a great need of solid waste management awareness among the employer. Lack of training and poor awareness is the major problems of the enterprises.

***Table 4.2.*** *The kind of solid waste activities carry out in the Enterprise*

| Activity/task/action | Valid | Frequency | Percent | Valid Percent | Cumulative Percent |
|---|---|---|---|---|---|
| What kind of solid waste activities do you carry out? | collector | 12 | 57.14 | 57.14 | 57.14 |
| | recycling | 0 | 0 | 0 | 0 |
| | Reuse | 2 | 9.52 | 9.52 | 9.52 |
| | No action | 7 | 33.33 | 33.33 | 33.33 |
| | Total | 21 | 100 | 100 | 100 |

Findings: The problems of waste management covers Collecting 57% Reuse9.52%, recycling0%, there no action to control waste is 33.33%. From the result shows that waste were not properly controlled. But there is no improvement in solid waste recycling.

***Table 4.3.*** *The type of solid waste material that found and collect in the Enterprise*

| Activity/task/action | Valid | Frequency | Percent | Valid Percent | Cumulative Percent |
|---|---|---|---|---|---|
| What kind of material does your firm collect? | Mixed | 7 | 33.33 | 33.33 | 33.33 |
| | Wood | 4 | 19.00 | 19.00 | 19.00 |
| | Metal | 7 | 33.33 | 33.33 | 33.33 |
| | Plastic | 3 | 14.29 | 14.29 | 14.29 |
| | Total | 21 | 100 | 100 | 100 |

From the above table it is seen that about 33.33 % Mixed waste is collected in the enterprises. But below 23 % waste is collected by there type. For proper waste management system, it collects by their type and size, otherwise the mechanism is poor. Hence from the result shows that small amount of wastes are collect by their type, then this small amount isolation is an indication of problems of solid waste management.

***Table 4.4.*** *The effect of conducting training and capacity building for solid waste management*

| Activity/task/action | Valid | Frequency | Percent | Valid Percent | Cumulative Percent |
|---|---|---|---|---|---|
| Do you expect that training and capacity building is the means of solving SWM problems? | yes | 5 | 23.81 | 23.81 | 23.81 |
| | no | 7 | 33.33 | 33.33 | 33.33 |
| | No idea | 9 | 42.86 | 42.86 | 42.86 |
| | Total | 21 | 100 | 100 | 100 |

From the above table shows that only 23% respondents know the importance training and capacity building as means of solid waste management but 33.33 % not agree and 42.84% of the respondent have not any idea. From the result shows that waste were training and capacity building the main problem of solid waste management in the enterprises.

***Table 4.5.*** *kaizen as a Solid waste management*

| Activity/task/action | Valid | Frequency | Percent | Valid Percent | Cumulative Percent |
|---|---|---|---|---|---|
| Do you expect that kaizen is solid management systems (swm)? | No idea | 21 | 100 | 100 | 100 |

From the above table it is seen that 100% of employer don't know kaizen as a Solid waste management. So there is a great need of solid waste management awareness among the employer. Lack of training and poor awareness is the major problems of the enterprises.

***Table 4.6.*** *Methods of collect solid waste*

| Activity/task/action | Valid | Frequency | Percent | Valid Percent | Cumulative Percent |
|---|---|---|---|---|---|
| What kind of technology does your firm use to collect the waste? | Local collector | 16 | 76.19 | 76.19 | 76.19 |
| | Employer | 5 | 23.81 | 23.81 | 23.81 |
| | Total | 21 | 100 | 100 | 100 |

The table shows that above 76% waste is collected by waste collector but the employer collect very small amounts about 24% waste from the total waste. This result shows that waste were not properly controlled for each working time

***Table 4.7.*** *The range of solid waste collect and stored in dust bin*

| Activity/task/action | Valid | Frequency | Percent % | Valid Percent % | Cumulative Percent % |
|---|---|---|---|---|---|
| Do you think that the waste is properly collect and stored in dust bin? | Yeas | 5 | 23.81 | 23.81 | 23.81 |
| | No | 16 | 76.19 | 76.19 | 76.19 |
| | Total | 21 | 100 | 100 | 100 |

From table shows that above 76% waste is not collected and stored within dust bin. The result implies that there are no waste management systems due to shortage of collecting facilities

***Table 4.8.*** *Waste Controlling techniques*

| Activity/task/action | Valid | Frequency | Percent | Valid Percent | Cumulative Percent |
|---|---|---|---|---|---|
| How many wastes are controlled per day? | Reuse | 3 | 14.30 | 14.30 | 14.30 |
| | prevention | 3 | 14.30 | 14.30 | 14.30 |
| | collecting | 4 | 19.00 | 19.00 | 19.00 |
| | No action is done | 14 | 52.38 | 52.38 | 52.38 |
| | Total | 21 | 100 | 100 | 100 |

The problems of waste management covers above 52% but from the response obtained that small amount of wastes were Collecting 19% Reuse 14.30 %, recycling0% and waste prevention is covers 14.30%. From the result shows that waste were not properly controlled.

***Table 4.9.*** *Different solid waste management mechanisms*

| Activity/task/action | Valid | Frequency | Percent | Valid Percent | Cumulative Percent |
|---|---|---|---|---|---|
| What do you think the government should do to make the solid waste situation in the enterprise better? | Crate recycling industry | 8 | 38.10 | 38.10 | 38.10 |
| | Pay monthly employer | 9 | 42.86 | 42.86 | 42.86 |
| | Training and awareness to the employer | 4 | 19.00 | 19.00 | 19.00 |
| | Total | 21 | 100 | 100 | 100 |

Different references book and literature explains that training, capacity building and awareness creation of the employer, is the best method of proper solid waste management. But the table shows, for waste controlling systems the responds have different ideas, 38% respondents told that solve the problem by Crate recycling industry, 42% Pay monthly employer, training and awareness to the employer is covers 19%.hence the enterprise needs training and capacity building to crate solid waste management mechanisms.

***Table 4.10.*** *The enterprise employers moral to implement solid waste management systems*

| Activity/task/action | Valid | Frequency | Percent | Valid Percent | Cumulative Percent |
|---|---|---|---|---|---|
| Will you encourage your worker/employer to give continuity to this activity? | No | 21 | 100 | 100 | 100 |

The table shows that 100% respondents did not encourage their worker to crate solid waste management systems.

### *4.3.2. Mid Assessment*

This (mid) assessment is conduction after training, implementation of kaizen and capacity building of the enterprises. The result should predict the problems of solid waste management in the enterprises.

***Table. 4.11.*** *The awareness of waste management at Mid Assessment stage*

| | Valid | Frequency | Percent | Valid Percent | Cumulative Percent |
|---|---|---|---|---|---|
| Activity/task/action | yes | 7 | 33.33 | 33.33 | 33.33 |
| | no | 14 | 66.66 | 66.66 | 66.66 |
| Do you know waste management? | Total | 21 | 100 | 100 | 100 |

From the above table it is seen that below 33.33 % of the responds know solid waste management but from pre assessment none of them don't know waste management's. Hence due to training and awareness creation there is a considerable positive change in the development of solid waste management.

***Table 4.12.*** *The kind and level of solid waste activities carry out in the Enterprise at Mid Assessment stage*

| Activity/task/action | Valid | Frequency | Percent | Valid Percent | Cumulative Percent |
|---|---|---|---|---|---|
| What kind of solid waste activities do you carry out? | collector | 13 | 61.90 | 61.90 | 61.90 |
| | recycling | 0 | 0 | 0 | 0 |
| | Reuse | 5 | 23.80 | 23.80 | 23.80 |
| | No action | 3 | 14.30 | 14.30 | 14.30 |
| | Total | 21 | 100 | 100 | 100 |

The table shows that there is in significance change is seen for waste reduction mechanisms from the first assessment time.

***Table 4.13.*** *The level and type of solid waste material that found and collects in the Enterprise at Mid Assessment stage*

| Activity/task/action | Valid | Frequency | Percent | Valid Percent | Cumulative Percent |
|---|---|---|---|---|---|
| What kind of material does your firm collect? | Mixed | 3 | 14.29 | 14.29 | 14.29 |
| | Wood | 5 | 23.80 | 23.80 | 23.80 |
| | Metal | 10 | 47.62 | 47.62 | 47.62 |
| | Plastic | 3 | 14.29 | 14.29 | 14.29 |
| | Total | 21 | 100 | 100 | 100 |

From the above table it is seen that about 14 % Mixed waste is collected in the enterprises. But above 40 % waste is collected by there type. For proper waste management system, waste collects by there type and size, otherwise the mechanism is poor. Hence from the result shows there was change from the first assessment period that large amount of wastes were collect by there type, then this is an indication of solid waste management mechanism.

***Table 4.14.*** *The effect of conducting training and capacity building for solid waste management at Mid Assessment stage*

| Activity/task/action | Valid | Frequency | Percent | Valid Percent | Cumulative Percent |
|---|---|---|---|---|---|
| Do you expect that training and capacity building is the means of solving SWM problems? | yes | 10 | 47.62 | 47.62 | 47.62 |
| | no | 6 | 28.57 | 28.57 | 28.57 |
| | No idea | 5 | 23.80 | 23.80 | 23.80 |
| | Total | 21 | 100 | 100 | 100 |

From the above table shows that only 48% respondents know the importance of training and capacity building as means of solid waste management but 28 % not agree and 23% of the respondent have not any idea. From the result shows that solid waste management mechanism were formed by conducting training and capacity building in the enterprises.

***Table 4.15.*** *Effect of kaizen as a Solid waste management at Mid Assessment stage*

| Activity/task/action | Valid | Frequency | Percent | Valid Percent | Cumulative Percent |
|---|---|---|---|---|---|
| Do you expect that kaizen is swm systems? | Yeas | 5 | 23.81 | 23.81 | 23.81 |
| | No | 6 | 28.57 | 28.57 | 28.57 |
| | No idea | 10 | 47.62 | 47.62 | 47.62 |
| | Total | 21 | 100 | 100 | 100 |

From pre assessment 100% of employer doesn't know kaizen as a Solid waste management. From the above table it is seen that 78% of employer don't know kaizen as a Solid waste management. So there is a change for the awareness of the employer .hence kaizen is one mechanism for suitable solid waste management systems of the enterprises.

***Table 4.16.*** *The level of solid waste collecting Methods at Mid Assessment stage*

| Activity/task/action | Valid | Frequency | Percent | Valid Percent | Cumulative Percent |
|---|---|---|---|---|---|
| What kind of technology does your firm use to collect the waste? | Local collector | 12 | 57.14 | 57.14 | 57.14 |
| | Employer | 9 | 42.82 | 42.82 | 42.82 |
| | Total | 21 | 100 | 100 | 100 |

The table shows that above 57% waste is collected by waste collector but the employer collecting capacity change from 24% to 43% waste from the total waste. This result shows that waste were well controlled for each working time. Employer training and encouragement is one mechanism for solid waste management.

***Table 4.17.*** *The amount of waste properly collect at Mid Assessment stage*

| Activity/task/action | Valid | Frequency | Percent | Valid Percent | Cumulative Percent |
|---|---|---|---|---|---|
| Do you think that the waste is properly collect and stored in dust bin? | Yeas | 8 | 30.10 | 30.10 | 30.10 |
| | No | 13 | 61.90 | 61.90 | 61.90 |
| | Total | 21 | 100 | 100 | 100 |

From table shows that above 62% waste is not collected and stored within dust bin. At the first result obtained that 76% of solid wastes were not properly collected. The result implies that there was management system that developed to control problems of solid wastes.

***Table 4.18.*** *The amount of waste manage by different mechanisms at Mid Assessment stage*

| Activity/task/action | Valid | Frequency | Percent | Valid Percent | Cumulative Percent |
|---|---|---|---|---|---|
| How many wastes are controlled per day? | Reuse | 4 | 19.05 | 19.05 | 19.05 |
| | prevention | 6 | 28.57 | 28.57 | 28.57 |
| | collecting | 7 | 33.33 | 33.33 | 33.33 |
| | No action is done | 4 | 19.05 | 19.05 | 19.05 |
| | Total | 21 | 100 | 100 | 100 |

The table shows that solid waste management systems change from Collecting 19%to 33.33%, Reuse 14.30%to19.05%, recycling0% and waste prevention 14.30%to 28.70%.

***Table 4.19.*** *The level of Government responsibility implemented to control waste at Mid Assessment stage*

| Activity/task/action | Valid | Frequency | Percent | Valid Percent | Cumulative Percent |
|---|---|---|---|---|---|
| What do you think the government should do to make the solid waste situation in the enterprise better? | Crate recycling industry | 4 | 19.00 | 19.00 | 19.00 |
| | Pay monthly employer | 9 | 42.86 | 42.86 | 42.86 |
| | Training and awareness to the people | 8 | 38.10 | 38.10 | 38.10 |
| | Total | 21 | 100 | 100 | 100 |

Different references book and literature explains that training, capacity building and awareness creation of the employer, is the best method of proper solid waste management. The above table shows, for solid waste controlling systems the perspective of the employer is change from the first. Hence 38% to change 19% respondents told that solve the problem by Crate recycling industry, from 42% to 42% Pay monthly employer, training and awareness to the employer is covers 19%to 38%. Hence conducting training and capacity building was one method of solid waste management mechanisms.

***Table 4.20.*** *The Enterprise worker moral apply Solid waste Management at Mid Assessment stage*

| Activity/task/action | Valid | Frequency | Percent | Valid Percent | Cumulative Percent |
|---|---|---|---|---|---|
| Will you encourage your worker/employer to give continuity to this activity? | Yes | 7 | 33.3 | 33.3 | 33.3 |
| | No | 14 | 66.67 | 66.67 | 66.67 |
| | Total | 21 | 100 | 100 | 100 |

The table shows that 33.33% respondents did not encourage there worker to crate solid waste management systems. Because employer encouragement is increase productivity and controls problems within team. Hence from the result shows that employee encouragement is changing from 0% to33.33%, for this reason wastes were properly collected and controlled. Then encouragement is one method of solid was management systems.

### *4.3.3. Post Assessment*

This (post) assessment is conduction after training, implementation of kaizen and capacity building of the enterprises. The result should predict the problems of solid waste management in the enterprises.

***Table 4.21.*** *Solid waste management awareness among the employer at Post Assessment Stage*

| | Valid | Frequency | Percent | Valid Percent | Cumulative Percent |
|---|---|---|---|---|---|
| Activity/task/action | yes | 16 | 71.20 | 71.20 | 71.20 |
| | no | 5 | 23.80 | 23.80 | 23.80 |
| Do you know waste management? | Total | 21 | 100 | 100 | 100 |

Table it is shows that above71% of the responds knows solid waste management but from pre assessment none of them don't know and at mid assessment covers 33.33% of solid waste management's. Hence due to training and awareness creation there is a considerable positive change in the development of solid waste management.

***Table 4.22.*** *The kind and level of solid waste activities carry out in the Enterprise at post Assessment stage*

| Activity/task/action | Valid | Frequency | Percent | Valid Percent | Cumulative Percent |
|---|---|---|---|---|---|
| What kind of solid waste management activities do you carry out? | collector | 15 | 71.42 | 71.42 | 71.42 |
| | recycling | 0 | 0 | 0 | 0 |
| | Reuse | 5 | 23.80 | 23.80 | 23.80 |
| | No action | 1 | 4.76 | 4.76 | 4.76 |
| | Total | 21 | 100 | 100 | 100 |

The table shows that there is in significance change is seen for waste reduction mechanisms from the first assessment time.

***Table 4.23.*** *The level and type of solid waste material that found and collects in the Enterprise at post Assessment stage*

| Activity/task/action | Valid | Frequency | Percent | Valid Percent | Cumulative Percent |
|---|---|---|---|---|---|
| What kind of material does your firm collect? | Mixed | 3 | 14.29 | 14.29 | 14.29 |
| | Wood | 5 | 23.80 | 23.80 | 23.80 |
| | Metal | 10 | 47.62 | 47.62 | 47.62 |
| | Plastic | 3 | 14.29 | 14.29 | 14.29 |
| | Total | 21 | 100 | 100 | 100 |

From the above table it is shows that about 14 % Mixed waste is collected in the enterprises. But above 40 % waste is collected by their type. For proper waste management system, waste collects by there type and size, otherwise the mechanism is poor. Hence from the result shows there was change from the first assessment period that large amount of wastes were collect by their type, then this is an indication of solid waste management mechanisms in the enterprises.

***Table 4.24.*** *The employer knowledge for Training and capacity building is the means of solving SWM problems at post Assessment stage*

| Activity/task/action | Valid | Frequency | Percent | Valid Percent | Cumulative Percent |
|---|---|---|---|---|---|
| Do you expect that training and capacity building is the means of solving SWM problems? | yes | 16 | 66.66 | 66.66 | 66.66 |
| | no | 3 | 14.29 | 14.29 | 14.29 |
| | No idea | 2 | 9.52 | 9.52 | 9.52 |
| | Total | 21 | 100 | 100 | 100 |

The above table shows that above 66.66% respondents know the importance of training and capacity building as means of solid waste management but 14.29 % not agree and9.52% of the respondent have not any idea. From the result shows that solid waste management mechanism were formed by conducting training and capacity building in the enterprises.

*Table 4.25. The employer knowledge for kaizen as the means of solving SWM problems at post Assessment stage*

| Activity/task/action | Valid | Frequency | Percent % | Valid Percent % | Cumulative Percent % |
|---|---|---|---|---|---|
| Do you expect that kaizen is swm systems? | Yeas | 15 | 71.43 | 71.43 | 71.43 |
| | No | 4 | 19.05 | 19.05 | 19.05 |
| | No idea | 3 | 14.29 | 14.29 | 14.29 |
| | Total | 21 | 100 | 100 | 100 |

From pre assessment 100% of employer doesn't know kaizen as a Solid waste management. From the mid assessment that 78% of employer doesn't know kaizen as a Solid waste management and the above table shows that below 15 % of employer don't know kaizen as a Solid waste management. So there is a change for the awareness of the employer .hence kaizen is grate mechanism for suitable solid waste management systems of the enterprises.

*Table 4.26. The level of solid waste collecting Methods at post Assessment stage*

| Activity/task/action | Valid | Frequency | Percent | Valid Percent | Cumulative Percent |
|---|---|---|---|---|---|
| What kind of technology does your firm use to collect the waste? | Local collector | 14 | 66.66 | 66.66 | 66.66 |
| | Employer | 7 | 33.33 | 33.33 | 33.33 |
| | Total | 21 | 100 | 100 | 100 |

The table shows that above66.66% waste is collected by waste collector but the employer collecting capacity change from 43% to 33.33% waste from the total waste. This result shows that waste were not controlled for each working time because of decreasing of employer incentives. Employer training and encouragement is one mechanism for solid waste management.

*Table 4.27. The amount of waste collect using dust bin*

| Activity/task/action | Valid | Frequency | Percent | Valid Percent | Cumulative Percent |
|---|---|---|---|---|---|
| Do you think that the waste is properly collect and stored in dust bin? | Yeas | 13 | 61.90 | 61.90 | 61.90 |
| | No | 8 | 30.10 | 30.10 | 30.10 |
| | Total | 21 | 100 | 100 | 100 |

From table shows that below 30% of solid waste is not collected and stored within dust bin. At the first result obtained that 76% and second 62% of solid wastes were not properly collected. The result implies that there was management system that developed to control problems of solid wastes. For proper collecting of solid waste and using waste collecting facility the amount of solid waste management problem is reduced from 76% to 30%.

*Table 4.28. The amount of waste manage daily by different mechanisms*

| Activity/task/action | Valid | Frequency | Percent | Valid Percent | Cumulative Percent |
|---|---|---|---|---|---|
| How many wastes are controlled per day? | Reuse | 4 | 19.05 | 19.05 | 19.05 |
| | prevention | 6 | 28.57 | 28.57 | 28.57 |
| | collecting | 9 | 42.86 | 42.86 | 42.86 |
| | No action is done | 2 | 9.52 | 9.52 | 9.52 |
| | Total | 21 | 100 | 100 | 100 |

The above table shows that solid waste management systems change from Collecting 19%to 42%, Reuse 14.30%to19.05%, and waste prevention 14.30%to 28.70%. But recycling still problem.

*Table 4.29. Governments Responsibility for SWM after the assesment*

| Activity/task/action | Valid | Frequency | Percent | Valid Percent | Cumulative Percent |
|---|---|---|---|---|---|
| What do you think the government should do to make the solid waste situation in the enterprise better? | Crate recycling industry | 6 | 28.57 | 28.57 | 28.57 |
| | Pay monthly employer | 2 | 9.52 | 9.52 | 9.52 |
| | Training and awareness to the people | 13 | 61.90 | 61.90 | 61.90 |
| | Total | 21 | 100 | 100 | 100 |

Different references book and literature explains that training, capacity building and awareness creation of the employer, is the best method of proper solid waste management. The above table shows, for solid waste controlling systems the perspective of the employer is change from the first. Hence 38% to change 28% respondents told that solve the problem by Crate recycling industry, from 42% to 9.52% Pay monthly employer, training and awareness to the employer is covers 19%to 62%. Hence conducting training and capacity building was one method of solid waste management mechanisms.

***Table 4.30.*** *Employer moral Apply Waste management after assessment*

| Activity/task/action | Valid | Frequency | Percent | Valid Percent | Cumulative Percent |
|---|---|---|---|---|---|
| Will you encourage your worker/employer to give continuity to this activity? | Yes | 14 | 66.66 | 66.66 | 66.66 |
| | No | 7 | 33.3 | 33.3 | 33.3 |
| | Total | 21 | 100 | 100 | 100 |

The table shows that 33.33% respondents did not encourage there worker to crate solid waste management systems. Because employer encouragement is increase productivity and controls problems within team. Hence from the result shows that employee encouragement is changing from 0% to66.66%, for this reason wastes were properly collected and controlled. Then encouragement is one method of solid was management systems.

According to the pre- and post-assessment, regardless of how enterprise employer ranked their awareness, comfort and regular use of solid waste management systems in there enterprise, all employer answered that they feel solid waste management system was extremely significant in applying production process and sustainable development.

My findings show a marked difference in the solid waste management systems of the enterprises at different assessment time/period. The difference is obtained due to training capacity building, material utilizations. These differences of pre and post results could be a reflection of the waste management gap in the enterprise. The result shows that from the beginning of the assessment solid waste management system is poor (there are no well defined solid waste management systems). After conducting training, implementation of kaizen the solid waste management system is change, Increase in training and capacity building about solid waste management system the employer was change solid waste management systems. My result show that problems of solid waste management is formed due to the factors like poor material utilization, waste prevention, recycling, resource reuse problems, storage and facility problems, improper waste collecting mechanisms are the common problems of solid waste managements in the enterprises

### *4.4. The Result of Observation, Interview and Questioner Data Interpretation and Discussions*

For these response the questionnaire result show that at different assessment period / time ,problems of solid waste management system is different ,before conducting training ,awareness, and capacity building to the enterprise there is no known and visible waste management systems. From the first assessment period the following factors were identified as problems of solid waste management.

#### *4.4.1. Poor Resource Utilization*

My result show that, Lack of skill man power most of the enterprise produce products wit out any engineering specification (drawings, mock-ups, material requirements of the product, were not available) causing Waste generations in the enterprises.

***(i) Waste Storage Problems***

In these enterprises solid wastes, which cannot be accommodated in the normal storage containers and need a special collection mechanism? The waste simply found in the working area and it is reduce production capacity of the enterprises.

***(ii) Poor Awareness, Training and Capacity Building***

There is lack of awareness, training, and education coupled with the negative attitude and lack of cooperation from government organizations. Therefore, awareness is crucial problem for sustainable solid waste management of the enterprises.

***(iii) Waste Prevention***

Waste prevention also called source reduction seeks to prevent waste from being generated. Waste prevention strategies include using less packaging, designing products to last longer, and reusing products and materials. Waste prevention helps reduce handling, treatment, and disposal costs and ultimately reduces the generation of methane. From the analysis observed that the waste prevention of the enterprises are very poor. From the total respondents, 86% did not consider the Waste Prevention mechanisms .due to this the problems controlled only 14% of the total of solid waste.

***(iv) Recycling***

Recycling is a process that involves collecting, reprocessing, and/or recovering certain waste materials (e.g., paint plastic& glass, metal scraps, concrete waste and plastics) to make new materials or products. Recycling generate many environmental and economic benefits. But the enterprises have not recycling mechanisms of solid waste because of shortage of facility, lack of knowledge and financial problems.

***(v) Reuse of Solid Waste***

Waste reuse means to find a new use for components that would be part of the Municipal solid waste, without the need for reprocessing them. These products can be either is reused for their original purpose, or for a different one. Examples of reuse are selling or donating used clothes, using empty glass containers for candy storage, turn empty jars into containers for leftover food, use of refillable pens, among others. But the enterprises have little knowledge for reusing of most solid wastes. Only small amount of plastic wastes use for secondary purpose.

***(vi) Collection Systems***

The proportion of solid waste generated to that collected is very low in the enterprises. The collection efficiency is also hampered by the inappropriate waste container. The solid waste is not collect properly to control problems of the enterprises production process. The enterprise has not appropriate storage containers to store and collect wastes together.

In general before the action is taken pre assessment result shows that (lack of training, capacity building, implementation of kaizen, poor collecting systems, lack of waste storage facility) the following problems were seen.

- Poor mobilization of resources.
- There are no viable economic incentives to encourage waste recycling, waste reduction to the employer
- Insufficient financial resources for recycling of wastes
- Lack of government and institutional support..
- Poor employer attitudes towards environmental cleanliness shifting responsibility
- Lack of general awareness of solid waste management
- Inadequate solid waste management (SWM) mechanisms at generation points including collection, storage.
- Over-reliance on imported and inappropriate technology and equipments enterprises.

***Fig. 4.1.*** *Working area of enterprises before action*

***4.4.2. Mid and Post Assessment Result Discussion***

The result shows after conducting of training, implementation of kaizen, visible solid waste management systems are seen in the enterprises and due to developing of solid waste management system through training, kaizen awareness and capacity building , most problems were reduced compare to before but some are still the problem like recycling.

***(i) Training Awareness and Capacity Building***

The lack of training and educational awareness on solid waste management by major enterprises constraint. Training and capacity building of the enterprises were conducted by trainer. It crate conducive environment to wards good attitude about solid waste management and there is good within the employers in the enterprises. Due to this training and capacity building the waste management system is change from 21% to56.83 %. My assessment shows that, Promoting capacity building of enterprises the key actions to crate proper solid waste management of the enterprises.

***Fig. 4.2.*** *conducting training Capacity building and awareness creation to the enterprises*

***(ii) Waste Prevention***

Currently, the degree of solid waste management is changed from the first time and before kaizen implementation. After kaizen the enterprises prevents waste barrier to understanding the benefits of solid waste management.

So as to reduce the amount of waste 14% from before action and increase solid waste prevention capacity from 14% to 24%.hence implementing of kaizen is one mechanism for solid waste management in the enterprises.

***(iii) Recycling***

Recycling refers to the removal of items from the waste stream to be used as raw materials in the manufacture of new products. But In the enterprises Recycling is not done because of the cost of system is difficult by the enterprises but collecting the recyclable material and to sell local collectors. Still this is the major problems.

***Fig. 4.3.*** *Recycling*

***(iv) Reused***

My result shows, for conducting training and capacity building increase the awkwardness of reuse of materials, because of these waste management system is are changed from the previous to this time. The Reuse of products is beast methods of solid waste prevention. They encouraging the customers to choose reusable products .they eliminate the production of waste at the source of usual generation and reduce the demands for large scale treatment and disposal facilities. The enterprise to regulate the types and amount of packaging used by manufacturers and make the reuse of materials the capacity increase from 9.52 to 21%.

***Fig. 4.4.*** *Reused*

***(v) Storage Facilities/ Storage Containers***

After proper training and capacity building the enterprises prepare dust bin for cost collect facility. The perspective of enterprises is change for waste collection system and they used the containers to collect waste. The storing facility changes from 23. % to 62% .proper solid waste storage facility is on mechanism for proper solid waste management systems.

***(vi) Collection Systems***

The Enterprises increase waste profile, so as to promote a general understanding of the significance of proper waste collecting mechanism. It is also possible that collect the waste by there type and size for manageable condition, Allows the use of same collection equipment as usual, they Improved waste collecting mechanisms.

- Preparing and designing of the waste collection sites
- Solid Wastes should be put in place
- There should apply daily and weekly schedule for shop cleaning and collection of wastes. The process improves from 21 % to 51%. Hence proper solid waste storage facility is other mechanism for proper solid waste management systems.

***Fig. 4.5.*** *waste collecting during kaizen*

In general after the action is taken (conducting training, capacity building, and implementation of kaizen, proper collecting systems, and waste storage facility) the following results are obtained

- Increase the capacity of the solid waste management in the enterprises.
- Separation of waste at source to reduce the amount disposed at the dumpsite
- Redesigning of the waste collection sites i.e. the transfer stations to be user friendly.
- There should be daily and a weekly schedule for collection of wastes
- Improve enterprise waste collection and disposal authority
- Reuse some wastes properly
- To identify the problems of solid waste management in the enterprises.
- The enterprises acquire knowledge about waste management systems.
- Insufficient financial resources for recycling of wastes

(still problem)

*Fig. 4.6.* *Working area of enterprises after action*

## *4.5. Comparisons of Solid Waste Management before to After Action*

*Table.4.31.* *The results of waste problems and improvements*

| Factor | Pre assessment/before action in% | Mid assessment (during Action) in% | Post assessment (after action)in% | Improvement/change in% |
|---|---|---|---|---|
| Prevention | 14 | 28 | 28 | 14 |
| Reuse | 11.76 | 21 | 21 | 10.34 |
| Collecting | 21 | 45 | 51 | 30 |
| Storage facility | 23 | 30 | 62 | 39 |
| Recycling | 0 | 0 | 0 | 0 |
| Trainings | 21 | 47 | 63 | 42 |
| Kaizen | 0 | 23.81 | 71.43 | 71.43 |
| Cumulative | | | | **29.54** |

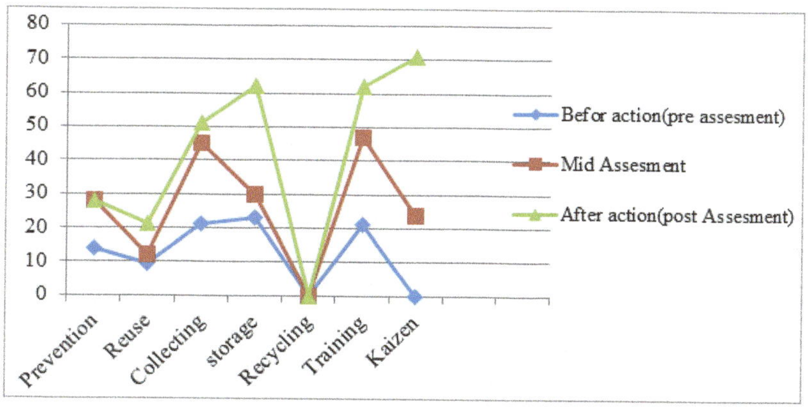


*Fig. 4.7.* *Line chart t analysis*

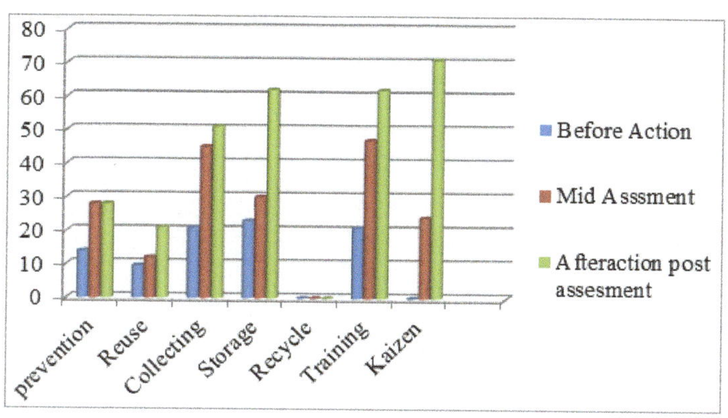


*Fig. 4.8.* *Pareto chart t analysis*

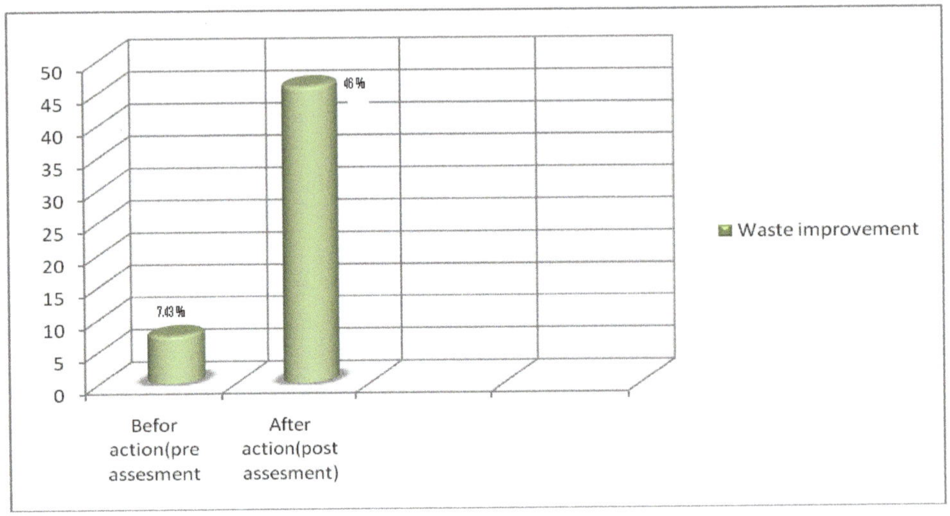


*Fig. 4.9.* *Problems of Waste management system improvement*

*Fig. 4.10.* *Working area of the enterprise's before and after action*

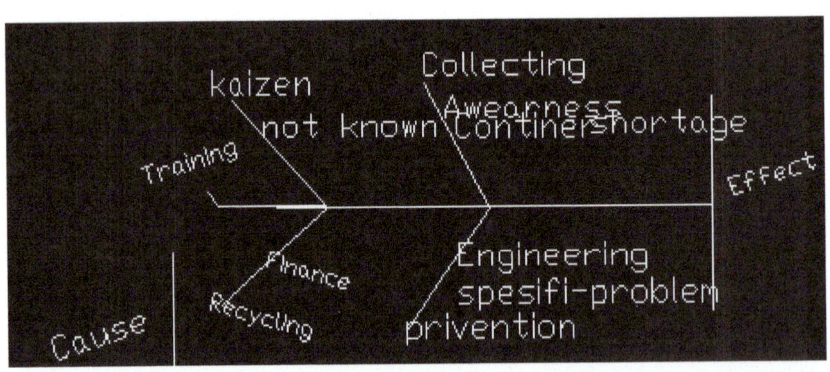


*Fig 4. 11.a)* *Fish bone) cause and effect diagram to see effects of Solid waste( before action)*

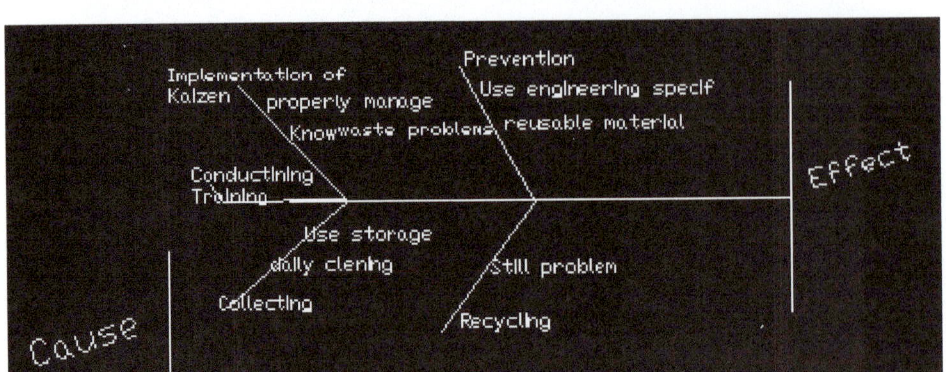


*Fig 4. 11b).* *Fish bone (cause and effect) diagram to develop of Solid waste management (after action)*

# 5. Summary

The aim of this research of problems of solid waste management was, to provide an efficient and sustainable solid waste management system and improving productivity of the enterprise through the development of solid waste management systems within participation stockholders.The conducted literature review revealed that, although similar studies were conducted in the previous the developed models were unable to consider simultaneously all the parameters influencing solid waste management this research were conducted using purposive sampling technique with sample size of one enterprise 1manager and 15 workers ,with total of 16 employers included in the study.

The study were used for analysis the problems of solid waste management's and actions taken to developed solid waste management systems in the enterprises. In the study using filed observation, unstructured interviews and questioners the research of problems of solid waste management of enterprises were conducted. The research was prepared as pre and post (after actions) analysis. The result shows that Solid waste management was characterize major problems of these enterprises. For assessing of the problem of solid waste management in Meserte Gabi metal and Machine work, the following findings were obtained like improper waste prevention, reusing, recycling, collecting, lack of training and awareness are the major problems.

Besides to reduce the problems we apply conduction training, capacity building to the enterprises and implementation of kaizen, solid waste management system were developed and there were positive change was seen in this enterprises. The result also shows that after these actions were taken the enterprise could adopt solid waste management systems and they reduce the amount of wastes in the following way. Achieving sustainable development of the enterprises, Improvement of waste prevention is increase from 14% to28%.Reuse of products is beast methods of waste prevention. Reuse of materials the capacity increase from 19.52 to 21%.on the other hand before and after action is done recycling is the major problems. There is no any recycling organizations are crated around Dessise. The enterprises use waste to collect and store wastes, the facility improves from 23 % to 62% and Collection of solid wastes is the most impact of waste management and using collection system the enterprises can increase the amount of waste from 21 % to 51%.

From the conducting questionnaires, interviews, and observation found that the production of the enterprise in the work shop were not consider the important factors of proper solid waste management's, however the researcher shown and considered important actions were done to develop proper solid waste systems in the enterprises. Hence from the action conducting training, capacity building, resource utilization, and implementation of kaizen were mechanisms that developed solid management systems in the enterprises.

# 6. Conclusion and Recommendation

## *6.1. Conclusion*

The problems of solid waste management result shows that Solid waste issues characterize major problems of Meserte Gabi metal and machine work enterprise. For assessing of the problem of solid waste management in the enterprises for, the following findings are obtained like waste prevention, reusing, recycling, collecting, storage facility and training and awareness are the major problems. The result also shows that after the action is taken (by conducting training, resources utilization, capacity building and implementation of kaizen) the enterprise should adopt solid waste management systems and they reduce the amount of wastes in the following way.

- Waste Prevention

Achieving sustainable development of the enterprises, Improvement of waste prevention is increase from 14% to28%.

- Reuse

Reuse of products is beast methods of waste prevention. Reuse of materials the capacity increase from 19.52 to 21%.

- Recycling

Before and after action is done recycling is the major problems. There is no any recycling organizations are crated around Dessise.

- Storage facilities/ Storage Containers

The enterprises use waste to collect and store wastes, the facility improves from 23 % to 62%

- Collection Systems

Collection of solid wastes is the most impact of waste management and using collection system the enterprises can increase the amount of waste from 21 % to 51%

Hence designing of solid waste management action research enterprise level through conducting training, capacity building, resource utilization, and implementation of kaizen were mechanisms that developed solid management systems in the enterprises, will help to enhance the solid waste reduction and improve productivity of the enterprises.

## *6.2. Recommendations for Action*

The findings of this study have important implications for interventions designed to enhance problems of Solid waste management in Meserte Gabi metal and Machine work enterprise i.e small and medium scale enterprise in Dessise and in similar sister enterprises and cities with in Ethiopia. Hence, policy makers and stakeholders could promote and facilitate the growth of small enterprises by accessing necessary infrastructural facilities.

Finally a number of concrete recommendations are done to improve the effectiveness of the enterprises solid waste management was proposed, covering the following areas.

Action From the enterprise

- Increase the amount of waste containers (in side and

outside the work shop).

- Encourage the employer by giving incentives
- Conduct workshops for stakeholders
- The enterprises should consider training the workers to upgrade their skills.
- Conduct continuous Trainings on solid and waste management technologies

Action from government

- Increase the enterprises awareness, education and participation in the waste management program.
- Conduct continuous Trainings on solid and waste management technologies
- Promote /crate recycling departments in educational institutions.
- The government/town and urban administrator) should design the waste collection sites of the enterprises.

## Acknowledgements

I would like to thanks Dr. Tafesse G/Senbet for giving the necessary documents and helpful guidance, for accomplishment of this research work. Next the author thanks very much for all my Paper Reviewers detail and wonderful reviewing and amazing feedback of the paper to increases scientific acceptance and strength of the paper. The authors also wish to thank the support from the sciencepublishinggroup.com/j/ijema under grants ISSN: 2328-7659 (Print); ISSN: 2328-7667 (Online).

## References

[1] Seadon (2006) Integrated Waste Management- Looking Beyond The Solid Waste Horizon.

[2] Ferguson, (1995). Managing and Minimizing Construction Waste, London.

[3] Yuping, (2003) Development Of The Nanshan Culture Tourism Zone In Hainan, China:

[4] Di Vita (2004). Renewable Resources and Waste Recycling. Environmental Modelling And Assessment

[5] Jones (2007).Environmental And Health Challenges of Municipal Solid Waste in China. China Environment Forum

[6] Flowerdew, Martin (2005). Methods in Human Geography: A Guide for Students Doing A Research Paper. Second Edition. Pearson Prentice Hall England.

[7] Forsyth, (2005). Enhancing Climate Technology Transfer through Greater Public-Private Cooperation: Lessons from Thailand and the Philippines.

[8] Furedy, (2002).Urban Waste and Rural Farmers: Enabling Low- Cost Organic Waste Reuse In Developing Countries. Retrieved August 2007, From City Farmers Website:

[9] (2004). International Solid Waste Association & United Nations Environment Program (ISWA & UNEP)

[10] By Solomon Worku Agaje July 2004 ,Socio Economic Determinants Of Growth Of Small Manufacturing Enterprises In Addis Ababa , Addis Ababa University School Of Graduate Studies

[11] By Camilla Louise Bjerkli Julu 2005 The Cycle Of Plastic Waste: An Analysis On The Informal Plastic Recovery System In Addis Ababa, Ethiopia, Norwegian University Of Science And Technology Depart

[12] Vivian W. Y. Tam(2009) Waste Management Parameters For Assessing Recycled Aggregate And Their Correlation Griffith School Of Engineering, Griffith University, Gold Coast, Queensland, Australia

[13] Juan Wary 2011Industry Extension Program No005/2003E.C Addis Ababa

[14] Christian Zurbrügg (2003) Urban Solid Waste Management In Low-Income Countries Of Asia How To Cope With The Garbage Crisis, Swiss Federal Institute For Environmental Science And Technology (EAWAG)

[15] By Dr Kalpana Sankar (2004),Environmental Protection That Benefits The Poor

[16] Prof. R. Rameezdeen (June 2010), Sustainability In Built Environment

[17] Gloria Elizabeth Ichim(2007) ,Assessing The Challenges And Potential Of Implementing Composting As Part Of A Municipal Solid Waste Management System, In Baisha,

# Spatio-Temporal Variations in Carbon Monoxide and Carbon Dioxide Levels in Three Motor Parks in Ibadan, Nigeria

**Godson Rowland Ana, Olusola Olabisi Ogunseye**

Department of Environmental Health Sciences, Faculty of Public Health, College of Medicine, University of Ibadan, Ibadan, Nigeria

**Email address:**

sholybliss4eva@gmail.com (O. O. Ogunseye)

**Abstract:** Carbon monoxide (CO) and carbon dioxide ($CO_2$) emissions arising from vehicles and combustion processes in motor parks predisposes to adverse health outcomes and associated health risks. There is dearth of studies and data on air quality in motor parks in Nigeria, hence, the need to assess the levels of CO and $CO_2$ and their relationship with meteorological parameters in three major motor parks in Ibadan, Nigeria. A cross-sectional comparative design was adopted for this study. Akinyele Motor Park (AMP) in Akinyele Local Government Area (LGA), Iwo Road Motor Park (IMP) in Ibadan North East LGA and New Garage Motor Park (NMP) in Ibadan South West LGA were purposively selected and levels of CO and $CO_2$ were monitored for 2 months with appropriate gas meters, meteorological parameters were also monitored using Ventus W155 wireless weather station. Values obtained were compared with WHO and ASHRAE guideline limits. Descriptive and inferential statistics were used for data analysis at $p=0.05$. CO concentrations (ppm) for AMP, IMP and NMP ranged from 2.0-106.0, 2.0 – 83.0 and 2.0 – 90.0 respectively while the mean $CO_2$ concentrations (ppm) were 395.4 ± 30.5, 356.3 ± 57.1 and 388.1 ± 42.1 respectively. There was a significant positive correlation between CO and $CO_2$ ($r=0.258$, $p=0.000$) and also with four meteorological parameters; temperature ($r=0.164$), rainfall ($r= 0.105$), heat index ($r= 0.134$) and dew point ($r= 0.127$) ($p<0.05$). A positive correlation was also found between $CO_2$ and four meteorological parameters; temperature ($r= 0.276$, $p=0.000$), rainfall ($r=0.125$), heat index ($r= 0.232$, $p=0.003$) and dew point ($r=0.028$). For the three motor parks, CO concentrations (ppm) were 80% higher than WHO guideline of 9 ppm for 8 hour monitoring. Mean $CO_2$ concentrations (ppm) were within ASHRAE guideline limit of 400 ppm. The study showed that meteorology has influence on CO and $CO_2$ concentrations and motor park users are exposed to high levels of CO. Routine monitoring of CO and $CO_2$ is recommended in order to ensure these emissions do not exceed guideline limits.

**Keywords:** Vehicles, Emissions, Meteorology, Motor Parks, Routine Monitoring

## 1. Introduction

The global vehicular fleet has increased ten-fold over the last 40 years, and it is predicted to increase even more over the next three decades. The United Nations estimated that more than 600 million people living in cities and towns all over the world are exposed to unhealthy and dangerous levels of air pollutants generated by vehicles [1]. Vehicular emissions account for 51% of carbon monoxide, 34% of nitrogen oxides and 10% of particulate matter released each year in the United States [2].

Carbon monoxide (CO) is a colourless, odourless, non-irritating gas produced as a byproduct of incomplete combustion of carbonaceous materials. These materials include petroleum products, coal, natural gas, wood, and plastics. CO can be produced at toxic levels by internal combustion engines, structural fires, industrial operations, and improperly vented heating or cooking appliances [3]. Most of CO emissions (95 - 98%) in a region can be related to anthropogenic activities [4].

When inhaled, CO binds to haemoglobin in red blood cells passing through the lungs, forming carboxyhemoglobin (COHb). Because CO binds to haemoglobin more tightly than oxygen does, CO occupies the sites normally used to bind and carry oxygen from the lungs to the tissues. One mechanism of

CO toxicity is that it decreases the oxygen content of arterial blood which reduces peripheral oxygen delivery [5].

Carbon dioxide ($CO_2$) is a dominant greenhouse gas [6]. Elevated levels of $CO_2$ in the atmosphere and its relationship with global climate change have attracted interests from scientific communities, becoming a topical issue posing both political and economic challenges worldwide [7]. $CO_2$ has drawn global attention due to its impact in terms of global warming as combustion of fuels in the power, transport and household sectors produce $CO_2$ and a wide range of short-lived air pollutants constituting directly or indirectly a substantial proportion of climate change and the bulk of adverse effects on human health from global energy use [8].

Road transportation is the second biggest source of greenhouse gas emissions in countries of the European Union (EU) where it contributes about one-fifth of the EU's total emission of $CO_2$ while passenger cars are responsible for 12% of EU's $CO_2$ emissions [9]. The most recent energy use and emission inventories indicate that China surpassed the United States as the world's largest carbon emitter in 2006 [10]. Most of the assessments carried out on anthropogenic $CO_2$ emissions were done in developed countries and emerging economies in Asia, which cumulatively account for 80% of $CO_2$ emissions [11]. Africa also contributed to emission levels as fossil fuel emissions for Africa were 285 Tg C (teragrams of carbon) accounting for 3.7% of the global emissions in 2005, however, fossil fuel emissions per capita in Africa are among the lowest in the world. Emissions from traffic are significant sources of air pollutants, especially in densely populated areas. $CO_2$ is the main product of fuel burning, but a wide range of other gases are emitted concurrently [12].

The weather of a place represents the state of the atmosphere over a brief period of time [13]. The weather conditions of any given location is often described in terms of meteorological elements which include the state of the sky, temperature, winds, pressure, precipitation, and humidity. These factors initiate and influence atmospheric processes [14]. The atmosphere is affected by direct and indirect energy releases *via* human activities such as burning of fossil fuel, industrial emissions and other phenomenon like El Nino and La-Nina [15]. It is believed that synoptic weather is an important driver of air pollution episodes via certain physical and chemical processes such as turbulent mixing, long-range transport, photochemical production and deposition [16-18]. Data are available on the influence of air pollutants on regional and global climate but very few studies have assessed the impact of air pollutants to every day weather [19].

The importance of transportation cannot be overemphasized as it is responsible for the movement of people in and around the city and it is also vital to the survival of such cities [20]. Motor parks are widespread and common public spaces in urban areas in Nigeria because many people use public transport systems, however, they differ in design, nature and services [21]. The goal of establishing of motor parks is to provide satisfaction to its users and commuters [21-22]. Several motor parks in developing countries are in deplorable state and their settings are not consistent with standards and best practices, hence incapable of delivering comfort and serenity to users [21]. Motor parks in Nigeria are public areas where various activities take place. In addition to vehicular emissions, there are other sources of emission in motor parks that release CO, $CO_2$ and other gases into the motor park ambient environment. Hence, this study was carried out to assess the levels of CO and $CO_2$ and their relationship with meteorological parameters in three major motor parks in Ibadan, Nigeria.

# 2. Materials and Methods

## *2.1. Study Area*

The study area was Ibadan. Ibadan is the capital of Oyo State, one of the thirty six states in Nigeria. Strategically, positioned on longitude 3°53' east of Greenwich Meridian and latitude 7° 23' north of the equator, this ancient city is located near the forest grassland boundary of south western Nigeria. The distance from Lagos is about 145 km North East [23]. As described by [24], Ibadan is situated close to the boundary between forest and grassland, which makes it a meeting point for people and products of both the forests and grassland areas.

This research was carried out in three major motor parks in Ibadan: Akinyele Motor Park (AMP) in Akinyele Local Government Area (LGA); Iwo Road Motor Park (IMP) in Ibadan North East LGA and the 'New Garage' Motor Park (NMP) located in the Ibadan South West LGA, within Ibadan municipal area. AMP caters for transport linking northern region which includes Kaduna, Abuja, Kano and other major cities. IMP caters for transport needs of passengers going towards eastern region comprising Benin, Warri, Port Harcourt, Uyo, Calabar and others. NMP links other major cities in the south west which includes Lagos, Abeokuta, Epe, Ijebu- Ode, and other coastal towns. These study sites were chosen purposively as they are the major exit points to reach northern, eastern and other western parts of the country. Many transportation devices such as taxis, mini-buses and buses originate and terminate at these points [25].

## *2.2. Description of Sampling Points*

Garmin GPS 60 was used to capture co-ordinates and elevation of three points in each motor park where monitoring of CO and $CO_2$ were done. Table 1 shows the sampling points in the motor parks and their corresponding co-ordinates and elevation.

***Table 1.*** *Co-ordinates and elevation of sampling points at the motor parks.*

| | Sampling points | Longitude | Latitude | Elevation |
|---|---|---|---|---|
| AMP, Ojoo | Point 1 | N07.46704° | E003.91339° | 243m |
| | Point 2 | N07.46731° | E003.91332° | 229m |
| | Point 3 | N07.46757° | E003.91288° | 231m |
| IMP, Iwo road | Point 1 | N07.40205° | E003.94246° | 239m |
| | Point 2 | N07.40185° | E003.94242° | 213m |
| | Point 3 | N07.40152° | E003.94212° | 230m |
| NMP, Apata | Point 1 | N07.39480° | E003.81088° | 213m |
| | Point 2 | N07.39448° | E003.81087° | 180m |
| | Point 3 | N07.39436° | E003.81103° | 175m |

### 2.3. Data Collection

CO monitoring was done using Extech CO 10 meter. 8-hour CO measurement was taken in each motor park twice a week for two months (August and September, 2014). The measurements were taken at 1hour intervals in three strategic points (front, back and centre) in order to obtain the average CO measurement from 8am-4pm. $CO_2$ monitoring was done using Telaire 7001 carbon dioxide and temperature monitor at the motor parks. 8-hour $CO_2$ measurement was taken in each motor park twice a week for two months (August and September, 2014). The measurements were taken at 1hour intervals in three strategic points (front, back and centre) in order to obtain the average $CO_2$ measurement from 8am-4pm. Meteorological parameters such as temperature, humidity, rainfall, heat index, dew point and sea level pressure were monitored using Ventus W155 wireless weather station. 8-hour monitoring of weather parameters was done twice a week in each motor park for 2 months (August and September, 2014) from 8am-4pm at 1hour interval.

### 2.4. Data Management and Statistical Analysis

Data was entered and analysed using statistical package for the social sciences (SPSS) version 20. Descriptive and inferential statistics were used in this study. Descriptive statistics was used to summarize data. Mean ± Standard Deviation (SD) and range was calculated for CO levels of the motor parks and compared with WHO guideline of 9ppm for 8-hour monitoring [26]. Mean ± SD and range was also calculated for $CO_2$ levels of the motor parks and compared with ASHRAE guideline of 400 ppm [27]. Mean ± Standard Deviation (SD) and range was calculated for meteorological parameters such as temperature, humidity, rainfall, heat index, dew point and sea level pressure. ANOVA was used to test for any significant differences in the levels of CO and $CO_2$ recorded at the motor parks. Pearson correlation test was carried out to check for relationships between CO, $CO_2$ and the meteorological parameters.

## 3. Results

### 3.1. Mean CO Concentration at the Sampling Points

Table 2 shows the mean CO concentration (ppm) of sampling points for morning and afternoon at the motor parks. Generally, afternoon mean CO concentrations were higher than morning. At AMP, Ojoo, for morning and afternoon, sampling point 3 recorded the highest concentration (15.4 ± 7.5 and 19.5 ± 10.9) while sampling point 2 recorded the lowest (11.7 ± 2.7 and 13.8 ± 6.6) respectively.

At IMP, Iwo road, during the morning period, sampling point 3 recorded the highest value (15.1 ± 8.3) while sampling point 2 recorded the lowest (12.0 ± 5.2). During the afternoon period, sampling point 1 recorded the highest concentration (24.0 ± 16.0) while sampling point 2 recorded the lowest (16.2 ± 11.1). At NMP, Apata, during the morning period, sampling point 3 recorded the highest concentration (15.5 ± 8.0) while sampling point 2 recorded the lowest (9.74 ± 3.13). During the afternoon period, sampling point 1 recorded the highest concentration (26.3 ± 16.9) while sampling point 2 recorded the lowest (9.6 ± 3.5).

For the three motor parks, sampling point 3 at NMP, Apata recorded the highest concentration (15.5 ± 8.0) for morning while sampling point 2 at NMP, Apata recorded the lowest (9.7 ± 3.1). For afternoon, sampling point 1 at NMP, Apata recorded the highest concentration (26.3 ± 16.9) while sampling point 2 at NMP, Apata recorded the lowest (9.6 ± 3.5). All the mean concentrations were higher than WHO guideline of 9 ppm.

### 3.2. Mean $CO_2$ Concentration at the Sampling Points

Table 2 also shows the mean $CO_2$ concentration (ppm) of sampling points for morning and afternoon at the motor parks. Generally, afternoon mean $CO_2$ concentrations were higher than morning. At AMP, Ojoo, during the morning and afternoon periods, sampling point 3 recorded the highest concentration (397.7 ± 38.9 and 410.9 ± 17.4) while sampling point 2 recorded the lowest (371.4 ± 37.4 and 396.0 ± 24.0) respectively.

At IMP, Iwo road, during the morning and afternoon periods, sampling point 1 recorded the highest concentration (361.4 ± 61.0 and 407.9 ± 47.6) while sampling point 2 recorded the lowest (327.7 ± 46.9 and 350.0 ± 51.6) respectively. At NMP, Apata, during the morning period, sampling point 3 recorded the highest concentration (381.1 ± 35.2) while sampling point 2 recorded the lowest (365.3 ± 26.6). During the afternoon period, sampling point 1 recorded the highest concentration (419.9 ± 44.4) while sampling point 2 recorded the lowest (388.8 ± 41.5).

For the three motor parks, sampling point 3 at AMP, Ojoo recorded the highest concentration (397.7 ± 38.9) for morning while sampling point 2 at IMP, Iwo road recorded the lowest (327.7 ± 46.9). During afternoon period, sampling point 1 at NMP, Apata recorded the highest concentration (419.9 ± 44.4) while sampling point 2 at IMP, Iwo road recorded the lowest (350.0 ± 51.6).

### 3.3. Temporal Variations in the Levels of CO

Fig. 1 shows the pattern of CO levels during the 8 hours of monitoring. For AMP, Ojoo, the chart shows that CO concentration (ppm) was lowest at 8am (11.3 ± 4.2) and highest at 12noon (17.2 ± 17.2). At IMP, Ojoo, CO concentration (ppm) was lowest at 8am (10.5 ± 3.1) and highest at 1pm (21.5 ± 15.0). At NMP, Apata, CO concentration (ppm) was lowest at 8am (9.4 ± 2.9) and highest at 4pm (19.3 ± 17.2). There was a statistically significant difference in the levels of CO across the 8 hours of monitoring at the three motor parks ($p<0.05$).

***Table 2.*** *CO and $CO_2$ mean±SD of the sampling points for morning and afternoon for the motor parks.*

| | | | | | | |
|---|---|---|---|---|---|---|
| | **AMP, Ojoo** | | | | | |
| | Point 1 | | Point 2 | | Point 3 | |
| | Morning | Afternoon | Morning | Afternoon | Morning | Afternoon |
| CO | 13.2±13.4 | 14.0±9.5 | 11.7±2.7 | 13.8±6.6 | 15.4±7.4 | 19.5±11.0 |
| $CO_2$ | 396.4±22.9 | 404.9±13.5 | 371.4±37.4 | 396.0±24.0 | 397.7±38.9 | 410.9±17.4 |
| | **IMP, Iwo Road** | | | | | |
| | Point 1 | | Point 2 | | Point 3 | |
| | Morning | Afternoon | Morning | Afternoon | Morning | Afternoon |
| CO | 14.8±7.0 | 24.0±16.0 | 12.0±5.2 | 16.2±11.1 | 15.1±8.3 | 18.7±13.1 |
| $CO_2$ | 361.4±61.0 | 407.9±47.6 | 327.7±46.9 | 350.0±51.6 | 338.5±58.9 | 362.3±41.4 |
| | **NMP, Apata** | | | | | |
| | Point 1 | | Point 2 | | Point 3 | |
| | Morning | Afternoon | Morning | Afternoon | Morning | Afternoon |
| CO | 14.7±9.3 | 26.3±16.9 | 9.7±3.1 | 9.6±3.5 | 15.5±8.0 | 14.8±5.0 |
| $CO_2$ | 380.0±44.0 | 419.9±44.4 | 365.3±26.6 | 388.8±41.5 | 381.1±35.2 | 402.8±41.4 |

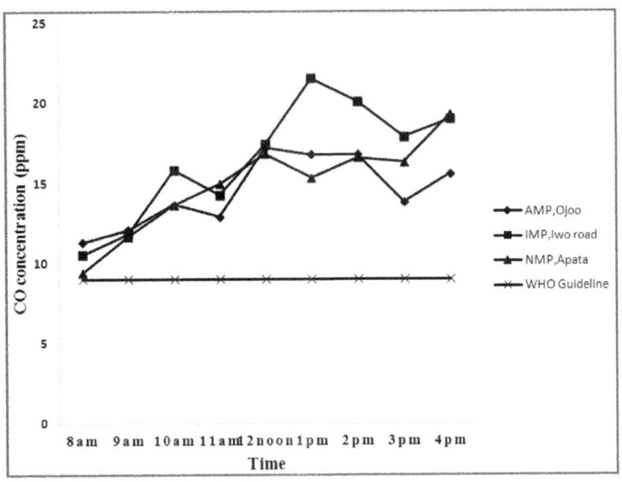


***Figure 1.*** *Concentration of CO during the 8 hours of monitoring.*

### *3.4. Temporal Variations in the Levels of $CO_2$*

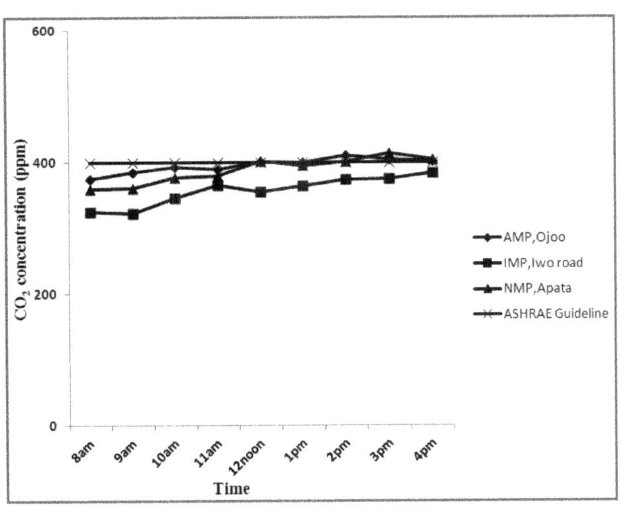


***Figure 2.*** *Concentration of $CO_2$ during the 8 hours of monitoring.*

Fig. 2 shows the pattern of $CO_2$ levels during the 8 hours of monitoring. For AMP, Ojoo, the chart shows that $CO_2$ concentration (ppm) was lowest at 8am (374.6 ± 37.2) and highest at 2pm (409.9 ± 12.7). At IMP, Iwo road, $CO_2$ concentration (ppm) was lowest at 9am (322.3 ± 55.0) and highest at 4pm (383.5 ± 52.2). At NMP, Apata, $CO_2$ concentration (ppm) was lowest at 8am (358.9 ± 23.6) and highest at 3pm (413.9 ± 55.9). There was a statistically significant difference in the levels of $CO_2$ across the 8 hours of monitoring at the three motor parks ($p<0.05$).

### *3.5. Spatial Variations in the Levels of CO and $CO_2$*

Table 3 shows the comparison of mean concentrations of CO and $CO_2$ at the three sampling points in each motor park, and the result of ANOVA shows that there were differences in the levels of CO and $CO_2$ at the sampling points in each motor park and the differences were statistically significant. ($p<0.05$).

***Table 3.*** *Comparison of levels of CO and $CO_2$ at sampling points of each motor park.*

| | | Point 1 | Point 2 | Point 3 | p-value |
|---|---|---|---|---|---|
| CO | AMP, Ojoo | 13.6±11.8 | 12.6±4.9 | 17.2±1.4 | 0.000 |
| | IMP, Iwo road | 18.9±12.6 | 13.9±8.5 | 16.7±10.8 | 0.000 |
| | NMP, Apata | 19.9±14.4 | 9.7±3.3 | 15.2±6.8 | 0.000 |
| $CO_2$ | AMP, Ojoo | 400.2±19.6 | 382.3±34.2 | 403.6±31.7 | 0.000 |
| | IMP, Iwo road | 382.1±59.7 | 337.6+49.9 | 349.1+52.8 | 0.000 |
| | NMP, Apata | 397.7+48.1 | 375.7+35.7 | 390.8+39.3 | 0.020 |

### *3.6. Mean and Range of CO and $CO_2$ for the Motor Parks*

Table 4 shows the overall mean and range for CO and $CO_2$ in the motor parks. The overall mean CO concentration (ppm) for AMP, Ojoo, IMP, Iwo road and NMP, Apata were 14.5 ± 9.3, 16.5 ± 11.0 and 14.9 ± 10.2 respectively ($p=0.009$) while the range were 2.0-106.0, 2.0-83.0 and 2.0-90.0 respectively. For AMP, Ojoo, IMP, Iwo road and NMP, Apata, CO (ppm) concentrations were 82%, 83% and 75% higher than WHO guideline of 9ppm for 8hour monitoring respectively. Generally, CO (ppm) concentrations were 80% higher than WHO guideline of 9 ppm for 8 hour monitoring across the motor parks.

Mean $CO_2$ concentration (ppm) for AMP, Ojoo, IMP, Iwo road and NMP, Apata were 395.4 ± 30.5, 356.3 ± 57.1 and 388.1 ± 42.1 respectively (p=0.000) while the range were 290.0-450.0, 255.0-524.0 and 320.0-560.0 respectively. These mean $CO_2$ concentrations are lower than ASHRAE guideline of 400 ppm.

Figures 3 and 4 show the mean±SD of CO and $CO_2$ across the motor parks respectively.

***Table 4.** Mean and range of CO and $CO_2$ for the motor parks.*

| | AMP, Ojoo | IMP, Iwo road | NMP, Apata | p-value |
|---|---|---|---|---|
| CO(ppm) | 14.5 ± 9.3 | 16.5 ± 11.0 | 14.9 ± 10.2 | 0.009 |
| | 2-106 | 2-83 | 2-90 | |
| $CO_2$(ppm) | 395.4 ± 30.5 | 356.3 ± 57.1 | 388.1 ± 42.1 | 0.000 |
| | 290-450 | 255-524 | 320-560 | |

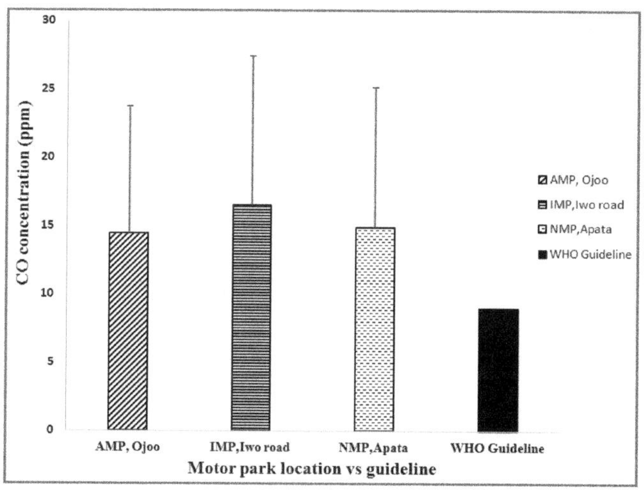


***Figure 3.** Mean CO concentrations across the motor parks compared with WHO guideline.*

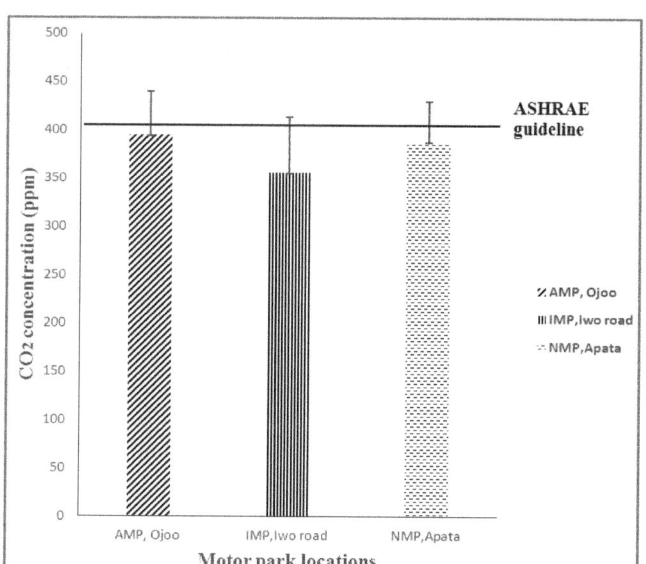


***Figure 4.** Mean $CO_2$ concentrations across the motor parks compared with ASHRAE guideline.*

### *3.7. Meteorological Assessment*

Table 5 shows the mean ± SD and range of meteorological parameters assessed at the motor parks. AMP, Ojoo recorded the highest mean temperature (28.5 ± 2.6) while IMP, Iwo road recorded the lowest (28.4 ± 2.0). NMP, Apata recorded the highest mean humidity (82.0 ± 9.3) while AMP, Ojoo recorded the lowest (79.3 ± 9.8). AMP, Ojoo recorded the highest mean rainfall (278.0 ± 404.8) while NMP, Apata recorded the lowest (144.4 ± 206.6). NMP, Apata recorded the highest mean heat index (33.8 ± 8.4) while IMP, Iwo road recorded the lowest (32.9 ± 4.1). NMP, Apata recorded the highest mean dew point (25.1 ± 1.5) while AMP, Ojoo recorded the lowest (24.4 ± 1.4). NMP, Apata recorded the highest mean sea level pressure (1012.6 ± 2.0) while IMP, Iwo road recorded the lowest (1011.7 ± 4.3).

***Table 5.** Mean and range of meteorological parameters for the motor parks.*

| | AMP, Ojoo | IMP, Iwo road | NMP, Apata |
|---|---|---|---|
| Temperature (°C) | 28.5 ± 2.6 | 28.4 ± 2.0 | 28.4 ± 2.3 |
| | 22.3-35.0 | 25.2-33.3 | 22.8-34.2 |
| Humidity (%) | 79.3 ± 9.8 | 79.5 ± 9.0 | 82.0 ± 9.3 |
| | 54.0-95.0 | 55.0-91.0 | 55.0-96.0 |
| Rainfall (mm) | 278.0 ± 404.8 | 205.1 ± 361.9 | 144.4 ± 206.6 |
| | 0.0-2309.0 | 0.0-2511.0 | 0.0-997.7 |
| Heat Index(°C) | 32.9 ± 5.5 | 32.9 ± 4.1 | 33.8 ± 8.4 |
| | 19.8-54.9 | 26.0-45.0 | 22.8-47.2 |
| Dew point(°C) | 24.4 ± 1.4 | 24.5 ± 1.3 | 25.1 ± 1.5 |
| | 21.4-30.5 | 22.3-28.0 | 19.3-28.8 |
| Sea level pressure (hpa) | 1012.1 ± 2.2 | 1011.7 ± 4.3 | 1012.6 ± 2.0 |
| | 1004.0-1018.0 | 993.0-1017.0 | 1005.0-1017.0 |

### *3.8. Relationships Between CO, $CO_2$ and Meteorological Parameters*

Pearson correlation test was carried out between CO, $CO_2$ and the meteorological parameters as shown in Table 6. The table shows that there was a significant positive correlation between CO and $CO_2$ (r=0.258, p=0.000). Fig. 5 shows the strength of the linear relationship between the levels of CO and $CO_2$ ($R^2$=6.7%). A significant positive correlation was also recorded between CO and four meteorological parameters namely; temperature (r=.0164), rainfall (r= 0.105), heat index (r= 0.134) and dew point (r= 0.127) ($p<0.05$) while there existed a negative correlation between CO and the remaining two meteorological parameters, i.e, humidity (r= -0.114, p=0.018) and sea level pressure (r= -0.002).

A positive correlation was also found between $CO_2$ and four meteorological parameters namely; temperature (r= 0.276, p=0.000), rainfall (r=0.125), heat index (r= 0.232, p=0.003) and dew point (r=0.028) while a negative correlation was recorded between $CO_2$ and the remaining two meteorological parameters i.e humidity (r= -0.286, p=0.000) and sea level pressure (r= -0.172, p=0.035).

There existed a significant negative correlation between temperature and two other meteorological parameters namely humidity (r= -0.795, p=0.000) and sea level pressure (r= -0.230, p=0.000), while temperature correlated positively with rainfall (r=0.155, p=0.001), heat index (r= 0.706, p=0.000) and dew point (r=0.413, p=0.000).

There was a significant positive correlation between humidity and two other meteorological parameters, namely; dew point (r= 0.210, p=0.000) and sea level pressure (r= 0.219, p=0.000), while a significant negative correlation existed between humidity and heat index (r= -0.431, p=0.000).

A significant positive correlation was observed between rainfall and two other meteorological parameters, namely; heat index (r= 0.163, p= 0.001) and dew point (r= 0.188, p=0.000). A positive correlation existed between heat index and dew point (r= 0.651, p= 0.000) while a negative correlation was noticed between heat index and sea level pressure (r= -0.223, p= 0.000).

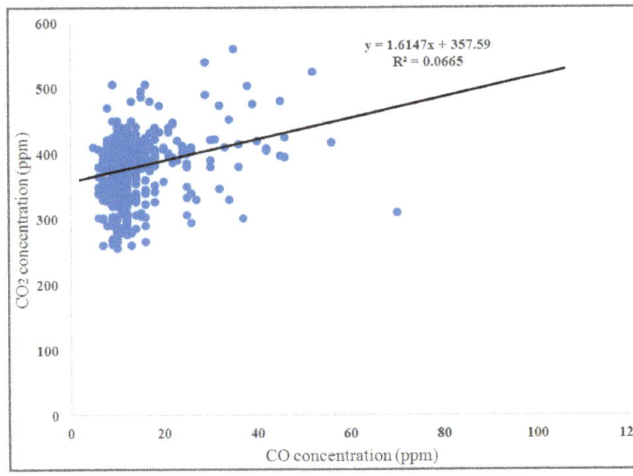


***Figure 5.*** *Relationship between levels of CO and $CO_2$ at the motor parks.*

***Table 6.*** *Correlation between CO, $CO_2$ and meteorological parameters.*

| Pearson Correlation coefficient Sig (2 tailed) | CO(ppm) | $CO_2$(ppm) | Temperature (°C) | Humidity (%) | Rainfall (mm) | Heat index(°C) | Dew point (°C) | Sea level pressure (hpa) |
|---|---|---|---|---|---|---|---|---|
| CO(ppm) | 1 | | | | | | | |
| $CO_2$(ppm) | 0.258**<br>0.000 | 1 | | | | | | |
| Temperature (°C) | 0.164**<br>0.001 | 0.276**<br>0.000 | 1 | | | | | |
| Humidity (%) | -0.114*<br>0.018 | -0.286**<br>0.000 | -0.795**<br>0.000 | 1 | | | | |
| Rainfall (mm) | 0.105*<br>0.029 | 0.125<br>0.113 | 0.155**<br>.155** | -0.048<br>0.319 | 1 | | | |
| Heat index(°C) | 0.134**<br>0.005 | 0.232**<br>0.003 | 0.706**<br>0.000 | -0.431**<br>0.000 | 0.163**<br>0.001 | 1 | | |
| Dew point (°C) | 0.127**<br>0.010 | 0.028<br>0.735 | 0.413**<br>0.000 | 0.210**<br>0.000 | 0.188**<br>0.000 | 0.651**<br>0.000 | 1 | |
| Sea level pressure (hpa) | -0.002<br>0.970 | -0.172*<br>0.035 | -0.230**<br>0.000 | 0.219**<br>0.000 | -0.003<br>0.945 | -0.223**<br>0.000 | -0.048<br>0.346 | 1 |

**. Correlation is significant at the 0.01 level (2-tailed)
*. Correlation is significant at the 0.05 level (2-tailed)

# 4. Discussion

Generally, afternoon mean CO concentrations were higher than morning. This may be due to increased influx of vehicles into the motor parks in the afternoon. This result is in line with the study by [28] in Enugu, Nigeria but at variance with studies by [29] in Lafia metropolis and [30] in Benin City where CO concentration was higher in the morning but lower in the afternoon.

There were variations in the mean CO concentration at the sampling points in the motor parks. This was due to the fact that, at each sampling point in the motor parks, activities that emit CO differ in degree and frequency. The pollution level at these sampling points varied as observed during CO monitoring in the motor parks and the difference in mean CO level across the sampling points was significant for the motor parks, hence, there were spatial variations in CO levels for the motor parks. The pattern of CO emissions during the 8 hours of monitoring revealed that there were temporal variations in CO concentrations across the three motor parks. This result is supported by [31] who stated that ambient CO concentrations have daily and seasonal variations, as well as complex spatial distributions.

IMP, Iwo road recorded the highest mean CO concentration (ppm) (16.5±11.0) while AMP, Ojoo recorded the lowest (14.5±9.3). There was significant difference in the mean CO concentration of the motor parks (p=0.009), hence, the motor parks vary in CO levels. The highest mean CO concentration recorded at IMP, Iwo road may be attributed to high vehicular density in the motor park than others. According to a study by [32], mean CO concentration at Iwo road near overhead bridge was 86 ppm while it was 32.5ppm at Ojoo trailer park. These mean CO concentrations were higher than the mean CO concentrations recorded in this study.

There were spatial and temporal variations in CO concentrations across the three motor parks and this is corroborated by the study by [30]. All the mean CO concentrations across sampling points, during the 8 hours of monitoring and overall mean for the motor parks were above the WHO guideline of 9 ppm [26].

The CO range (ppm) for AMP, Ojoo, IMP, Iwo road and NMP, Apata were 2.0-106.0, 2.0-83.0 and 2.0-90.0 respectively and generally, these CO concentrations (ppm) were 80% higher than WHO guideline of 9 ppm for 8hour monitoring across the motor parks. Hence, motor park traders are exposed to high concentration of CO in the motor parks. This finding is supported by a study that investigated the ambient carbon monoxide and carboxyhaemoglobin levels in Ibadan City, Nigeria by [33], it was reported that ambient carbon monoxide levels in Ibadan city were between 3.0 and 55.0 ppm with a mean value and standard deviation of 20.12 ± 1.40ppm. The study indicated that Ibadan city dwellers were generally exposed to higher levels of CO than WHO permissible guideline limit. Vocations and occupations requiring workers to spend long periods on the highways are at higher risk of exposure. Potential exposures that exceed the existing guideline may be of greater concern to public health because they increase the total body burden for CO [34].

Generally, afternoon mean $CO_2$ concentrations were higher than morning. This may also be due to increased influx of vehicles into the motor parks in the afternoon, and also increase in air pollution related activities in the motor parks. This could also be explained by stating that CO is co-emitted with $CO_2$ from combustion sources [7], hence, increase in CO concentration in the afternoon when compared to morning will result in increased $CO_2$ concentration in the afternoon as compared to morning.

There were variations in the mean $CO_2$ concentration at the sampling points in the motor parks. This is due to the fact that at each sampling point in the motor parks, $CO_2$ emitting activities vary. The emission levels at these sampling points differ as observed during $CO_2$ monitoring in the motor parks and the difference in mean $CO_2$ levels across the sampling points was statistically significant for the motor parks, hence a spatial variation in $CO_2$ level was observed for the motor parks. The pattern of $CO_2$ emission during the 8 hours of monitoring revealed that there were temporal variations in the $CO_2$ concentration across the three motor parks.

AMP, Ojoo recorded the highest mean $CO_2$ concentration (395.4 ± 30.5) while IMP, Iwo road recorded the lowest (356.3±57.1). This result is similar to that reported in a study by [8] where the mean $CO_2$ concentration in traffic area TA (TA comprised of Ojoo park, Mokola round about, Agbowo complex area, Challenge park and Oluyole Industrial area) was highest at 390 ± 64 ppm during the rainy season. However, the result of this study is lower than that reported by [35], in which $CO_2$ concentration was as high as 1710 ppm at Kawo motor park in Kaduna.

There was significant difference in the mean $CO_2$ concentration across the motor parks (p=0.000), hence, the motor parks vary in $CO_2$ concentrations. All the mean $CO_2$ concentration of the motor parks are below ASHRAE guideline of 400 ppm [27]. There was a significant positive correlation between CO and $CO_2$ (r=0.258, p=0.000, $R^2$=6.7%). CO is co-emitted with $CO_2$ from combustion sources, leading to a significant positive correlation between them when combustion is a significant source of observed $CO_2$ [7]. There was a significant positive correlation between CO and temperature (r= 0. 164, p= 0.001). This could be attributed to the fact that internal combustion engines of vehicles perform better in cold temperature and emit less CO but as temperature increases, the engine efficiency reduces, hence, releasing more CO into the motor park ambient environment. A study by [36] at Gaza strip reported a positive correlation between outdoor CO and outdoor temperature during winter.

CO concentrations in the motor parks have significant relationships with meteorological parameters, hence, meteorology has influence on the levels of CO in the motor parks. This assertion is corroborated by [31] who noted that meteorological conditions highly influence CO levels in urban regions. There was also a significant positive correlation between $CO_2$ and temperature. This is expected as CO is co-emitted with $CO_2$ from combustion sources [7]. However, in a study by [37] in Poland, significant negative correlation was found between $CO_2$ and air temperature. This is further corroborated by a study by [19] in Eastern China that showed that intense air pollution significantly affects local synoptic weather by influencing certain weather parameters indicating clear air pollution–weather interactions. Generally, for the motor parks, temporal variations were observed for all meteorological parameters and there were marked relationships between them.

# 5. Conclusion and Recommendations

This study assessed levels of CO and $CO_2$ and their relationships with meteorological parameters in three major motor parks in Ibadan, Nigeria. Mean CO concentration for afternoon was higher than that of morning at the motor parks due to increased influx of vehicles into the motor parks in the afternoon, and also increase in air pollution related activities. There were temporal variations in CO concentration across the 8 hours of monitoring and spatial variations across the sampling points for the motor parks. IMP, Iwo road recorded the highest mean CO concentration while AMP, Ojoo recorded the lowest. Generally, CO (ppm) concentrations were 80% higher than WHO guideline of 9 ppm for 8 hour monitoring across the motor parks. Therefore, motor park users are exposed to high concentrations of CO and are susceptible to effects of CO exposure.

Mean $CO_2$ concentration for afternoon was higher than that of morning at the motor parks due to increased influx of vehicles into the motor parks in the afternoon, and also increase in air pollution related activities. There were temporal variations in $CO_2$ concentration across the 8 hours of monitoring and spatial variations across the sampling points for the motor parks. AMP, Ojoo recorded the highest mean $CO_2$ concentration while IMP, Iwo road recorded the lowest. The mean $CO_2$ concentrations for the motor parks were lower than ASHRAE guideline of 400 ppm.

There was a significant positive correlation between CO and $CO_2$. Generally, CO and $CO_2$ concentrations in the motor parks have significant relationships with the meteorological parameters, hence, meteorology has influence on the levels of CO and $CO_2$ in the motor parks. Temporal variations were observed for all the meteorological parameters across the 8 hours of monitoring at the motor parks and there were marked relationships between them.

Fixed site CO and $CO_2$ monitoring equipment should be installed at the motor parks to ensure regular assessment of these gases at the motor parks so that they do not exceed guideline limits.

## Acknowledgements

The authors are grateful to staff and students in the Department of Environmental Health Sciences, Faculty of Public Health, College of Medicine, University of Ibadan, who contributed to the successful completion of this research. The technical support from concerned individuals within and outside the University is also gratefully acknowledged.

## References

[1] Cacciola, R.R., Sarva, M., and Polosa, R. 2002. Adverse respiratory effects and allergic susceptibility in relation to particulate air pollution: flirting with disaster. *Allergy* 57:281–286.

[2] Olamijulo, J. O. and Ana, G. R. E. E. 2013. Air Quality in Relation to Vehicular Traffic-Related Gaseous Emissions in two Selected Local Government Areas in South-Western Nigeria. *International Journal of Environmental Monitoring and Analysis.* Vol. 1, No. 4, 2013, pp. 121-127.doi: 10.11648/j.ijema.20130104.12.

[3] Centers for Disease Control and Prevention. 1997. Carbon Monoxide Intoxication and Death in a Newly Constructed Sewer Manhole. Hazard Evaluation and Technical Assistance Report HETA 98-0020. http://www.cdc.gov/niosh/hhe/.

[4] Boudaghpour, S. and Jadidi A. 2009. Investigation of the effect of outlet pollutants of cement production industries around Tehran and approaches to control and eliminate pollutants. *International Journal of Physical Sciences* 4:486-495.

[5] Piantadosi, C.A. 2004. Carbon monoxide poisoning. *Undersea and Hyperbaric Medicine Journal* 31.1:167–177.

[6] Canadell, J. G., Raupach, M. R. and Houghton, R. A. 2009. Anthropogenic $CO_2$ emissions in Africa. *Biogeosciences,* 6: 463–468.

[7] Wang, Y., Munger, J. W., Xu, S., McElroy, M. B., Hao, J., Nielsen, C. P., and Ma, H. *et al.* 2010. $CO_2$ and its correlation with CO at a rural site near Beijing: Implications for combustion efficiency in China. *Atmospheric Chemistry and Physics* 10:8881–8897.

[8] Ana, G.R., Ojelabi, Peju and Shendell, Derek, G. 2014. Spatial-temporal variations in carbon dioxide levels in Ibadan, Nigeria. *International Journal of Environmental Health Research* 2014: 1-12. *http://dx.doi.org/10.1080/09603123.2014.938024*

[9] De Wilde, H.P.J. and Kroon, P. 2013. Policy options to reduce passenger cars emissions after 2020. Energy Research Centre of the Netherland, February, 2013.

[10] Gregg, J. S., Andres, R. J. and Marland, G. 2008. China: Emissions pattern of the world leader in $CO_2$ emissions from fossil fuel consumption and cement production. *Geophysical Research Letters* 35:10.

[11] Raupach, M. R., Marland, G., Ciais, P., Le Quere, C., Canadell, J. G., Klepper, G., and Field, C. B. 2007. Global and regional drivers of accelerating $CO_2$ emissions, *Proceedings of the National Academy of Sciences*, 0700609104.

[12] Popa, M.E., Vollmer, M. K., Jordan, A., Brand, W. A., Pathirana, S. L., Rothe, M., and Röckmann, T. 2014. Vehicle emissions of greenhouse gases and related tracers from a tunnel study: CO: $CO_2$, $N_2O$:$CO_2$, $CH_4$:$CO_2$, $O_2$:$CO_2$ ratios, and the stable isotopes $^{13}C$ and $^{18}O$ in $CO_2$ and CO. *Atmospheric Chemistry and Physics*, 14, 2105–2123.

[13] Aiyelabegan, A. T. 2014. Latitudinal dependence of some meteorological parameters in Nigeria. *International Journal of Technical Research and Applications,* Volume2, Special Issue 6 (Nov-Dec 2014), pp.7-10.

[14] Ayoade, J.O. 1993. *Introduction to the Tropics*. Lagos: Spectrum Books Ltd.

[15] Ogolo, E. O. and Adeyemi, B. 2009. Variations and Trends of Some Meteorological Parameters at Ibadan, Nigeria. *The Pacific Journal of Science and Technology* 10.2:981-987.

[16] Hegarty, J., Mao, H., and Talbot, R. 2007. Synoptic controls on summertime surface ozone in the northeastern United States, *Journal of Geophysical Research*, 112, D14306, doi: 10.1029/2006JD008170.

[17] Ding, A. J., Wang, T., Xue, L. K., Gao, J., Stoh, A., Lei, H. C., Jin, D. Z., Ren, Y., Wang, Z. F., Wei, X. L., Qi, Y. B., Liu, J., and Zhang, X. Q. 2009. Transport of north China mid-latitude cyclones: Case study of aircraft measurements in summer 2007, *Journal of Geophysical Research*, 114, D08304, doi: 10.1029/2008JD011023.

[18] Zhang, Y., Mao, H. T., Ding, A. J., Zhou, D. R., and Fu, C. B. 2013. Impact of synoptic weather patterns on spatio-temporal variation in surface $O_3$ levels in Hong Kong during 1999–2011, *Atmospheric Environment*, 73: 41–50.

[19] Ding, A. J, Fu, C. B., Yang, X. Q., Sun, J. N., Petäjä, T., Kerminen, V.M., Wang, T., Xie, Y., Herrmann, E., Zheng, L. F., Nie, W., Liu, Q., Wei, X. L, and Kulmala, M. 2013. Intense atmospheric pollution modifies weather: a case of mixed biomass burning with fossil fuel combustion pollution in eastern China. *Atmospheric Chemistry and Physics*, 13, 10545–10554.

[20] Onokala, P.C. 2001. Urbanization and Urban transportation problems in Nigeria. *Issues in urbanization and urban administration in Nigeria*. E.O. Ezeani and N.N. Elekwe. Eds. Enugu: Jamoe Enterprises (Nigeria) Publishers. 168-186.

[21] Adedayo, O. F. and Zubairu, S. N. 2013. An assessment of facilities in motor parks in Minna, Niger State, Nigeria, through post-occupancy evaluation. *Management* 3.7:360-367.

[22] Anable, J. 2005. Complacent Car Addicts' or 'Aspiring Environmentalists'? Identifying travel behaviour segments using attitude theory. *Elsevier Journal of Transport Policy* 12:65-78.

[23] Ayeni, M.O.A. 1994. The metropolitan area of Ibadan; its growth and structure. *Ibadan Region.* M.O Filani, *et al.* Eds. Nigeria: Rex Charles Publishers. Chapter 7:72-84.

[24] Ajayi, O. *et al.* 2012. Flood Management in an Urban Setting: A Case Study of Ibadan Metropolis. *Nigerian Association of Hydrological Sciences.* 65-81. www.unaab.edu.ng.

[25] Titus, S.I. *et al.* 2010. Refuse disposal practices in three major motor parks in Ibadan municipality, Nigeria. *Journal of Public Health and Epidemiology,* 2.4:82-86.

[26] World Health Organization 1999. *Environmental Health Criteria 213: Carbon Monoxide.* 2nd ed. Geneva: World Health Organization.

[27] American Society for Heating Refrigerating and Air-conditioning Engineers Incorporated (ASHRAE) 2008. *Selecting $CO_2$: Criteria for Outdoor Air Monitoring.* ASHRAE Journal, www.ashrae.org.

[28] Nwadiogbu J.O. *et al.* 2013. Contributions of vehicular traffic to carbon monoxide emissions in Enugu Metropolis, Nigeria. *Archives of Applied Science Research* 5.6:189-192.

[29] Asheshi O.O. 2012. Measurement of Traffic Emission in Lafia Metropolis. *Journal of Science and Multidisciplinary Research.*4:34-46.

[30] Ukpebor, E.E. *et al.* 2010. Spatial and Diurnal Variations of Carbon Monoxide (CO) Pollution from Motor Vehicles in an Urban Centre. *Polish Journal of Environmental Studies* 19.4:817-823.

[31] Han, X. and Naeher, L.P. 2006. A review of traffic-related air pollution exposure assessment studies in the developing world. *Elsevier Environment International* 32:106 – 120.

[32] Osuntogun, B.A. and Koku, C.A. 2007. Environmental Impacts of Urban Road Transportation in South-western states of Nigeria. *Journal of Applied Sciences* 7.16:2356-2360.

[33] Banjoko S.O. *et al.* 2008. Ambient carbon monoxide and carboxyhaemoglobin levels in Ibadan City, Nigeria: A source of health inequality between developed and developing nations? *Journal of Environmental Health Research* 1.7. http://www.cieh.org/jehr/jehr3.aspx?id=11484

[34] United States Environmental Protection Agency 1999. *Air Quality Criteria for Carbon Monoxide (External Review Draft).* National Center for Environmental Assessment, Research Triangle Park, North Carolina: USEPA.

[35] Ndoke, P.N. *et al.* 2006. Contributions of Vehicular Traffic to Carbon Dioxide Emissions in Kaduna and Abuja, Northern Nigeria. *Leonardo Electronic Journal of Practices and Technologies* Issue 9:81-90.

[36] Elbayoumi, M. *et al.* 2014. The effect of seasonal variation on indoor and outdoor carbon monoxide concentrations in Eastern Mediterranean climate. *Atmospheric Pollution Research* 5:315-324.

[37] Napiorkowska, M. and Tomaszewska, M. 2013. The relationship between Carbon dioxide ($CO_2$) /derived/ from SCIAMACHY.ENVISAT-1, meteorological parameters, and vegetation indices – case study of Poland. *Geoinformation Issues.* 5.1.5:43–56.

# 6

# The Effectiveness of using MODIS Products for Monitoring Climate Change Risks over the Nile Delta, Egypt

**Hossam Ismael**

Geography and GIS Department, Faculty of Arts., Assiut University, New Valley Branch, Egypt

**Email address:**
hosam.ismael@artnv.au.edu.eg, hossam_geography@yahoo.com

**Abstract:** Climate change is the one of greatest challenges that faces the human being nowadays as the Earth's climate is getting warmer. The National Oceanic and Atmospheric Administration (NOAA) and the National Aeronautics and Space Administration (NASA) have indicated that the temperature average of the Earth's surface has increased about 1.2 to 1.4 C since 1900. Other climatic aspects are exposed to change as well such as patterns of precipitation and storms. The most common reason that leads to climate change is very likely human activities (e.g. fuel combustion and pollution). The Study area is the most affected region in the world by climate change impacts according to the fourth report of the Intergovernmental Panel on Climate Change [4]th Report of IPCC, 2007. This report presents a scenario of destruction of the settlement centers in Nile Delta, Port Said in the east and Alexandria in the west (10 million people are at risk), besides, losing more than 86 square kilometers of the northern lakes, about 200,000 acres of the most valuable agricultural land as a result of high temperature and the consequent rise in average sea level. In Egypt, air pollutants (e.g. $SO_2$ and $CO_2$) gave rise to high concentrations of air pollutants especially in Nile delta, due to bio mass fire which is called 'Black Cloud' phenomenon. The main aim of this study was to present the effectiveness of using both the MODIS atmosphere data produced by the Terra mission and to describe differences with comparable products to be produced by Aqua. To achieve this aim the study used the HYDRA visualization software with the characteristics of the MODIS climatic data. Results obtained from MODIS data are validated by using the previously mentioned data sets to reveal the nature and the characteristics of the climate change. Fire, dust Detection with MODIS, AIRS, and AOD analysis clearly indicates large amounts of aerosols that form the black cloud events over various locations within the Nile delta region. Also the results agreed with the observed values in the study area, and highly required for many applications related to integrated remote sensing techniques with actual field measurements and data Meteorological Authority in different periods to reduce the risk of climate.

**Keywords:** HYDRA Visualization, Heat Island Impacts, MODIS Images, Terra and Aqua

## 1. Introduction

Rapid global climate change is induced by increased atmospheric greenhouse gas concentration, especially $CO_2$, which results from human activities such as fossil fuel combustion and deforestation. This directly impacts terrestrial, which continues to change in both space and time (Melillo et al. 1993; Prentice et al. 2001; Nemani et al. 2003), and ultimately impacts in turn greatly affects the human society (Milesi et al. 2005). It has been shown that the oceans and the biosphere, especially terrestrial ecosystems, currently play a major role in reducing the rate of the atmospheric $CO_2$ (Prentice et al. 2001; Schimel et al. 2001).

The Nile delta is suffering from high ambient concentrations of atmospheric pollutants including particulates (PM), carbon monoxide, nitrogen oxides, ozone and sulfur dioxide (Abu- Allaban et al., 2007, Abu-Allaban et al., 2002, El-Metwally et al., 2008). Pollution in this area is represented in what is called "Black cloud" over Nile delta. The existence of this black cloud could be attributed to many reasons among which are biomass burning, local emission and long rang of transport during the fall season. Several studies have been conducted to investigate this phenomenon using ground-based and satellite air quality data as compared to other megacities (Abu- Allaban et al., 2002, 2007, 2009; Alfaro and Wahab, 2006; El-Askary and Kafatos, 2008; El-Askary, 2006, 2008, 2010); El- Metwally et al., 2008; Favez et al., 2008; Kanakidou et al., 2011; Mahmoud et al.,

2008; and Marey et al., 2010; Prasad et al., 2010 and Zakey et al., 2004). Marey et al. (2010) have used a multi- sensor approach using the Moderate Resolution Imaging Spectrometer (MODIS) and the Multi-angle Imaging Spectro Radiometer (MISR) with meteorological data and trajectory analysis to examine the reasons for the black cloud formation over Cairo.

MODIS fire detection has identified the aerosol source which is burning the agricultural waste after harvest season in the Nile Delta region. Prasad et al. (2010) suggest that the long range of dust transport at high altitude (2.5-6 km) from the Western desert and its deposition over the Nile Delta region is one of the major contributors to air pollution episodes during this season. Nile delta and Great Cairo have witnessed different seasonal climate systems that are highly influenced by sand and dust storms coming from the western desert as well as local, increasing anthropogenic activities (El-Askary et al, 2010). As a result, it is characterized by a complicated meteorology that varies in the different times of the year. This study suggests that the pollution during the black cloud episode is the worst on human beings in addition to its effect on the local climate. During winter the climate is generally cold, humid and rainy; while during the summer season the weather is hot and dry (Zakey et al, 2008).

Several studies have been conducted to examine the main reasons for the increasing pollution levels in Cairo and Nile delta using ground-based and satellite air quality data (Zakey et al., 2004; Abu-Allaban et al., 2002, 2007; Alfaro and Wahab, 2006; El-Askary and Kafatos, 2008; Favez et al., 2008; Mahmoud et al., 2008; and El-Metwally et al., 2008; Marey et al., 2010 and Prasad et al., 2010). However, the genesis of the fall episodes (the black cloud season) is still under discussion and needs more investigations. In order to acquire further synoptic information and visualization of the aerosol and clouds characteristics, it is useful to analyze satellite data as well, since these data provide horizontal and vertical covering with fairly high (e.g. daily and monthly) temporal resolution. (Marey et al, 2010, 2011) utilized data from several satellite instruments to examine the most likely reasons for the black cloud formation over Nile delta.

In addition to the above, several studies have proved that the general behavior of MODIS retrieved aerosol optical depth (t) and has been validated against ground-based measurements, and good agreements were reported between MODIS and Sun photometer (SP) t over both oceans and land regions [e.g., Remer et al., 2002, 2005]. Reported in these studies, the t retrieved from MODIS and SP share a high degree of correlation, and the uncertainty of MODIS t over oceans was estimated to be 0.03 ± 0.05t [e.g., Remer et al., 2002, 2005]. On other hand (Jianglong Zhang and Jeffrey S. Reid, 2006) found fairly similar results using 1 years' worth of Terra (MOD04) and AQUA (MYD04) MODIS level II over-ocean aerosol product.

The impacts of pollution on aerosol optical properties and the long range of dust transport have been widely studied over Asia and other regions. Recent studies, using different sensors over optical and microwave spectra, have analyzed aerosol optical properties during dust and pollution events over Northern India and East Asia (Uno et al. 2001, Carmichael et al. 2002, Di Girolamo et al. 2004, Singh et al. 2004, El-Askary et al. 2006).

Dust storms have also been widely investigated over Egypt through a multisensory approach utilizing the differing dust behaviour over ultraviolet, visible and microwave regions of the spectrum (El-Askary et al. 2003), as well as Saharan dust transportation (Kallos et al. 2006). However, a lot still needs to be done for Africa and especially for North Africa, where one of the highest populated cities of the world is present. El-Askary (2006) has recently presented a detailed seasonal analysis of the aerosol parameters over Cairo and the Greater Delta region from 2000 to 2005. High AOD values were found over Cairo and the Greater Delta region during the spring and autumn times of the year using MODIS data. These times represent the periods of dust storms occurrence (April and May) and, more recently, the occurrence of black cloud (September and October).

Also, El-Askary et al (2008) have found that, since 1999, a black cloud appeared, above the Nile Delta and Cairo, during October and was reported by the media. During that time of the year, farmers have burned rice straw, a by-product from the rice harvest, to clear fields for the next crop. The main aim of this study id to present the effectiveness of using each of the MODIS atmosphere data products, that are being produced by the Terra mission and to describe differences with comparable products produced by Aqua. To achieve these objectives, the study has applied some threshold tests, such as MODIS fire detection, investigating Aerosols with MODIS, and dust detection with MODIS and AIRS.

## 2. Study Area

The Nile delta covers only 2% of Egypt's area but hosts 47% of the country's population and comprises 65% of its agricultural land. According to (Hereher, 2009), the Nile Delta was formed during flood seasons by Nile sediments. After passing Cairo, the Nile bifurcates into two branches: Rosetta in the west and Damietta in the east. Between and beyond these two promontories a wide coastal plain exists that was built up during the last 8–6 ka (Stanley and Warne 1994), forming one of the most famous deltas in the world in earlier times there were seven branches each one has its own lobe into the Mediterranean Sea. Five of them have silted up, leaving only the Rosetta and Damietta tributaries (Sneh and Weissbrod 1973).

The Nile Delta occupies an area of 22,000 km$^2$, and generally flat; its apex at Cairo (160 km upstream) lies at an elevation of +18 m (Sestini 1992). The climate in the Nile Delta is Mediterranean: hot in the summer and mild in the winter. Temperature averages are 19°C in the winter and 33°C in the summer. Precipitation ranges from 200 mm/year along the Mediterranean coast to only 22 mm/year at Cairo. Excluding Great Cairo, the Nile Delta contains another 10 administrative governorates: Alexandria, Kafr El-Sheikh, El-Dakahlia, Damietta, Port Said, El-Behira, El-Gharbia,

El-Minofia, El-Sharkia, and Ismailia. Government statistics (the Central Agency for Public Mobilization and Statistics, http://www.capmas.gov.eg/) report that these governorates host 47% of Egypt's 88 million populations in the 2015 census. The delta coast possesses many highly populated cities, such as Alexandria, Damietta, Port Said, and Rosetta. Delta coast is characterized by its urban, industrial, touristic and international trading importance, In addition, the region contains 63% of Egypt's agricultural land which constitutes 8.6 million acres in 2005 (Hereher, 2009), and 25% of the total Mediterranean coastal wetlands, that are not only a fishing resource but also a good shelter for migratory birds (Sestini 1992).

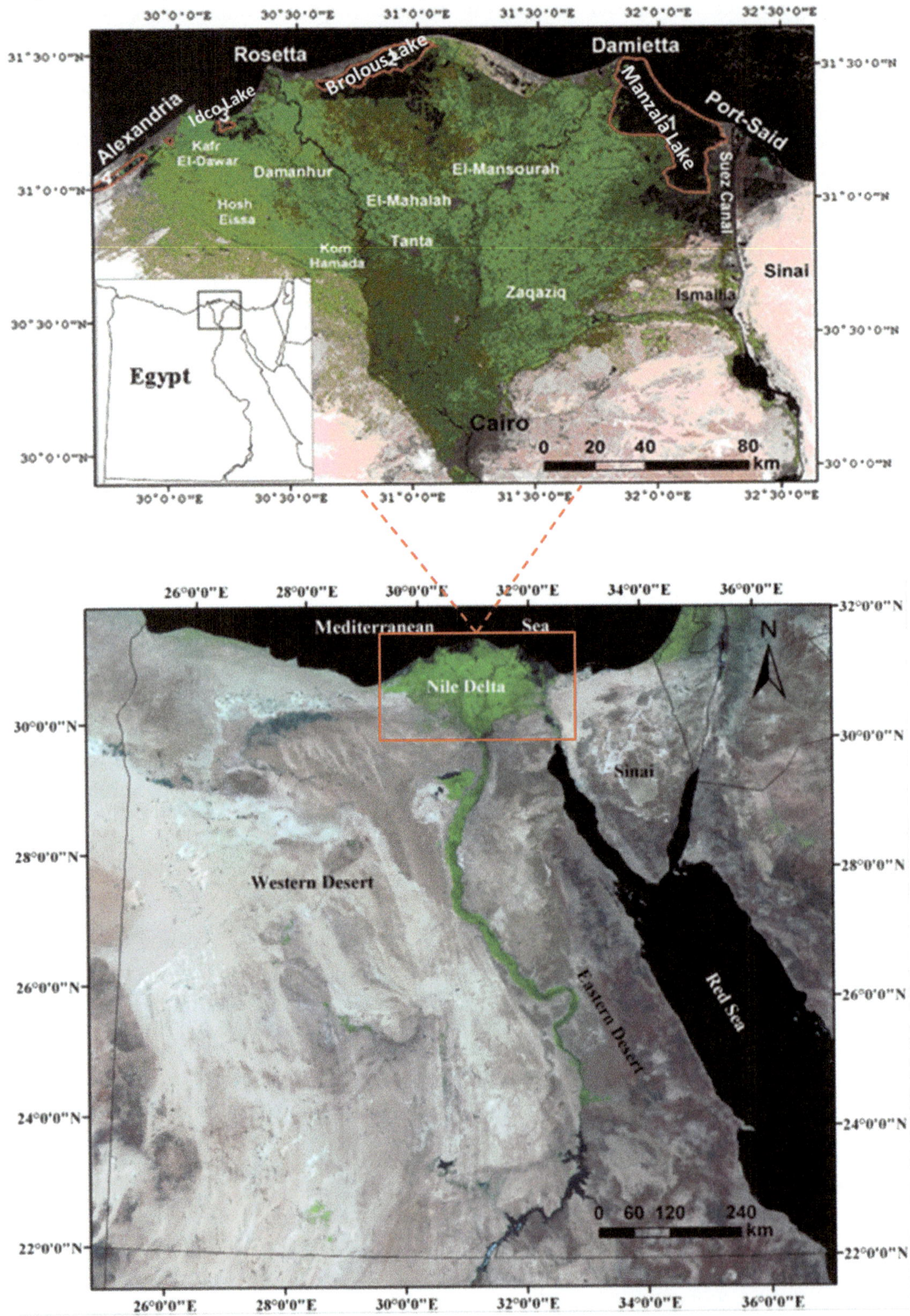


*Figure 1. An image of the Nile delta on 12th march 2013 compiled from 4 MODIS tiles.*

## 3. MODIS and Its Products

The MODerate-resolution Imaging Spectroradiometer (MODIS) is one of five scientific instruments onboard the satellite platform, Terra, part of NASA's Earth Observation System (EOS), It provides data for retrieving land surface temperature (LST) at 1-km resolution with almost daily coverage of the Earth, which is invaluable for both local and global change research. The Moderate Resolution Imaging Spectroradiometer (MODIS) is an Earth Observing System (EOS) and a facilitating instrument that is currently flying aboard the Terra and Aqua spacecraft. It suits the global monitoring of atmospheric properties from space, and is based on heritage sensors such as the Advanced Very High Resolution Radiometer (AVHRR), Landsat Thematic Mapper (TM), High-resolution Infrared Radiation Sounder (HIRS), and the Nimbus-7 Coastal Zone Color Scanner (CZCS). The wide spectral range (0.41–14.24 m), frequent global coverage (one to two days revisit), and two high spatial resolution bands (250 m), permit state of the art global monitoring of atmospheric profiles, column water vapor amount, aerosol particles, and the subsequently formed clouds (King et al, 1999). MODIS scans a swath width of 2330 km that is sufficiently wide to provide nearly complete global coverage every two days from a polar-orbiting, sun-synchronous, platform at an altitude of 705 km. MODIS provides images in 36 spectral bands between 0.415 and 14.235 m with spatial resolutions of 250 m (two bands), 500 m (five bands), and 1000 m (29 bands) (King et al 2003).

MODIS twin sensors were launched under the auspices of the NASA Earth Observing System (EOS) program: the first on 18 December 1999 aboard the Terra satellite, and the second on 4 May 2002 aboard the Aqua satellite; and both have measured the reflected and emitted radiance from the Earth and the atmosphere, day and night. Terra and Aqua, which are both polar-orbiting satellites, cross the equator during the daytime at approximately 1030 (morning) and 1330 (afternoon) local time (LT), respectively. Radiance data are acquired by MODIS in 36 spectral bands, spanning 0.645 – 14,385 μm four focal plane assemblies (Table.1), wavelengths, which range from the visible (VIS) through the near-infrared (NIR) and mid-infrared (MIR) up to the thermal infrared (TIR) regions of the electromagnetic spectrum. They are acquired in one of three spatial resolutions at nadir: 0.25 km (bands 1–2: VIS), 0.5 km (bands 3–7: VIS-MIR), and 1 km (bands 8– 36: VIS-TIR) (King et al, 1992).

***Table 1.*** *MODIS Channel Number, Wavelength (m), and Primary Application. From (Manzel, 2009).*

| MODIS Reflected Solar Bands | | | MODIS Thermal Emissive Bands | | |
|---|---|---|---|---|---|
| **Band** | **Wavelength (micron)** | **Primary application** | **Band** | **Wavelength (micron)** | **Primary application** |
| 1,2 | 0.645, 0.866 | Land, Clouds | 20-23 | 3.75(2), 3.95, 4.05 | Surface / cloud temperature |
| 3,4 | 0.470, 0.555 | Land, Clouds | 24,25 | 4.46, 4.51 | Atmospheric temperature |
| 5-7 | 1.24, 1.64, 2.13 | Land, Clouds | 27,28 | 6.71, 7.32 | Water vapor |
| 8-10 | 0.415, 0.443, 0.490 | Ocean color | 29 | 8.55 | Surface / cloud temperature |
| 11-13 | 0.531, 0.565, 0.653 | Ocean color | 30 | 9.73 | Ozone |
| 14-16 | 0.681, 0.750, 0.865 | Ocean color | 31,32 | 11.03, 12.02 | Surface / cloud temperature |
| 17-19 | 0.905, 0.936, 0.940 | Water vapor | 33,34 | 13.33, 13.63 | Cloud top properties |
| 26 | 1.375 | Cirrus clouds | 35,36 | 13.93, 14.23 | Cloud top properties |

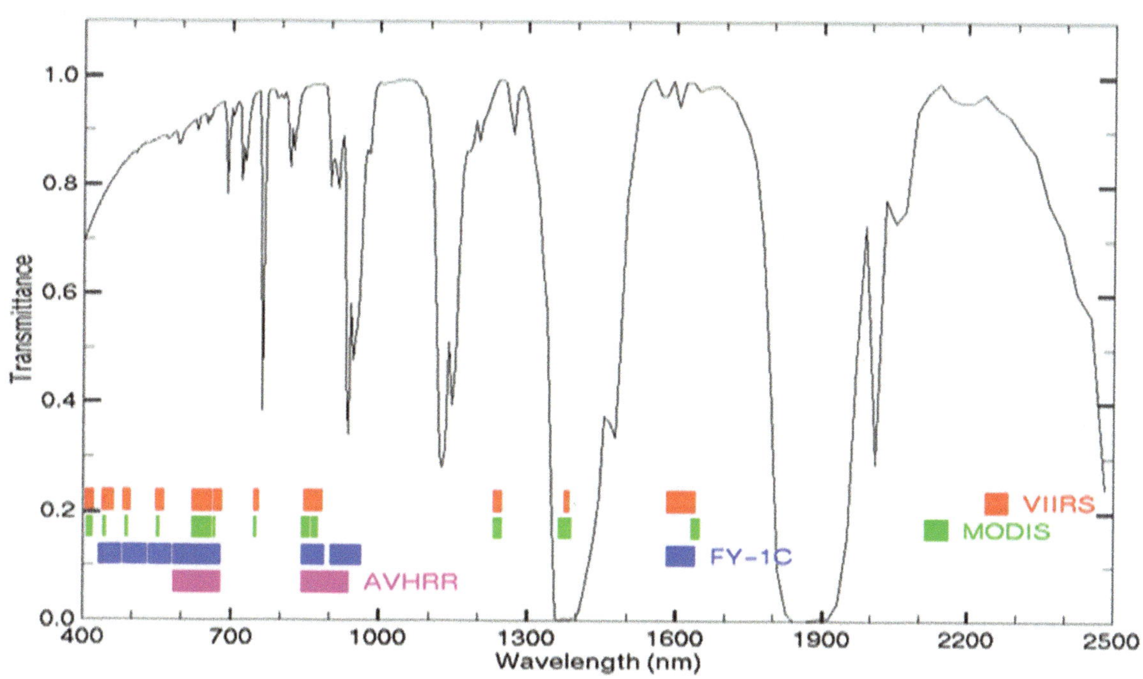

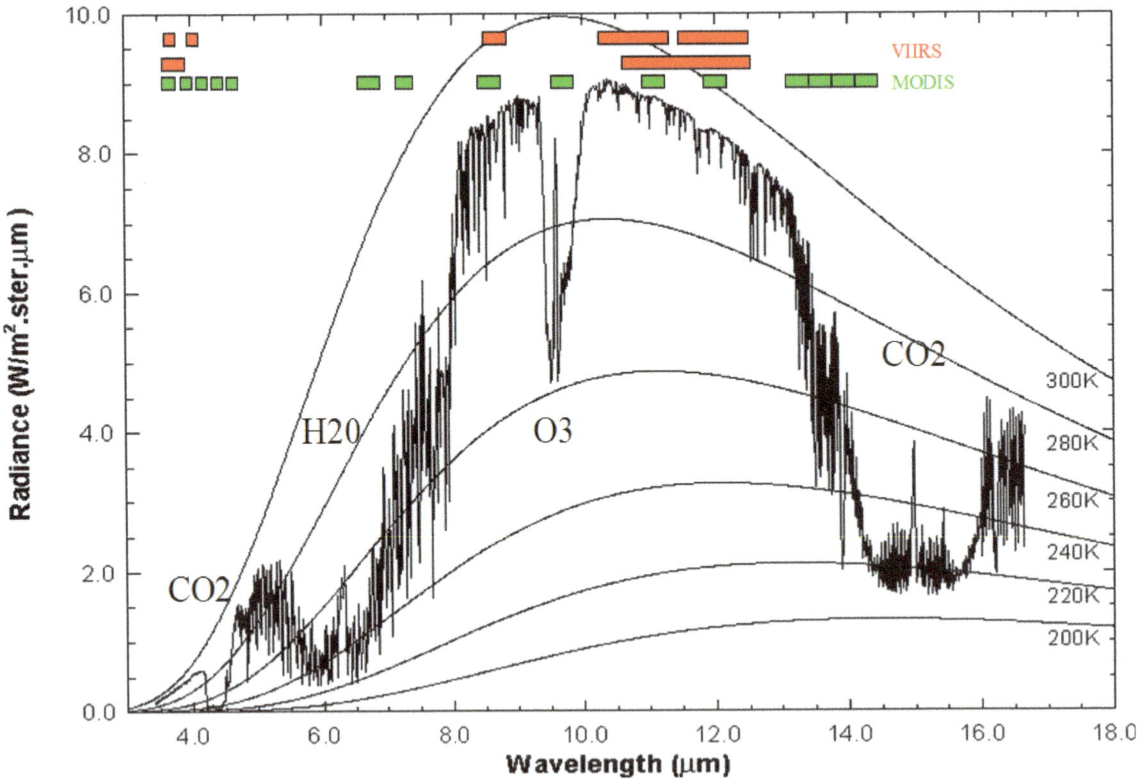


***Figure 2.*** *Visible and Infrared absorption features and MODIS spectral band locations (Manzel et al, 2006).*

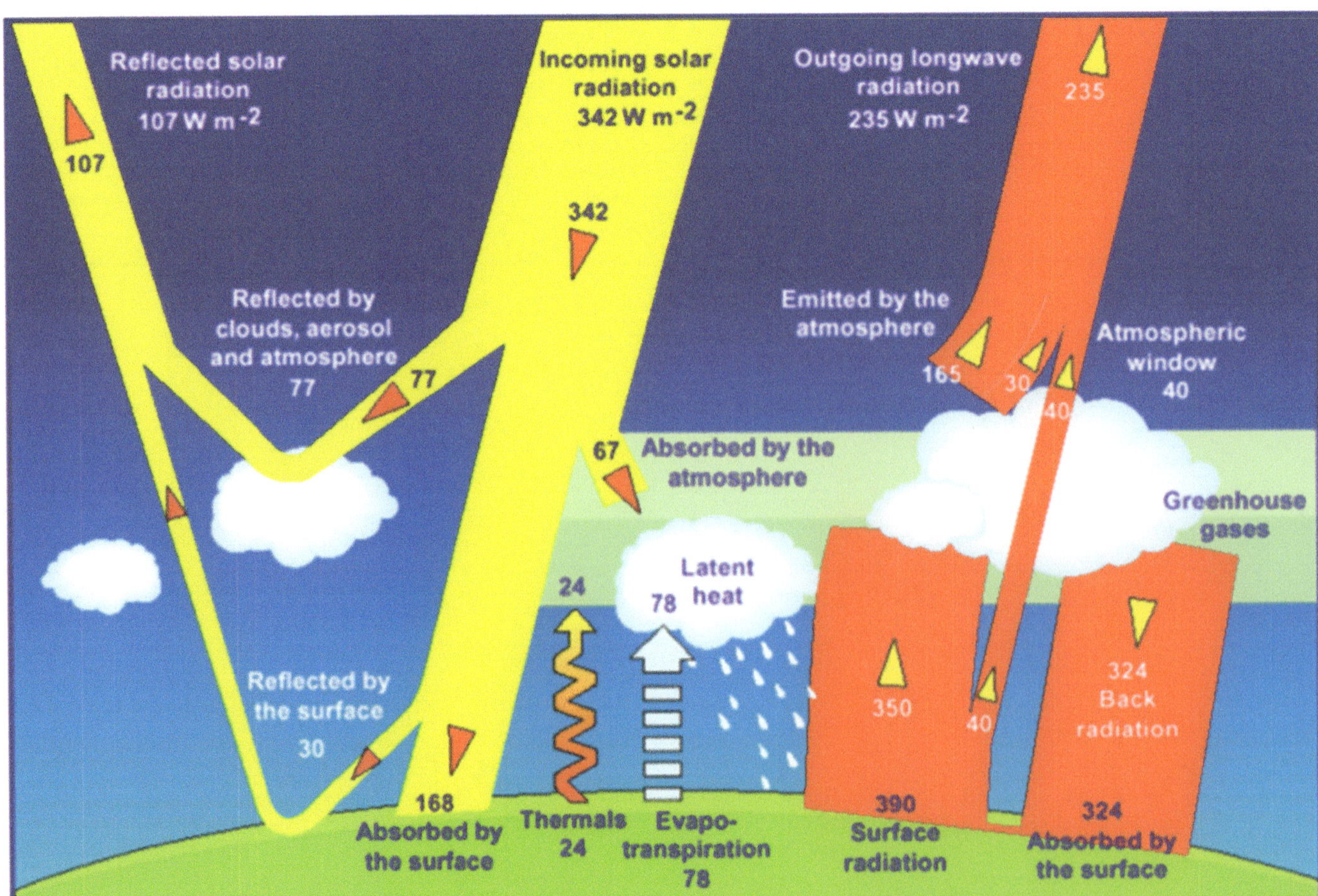


***Figure 3.*** *Climate system energy balance with role of clouds, evaporation, and other radiative processes indicated in watt per meters squared (Parkinson and Greenstone, 2000).*

The products generated from are archived continuously by appropriate NASA data centers and are distributed freely. The algorithms used to generate these products undergo periodic revisions, and the data users are not always sure about the version and quality of the products they are using at any given time. It is, therefore, necessary to conduct periodic calibration and evaluation of the products to keep track of their evolution and make the information available to users (Antonili et al, 2008). (Figure.2) shows MODIS emissive bands superimposed on high resolution earth atmosphere emission spectra with Planck function curves at various temperatures superimposed (top) and MODIS reflective spectral bands superimposed on the earth reflection spectra (bottom). (Figure.3) illustrates the annual mean of global energy balance for the earth-atmosphere system. Tracking the global energy balance trends is a vital part of climate change research.

The MODIS spectral bands have been chosen to be sensitive to various reflection, absorption, and scattering spectral signatures. (Figure.4) shows the processing schema used for deriving atmospheric products from level 0 (raw instrument data) to level 1 (calibrated earth located radiances) to level 2 (5x5 field of view derived atmospheric products) to level 3 (time composited global product fields in even interval lat-lon grid boxes).

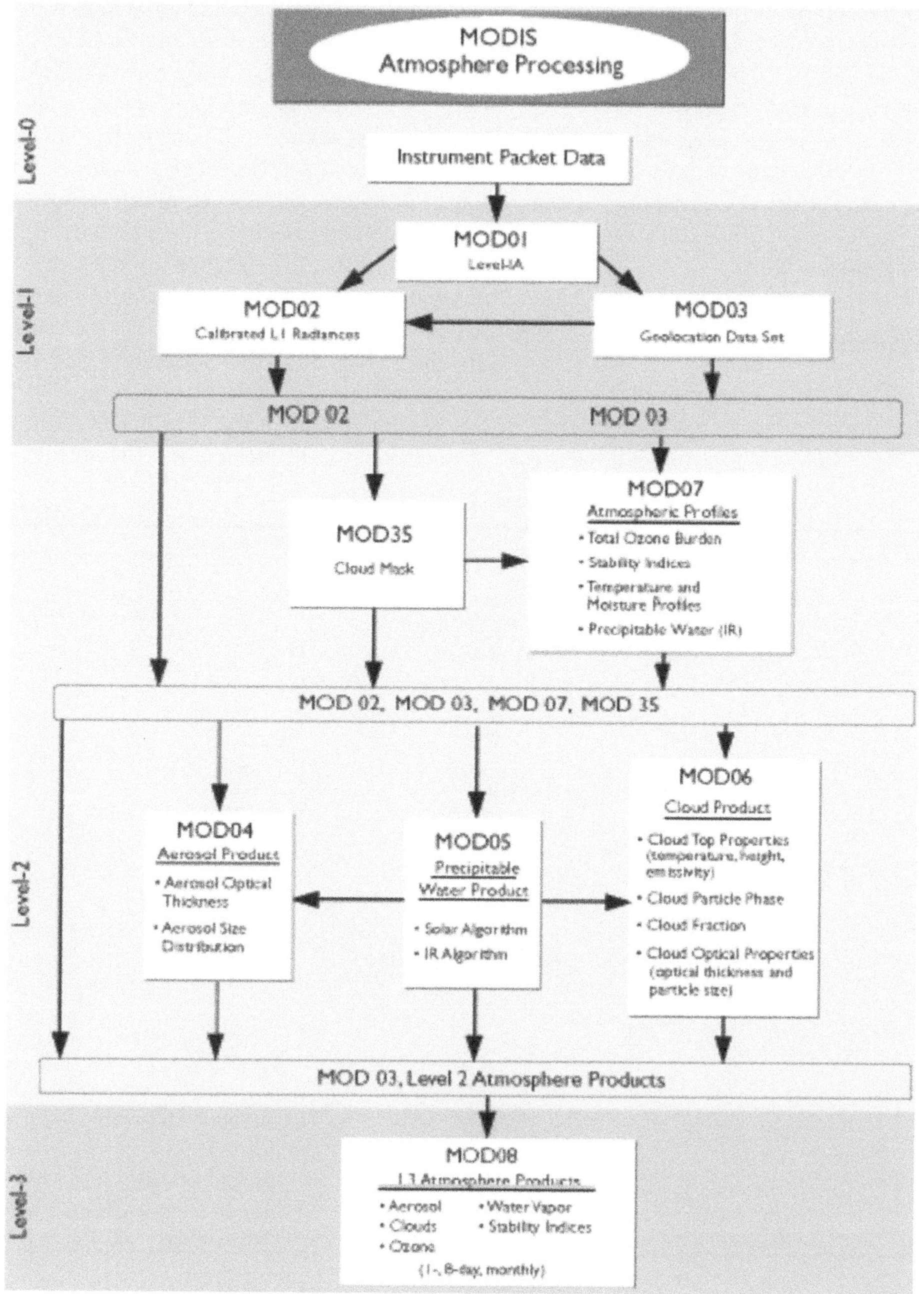


***Figure 4.*** *MODIS Atmospheric Processing schema from level 0 (raw instrument data) to level 1 (calibrated earth located radiances) to level 2 (5x5 field of view derived atmospheric products) to level 3 (time composited global product fields in even interval lat-lon grid boxes) adapted from the EOS Data Products Handbook.*

The products generated from MODIS are archived continuously by appropriate NASA data centers and are distributed freely. The algorithms used to generate these products undergo periodic revisions, and data users are not always sure about the version and quality of the products they are using at any given time. Therefore, it is necessary to conduct periodic calibration and evaluation of the products to keep track of their evolution and make the information available to users.

The MODIS spatial resolution is 250-m bands are centered at 0.65 and 0.86 μm with the 500-m bands at 0.47, 0.56, 1.24, 1.63, and 2.13 μm. Each band's spectral response is determined by an interference filter overlying a detector array imaging a 10-km along-track scene for each scan (i.e., 40, 20, and 10 element arrays for the 250, 500, and 1000-m bands, respectively. The MODIS-derived aerosol properties (t) and the fraction of fine mode particle over land [Kaufman et al., 1997a; Chu et al., 1998, 2002] in tandem with those (t), the fraction of fine mode particle, and effective radius) over ocean [Tanre´ et al., 1997, 1999; Remer et al., 2002,2005] enable to study the global aerosols comprehensively.

## 4. Materials and Methods

The study is based on MODIS fire algorithm and uses a series of tests to determine if a pixel contains a fire in order to measure the impacts of the black cloud over Nile delta (table.2). It has also used the visible thresholds tests and spectral cloud mask (MOD35) which serves as a primary ancillary input to the other cloud algorithms.

The focus of the product is to prove whether the pixel is unobstructed between the surface and satellite or not. In addition to the potentiality for obstruction in the line of sight due to clouds, heavy aerosols (e.g., fire and dust detection) and will act also to decrease the likelihood of finding clear-sky conditions. The data of this study are collected during two years of monthly average observations of AOD over the Delta region derived from MODIS level-2 AOT atmospheric daily products data at a 10 km resolution.

Finally, the study is based on HYDRA. HYper-spectral data viewer for Development of Research Applications. HYDRA enables interrogation of multispectral (including hyperspectral) fields of data so that (a) pixel location and spectral measurement values (radiance or brightness temperature) can be easily displayed; (b) spectral channels can be combined in linear functions and the resulting images are displayed; (c) false color images can be constructed from multiple channel combinations; (d) scatter plots of spectral channel combinations can be viewed; (e) pixels in images can be found in scatter plots and vice versa; (f) transects of measurements are displayed, and (g) temperature and moisture as well as spectra from selected pixels are compared. HYDRA software has become a part of the World Meteorological Organization Virtual Laboratory for Satellite Meteorology.

***Table 2.*** *Summary of some MODIS Cloud Tests.*

| Scene | Solar/Reflectance | Thermal | Comments |
|---|---|---|---|
| Low cloud over water | $R_{0.86}$, $R_{0.67}/R_{0.86}$, $BT_{11}$-$BT_{3.7}$ | Difficult, Compare $BT_{11}$ to daytime mean clear-sky values of $BT_{11}$; $BT_{11}$ in combination with BT difference tests; Over oceans, expect a relationship between [$BT_{11}$-$BT_{8.6}$], [$BT_{11}$-$BT_{12}$] due to water vapor correlation to SST | Spatial and temporal uniformity tests sometimes used over water scenes; Beware of Sun-glint regions |
| High Thick cloud over water | $R_{1.38}$, $R_{0.86}$, $R_{0.67}/R_{0.86}$, | $BT_{11}$; $BT_{13.6}$; $BT_{6.7}$ [$BT_{11}$-$BT_{8.6}$], [$BT_{11}$-$BT_{12}$] | Easy, good contrast in vis and IR |
| High Thin cloud over water | $R_{1.38}$, $\Delta R_{0.86}$, [$R_{0.65}$–$R_{1.6}$]/[$R_{0.65}$+$R_{1.6}$]; | $BT_{6.7}$; $BT_{13.9}$ [$BT_{11}$-$BT_{12}$], [$BT_{3.7}$-$BT_{12}$], [$BT_{8.6}$-$BT_{11}$] | For $R_{1.38}$, be careful of surface reflectance for atmospheres with low total water vapor amounts. For BT difference tests, beware of variations in surface emissivity. |
| Low cloud over vegetation | $R_{0.86}$, $R_{0.67}/R_{0.87}$, $BT_{11}$-$BT_{3.7}$; [$R_{0.86}$–$R_{0.65}$] /[$R_{0.86}$+$R_{0.65}$] | Difficult, $BT_{11}$ in combination with BT difference tests. | NDVI (Normalized Difference Vegetation Index). Other ratio tests also are good. Can make tests a function of ecosystem. |
| High Thick cloud over vegetation | $R_{1.38}$, $R_{0.86}$, $R_{0.67}/R_{0.86}$, [$R_{0.86}$–$R_{0.65}$] /[$R_{0.86}$+$R_{0.65}$] | $BT_{11}$; $BT_{13.9}$; $BT_{6.7}$ [$BT_{11}$-$BT_{8.6}$], [$BT_{11}$-$BT_{12}$] | Can make tests a function of ecosystem to account for variations in surface emittance and reflectance. |
| High Thin cloud over vegetation | $R_{1.38}$, $R_{0.86}$, $R_{0.67}/R_{0.86}$, [$R_{0.86}$–$R_{0.65}$] /[$R_{0.86}$+$R_{0.65}$] | $BT_{13.9}$; $BT_{6.7}$ [$BT_{11}$-$BT_{8.6}$], [$BT_{11}$-$BT_{12}$] | Beware of variations in surface emittance and reflectance. |
| Low cloud over bare soil | $R_{0.86}$, $R_{0.67}/R_{0.86}$, [$BT_{11}$-$BT_{3.7}$] | $BT_{11}$ in combination with brightness temperature difference tests. [$BT_{3.7}$-$BT_{3.9}$] | Difficult due to brightness and spectral variation in surface emissivity. Surface reflectance at 3.7 and 3.9 μm is similar. |

MODIS data are stored in the Hierarchical Data Format (HDF) (cf ,*.hdf.ncsa.uiuc.edu* .(Within this HDF file, each product has an associated Science Data Set (SDS) name. For example, the cloud-top pressure product is given the SDS name" Cloud_Top_Pressure ,"and for Terra is stored in the HDF data designation file "MOD_06L) "2which contains all Level 2-cloud-top property and optical/microphysical retrieval SDS's). Similarly, cloud mask results are found in the

"MOD_35L "2HDF file.

## 5. The Results

MODIS fire algorithm is used to differentiate fire pixels from nearby non-fire (background) pixels. It is used as an indicator for the black cloud impacts. This involves comparing the potential fire pixel to the surrounding pixels. This contextual approach uses the average brightness temperature of the non-fire pixels in Band 21(3.99 m) that surround the potential fire pixel and also the Mean Absolute Deviation (MAD). MAD is the mean of the absolute value of the differences between the average and the deviation from the average.

MODIS fire algorithm has used to differentiate fire pixels from nearby non-fire (background) pixels. It was used as an indicator for the black cloud impacts. This involves comparing the potential fire pixel to the surrounding pixels. This contextual approach uses the average brightness temperature of the non-fire pixels in Band 21(3.99 μm) that surround the potential fire pixel and also the Mean Absolute Deviation (MAD). MAD is the mean of the absolute value of the differences between the average and the deviation from the average:

MAD $= \frac{1}{N}\sum_{i} | x_i - \bar{x} |$ where i is the number of points you are summing.

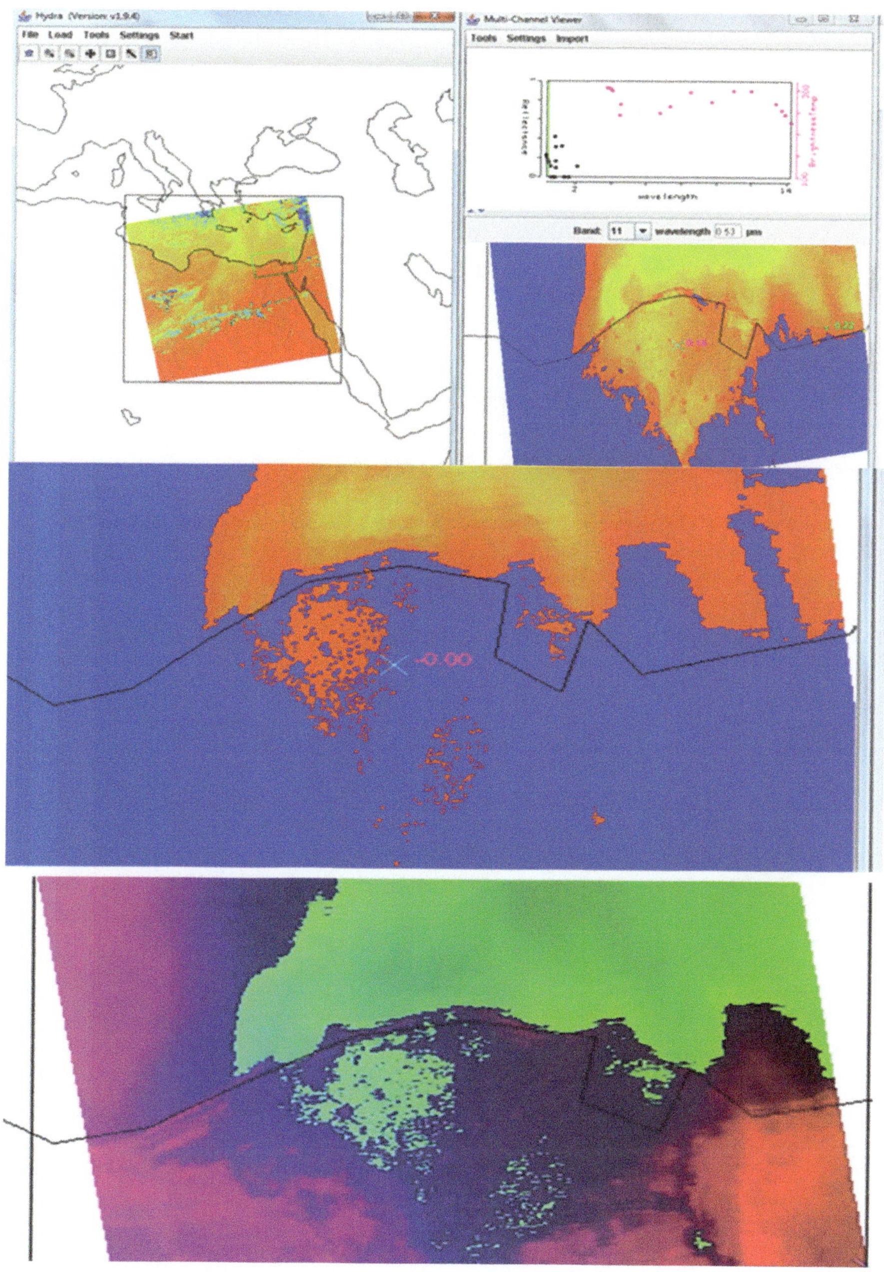


***Figure 5.*** *Black cloud display with 11 μm brightness temperature image at 10 km from Aqua MODIS on 22 March 2010 at 1110 UTC (top), 1 km resolution Band 3 (0.47 μm), Band 13io(0.67μm) reflectance image in the Multi-Channel Viewer (middle and down) respectively.*

For pixels that did not pass the first BT4 test, they must pass the next 3 dynamic thresholds to be labeled a fire pixel (Figure 5).

| Threshold tests | Thresholds for data selected in (Figure. 5) |
|---|---|
| BTDIF4-11 > Average | BTDIF4-11 > 3.2 K + 3.5 * (.63) K |
| BTDIF4-11 + 3.5 (MAD BTDIF4-11) | BTDIF4-11 > 5.40 |
| BTDIF4-11 > Average | BTDIF4-11 > 6K + 3.2 K |
| BTDIF4-11 + 6 K | BTDIF4-11 > 9.2 K |
| BT4 (Band 21 3.99 μm BT) > | BT4 > 298.52 K + 3 * 16.57 K |
| Average BT4 + 3(MAD BT4) | BT4 > 348.23 K |

Manzel et al (2009) have concluded that the high reflectance measurements result from thick clouds at all levels, and the infrared window brightness temperature provides a good indication of the cloud level. Intermediate reflectance data are subject to ambiguous interpretations since they result from a mixture of cloud and surface contributions. Visible data used in the determination of the cloud mask must be uncontaminated by sun glint. Algorithms which include solar reflectance data are constrained to solar zenith angles less than 85°. Sun glint occurs when the reflected sun angle, θr, lies between 0° and approximately 36°.

$$\cos\theta r = \sin\theta \sin\theta o \cos\varphi + \cos\theta \cos\theta o$$

Where θo is the solar zenith angle, θ is the viewing zenith angle, and φ is the azimuthal angle. Sun glint is also a function of surface wind and sea state. The entire single pixel tests mentioned so far relies on thresholds. Thresholds are never global. There are always exceptions and the thresholds must be interpreted carefully.

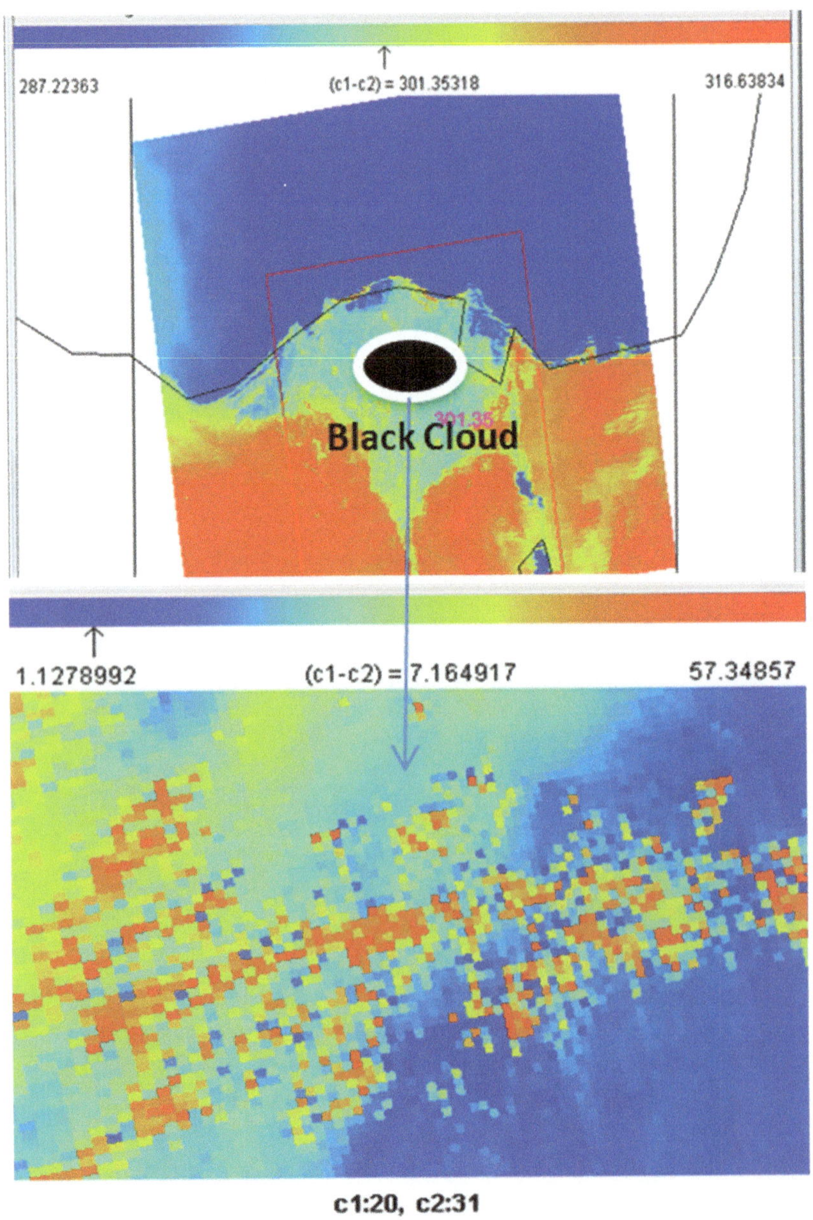


*Figure 6. The potential fire of black cloud over Nile delta region a BT4 average and BT4 MAD and a BTDIF4-11 average and BTDIF4-11 MAD, showing biomass fir occurrence at lower altitudes. 900 mb.*

By loading Band 22 (3.99 μm) into the Multi-Channel Viewer and overlay the MODIS fire Level 2 product. The fire product has displayed over the image and a new MODIS Fire Mask Product window will be displayed, with a button at the bottom that allows you to toggle the mask. After that, we compute and select a potential fire of black cloud over Nile delta region and determine a BT4 average and BT4 MAD and a BTDIF4-11 average and BTDIF4-11 MAD (figure.6).

By loading Band 22 (3.99 m) into the Multi-Channel Viewer and overlay the MODIS fire Level 2 product. The fire product is displayed over the image and a new MODIS Fire Mask Product window will be displayed, with a button at the bottom that allows toggling the mask. After that, we compute and select a potential fire of black cloud over Nile delta region and determine a BT4 average and BT4 MAD and a BTDIF4-11 average and BTDIF4-11 MAD (figure.6).

The black cloud over Nile delta was detected with threshold tests that rely on brightness temperatures in three infrared spectral bands; they are BT11, BT13.9, and BT6.7. Infrared window thresholds, BT11, are practical in certain conditions; however, they will vary in moisture content of the atmosphere. Over land, BT11 is further complicated by the fact that the surface emissivity varies appreciably according to soil and vegetation type. Thus, BT11 is used primarily to detect high, thick clouds and thresholds are set accordingly.

In addition, black clouds are likely present when BT11 is less than 300 K over Nile delta. BT13.9 provides good sensitivity to the relatively semi-arid regions of the atmosphere because of CO2 absorption. The same is true for BT6.7 because of $H_2O$ absorption. These spectral bands receive most of their radiation near 300 hPa and only clouds above 500 hPa make strong radiance contributions; negligible contributions come from the earth surface. Thus a threshold for BT13.9 and BT6.7 can isolate clouds above 500 hPa (Figure.6).

According to dust detection affecting Nile delta with MODIS and AIRS, it was illustrated that by using the main menu load MOD021KM.A2008055.1130.005.2014056124414.hdf. This granule shows a dust storm over Egypt. The temperature brightness differences show a larger range over Nile delta by representing the Band 31 (11 m) on the x-axis and [Band 29 (8.6 m) minus Band 31 (11 m)] on the y-axis. In addition to the capability of distinguishing dust storm from clouds using these channels (Figure.7)

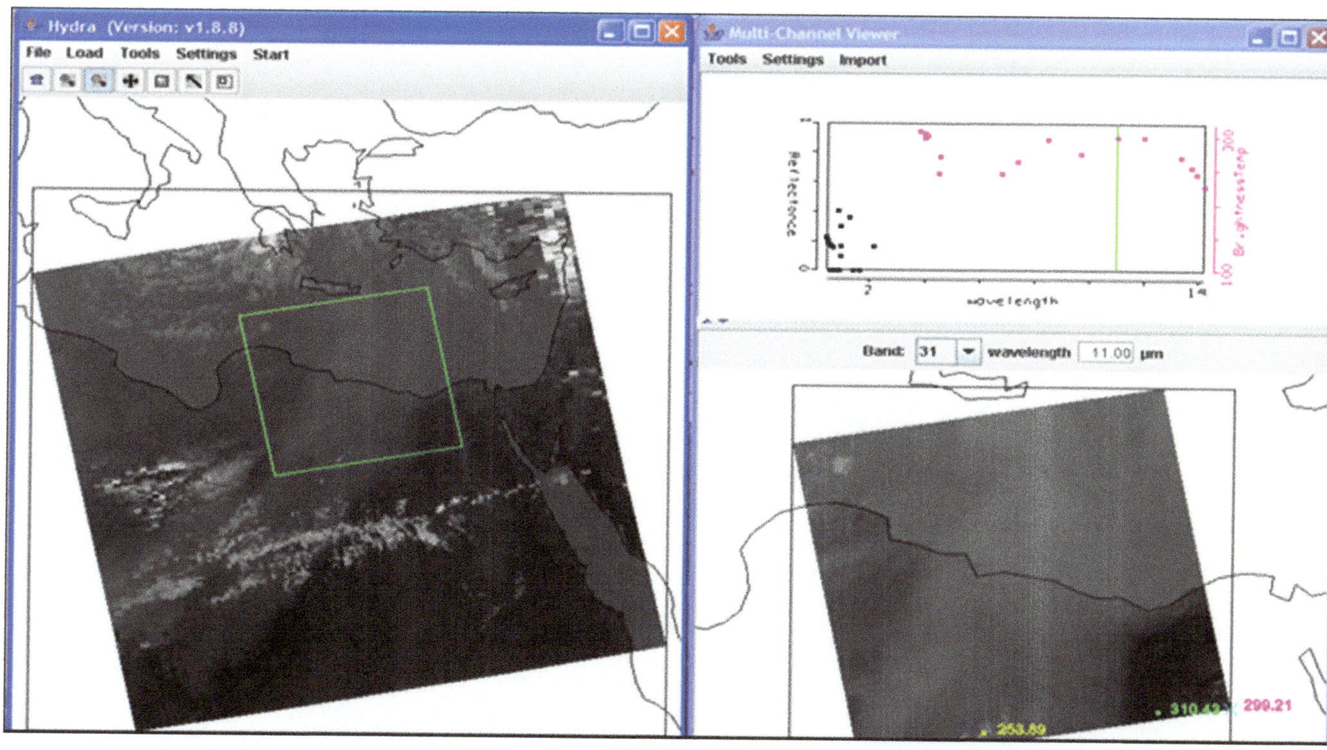


***Figure 7.*** *Dust storm scene over Egypt affecting Nile delta viewed by MODIS.*

The MODIS aerosol product is used to study aerosol climatology, sources and sinks of specific aerosol types (e.g., biomass burning aerosol or black cloud over Nile delta), interaction of aerosols with clouds, the hydrological cycle and atmospheric dynamics, and atmospheric corrections of remotely sensed surface reflectance over the land (Figure. 8). MODIS measures aerosols, which include dust, sea salt, smoke from agriculture fires, and some kinds of pollution, directly affect the amount of sunlight that reaches the earth by scattering and absorbing the incoming radiation.

The MODIS fire and thermal anomalies products contain information unique to understanding the timing and distribution of fires and characteristics such as the energy emitted from the fire and is available for both day and night periods. Temporal composites include an eight-day and monthly day and night fire occurrence aggregation and a summary of the number of fires in classes related to the strength of the fire.

El-Askary et al (2008) have used aerosol product seasonally over Nile delta and Cairo, and they had found that the dust aerosols during MAM are found at higher altitudes, yet are low compared to the anthropogenic aerosols revealed from the CTP values shown in (figure.9a). The reason that they are found at higher altitudes as compared to SON aerosols is due to the normal temperature gradient during this particular time of the year over Cairo.

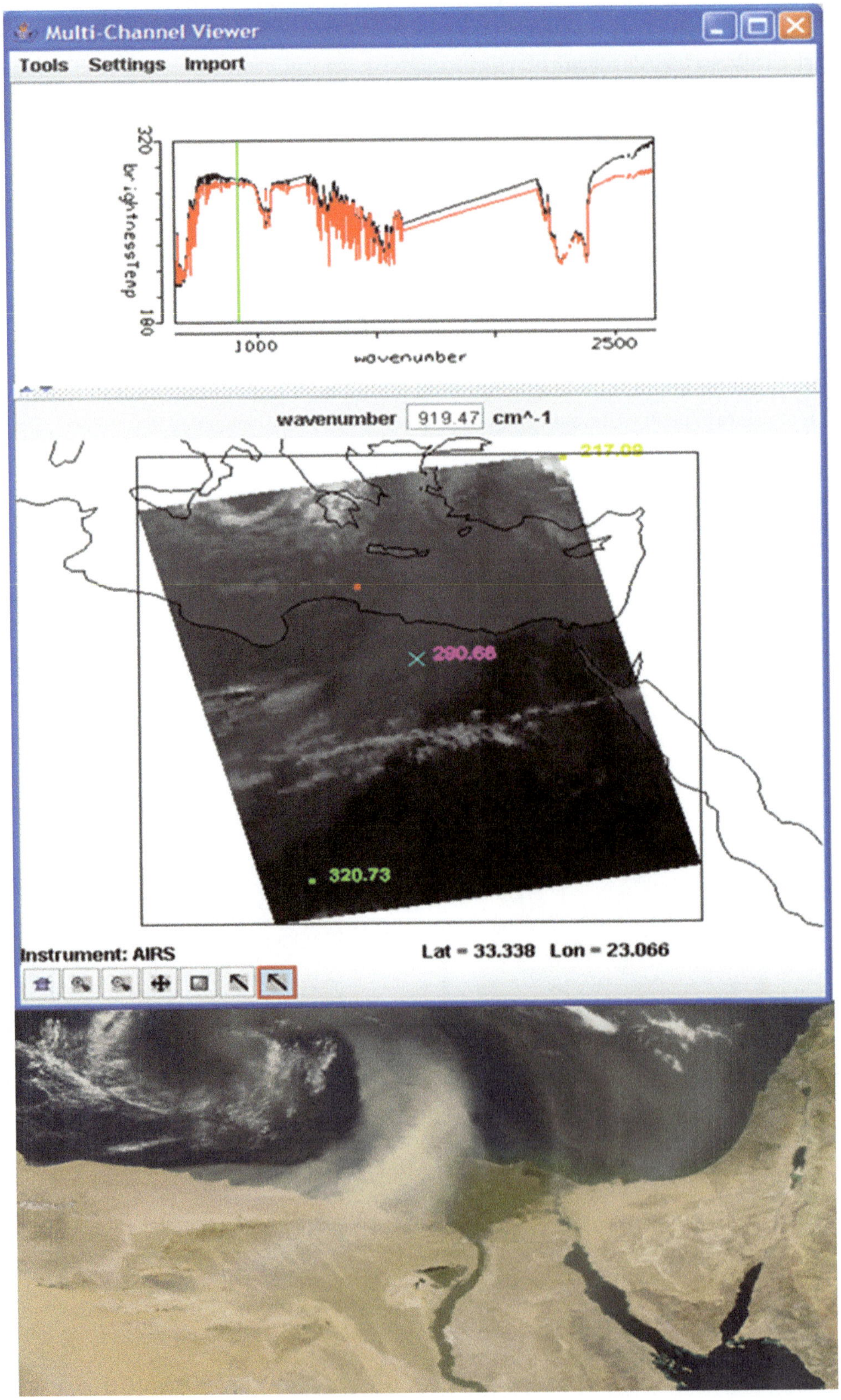


***Figure 8.*** *AIRS measurements over the same dust storm.*

However, the CTT during SON of the three years shows a decreasing trend from September to November, reflecting the higher levels of aerosols at the beginning and decreasing in concentration towards November, as shown in (Figure.9b). It is evident that, during the pollution episodes, lower CTTs were achieved due to the decreasing trend, as compared to the dust episodes that show an increasing trend in the CTT values (Devasthale et al. 2005). This implies that anthropogenic aerosols have an indirect impact in the thermal infrared, compared to the dust aerosols. Anthropogenic pollutants during SON are found at a very low altitude, less than1 km closer to the earth's surface, as revealed from the CTP values (.920 mb). One important issue that needs to be addressed is whether MODIS aerosol retrievals (at 10 * 10 km2 grid size) are applicable to regional or local air pollution. Presented here are two case studies: (1) the Nile delta (200 *200 km$^2$), and (2) Greater Cairo (40 *50 km$^2$).

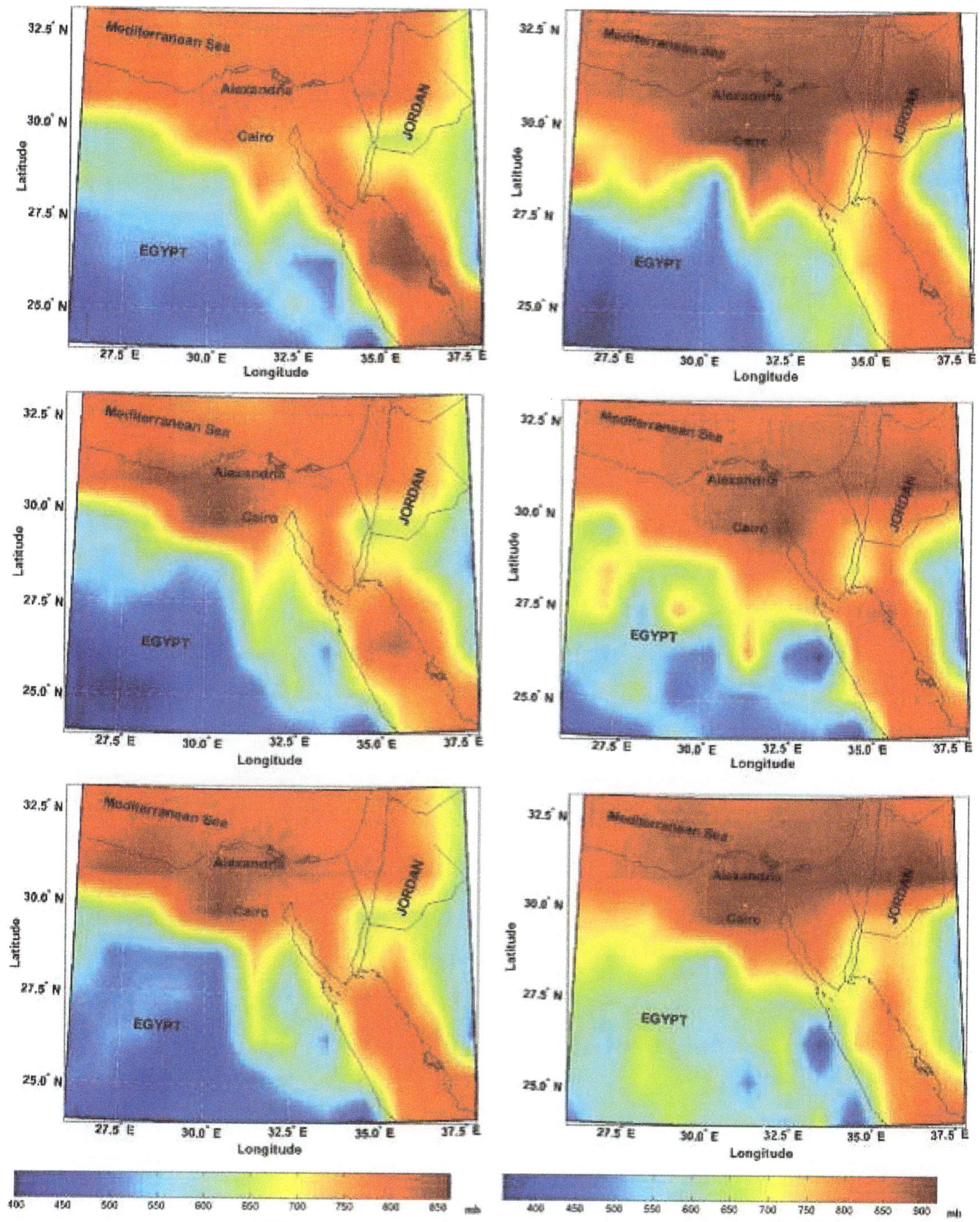


(a) (b)

*Figure 9. The cloud top pressure CTP over Cairo and the Greater Delta region during (a) spring (MAM) of 2004, 2005 and 2006, showing dust occurrence at higher altitudes, 750–800mb and (b) autumn (SON) of 2004, 2005 and 2006, showing anthropogenic aerosols occurrence at lower altitudes. 900 mb. (El-Askary et al, 2009).*

# 6. Conclusion

MODIS is ideal for monitoring large-scale changes in the biosphere that will yield new insights into the workings of the polluted area by the black cloud. While no current satellite sensor can directly measure that anthropic phenomenon concentrations in the atmosphere. MODIS from both the Terra and Aqua platforms can be successfully used as a climate model to integrate with climate data from stations for linear regression estimates and measure the change impacts of some elements such as daily maximum and minimum air temperatures changes, clouds cover, aerosols and carbon dioxide concentrations changes at a local scale on Nile Delta. MODIS products retrievals provide a useful perspective to the capability for monitoring regional and local air pollution and to identify the aerosol source as the burning of agricultural waste after harvest season in the Nile Delta region.

It was illustrated that the variability and similarity in the aerosol patterns over the entire region are associated with the presence of a clear annual component for the locations under investigation, this is in line with (El-Askary et al, 2009). The study has concluded that the Delta region is a strong source of pollution airborne particles. These particles can travel long distances to other cities and neighboring Mediterranean countries and thus influence the aerosol radiative forcing at regional scales. Moreover, aerosols are likely to affect the water cycle by suppressing precipitation in a region of the world which is already very dry

AOD analysis clearly indicates large amounts of aerosols forming the black cloud events over various locations within the Delta region. The main reasons of the black cloud phenomena over Nile delta are the natural factors, such as desert dust and the anthropic factors, such as Biomass Burning especially for rice crop (Figure.10).

It should be noticed that aerosol from natural sources, such as sea salt and desert dust; contain larger particles than aerosols were emanating from human-produced combustion sources such as agricultural and deforestation burning or urban/industrial pollution. The mature MODIS algorithm includes aerosol optical thickness at several wavelengths, information on particle size, and aerosol-reflected flux at the top of the atmosphere, which is expected to be more accurate than the optical thickness retrievals.

***Figure 10.*** *The anthropic factors affecting black cloud over Nile delta.*

# Acknowledgements

This research was supported through training course organized by FORGEA international and Twas in Monteponi, Italy. The author would like to thank W. Paul Menzel, NOAA Satellite and Information Service University of Wisconsin, Madison, WI and Paolo Antonelli, UW/SSEC for their contributions to this work.

# References

[1] Abu-Allaban, M.; Lowenthal, D.H.; Gertler, A.W. & Labib, M. (2009). Sources of volatile organic compounds in Cairo's ambient air. *Environmental Monitoring and Assessment*, Vol.157, No.1-4, (October 2009), pp. 179–189, ISSN 0167-6369.

[2] Alfaro, S.C. & Abdel Wahab, M.M. (2006). Extreme variability of aerosol optical properties: The Cairo aerosol characterization experiment case study. *Proceedings of the NATO Advanced Research Workshop on Remote Sensing of the Atmosphere for Environmental Security*, pp. 285–299, ISBN: 978-1-4020-5089-3, Rabat, Morocco, 16-19 November, 2005.

[3] Astitha, M.; Kallos, G. & Katsafados, P. (2008). Air pollution modeling in the Mediterranean Region: Analysis and forecasting of episodes. *Atmospheric Research*, Vol.89, No.4, (September 2008), pp. 358-364, ISSN 0169-8095

[4] B. H. Tang, Z.-L. Li, and Y. Y. Bi, "Estimation of land surface directional emissivity in midinfrared channel around 4.0 μm from MODIS data," Opt. Express 17(5), 3173–3182 (2009), http://dx.doi.org/10.1364/OE.17.003173.

[5] Becker. F, and Z.-L. Li, "Towards a local split window method over land surfaces," Int. J. Remote Sens. 11(3), 369–393 (1990), http://dx.doi.org/10.1080/01431169008955028.

[6] Barnaba, F. & Gobbi, G.P. (2004). Aerosol seasonal variability over the Mediterranean region and relative impact of maritime, continental and Saharan dust particles over the basin from MODIS data in the year 2001. *Atmospheric Chemistry and Physics Discussions,* Vol.4, No.4, (August 2004), pp. 4285-4337, ISSN 16807367.

[7] Christopher D. G et al, 2006. The impacts of climate change in coastal marine systems. Ecology Letters, (2006) *9*: 228–241.

[8] Dubovik, O., B. N. Holben, T. F. Eck, A. Smirnov, Y. J. Kaufman, M. D. King, D. Tanre´, and I. Slutsker (2002), Variability of absorption and optical properties of key aerosol types observed in worldwide locations,J. Atmos. Sci., 59, 590–608.

[9] El-Askary, H.; Farouk, R.; Ichoku, C. & Kafatos, M. (2009). Transport of dust and anthropogenic aerosols across Alexandria, Egypt. *Annales Geophysicae,* Vol.27, No.7, (July 2009), pp. 2869-2879, ISSN 0992-7689

[10] El-Askary H. & Kafatos, M. (2008). Dust Storm and Black Cloud Influence on Aerosol Optical Properties over Cairo and the Greater Delta Region, Egypt. *International Journal of Remote Sensing*, Vol.29, No.24, (December 2008), pp. 7199 – 7211, ISSN 0143-1161

[11] El-Askary, H. (2006). Air pollution Impact on Aerosol Variability over mega cities using Remote Sensing Technology: Case study, Cairo, Egypt. *Egyptian Journal of Remote Sensing & Space Science*, Vol.9, (July 2006), pp. 31-40

[12] El-Askary, H.; Sarkar, S.; Kafatos, M. & El-Ghazawi, T. (2003). A multisensor approach to dust storm monitoring over the Nile Delta. *IEEE Transactions on Geoscience & Remote Sensing,* Vol.41, No.10, (October 2003), pp. 2386 – 2391, ISSN 0196-2892.

[13] El-Metwally, M.; Alfaro, S.C.; Abdel Wahab, M.M. & Chatenet, B. (2008). Aerosol characteristics over urban Cairo: Seasonal variations as retrieved from Sun photometer measurements. *Journal Geophysical. Research,* Vol.113, No.D14219, (July 2008), pp. 1-13 ISSN 0148-0227.

[14] Jiménez-Muñoz. J, and J. A. Sobrino, "A generalized single-channel method for retrievingland surface temperature from remote sensing data," J. Geophys. Res. 108(D22), 4688–4697 (2003), http://dx.doi.org/10.1029/2003JD003480.

[15] Jiang. G -L. Li, and F. Nerry, "Land surface emissivity retrieval from combined mid-infrared and thermal infrared data of MSG-SEVIRI," Remote Sens. Environ. 105(4), 326–340 (2006), http://dx.doi.org/10.1016/j.rse.2006.07.015.

[16] Kaufman, Y. J., B. N. Holben, D. Tanre´, I. Slutsker, A. Smirnov, and T. F. Eck (2000), Will aerosol measurements from Terra and Aqua polar orbiting satellites represent the daily aerosol abundance and properties, Geophys. Res. Lett., 27, 3861– 3864.

[17] Kaufman, Y. J., D. Tanre´, and O. Boucher (2002), A satellite view of aerosols in the climate system, Nature, 419, 215– 223.

[18] King, M. D., Y. J. Kaufman, W. P. Menzel, and D. Tanre, 1992: Remote Sensing of Cloud, Aerosol and Water Vapor Properties from the Moderate Resolution Imaging Spectrometer (MODIS). IEEE Trans. and Geoscience and Remote Sensing, 30, 2-27.

[19] King, M. D., W. P. Menzel, Y. J. Kaufman, D. Tanré, B. C. Gao, S. Platnick, S. A. Ackerman, L. Remer, R. Pincus, and P. A. Hubanks, 2003: Cloud and aerosol properties, precipitable water, and profiles of temperature and humidity from MODIS. IEEE Trans. Geosci. Remote Sens., 41, 442-458.

[20] King, M. D., W. P. Menzel, Y. J. Kaufman, D. Tanré, B. C. Gao, S. Platnick, S. A. Ackerman, L. A. Remer, R. Pincus, and P. A. Hubanks, 2003: Cloud, Aerosol and Water Vapor Properties from MODIS., *IEEE Trans. Geosci. Remote Sens.*, 41, pp. 442-458.

[21] Marey, H. S., Gille, J. C., El-Askary, H. M., Shalaby, E. A., and El- Raey, M. E.: Study of the formation of the "black cloud" and its dynamics over Cairo, Egypt, using MODIS and MISR sensors, J. Geophys. Res., 115, D21206, doi:10.1029/2010JD014384, 2010.

[22] Marey, H. S., El-Askary, H. M., Shalaby, E. A., and El- Raey, M. E: Aerosol climatology over Nile Delta based on MODIS, MISR and OMI satellite data, Atmos. Chem. Phys., 11, 10637–10648, 2011.

[23] Menzel W. P., D. P. Wylie, and K. I. Strabala, 1992: Seasonal and diurnal changes in cirrus clouds as seen in four years of observations with the VAS, *J. Appl. Meteor.*, 31, 370-385

[24] Menzel W. P. and J. F. W. Purdom, 1994: Introducing GOES-I: The first of a new generation of Geostationary Operational environmental Satellites. *Bull. Amer. Meteor. Soc.,* Vol. 75, No. 5, pp. 757-78.

[25] Menzel W. P. et al, 2009: Laboratory Exercises for remote sensing applications with meteorological satellite, Space Science and Engineering Center, University of Wisconsin Madison, WI.

[26] Moeller, C. C., H. E. Revercomb, S. A. Ackerman, W. P. Menzel, and R. O. Knuteson, 2003: Evaluation of MODIS thermal IR band L1B radiances during SAFARI 2000. *J. Geophys. Res.*, 108, D13, 8494.

[27] Mu Q, Heinsch FA, Zhao M, Running SW, Cleugh HA, Leuning R (2006b) Development of a global evapotranspiration algorithm based on MODIS and global meteorology data. Remote Sens Environ 111(4):519–536.

[28] Nerry. F: Petitcolin. F, and M. P. Stoll, "Bidirectional reflectivity in AVHRR channel 3: application to a region in North Africa," Remote Sens. Environ. 66(3), 298–316 (1998), http://dx.doi.org/10.1016/S0034-4257(98)00066-2.

[29] Platnick S, M. D. King, S. A. Ackerman, W. P. Menzel, B. A. Baum, J. C. Riédi, R. A. Frey, 2003: The MODIS cloud products: algorithms and examples from Terra. *IEEE Trans. Geosci. Remote Sens.*, 41, pp. 459-473.

[30] Qin. Z, A. Karnieli, and A. Berliner, "A mono-window algorithm for retrieving land surface temperature from landsat TM and its application to the Israel-Egypt border region," Int. J. Remote Sens. 22(18), 3719–3746 (2001), http://dx.doi.org/10.1080/01431160010006971.

[31] Remer, L. A., et al. (2005), The MODIS aerosol algorithm, products and validation, J. Atmos. Sci., in press.

[32] Sobrino. J, et al., "Multi-channel and multi-angle algorithms for estimating sea and land surface temperature with ATSR," Int. J. Remote Sens. 17(11), 2089–2114 (1996), http://dx.doi.org/10.1080/01431169608948760.

[33] Smirnov, A., B. N. Holben, T. F. Eck, O. Dubovik, and I. Slutsker (2000), Cloud screening and quality control algorithms for the AERONET data base, Remote Sens. Environ., 73, 337– 349.

[34] Tang. T, et al., "Generalized split-window algorithm for estimate of land surface temperature from Chinese geostationary FengYun meteorological satellite (FY-2C) data," Sensors 8(2), 933–951 (2008), http://dx.doi.org/10.3390/s8020933.

[35] Zakey, A.S.; Abdel-Wahab, M.M.; Pettersson, J.B.C.; Gatari, M.J. & Hallquist, M. (2008). Seasonal and spatial variation of atmospheric particulate matter in a developing megacity, the Greater Cairo, Egypt. *Atmósfera* Vol.21, No.2, (January 2008), pp. 171- 189, ISSN 0187-6236

[36] Zhang. H, "A physically based algorithm for land surface emissivity retrieval from combined mid-infrared and thermal infrared data," Sci. China Ser. E 43(Suppl 1), 22–33 (2000), http://dx.doi.org/10.1007/BF02916575.

[37] Z.-L. Li et al., "Evaluation of six methods for extracting relative emissivity spectra from thermal infrared images," Remote Sens. Environ. 69(3), 197–214 (1999), http://dx.doi.org/10.1016/S0034-4257 (99)00049-8.

# Environmental Evaluation of Small Scale Gold Mining in the Bonkani Region in the Upper East Area of Côted'Ivoire

**Eboua Narcisse Wandan[1], Gnammytchet Barthelemy Koffi[2], Kouakou Bob Kouadio[3], Zoly Desiré Gomba[4]**

[1]Environmental Management Research Laboratory, Department of Water, Forestry and Environment, INP–HB, Yamoussoukro, Côte d'Ivoire
[2]Civil Engineering Research Laboratory, Geosciences and Geographical Sciences, Department of Earth Sciences and Mineral Resources, INP–HB, Yamoussoukro, Côte d'Ivoire
[3]Sustainable Forest Management Research Laboratory, Department of Water, Forestry and Environment, INP–HB, Yamoussoukro, Côte d'Ivoire
[4]Departmental Directorate of Water and Forests, Bouna, Côte d'Ivoire

**Email address:**
wandaneb@yahoo.fr (E. N. Wandan), kofgnam@yahoo.fr (G. B. Koffi), kouadio_bob@yahoo.fr (K. B. Kouadio), zgombadesire@gmail.com (Z. G. Desire)

**Abstract:** Artisanal mining is usually done by workers with limited understanding of the long term impacts of their activities on the environment and on their health and with limited capacity to mitigate the risks. The aim of this study was to evaluate the environmental impact of this activity in the Department of Bouna where this activity has gained Importance. This study was done through data collection, interviews of miners, traders, and government services' agents, and on site diagnosis. Six steps were inventoried on the mining sites: site prospection, digging holes, crushing mined ore, grinding of crushed ore, washing the flour after milling, and gold recovery with mercury. Among the activities on mining sites, 24% were found to present potential impact of major importance, 39% have potential impact of medium importance, and 37% have minor potential impacts. The major impacts are those impacting soil, landscape, and vegetation and water resources. These results showed that small scale gold mining significantly impacts the environment. As recommendation, it was proposed site rehabilitation by reforestation, the help of miners to use retors which are mercury traps or eventually mercury-free gold extraction technologies such as chlorination. A simple policy of solid and liquid waste management must be developed by site owners. They must ensure the education of miners and of the population living on mining sites on the importance of using and building effective systems for on-site sanitation. Besides that, programs focusing on HIV/STI prevention, with targeted interventions for high risk women must be implemented under private/public/NGO partnership. Since the resilience of the gold mining activities has proven to be one of the main sources of livelihoods in local communities of the extraction sites, the state should define a consistent policy of artisanal gold mining taking into account the environmental, social and health.

**Keywords:** Artisanal, Mining, Bouna, Cote d'Ivoire, Environnemental Impact

## 1. Introduction

Gold mining on a small scale which refers to both to the rudimentary process of ore extraction, is a key livelihood activity employing over 13 million workers and maintaining 80 to 100 000 000 people worldwide. It produces about 350 to 800 tons of gold per year and contributes to about 20-30% of world production.

People who depend on artisanal gold mining are generally from poor rural areas in developing countries [1]. Some people living in rural areas are dependent on mining as the main source of income or as an essential complement to their meager farm income. In many African countries such as Ghana, Mali and Burkina Faso, gold mining has become important because of the escalation of poverty and the lack of employment opportunities in the formal sector [2].

In Côte d'Ivoire, following the socio-political crises, and the economic difficulties faced by farmers in recent decades, several gold mining strategies have been implemented by population to address the situation. According to the Ministry of Mines, Côte d'Ivoire abounds over 1,000 artisanal gold mining sites and more than 500,000 people live in this rural activity (http://intellivoire.net/cote-divoire-le-gouvernement-sengage-a-organiser-lorpaillage/). This is largely observed in the department of Bouna, in the Bonkani region, where there has been in recent years, a "gold rush"[3].

Driven by poverty, artisanal mining is usually done by workers with limited understanding of the long term impacts of their activities on the environment and on their health and with limited capacity to mitigate the risks. The absence of officials responsible for law enforcement and poor medical arrangements perpetuate the chaos and insecurity in the areas of gold mining [1]. Small-scale gold mining activities have been widely criticized for being wasteful, inefficient, ineffective, and harmful to the environment in addition to the illegal trade in minerals. It is commonly accepted that it is largely informal and a greater proportion of activities are illegal [4].

Social problems caused by gold mining such as crime, alcohol, sex abuse and land conflicts are also mentioned. In the rural community of Kompti in Burkina Faso for example, for the only month of July 2006, the social service has been asked to solve five rapes, six cases of drugs and eleven cases of marital conflict [5].

The aim of this study was to evaluate the environmental impact of artisanal gold mining in the Bonkani in order to improve the environmental performance of gold mining in Côte d'Ivoire. The environment being defined as "*the dynamic system defined by the physical interactions, biological and cultural, perceived or not, between humans, other living beings and all components of the environment, whether natural or processed or created by man*" [6].

## 2. Methodological Approach

### *2.1. The Study Area*

The study was done in two small scale mining sites located in the Department of Bouna in the Boukani region, eastern Côte d'Ivoire (Figure 2). The population is 96,241 inhabitants [7] and the economic activities are based on agriculture. The main cash crops are cashew nuts, shea nuts and cotton whose production is almost abandoned. In terms of staple crops: yam, maize, millet, sorghum, lowland rice, beans and locust bean are grown in this region. The Department of Bouna is an area conducive to livestock which constitutes the second pillar of the economy of the Department.

The study area consists of two gold mining sites located on the lands of two villages: Lomidouo and Kintan at about 120 and 45 km south-east of the city of Bouna respectively (Figure 3).

### *2.2. Methods*

#### *2.2.1. Survey-Diagnosis*

Since artisanal gold mining, as a socio-economic fact, cannot be understood alone, a qualitative method that favors any or all relative to the elements was used [8]. We resorted to several techniques to collect qualitative data: direct observation, data collection, and individual or sometimes grouped semi-directive (requiring open answers) or unstructured (leaving room for digressions and spontaneous conversations) interviews. We also obtained first-hand information from various informants (artisanal miners, traders, government services, etc.).

***Figure 1.*** *Localization of the mining sites in the Department of Bouna in the Boukani region.*

#### *2.2.2. Impact Assessment*

The impact assessment can be divided into three partially overlapping phases [9]: identification (identify the impacts associated with each phase of the project and activities); forecast (predict the nature, magnitude, extent and duration of the main impacts); and assessment (determining the absolute importance of impacts).

#### *2.2.3. Identification, Description and Analysis of Impacts Detailed*

After the description and analysis of the initial state of the mining sites, we proceeded with the identification of the impacts from the checklist coupled to a matrix of potential interactions [10]. This help to find the different relationships between the sources of impacts (the various gold mining activities) and the receivers (the components of the environment undergoing disturbances compared to the initial state of the gold mining area) [11]. We used for this purpose the matrix of Leopold for potential interactions and the network diagram for induced impacts.

#### *2.2.4. Impact Assessment*

The aim of the impact assessment was to assign absolute importance to expected impacts, associated with mining activities, and determine the order of priority in which these impacts must be avoided, mitigated or offset [12]; [13]. This assessment is generally based on the key factors of the environment. In this study, the factors considered are ecological (effect on wildlife habitat, tolerance, sensitivity, biodiversity and ecosystem carrying capacity, the sustainability of local species populations) and social (effect on the health and safety of humans, loss or gain commercial value, aesthetic value, etc.). The methodology was based on the matrix of Fecteau [14].

## 3. Results and Discussion

The different stages of mining and processing of gold identified in both sites of gold mining were:

1) The prospection which consists on identifying gold-rich sites in the area; it is generally carried out by former miners who have gained experience.
2) Sinking or digging holes in the site that has been found to be rich in gold. This is done using rudimentary instruments and the holes are of varying sizes, depending on the richness of the site.
3) The crushing or reduction of the size of the mined ore; it is also done manually with the aid of hammer or anvil (granite stone). The crashers usually use knot bag to prevent the flying of particles and to protect their fingers (Photo 1).

***Figure 2.*** *Crushing of ore with a hammer or in a mortar.*

4) The grinding which consists on reducing the ore in powder (flour) in mechanical mills (Photo 2).

***Figure 3.*** *Ore grinding.*

5) The washing process involves mixing the flour after milling with water and washing this mixture gravimetrically on a carpet ramp in order to trap gold. The mud recovered from the carpet is cleaned until a black powder was obtained (Photo 4).

*Figure 4. Washing the ground ore and recovering the sludge containing gold.*

6) Gold recovery involves mixing the obtained black sludge with mercury. The gold-mercury mixture is then burned with a blowtorch until obtaining pure gold which is immediately weighed and sold locally to a buyer who is actually an employee of the site owner.

*Figure 5. Recovery of gold by mercury amalgamation and heating for purification and weighing system.*

### *3.1. Identification of Observed Impacts on Environment*

By considering the matrix of Leopold, the potential impacts of gold mining on the environment identified in this study were classified into two categories: the impacts on the biophysical environment and the impacts on the socio-economic environment. Concerning the biophysical environment, the identified matrices were the fauna, the flora, soil and water resources. The socioeconomic environment takes into account the health, the economic and social well-being.

Twenty four percent of the potential impacts were of major importance, 39% have impacts of medium importance, and 37% are minor impacts. The major impacts are those impacting soil, landscape, and vegetation and water resources. These results showed how that small scale gold mining significantly impacts the environment.

#### *3.1.1. Potential Impacts on the Biophysical Environment*

***(i). Wildlife***

Throughout the duration of this study on the two sites, we have not identified the presence of wildlife. Similarly, local residents and hunters that we met reported the scarcity of wildlife in the gold mining areas.

Gold mining has a significant negative impact on wildlife (terrestrial mammals, soil microfauna, aquatic fauna, insects or even birds). This impact was due to noise from miners, disruption and destruction of animals' habitats due to the cutting of trees used to support wells, and for the construction of shelters, etc. This has led to the scarcity of certain species or their loss, and the erosion of biodiversity. The destruction of nesting sites and their niche has led to the disappearance or relocation of birds. The waterway adjacent to the sites has become totally turbid causing the depletion of aquatic species.

***(ii). The Flora and Vegetation***

The whole forest resources is under severe human pressure manifested by excessive deforestation for fuel wood, wood for the building of temporary shelters. There are also the clear-cut of trees for the support of mine shafts; it takes about 500 feet of treeto support wells (Fig. 7).

Two species have been identified as the most cut because of their hardness; and their interesting biological quality; shea (*Vitellaria paradoxa*) and nere (*Parkia biglobosa*). The fat extracted from shea seeds after cooking is a multipurpose product. It is used locally in food and cosmetics industry, food (chocolate), pharmaceutical. The pulp of néré rich in vitamins is an excellent food for both humans and livestock. Porridge and fermented grains are an excellent seasoning for sauces (soumbala) and would have antihypertensive properties [15].

This cutting has resulted in the reduction of forest areas and the density of plant species. We observed, moreover, a clear difference between the appearance of the once closed landscape, dense and lush and today landscape characterized by increased degradation (Fig. 6).

***Figure 6.*** *Original savannah (left) and cleared savannah (right).*

***Figure 7.*** *Trees cut from the savannah used to stabilize the walls of wells.*

### *(iii). Water resources*

Water is involved in almost all the activities of artisanal gold mining. During sinking, workers evacuate large quantities of water from wells with motor pumps. Milling uses in continuous water to cool engines. Washing ore consumes a lot of water; it takes about 200 liters of water to wash a bag of 50 kg of ore "flour". We can include the lives of miners, which require daily water needs (nutrition, laundry, shower,

(a) (b)

(b) (d)

***Figure 8.*** *Water use on the mining sites. (a) stagnant water; (b) tricycle motor transporting water from the river to the sites; (c) water use for the cooling of grinding machine; (d) river water pumped to the sites.*

All these activities contribute to lower the level of the water table and the scarcity of water as resource. These activities lead also to the disturbance of the quality of the river water causing the turbidity and lowering of the river bed during the dry season flow.

Used batteries from lighting torches (the sites are not illuminated) that are often thrown into wells, mercury used to amalgamate gold, spilled oils, hydrocarbons and diesel that are seeping into ground may be carried into the nearby Binada river, adjacent to the sites or may infiltrate into groundwater. This may impact the sustainably of aquatic environment and its biodiversity.

The impacts on water are: the depletion of water resources (massive use of water, water discharges during sinking), pollution of surface water and / or groundwater

### *(iv). Soil and Landscape*

(a)

(b)

(c)

(d)

***Figure 9.*** *Impacts of gold mining on the ground - (a) open wells abandoned, (b) bare soil eroded by trickling; (c) waste scattered on the ground; (d) earth mounds) disfiguring the landscape.*

Soil is affected in one way or another at every stage of artisanal gold mining. The physical impacts left following the completion of the wells, forming heaps and deforestation that results negatively influenced the quality of soil. We can add the presence of solid and liquid waste on the site (household waste, oils, hydrocarbons, human faeces, plastic, wastewater, etc.). All these result on the erosion and loss of soil fertility, the contamination by solid and liquid waste, and the degradation of the landscape.

### *(v). Air*

Several activities are source of air pollution particularly grinding the ore. Indeed, workers have no protective mask while the fine powder is easily transported by wind. Huge emissions of gases were observed, fumes but also noise (mills, pumps and motors). Air may be polluted by mercury vapors from burning amalgamated gold. Mercury is used to extract gold from ore by forming a mixture of approximately equal parts of gold. Mercury is then heated, leading to its evaporation and gold is collected. This emitted Mercury may represent a significant source of air pollution. The air content of mercury vapor around the points where the amalgam is heated can reach very high values and can almost always exceed the limit set by the [16](1.0 $\mu g / m^3$) for the Exposure of the general population.

The decomposition of solid and liquid wastes produced by miners generates strong odors. These wastes are dumped into the wild without any sanitation measures (Photo 9). Despite the facts that such pollution is felt by the concerned population, no protective measures are taken by small-scale miners to reduce or control them.

### *(vi). Health*

Miners are mostly exposed to a number of diseases that are related to their living and working condition on the mining sites. According to a private medical office located on one site, miners presented a number of diseases that are related to the lack of sanitation and hygiene in both sites, but also many accidents occurred. Malaria is due to the presence of stagnant water from the toilet. Diarrheal diseases, typhoid and cholera, are caused by poor hygiene. Skin diseases are caused by abrasions and ore jets during the sinking and crushing operations. Respiratory diseases (cough, pneumonia, angina, etc.) are caused by dust exposure especially during grinding and drilling. This can affect children who accompany their mothers on the sites (Photo 9).

Water used in the sites is supplied from the river or from stagnant water. This lack of hygiene characterized site restoration, both individuals and providers, in terms of drinking water quality and the management of household waste. Meals served in restaurants: cereal, rice, pasta and beans or canned are often poorly conditioned. The individual food ration is insufficient and does not compensate for the efforts made by the miners during the day.

Exposure to mercury during the amalgamation represents a real public health problem. Studies on the use of mercury in mineral processing conducted on 11 gold mining sites in Burkina Faso [17] illustrated the impact of artisanal mining on health. The average value of mercury is found to be 194.5 mg/g in the urine. More than 98.9% have urinary concentrations of mercury beyond the reference values for the general population; 68.8% have concentrations beyond the professional reference values (35 mg Hg/g), and almost half (49.5%) presented values above 100 mg Hg/g.

(a)

(b)

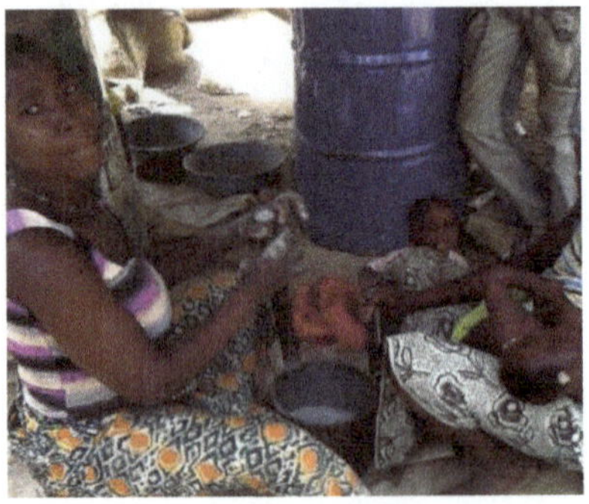

(c)

(d)

***Figure 10.*** *Waste dumping site, wastewater from latrine (haut) and women with their baby, crushing ore in a pestle.*

Several cases of STI/AIDS were reported by the private health center located on the sites, even if there were no screening session. Prostitutes encountered on sites reported using adequate methods of protection. The STI/AIDS come from the luxury prostitution of salesgirls who become prostitutes at night, lured by the easy money especially during good production period. This is favored by alcohol and drugs and these girls will go with miners unprotected and who constantly are changing partners.

Studies elsewhere indicated that the number of women with STIs identified was more than 4 times higher than that of men in the same age group [18];[4].

Noise pollution is also encountered in the sites, due to vibration produced by the digging of wells, ore crushing and grinding, the operation of some machines such as mills, power generators, and exhaust pipes, etc. This noise that exceeds the regulatory standards has very negative consequences, not only on human health but also on wildlife. Local people have no means of protection against noise and vibration.

### 3.1.2. Potential Impacts on the Socio-economic Background

#### (i). The Economy

Gold mining is important for the economy of miners and the local communities. This activity mostly has a very positive impact on jobs. When the activity is profitable, according to the authorities of the sites, up to100 - 150 workers excluding daily laborers can be found. This number can drop to 50 or even 20 when another site is found more productive. Workers on the sites are very mobile and most workers on the sites are from the countries of the sub-region where the activity of gold extraction is a tradition. Therefore money obtained is often repatriated to their countries of origin.

The sale of gold extracted from a bag of ore can earn up to 10 million CFA (17 200 euros). Although we have been unable to obtain information on the income of workers in the whole chain, gold panning help to increase the incomes of miners through direct sales of gold. Ofthe76 workers interviewed, about 30.3% said they have sent money to theirs families and that this money have been used to deal with household expenses (children's education, food and health), 10, 57% spent on food in working sites, 26.3% of respondents said they spent their income for food, for clothing (23.7%), and10.5% for personal items (Table 1). This study found that most miners have ventured into mining as a survival strategy rather than as a long term investment.

***Table 1.*** *Goods and services purchased using money from the sale of gold.*

| Expenditures | Number of Respondents | Percentage |
|---|---|---|
| Electrical gadgets | 7 | 9,2 |
| Schooling and hospitalization | 5 | 6,6 |
| Fun and recreation | 4 | 5,3 |
| personal items | 8 | 10,5 |
| Shipping money to the family | 23 | 30,3 |
| Food | 15 | 19,7 |
| Clothing | 7 | 9,2 |
| Investing in real estate | 2 | 2,6 |
| Gear wheel | 3 | 3,9 |
| Other | 2 | 2,6 |
| Total | 76 | 100 |

(a)

(b)

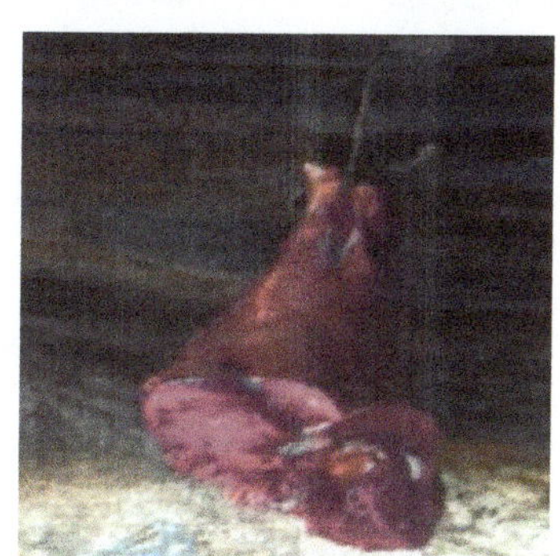

(c)

(d)

*Figure 11. Commercial activities on the mining sites- (a)food; (b) grocery, (c) butchery; video club.*

Gold mining contributed also to the creation of additional activities that have developed around the sites. On both sites, we have identified restoration, butchers, bars, shops, blacksmith and mechanics, clothes sellers, clubs and video games, and fruit and vegetables sellers.

***(ii). Social***

Talks we have had with the site leaders and traditional authorities, it appears that the holders of the sites are very involved in local communities. Thus, a percentage (which has not been given to us) from the sale of gold would be returned to the sites, the villages and to the royalty of Bouna. This commitment also involves the construction of school infrastructure and sponsorship of sporting activities.

But gold mining also causes loss of agricultural land and deforestation, which negatively impact on the local economy.

***Figure 12.*** *Children working on the mining sites.*

Gold mining at both sites also has a negative impact on people especially women and children. Indeed, on both sites we found the use of children for certain tasks such as crushing and transporting the ore. The girls say they work for parents who live in the surrounding villages (Photo 11). We have not been able to obtain information about the level of their remuneration, but they are paid by task. They come every morning to go back the evening, often without finding work and without enough money to afford a meal. In these situations they are at the thank you of any worker who makes them advances by paying their meal. According to indiscretions, this resulted in some cases of sexual abuse.

***(iii). On Security***

The impact of gold mining on safety is related to the working conditions on the site and facilities. The lack of personal protective equipment increases the risk of injury. Possible electrocution due to the outdoor exposure to power lines (Kintan site) and which may affect the safety of people on the site (fire hazard). Banditry and insecurity are virtually present. The presence of many two-wheeled vehicles (carts, motorcycles) often causes accidents that can in some cases cause irreversible trauma.

Crime and violence are common currency so that the holders of the sites use the police (Republican Security Force) to provide security at the sites. Traditional authorities as well as officials at the sites do not have the means to control deviant behavior among migrant minors [19]; [20].

## 4. Recommendations

The recommendations aimed at the rehabilitation of the sites after the closing of the sites by reforestation and helping miners master the new technologies through technical supervision and monitoring. This requires helping miners to use mercury trap which is a device uses that reduce the release of mercury vapor in the atmosphere. It is an assembly of plumbing parts in which the amalgam is burnt; mercury vapor passes through a tube and is condensed in a cooler [21].

It is also possible to teach the miners how to use mercury-free gold extraction technologies such as chlorination. This has been tested in Guyana and has resulted in a recovery rate of over 93% in the laboratory.

In terms of hygiene and security, we must develop a simple policy of solid and liquid waste management. Household waste can be buried in pits and used to make the traditional composting beneficial to farmers in village guardian's sites.

In terms of health, we propose that the owners of the sites undertake campaign of education of miners and the population on the importance of using and building effective systems for on-site sanitation: namely to equip the toilet for a good management of their excreta. This type of installation can also be built by the owners of the sites to ensure the health of users. Besides that, programs focusing on HIV/STI prevention, with targeted interventions for high risk women such as the one found in both sites, will be extremely important in such high transmission communities where there is substantial recent migration of men and women seeking work. Such programs can be initiated by a private/public/NGO partnership.

Institutionally, the state should define a consistent policy of artisanal gold mining taking into account the environmental, social and health.

## 5. Conclusion

The results of the studies have shown that working conditions are generally dangerous and unsanitary and deplorable living conditions in traditional gold mining systems. Internal costs of small-scale mining borne directly by minors include the poor conditions of health and safety in mines. External costs are supported by the surrounding communities, such as the degradation of the environment around the mining area and social disruption in the form of prostitution and alcoholism. Riparian Communities mines suffer environmental degradation and social problems. These negative impacts will remain for a long time in the areas of mining sites long after the miners have abandoned sites.

However, the resilience of the gold mining activities has proven to be one of the main sources of livelihoods in local communities of the extraction sites because they provide income and employment for survival from day to day. Panning for gold is proving to be a very important activity; it also creates many direct and indirect jobs such as catering and distribution. Unfortunately, these activities do not generate revenue for the national treasury, although they participate in local development through donations and infrastructure construction.

If gold mining to continue, there must be reconciled with the preservation of nature and its occupants. This should be a concern for policy-makers, technicians, civil society, traditional and religious leaders, miners who must work each for good organization of this carrier activity. The State shall for its part, promote research and made available on fair and necessary information to establish appropriate policies for the mining sector and train human resources with the ability to monitor compliance with environmental standards.

In overall, the sustainability of small-scale gold mining is a contentious issue given the level of environmental degradation and the disruption of social relations taking place in the district. However, the sustainability of the sector mainly rests upon proper regulation by the national government in order to improve monitoring mining activities and remitting of revenue to the fiscal.

## Acknowledgements

We would like to thank the regional District of Water and Forestry, and Mine. We would like also to thank the OIPR of Bouna for their logistic support during this study. We also thank Issiaka, the manager of the mining sites who guide us during the visit of the sites and did the translation during the interview.

## References

[1] Heemskerk, M.; Oliviera, M. (2003). "Perceptions of Small-scale Gold Mining Impacts: Results from Focus Group Discussions in Mining Camps and Affected Communities. World Wildlife Fund Report.

[2] LOGAN, M. (2004). Making Mining Work: Bringing Poverty Stricken Small-scale Miners into the Formal Private Sector. International Research Centre.

[3] LATAPIE D. (2001). La Fabuleuse Histoire de la ruée vers l'or, Californie - XIXe siècle, Mémoire vive. Privat, p. 10.

[4] Shoko, D. S. M; Veiga, M. M. (2004). Removal of Barriers to Introduction of Cleaner Artisanal Gold Mining and Extraction Technologies. Global Mercury Project Report.

[5] Thiobane M. (2012). Orpaillage dans la région de Kédougou : La malédiction de l'or. Les «Cahiers du GREP» N° 05 - Mai 2012 p 3-5.

[6] Albarre, G., André, H., Bochkoltz, F., Doulliez, J., Goffin, L., Jeanfils, E., Offergeld-Luxen, J., Pauwels, D., et Tibesar, A., 1978. Vers une transdiscipline de l'environnement. Fondation universitaire luxem- ourgeoise, Série « Notes de recherche » 14. Arlon, Belgique, 96 p.

[7] INS, 1998. Recensement général de la population et de l'habitation 1998. Volume III, Tome 1, Région des lacs, 33p.

[8] Aktouf, O (1987). Méthodologie des sciences sociales et approche qualitative des organisations : Une introduction à la démarche classique et une critique. Les Presses de l'Université du Québec, 1987, 213 pp.

[9] Programme des Nations Unies pour l'environnement [9] (UNEP/PNUE). 2012. Réduire l'utilisation du mercure dans le secteur de l'orpaillage et de l'exploitation minière artisanale. Guide pratique. Disponible sur le site suivant: http://www.unep.org/hazardoussubstances/Portals/9/Mercury/Guide%20Practique%Reduire%20L'Utilisation.

[10] Galvez, R. and Guesdon, G-A.2011 Le génie environnementale de la théorie à la pratique : la gestion de déchets et sites contaminés, l'évaluation des impacts environnementaux. Editions Multimondes.

[11] FEKOUA D, 2010. Anthropisation et risques environnementaux sur les collines de Yaoundé. Master professionnel en études d'impacts environnementaux, Centre régional d'enseignement spécialisé en agriculture forêt/ bois Cameroun.

[12] Sadar M. H.: Evaluation des impacts environnementaux, 2è Ed., Carleton University Press, 1996.

[13] André P, Delisle C-E., Revéret J-P et Sene A. (1999). L'évaluation des impacts sur l'environnement : Processus, acteurs et pratique. PresseInternationalesPolytechnique, 416 p.

[14] Tchindjang, M. (2009) : Cours en ligne sur les EIE. www.cm.refer.org.

[15] Louppe, D. et Ouattara N. 1997. Influence du Karité sur les productions agricoles du Nord de la Côte d'Ivoire. Mémoire présenté au XIè Congrès forestier mondial à Antalya (Turquie) du 13 au 22 octobre 1997.

[16] WHO 1980. Recommended health-based limits in occupational exposure to heavy metals. Tech. Report Series 647: Inorganic Mercury, WHO Environmental Health Criteria, 118. Geneva: WHO, 1980:168.

[17] Ouedraogo, A. H. 2006. L'impact de l'exploitation artisanale de l'or (orpaillage) sur la santé et l'environnement. Médiaterre.(http://www.mediaterre.org/afrique-ouest/actu,20061121095625.html); p2.

[18] Sawadogo, E, 2011..L'impact de l'exploitation artisanale de l'or: cas du site de Forafora dans la province de Poni. Université de Ouagadougou. [Rapport]: Mémoire de maîtrise en géographie.

[19] Metcalf S. M. (2008). Identifying Strategies for Effective Artisanal and Small-scale Gold Mining Interventions in Kadoma-Chakari: Zimbabwe. Unpublished MSc Thesis, University of British Columbia.

[20] Hollaway, J. (1999). “The Button Gold Mine, Zimbabwe”, Jennings, S. N. (Ed). Small-scale Gold Mining: Examples from Bolivia, Philippines and Zimbabwe, Geneva: International Labour Organisation. Geneva.

[21] NGO MINYEM, K (2012). Réalisation d’un diagnostic environnemental pour l’amélioration de la performance environnementale de l’orpaillage au Burkina Faso ; Cas du site de Goumbeledougou. Mémoire de fin d’études de Master d’Ingénierie de l’eau et de l’environnement. Institut International de l’Eau et de l’Environnement, Burkina Faso. 85p.

# Assessment and Mapping of the Vulnerability of Soils in Imo State, Nigeria to Erosion Hazard using Geographic Information System

**Emeka Udokporo[1], Martin Atu Ngozika Anikwe[2, *], Kevin Ejike Chukwu[1]**

[1]Department of Geography and Meteorology, Enugu State University of Science and Technology, Enugu, Nigeria
[2]Department of Agronomy and Ecological Management, Enugu State University of Science and Technology, Enugu, Nigeria

**Email address:**
anikwema@yahoo.co.uk (M. A. N. Anikwe)

---

**Abstract:** Soil erosion by water is a critical problem in Imo State terrain due to anthropogenic pressure on its landscape. Assessment and mapping of erosion prone area is essential for soil conservation and watershed management. The objectives of the study are to use satellite imageries to assess and map the land cover, vegetation, land use, topography and hydrology; and to use field sampling to map soil properties and soil erodibility of Imo State with a view to assessing soil erosion hazard in Imo State. The DEM was generated to prepare slope map, elevation map and hill shading map. Results from the study show that 45.4 % of the land cover of Imo State representing 2256.7 $Km^2$ is covered by light vegetation whereas 24.7 % of the land area representing 1229 $Km^2$ is covered by thick vegetation. The result also showed that a large percentage of the entire land area (21%) representing 1046.3 $Km^2$ is cultivated whereas 6.4% representing 319 $Km^2$ is built up with houses and infrastructure like roads, bridges etc. The area covered by thick vegetation (24.7%) is categorized under slightly vulnerable to stable. A larger percentage (45.4%) is covered by light vegetation, and therefore, classified as moderately vulnerable to erosion whereas 27.4% comprising cultivated and built up areas are classified as highly to extremely vulnerable to erosion. Results also show that most parts of the state (4133$Km^2$) representing 84% of the entire landscape fall into slightly to less vulnerable to erosion with a slope class between 1 – 4% slope. The results on hill shading concur with that of slope above. As expected, the dispersion ratios were affected by the type of underlying parent material. Results from the characterization of soil properties of study area show since most soils in the study area have high dispersion rates, depicting high vulnerability to soil erosion and 84% of the entire landscape fall into slightly to less vulnerable to erosion with a slope class between 1 – 4% slope, it then follows that high rates of erosion recorded for the area may be due to land use and management as light vegetation, cultivation and built areas cover 73% of the land area and classified as moderately to extremely vulnerable to erosion. The result at landscape and plot scale varies. Therefore, model results need to be carefully used for local level soil conservation planning.

**Keywords:** Assessment, Mapping, Vulnerability, Gully Erosion, Nigeria

---

## 1. Introduction

Globally, environmental issues have become major concern to governments and citizens of various nations, including Nigeria. The environment, which is at the heart of economic, social, cultural and human activities, has been disrupted by man's neglect and abuse of the environment. Pollution, deforestation, erosion, landslides, global warming etc. are the aftermaths of this abuse in the ecosystem. The issue of protection, mapping and monitoring becomes paramount in the face of the increasing population of residents in the southeast geopolitical zone of the country. They are severely impoverished, particularly, the rural dwellers due to environmental degradation and increasing population impact on the environmental resources.

Southeastern Nigeria is a typical gully erosion region in Nigeria. The presence of gully sites is one of the hazard features that characterize this zone as well as other States that adjoin them (Ofomata, 1985). Agulu-Nanka in Anambra State is an area badly affected by water erosion. Up to 250 tonnes per hectare of soil have been lost in severe storms. Similarly, Asiabaka and Boers (1988) had estimated that over 1,970 gully sites occur in Imo and Abia States. A conservative assessment shows the distribution of known gully sites, in different stages of development are as follows; Abia (300), Anambra (700), Ebonyi (250), Enugu (600), Imo (450) (Igbokwe et al. 2003) (Egboka, 2004).

In Imo State, the impact of gully erosion has been unimaginable, displacing communities and causing destruction of farm lands. However, human activities like land-clearing, and deforestation, overgrazing, as well as the creations of firewood tracks accelerate the natural rates of these processes (Ogbonna, 2012). The immediate site effect is the loss of soil while the off-site effects include the yield of sediment on the river network, which results in declining water quality and damage to the hydraulic structures. It is, therefore, necessary to assess and to manage areas that are vulnerable to gully erosion in order to mitigate any damage associated with it. Gullies are triggered by man's activities, rainfall erosivity and soil erodibility (Ogbonna, 2012).

Gully Erosion is, however, a prominent feature in the landscape of Imo State. The topography of the area as well as the nature of the soil contributes to speedy formation and spreading of gullies in the area.

The resultant need to predict gully occurrences has led to the development of numerous stochastic and process-based models, with increasing emphasis on the use of the Geographic Information Systems (GIS), Satellite Remote Sensing (SRS) and Geostatistical models.

Inspite of technological advancement, erosion menace still remains a major problem in Nigeria (especially in South Eastern Nigeria). The yearly heavy rainfall has very adverse impacts altering existing landscape and forms. Such landforms create deep gullies that cut into the soil. The gullies spread and grow until the soil is removed from the sloping ground. Gullies when formed expand rapidly coupled with exceptional storm or torrential rain down the stream by head ward erosion gulping up arable lands, economic trees, homes, lives, sacking of families, valuable properties that are worth millions of naira. As a matter of fact, there is a direct correlation between development and the effect of gully erosion.

In order to protect the environment and the natural resources for the future generations, it is important to assess and map areas of Imo State that are vulnerable to gully erosion. There is dearth of information on the number, types and stages of gully erosion in different parts of Imo State. More importantly, very little information exist on the vulnerability of soils in Imo State to gully erosion. Much model-derived information on rates of erosion and on erosion risk of cultivated land may be of dubious value, and thus, there is need for other ways to assess vulnerability of soils in Imo State to gully erosion. In developing a sustainable remediation procedure, an assessment of risk of erosion through a spatial mapping of the areas vulnerable to gully formation is necessary.

Hence, in order to plan for a better environmental decision-making policy, the economic evaluation of these environmental problems is important, hence the decision to carry out this research.

The main objectives of the study is to assess and map the vulnerability of parts of Imo State, Nigeria to gully erosion hazard using geographic information system, remote sensing techniques as well as field sampling techniques.

The specific objectives are:

1. To use satellite imageries to map the land use and management, vegetation and land cover, topography and hydrology of Imo State, Nigeria with a view to assessing soil erosion hazard in Imo State.
2. To use field sampling to map soil properties and soil erodibility with a view to assessing soil erosion hazard in Imo State.

## 2. Materials and Methods

### *2.1. Study Site*

Imo State occupies the area between the lower River Niger and the upper and middle Imo River. Imo State is bounded on the east by Abia State, on the west by the River Niger and Delta State; and on the north by Anambra State, while Rivers State lies to the south. Imo State covers an area of about 5,100sq km. Imo state lies within latitudes 4°45′ N and 7°15′ N and longitudes 6°50′E and 7°25′ E (Okoro et al. 2014)

Imo State is underlain by the Benin Formation of coastal plain sands. This formation, which is of late Tertiary age, is rather deep, porous, infertile and highly leached. In some areas, impermeable layers of clay occur near the surface, while in other areas, the soil consists of lateritic material under a superficial layer of fine grained sand. Ogbonna (2012) noted that seven (7) geologic formations cover the old Imo State, these are:

(1) *Niger Delta Formation* – consist of sands, gravels, clays mainly arenaceous with very low dip structure to the SSW;

(2) *Coastal Plain Sands Formation* – consists principally of sands, gravels, clays, shales (lignites). Topographycally, it consists of gently sloping plain;

(3) *Bende Ameke Group Formation* – lithologies are clastic sedimentary rocks consisting mainly of lignites conglomerates with interbedded sandstones, and shales.

(4) Imo Shales Formation - *consists* of clay-shales with intra-formational sand bodies. It is typified with lowland ridges.

(5) This formation is generally called The *Plateau and Escarpment* made up of three groups: *Upper Coal Measure, False Bedded Sandstone, and Lower Coal Measures* – lithologies are sands, gravels, siltstones, sandstones and grits. The topography consists of minor cuesta, mesas and escarpment.

(6) This formation consists of – *Asata-Nkporo Shale, Awgu-Ndeaboh Shale Group, Ezeaku Shale Group, and Asu River Group.* Lithologies are shale with lenticular sand bodies, thin limestones, and minor sandstone lenses with igneous bodies.

(7) *Asu River Group Formation* – consist of crystalline rocks, with topography of mountain terrain with plan ted

margins.

The type of geology and soil of an area plays an important factor in the development of gully erosion. Rivers are few with vast inter fluves which are characterised by dry valleys that carry surface drainage in periods of high rainfall. The phenomenal monotony of the terrain may be accounted for by the absence of any tectonic disturbances and by the homogeneity of the rock structure (Udo, 1970). The main streams draining the state are Imo, Otamiri, Njaba and Ulasi rivers, all of which have very few tributaries. With the exception of Imo River, which runs through the area underlain by the Imo Shales, other rivers rise within the coastal plain sands. Generally, river valleys constitute the major physical features, which are often marshy.

The undulating nature of the interfluves gives rise to numerous depressions especially in the northeast Rainfall distribution is bimodal, with peaks in July and September and a two week break in August. The rainy season begins in March and lasts till October or early November. From March to May, there are violent storms which destroy crops and houses. Rainfall is often at its maximum at night and during the early morning hours.

However, variations occur in rainfall amount from year to year. Annual rainfall varies from 1,990 to 2,200 mm. Temperatures are similar all over the state. The hottest months are January to March, with the mean annual temperature above 20°C. The influence of the harmattan lasts for about nine weeks (i.e. from late December to late February).

Imo State has an average annual relative humidity of 75 percent which is highest during the rainy season, when it rises to about 90 percent. The high temperature and humidity experienced in the state favour luxuriant plant growth, which ideally should produce the climax vegetation of the tropical rain forest.

Economic trees like the iroko (*Chlorophora excelsa*), mahogany (*Swietenia macrophylla*), obeche (*Triplochiton scleroxylon*), Gmelina (*Gmelina aborea)*, bamboo (*Phyllostachys edulis*), rubber (*Ficus elastica*) and oil palm (*Elaeis guineensis*) pre dominate. But due to high population density, most of the state has been so farmed and degraded that the original vegetation has disappeared. Thus farmers are forced into marginal lands, a situation aggravated by the rising demand for fuelwood. Deforestation has triggered off acute soil erosion in the study area.

### 2.2. Data Acquisition

The data collected for the study were classified into primary and secondary data. Secondary data source includes information on rainfall distribution from January to December, 1982-2013 collected from records of Nigeria Meteorological Agency (NIMET). Existing road network, settlement distribution, administrative map, drainage patterns and vegetation/land use was obtained from ASTAL Uyo whereas geology, rainfall, relief, and soil maps printed and published by Ministry of Lands, Survey and Urban Planning, Owerri, Imo State from 1980-date were collected. The primary data for this research work is a *satellite image* and *boundary map* of Imo State. The States Ministry of Lands and Urban Development provided the administrative map of Imo State as compiled by the States survey division. The analogue map was scanned and the boundary extracted using ArcGIS software. Landsat 7 ETM+ and Landsat 4-5 satellite imagery were adopted for this research after due consideration because of the suitability of the resolution of the satellite and their spectral characteristics. Landsat imageries were stored using aclassification method of paths and rows. United States Geological Survey (USGS) website serves as a repository for archived Landsat satellite imagery. This research was carried out using Global Mapper v12, a raster based Geographical Information System, ArcGis 9.3, a vector based Geographical Information System.

A total of 40 systematic sampling polygons were established throughout the State boundary. The downloaded satellite imagery band 1, 2, 3,4,5,7 were imported in Arc Map 9.3 satellite imagery and the *composite tool* was used to combine all the bands (panchromatic in appearance) to form a multi-spectral imagery. The Red, Green, Blue (RGB) of the image was set at R-Band 7, Blue-Band 5 and Green-Band 3 to form a true colour imagery.

The 40 systematic sampling polygons created were used as input to create unique signature in Arc Map by using GIS facilities. Then, all sampling plots located with either of the 3-uniques classification of Thick Vegetation, Light Vegetation and Non Vegetation (includes roads, buildings, and hills).

*Image enhancement and classification*

In this research, principal components and direct adjustment of brightness were used in order to obtain strong contrast between pixels and to create new images. These new images were used, as extra bands, in order to improve image classification techniques. In order to classify land cover within the study area, Image classification was made using supervised techniques, which require that the analyst have distinctive spectral signatures for all types of land cover classes. After accuracy assessment of all used methods maximum likelihood method was used for the classification.

### 2.3. Fieldwork

A reconnaissance survey was carried out in the in the study area to identify sampling points. Sampling points were chosen in each selected site using free survey technique (observation points that are representative of the site are chosen by the surveyors based on personal judgment and experience) (Mulla and McBratney, 2000).The 5 soil groups representing soils formed over alluvium, coastal plain sands, Imo shale clay, and upper coal measures were selected. Three auger and three core samples were collected from 0-20 cm depth in each of the site.

The auger samples were composited and its sub-samples used for analysis whereas the core samples were analyzed separately and mean results computed and used for statistical analysis. Differences in management practices and edaphoclimatic properties of the soils influenced the choice of the different sites. Sampled locations are shown in Fig. 1.

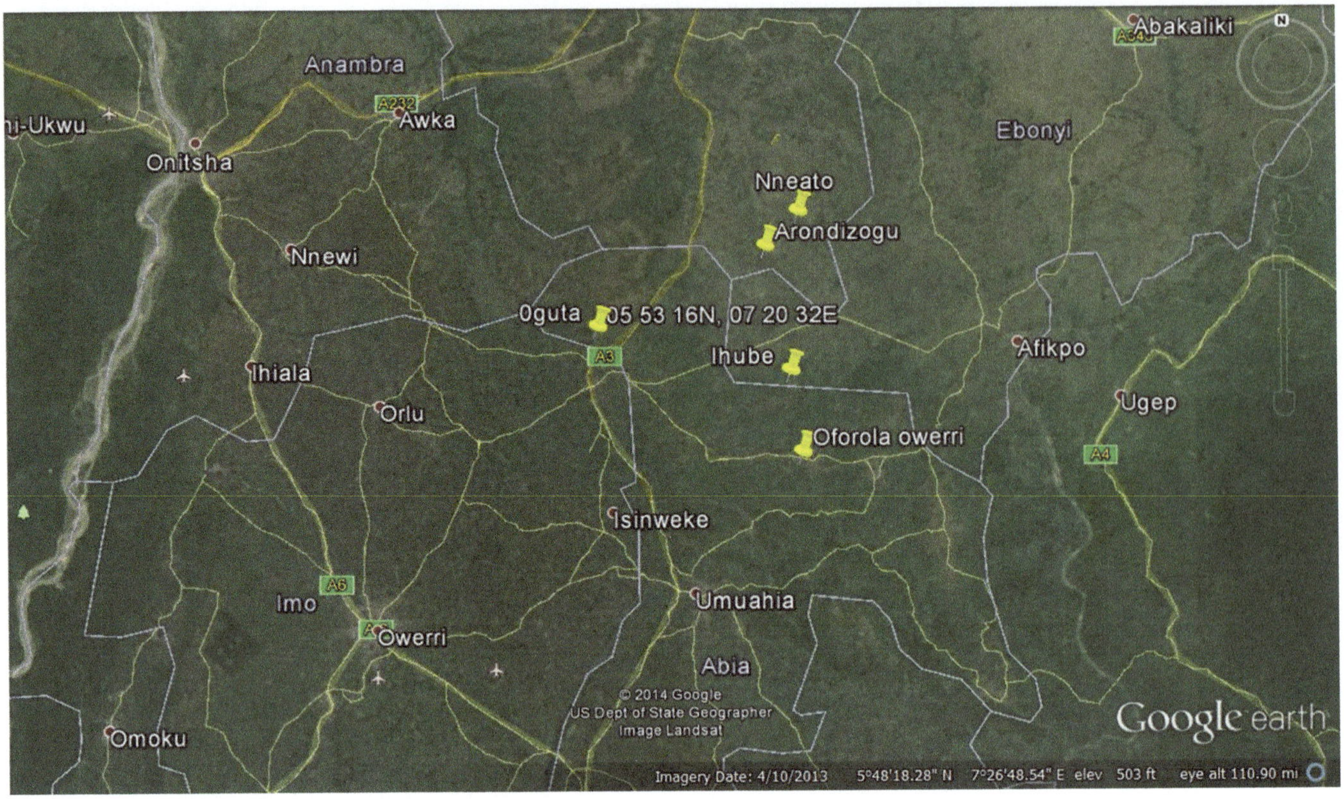


*Fig. 1. Sampling Locations in the Study area (from Google Earth, 2014).*

### *2.4. Laboratory Analyses*

Soil samples were air-dried and sieved using 2-mm mesh sieve before laboratory determinations. Particle size distribution was determined in both distilled water and calgon by hydrometer method (Gee and Bauder, 1986). Thereafter dispersion ratio, used as indirect measures of erodibility was computed as follows:

$$\text{Erosion index (EI)} = \frac{DR}{\%\,CLAY\ {}^{1}/_{2}WHC}$$

Where DR =dispersion ratio

WHC =Water holding capacity

$$\text{Dispersion ratio (DR)} = \frac{\%\ \text{silt}+\%\ \text{clay}\ (H_2O)}{\%\ silt+\%\ clay\ (calgon)}$$

Particle size distribution (textures) was obtained by the hydrometer method (Gee and Bauder, 1986). Soil water holding capacity was determined on undisturbed samples as the difference of water contents at –0.03 MPa determined by pressure plate and water content at –1.5 MPa determined by pressure membrane (Dane and Hopmans, 2002).

*Erodibility Classes*

For the purpose of this study, five erodibility classes were used to depict the vulnerability of the soil to erosion as follows:

1. Extremely vulnerable
2. Highly vulnerable
3. Moderately vulnerable
4. Slightly vulnerable
5. Less vulnerable

## 3. Results and Discussion

### *3.1. Land Use and Management*

The results of the study show that 33.6 % of the total land cover of Imo State is cultivated, and therefore, highly vulnerable to soil erosion depending on the management of the soils (Fig. 5). However, only a small portion (9.1%) actually contains human settlements and buildings including road networks (Fig. 3 and 4).The area coveringsettlements and road networks are classified as highly vulnerablebecausehumanactivities increase soil erodibility rates.Human intervention increases with population growth and pressure. For example, based on his study in the Gauche catchment area (234 ha) in West Hararghe zoneKenya, Thomas (1991) reported that of all the visible erosion incidents, 81% were caused by human induced factors while the rest were by natural factors. The most common human-induced factors that cause accelerated erosion include deforestation, inappropriate agricultural practices such as over cultivation and overgrazing. Croplands and pastures are susceptible to erosion but croplands are more vulnerable because the soil is repeatedly tilled and left without a protective cover of vegetation. The socio-economic situation in rural areas often leads people to use their environment inappropriately which induce land degradation. In any area the type of land use affects the level of soil protective cover and consequently the rate of erosion and erodibility. Deforestation and the removal from the fields of dung and crop residues for fuel and feed causes a steady reduction in the organic matter content of highland soils, rendering them less productive and more easily erodible.

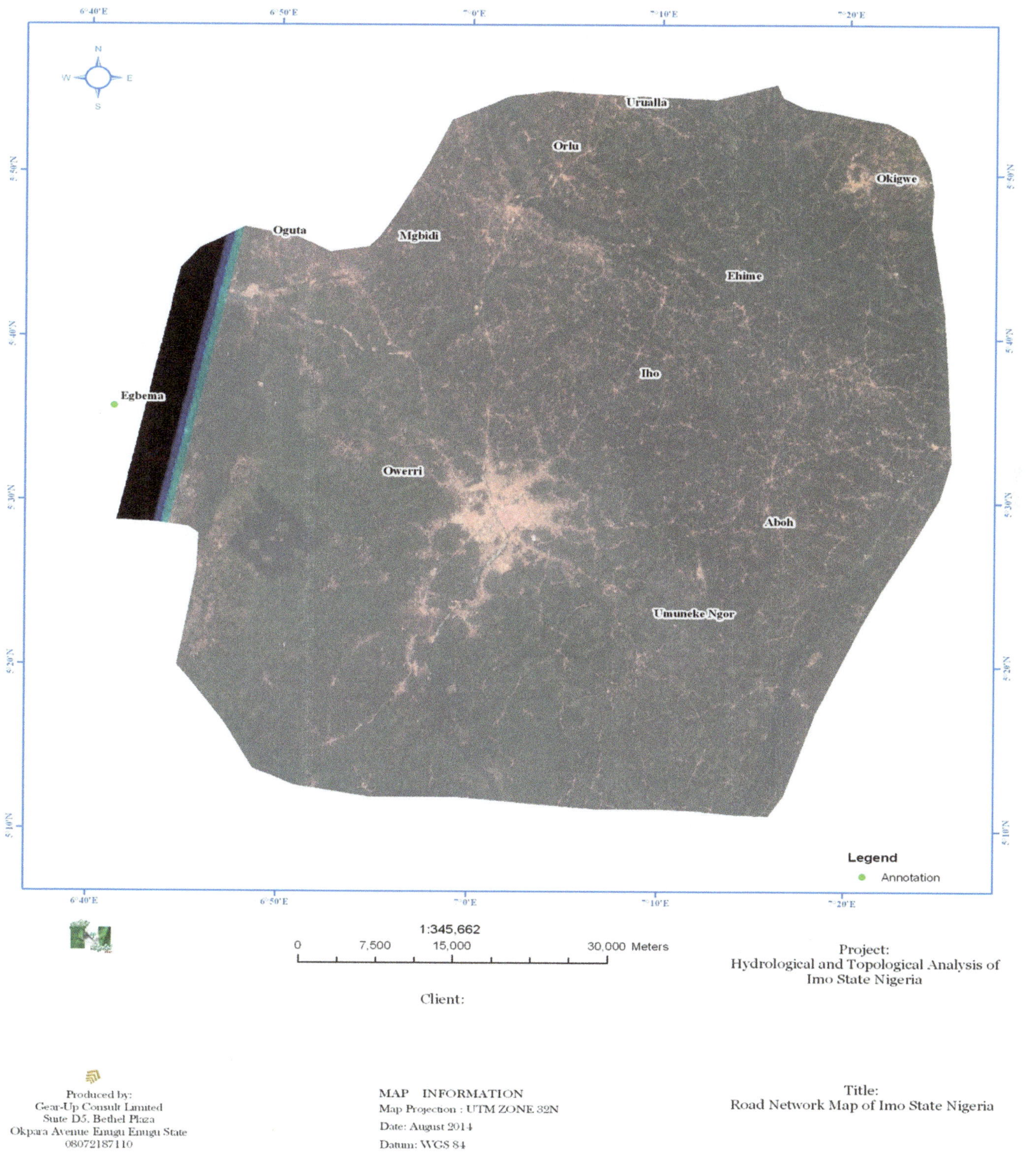


***Fig. 2.*** *Satellite imagery of Imo State.*

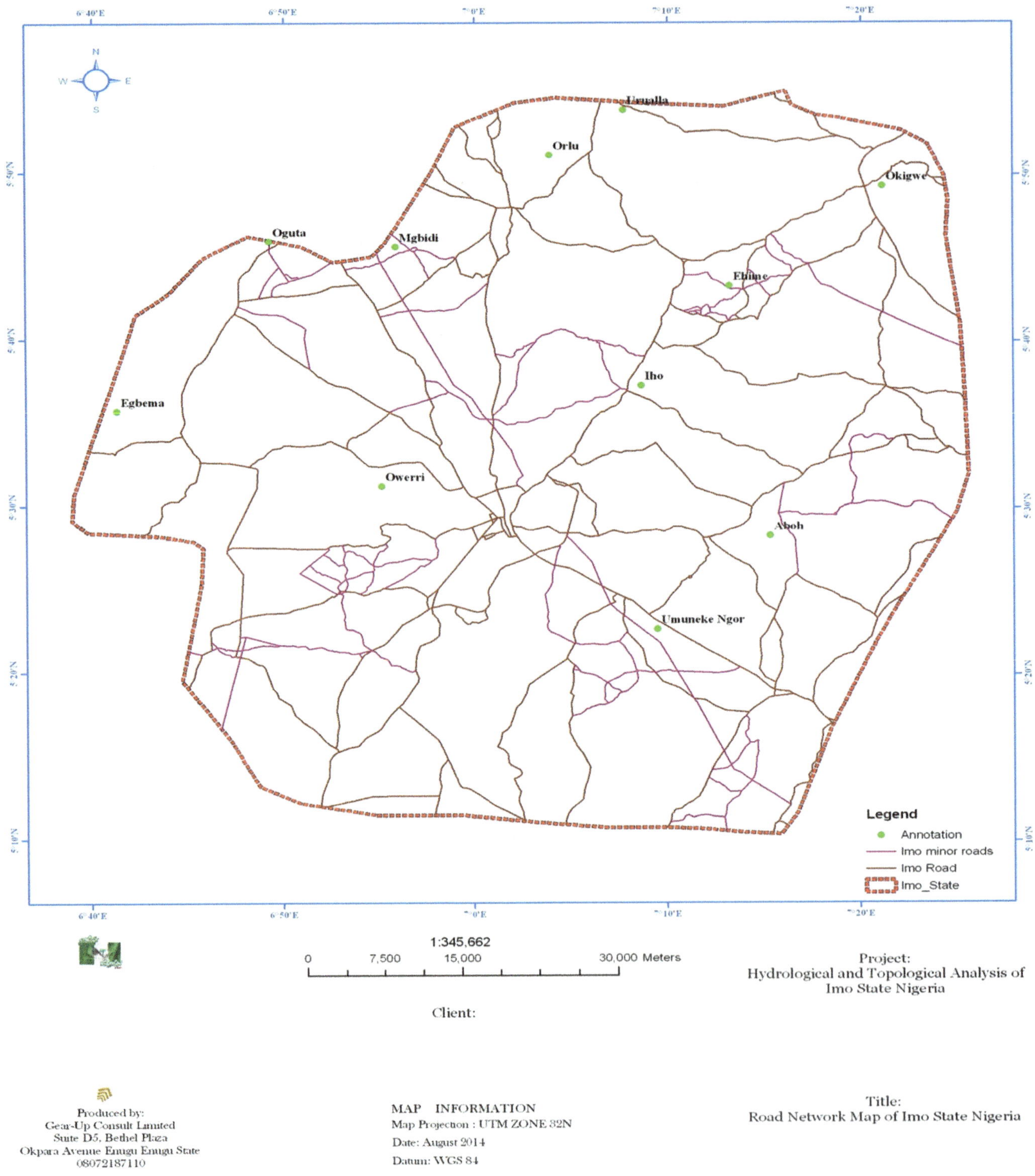


***Fig. 3.*** *Road network map of Imo State.*

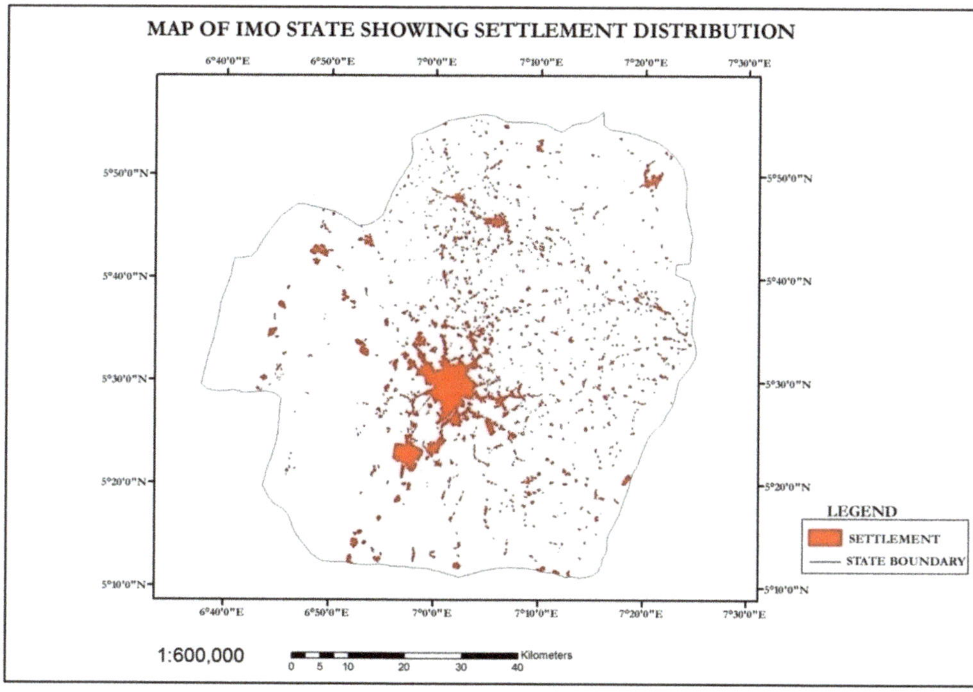


*Fig. 4. Settlement map of Imo State from NigSat1 (Source ASTAL Uyo).*

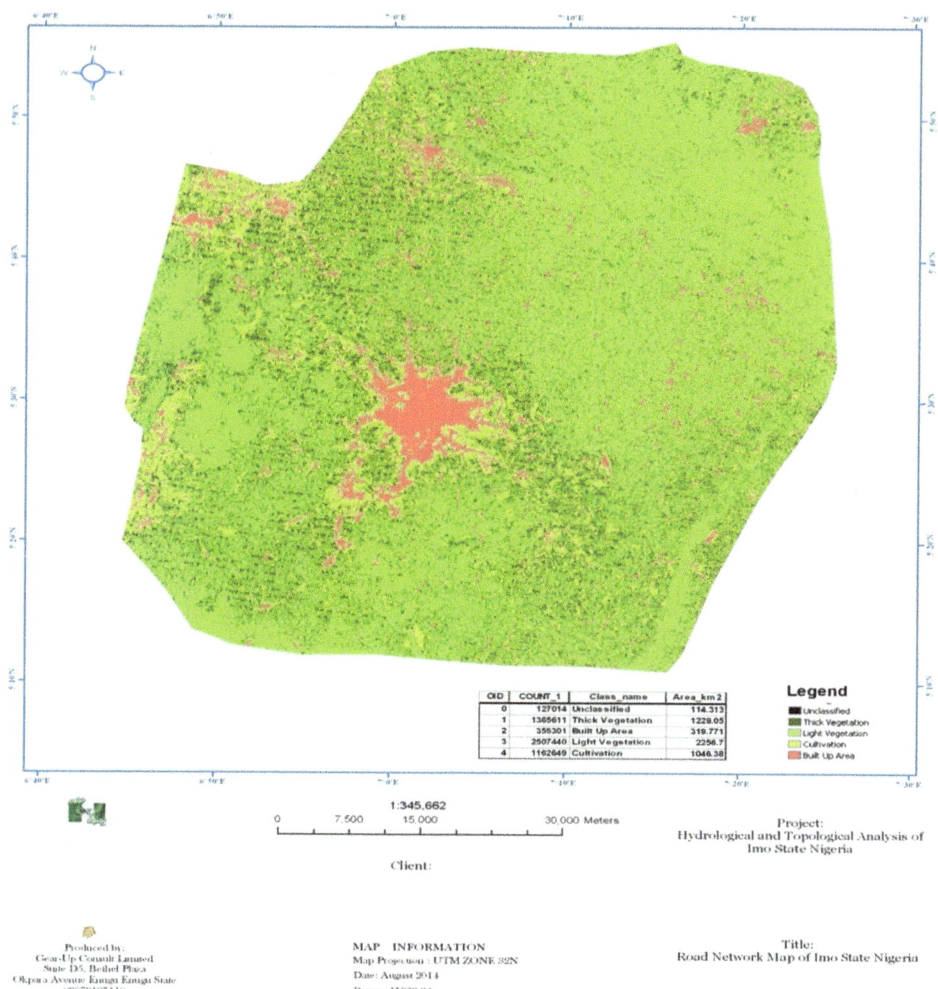


| OID | COUNT_1 | Class_name | Area_km2 |
|---|---|---|---|
| 0 | 127014 | Unclassified | 114.313 |
| 1 | 1365611 | Thick Vegetation | 1229.05 |
| 2 | 355301 | Built Up Area | 319.771 |
| 3 | 2507440 | Light Vegetation | 2256.7 |
| 4 | 1162649 | Cultivation | 1046.38 |

*Fig. 5. Land use map of Imo State.*

Fallowing has been traditionally used as a soil management and fertility restoration strategy as vegetative regrowth during fallowing helps these processes.Where there has been persistent population pressure on arable land, the length of the fallowing period has shortened over time leading to continuous cropping.

Tillage operations are sometimes carried out along slopes. Furrows formed along slopes cannot slow down runoff compared to those made along contours. Cultivation into fine soil tilts also makes the soil vulnerable to erosion during the early part of the main rainy season. Management of animals and their feed sources is also a major contributor to soil erosion in parts of the region. Mass movement of soil can be caused by human activity and land use change.

### *3.2. Vegetation and Land Cover of Imo State*

Results from the study show that 45.4 % of the land cover of Imo State representing 2256.7 $Km^2$ is covered by light vegetation whereas 24.7 % of the land area representing 1229 $Km^2$ is covered by thick vegetation (Fig. 5). The result also showed that a large percentage of the entire land area (21%) representing 1046.3 $Km^2$ is cultivated whereas 6.4% representing 319 $Km^2$ is built up with houses and infrastructure like roads, bridges etc.

The area covered by thick vegetation (24.7%) is categorized under slightly vulnerable to stable. A larger percentage (45.4%) is covered by light vegetation, and therefore, classified as moderately vulnerable to erosion whereas 27.4% comprising cultivated and built up areas are classified as highly to extremely vulnerable to erosion.

Throughout the world, the lowest erosion rates, ranging from 0.004 to 0.5 t/ha per year, are found in undisturbed forests (Pimentel et al. 1998). However, once forest land is converted to agriculture, erosion rates increase because of vegetation removal, over-grazing and tilling. Vegetation cover reduces erosion. Living and dead plant biomass reduces soil erosion by intercepting and dissipating raindrops and wind energy. Above-ground foliage slows the velocity of water running over the soil decreasing the volume of water and soil lost in surface runoff. Plant roots physically bind particles, thus stabilising the soil and increasing its resistance to erosion. Plant roots also enhance water conservation by creating pores in the soil surface that enable water to enter easily into the soil matrix. The uptake of water by plant roots also depletes the soil water content and thereby further increases infiltration rates.

In recent times, felling of trees for fuel, wood and charcoal without replacement has become a serious problem contributing to the loss of vegetation and hence to increased soil erosion. The increase in the human population has reduced land holding per capita and created pressure on limited land for agricultural production. Also, the removal or destruction of vegetation cover through overgrazing and bush burning etc. leads to land degradation as such practices leave soils bare and exposed to erosion and other degradation processes.

### *3.3. Soil Properties*

The major soil class found in Imo State is Ultisols. However, several different types of Ultisols are found in different places at lower classification levels and that infers that each of these has properties that affect soil degradation differently. Soils vary in their resistance to erosion partly based on texture and amount of organic matter. The resistance also depends on soil condition and depth. Soils high in silt and low in clay and sand are highly erodible (Nill et al. 1996). The high erodibility of silty soils is explained by their weak structural stability. They rapidly form surface seals upon the impact of rain drops. Erosion is less on clayey soils due to better aggregation and on sandy soils due to the non-sealing surface. Most of the differences in soil characteristics highly depend on the type of parent material found in the study area.

A total of seven (7) different parent materials were found in the study area. As earlier mentioned, five sites representing four of the seven parent materials were identified in the study area and soil samples and field measurements taken from the sites to complement GIS studies. The results (Table 2) showed that all the soils exhibited very high dispersion ratios ranging from 58-82%. As expected, the dispersion ratios were affected by the type of underlying parent material. The highest dispersion ratios were found in *Nneato* (82%) and *Ihube* (78%) representing sites in Upper Coal measures. *Oguta* site representing Alluvium (Lacustrine Deposit) had a dispersion ratio of 71% whereas Oforola Owerri site representing Coastal Plain Sandsand Aronizogu representing Clay Shale had dispersion ratios of 63 and 58 % respectively. According to Middleton (1930) soils having dispersion ratio greater than 15% are erodible in nature. Dispersive soils are structurally unstable and therefore highly susceptibility to erosion mainly due to the high sodium content. A dispersive soil can become rapidly deflocculated even in still water with low dissolved salt concentration (Shoghi et al., 2013).

### *3.4. Topography and Hydrology of Imo State*

Imo State's topography consists of high altitudes and rugged landscapes, as described earlier. The rugged topography and steep slopes affect soil erosion rate through its morphological characteristics. Two of these, namely gradient and slope length, are essential components in quantitative relationships for estimating soil loss (Wischmeier and Smith 1978).

Results of the study show that most parts of the state (4133$Km^2$) representing 84% of the entire landscape fall into slightly to less vulnerable to erosion with a slope class between 1 – 4% slope. However, 569.3 $Km^2$ of land representing 12% of the land area are moderately vulnerable to erosion whereas 4.5% of the entire land area is highly to extremely vulnerable to erosion with slope class of between 7- >12% (Fig. 6). These results concur with that on Fig. 7 showing analysis of hill shading for Imo state which showed that 87% of the area shaded fall within the highest heights of

176 – 187m and above whereas the remaining part (13%) fall within heights lower than 176m. Hill shading is the shadows drawn on a map to simulate the effect of the sun's rays over the varied terrain of the land.It is a hypothetical illumination of a surface according to a specified azimuth and altitude for the sun. Hill shading creates a three-dimensional effect that provides a sense of visual relief for cartography, and a relative measure of incident light for analysis.

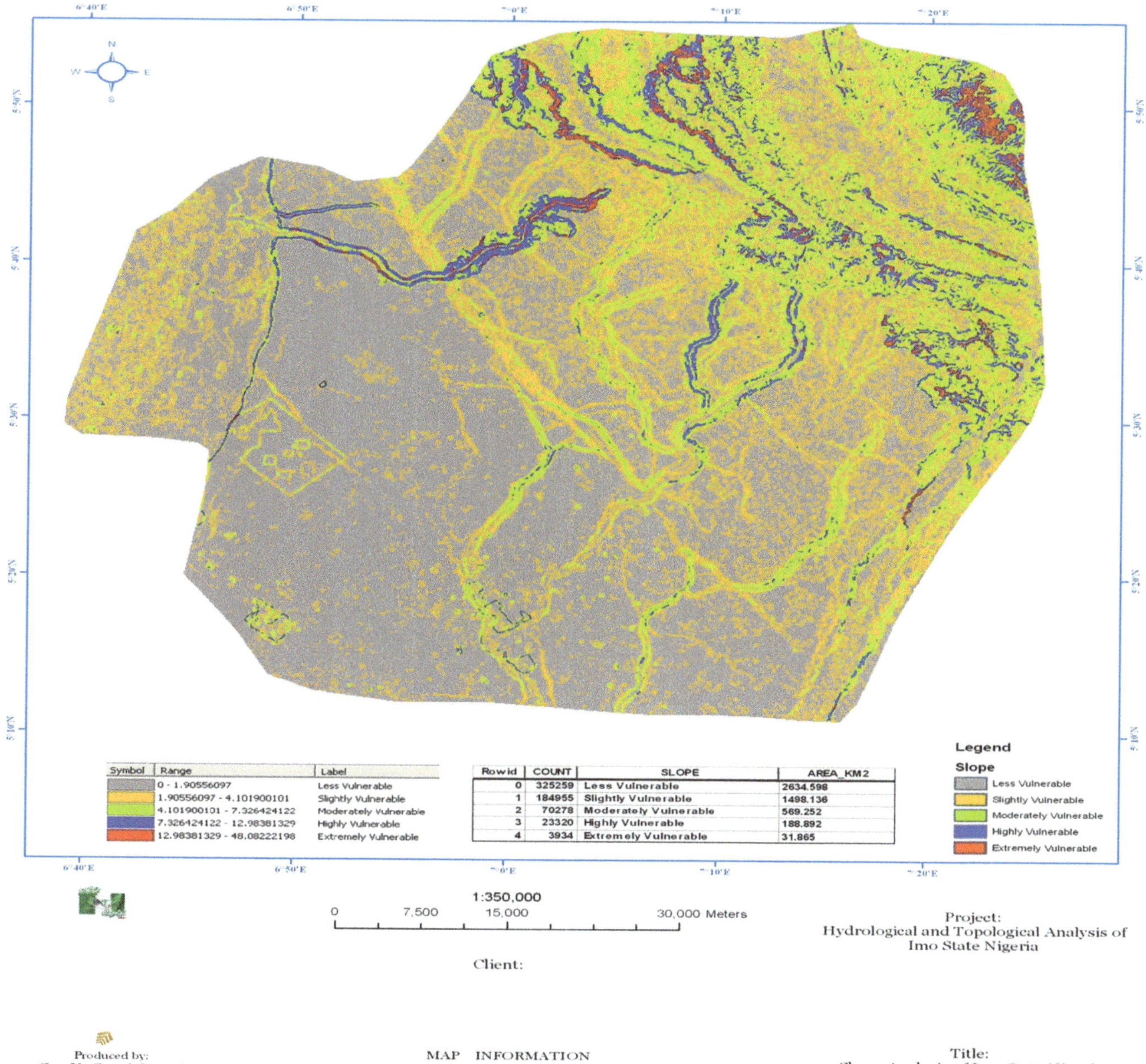


***Fig. 6.*** *Slope Model of Imo State, Nigeria.*

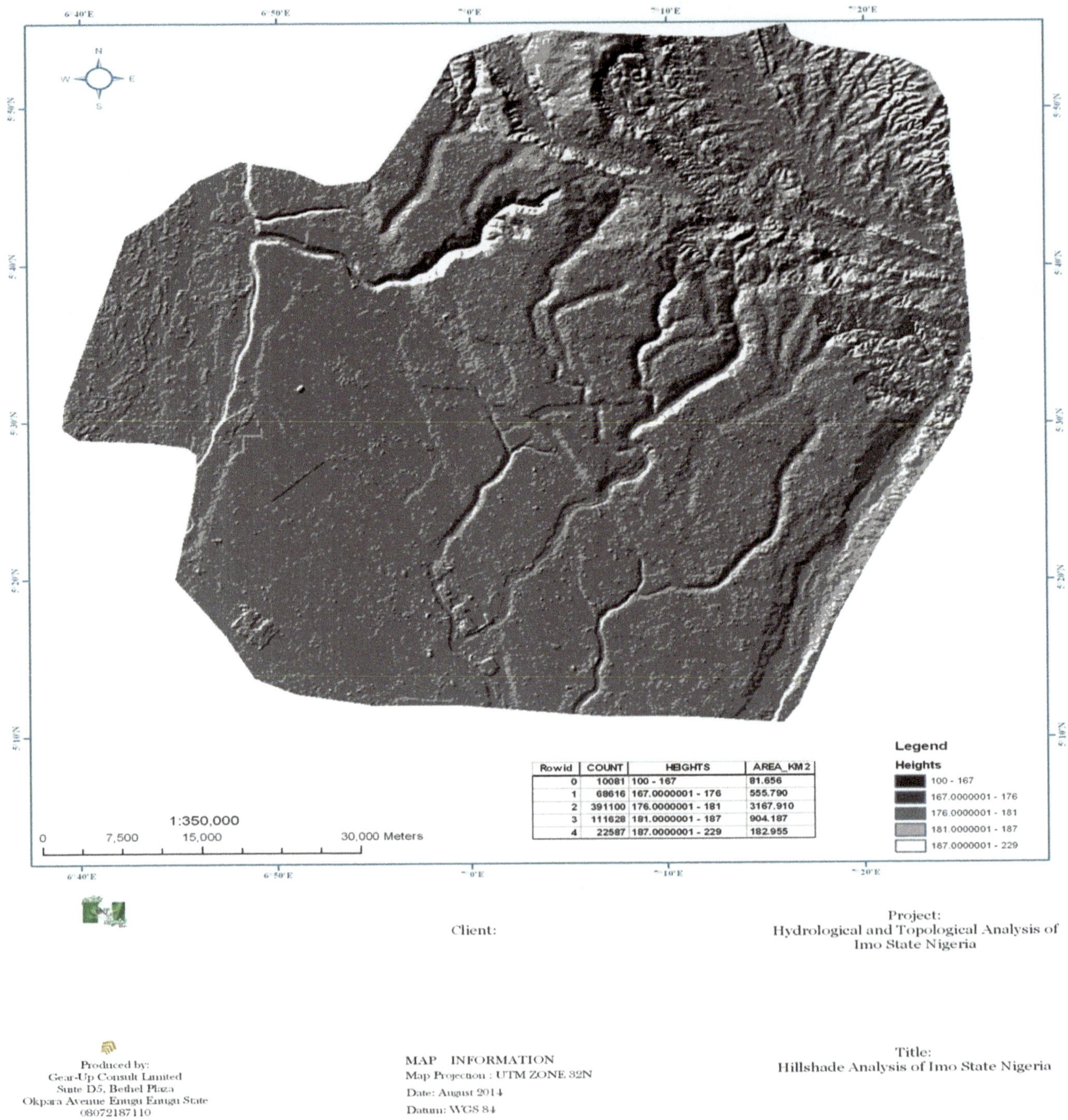


| Rowid | COUNT | HEIGHTS | AREA_KM2 |
|---|---|---|---|
| 0 | 10081 | 100 - 167 | 81.656 |
| 1 | 68616 | 167.0000001 - 176 | 555.790 |
| 2 | 391100 | 176.0000001 - 181 | 3167.910 |
| 3 | 111628 | 181.0000001 - 187 | 904.187 |
| 4 | 22587 | 187.0000001 - 229 | 182.955 |

***Fig.** 7. Hill Shade Model of Imo State, Nigeria.*

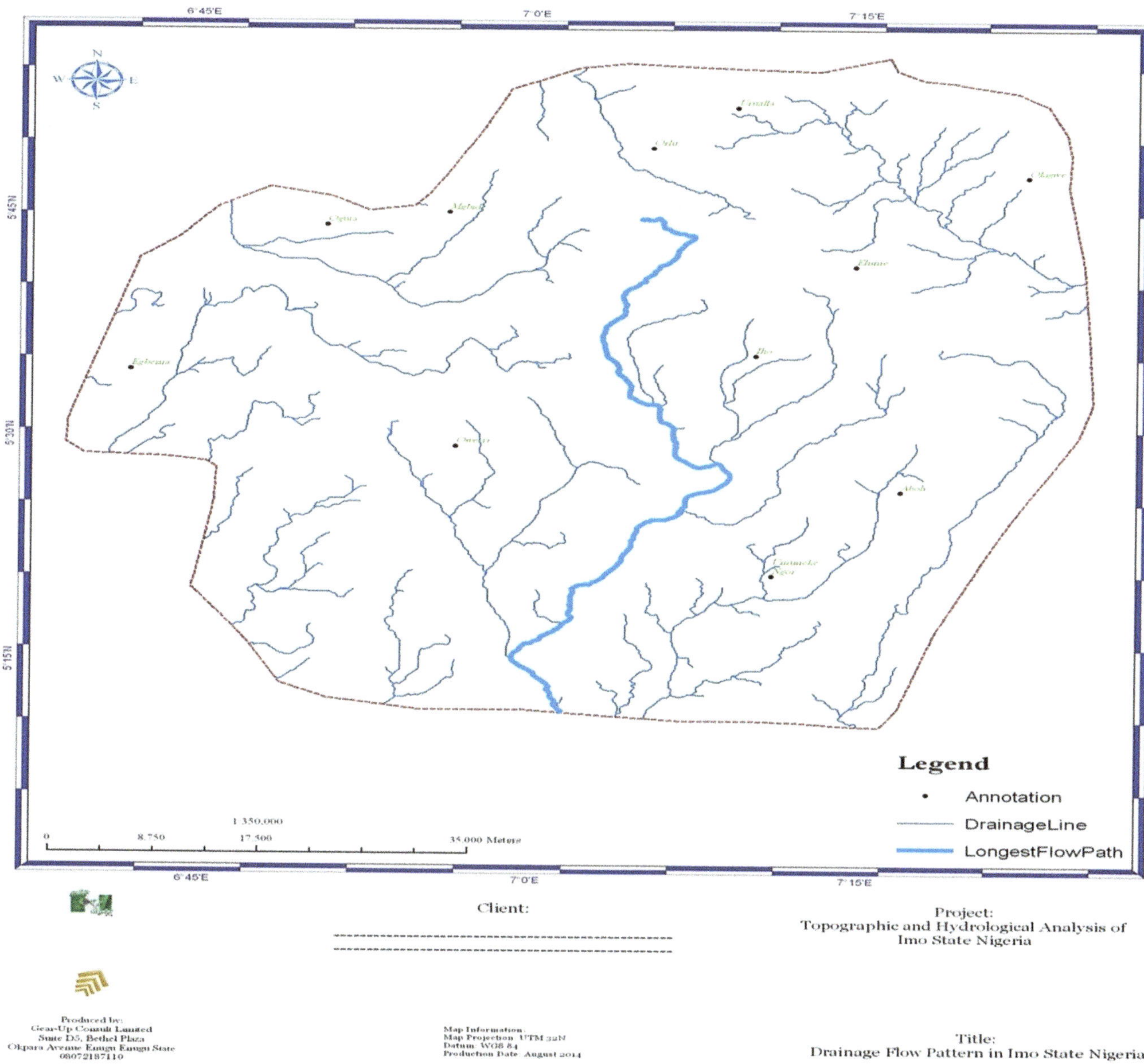


***Fig. 8.*** *Drainage Model of Imo State, Nigeria.*

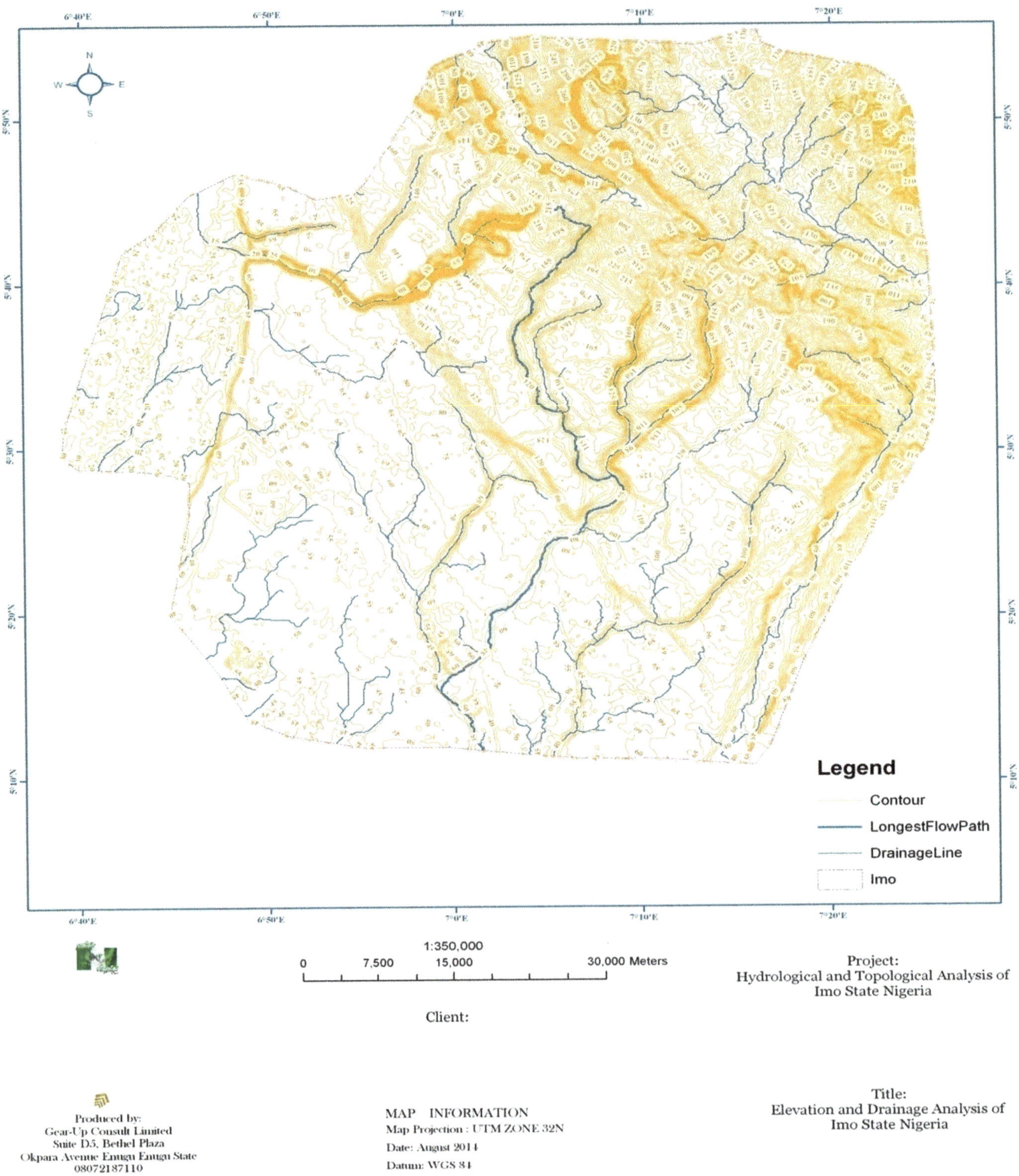


***Fig. 9.** Elevation and Drainage overlay model ofImo State.*

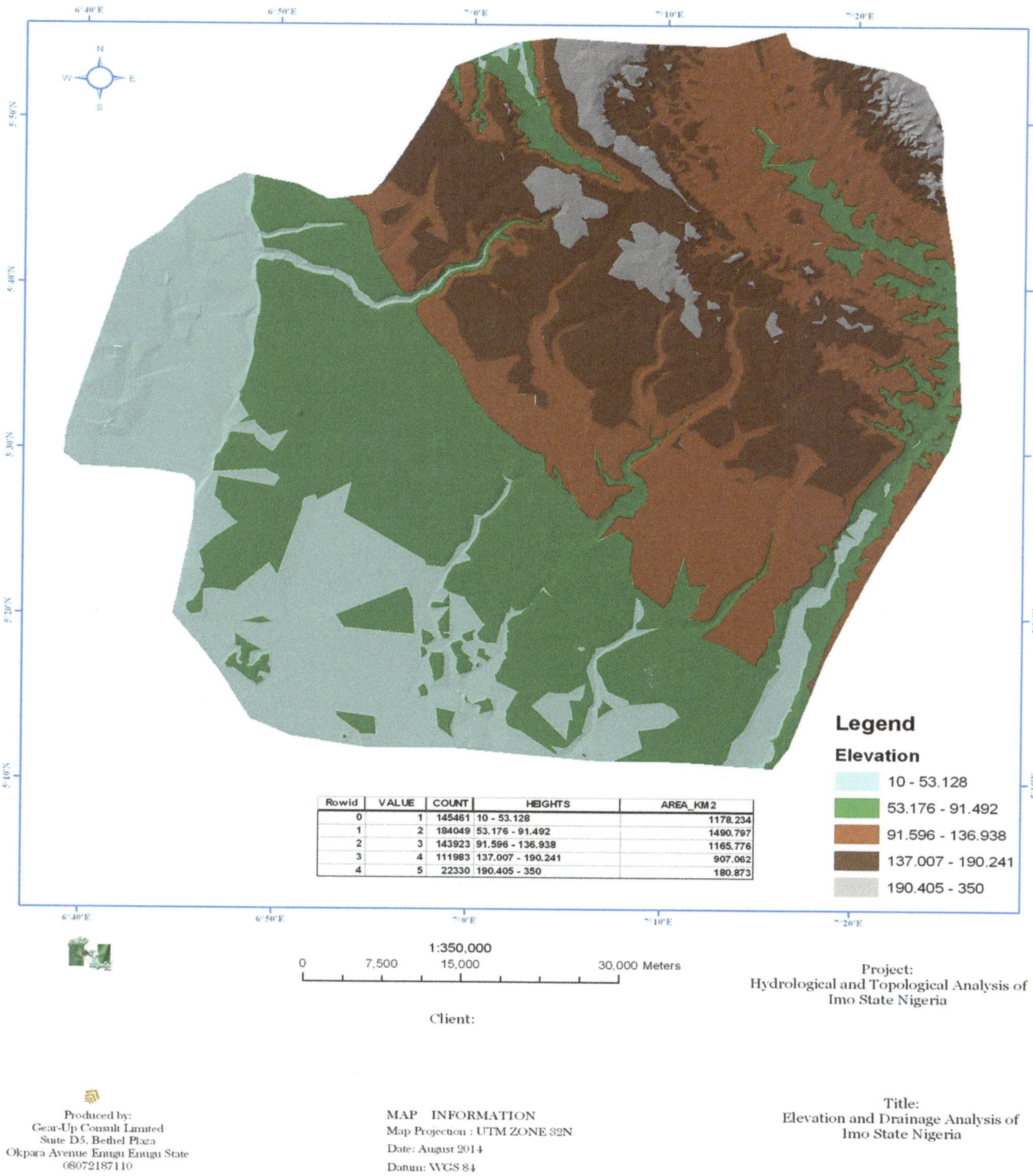


| Rowid | VALUE | COUNT | HEIGHTS | AREA_KM2 |
|---|---|---|---|---|
| 0 | 1 | 145461 | 10 - 53.128 | 1178.234 |
| 1 | 2 | 184049 | 53.176 - 91.492 | 1490.797 |
| 2 | 3 | 143923 | 91.596 - 136.938 | 1165.776 |
| 3 | 4 | 111983 | 137.007 - 190.241 | 907.062 |
| 4 | 5 | 22330 | 190.405 - 350 | 180.873 |

***Fig. 10.*** *Elevation Model/Drainage Overlay analysis of Imo State, Nigeria.*

Results of the elevation and drainage analysis (Fig. 8, 9 and 10) follow the previous trend depicting that 78% of the entire land area representing 3833 $Km^2$ fall within less to moderately vulnerable erodibility class.

On sloping lands, more than one-half of the soil particles that are dislodged by raindrops during rainfall are carried downhill. Erosion increases dramatically because the increased angle facilitates water flow and soil movement. It is not surprising, therefore, that areas on the highest elevations suffer some of the state's highest erosion rates.

Data for assessment of the effect of slope gradient and length on soil erosion is limited. However, it is generally accepted that an increase in slope and slope length will increase erosion because they lead to an increase in overland flow volume and velocity. Runoff on low slopes flows slowly and quickly forms a water layer deep enough to act as surface mulch. Increasing slope length enhances soil loss as more runoff can accumulate on long slopes. Thomas (1991) identified that slope shape together with ground/field attributes exercise a strong influence on the nature and extent of visible erosion damage.

On steep slopes, soils are generally shallower and their nutrient and water storage capacities are limited. Thus, soils in these areas, when exposed to soil eroding agents, face greater degradation consequences compared to soils in flat areas. Since most of the terrain of the highlands of the region is undulating and hilly, most agricultural land is situated on sloping ground. Increasing population has resulted in an increasing demand for cultivable land which has increasingly moved on to steeper slopes previously covered (through cutting and burning) by forests.

## 4. Conclusions

Results show that most of the soils in the study (79.9 %) are classified as highly to moderately vulnerable to erosion based on the RUSLE soil erodibility classification. This means that the combined effect of cultivation (33%), light vegetation (37.2%) and human settlements (9.1%) combine to expose the soils to greater risk of gully erosion. Thus 80% of the soil of Imo State will be classified as highly vulnerable to erosion. However, the total percentage area under thick vegetation is 18.3% and this area fall under slightly vulnerable classification. The natural vegetation protects the soil against the impacts of rainfall and it is a source of organic matter to the soil. These factors improve infiltration and enhance the recharging of groundwater reservoirs. When vegetation cover is displaced, infiltration capacity is decreased, resulting in surface runoff, which will carry sediments and nutrients into rivers (Van Oost et al., 2000; Zuazo and Pleguezuelo, 2008). Moreover, changes in precipitation volume and intensity caused by climate changes may increase the energy available in rainfall for detaching and carrying sediments. According to Yang et al. (2003), the global average soil erosion is projected to increase approximately 9% by 2090 due to climate changes.

Although soil erosion is a natural and inevitable process, the accelerated rates of soil loss, caused by the factors mentioned above, represent a serious environmental problem. For instance, increased rates of soil erosion are directly associated with nutrient loss, which may reduce agricultural productivity (Bakker et al., 2007) and cause water bodies' eutrophication (Istvánovics, 2009). In some cases, advanced stages of soil erosion, such as rill and gully erosions, can devastate entire areas, turning them unusable for agricultural purposes (Valentin et al., 2005; Kirkby and Bracken, 2009).

## References

[1] Asiabaka, C. C., & Boers, T. M. (1988). An Analysis of existing traditional methods of Farming and Erosion Control among Farmers in South Eastern Nigeria. In *Proceeding of First International Symposium on Erosion in South Eastern Nigeria* (Vol. 1, No. 1, pp. 109-114).

[2] Bakker, M. M., Govers, G., Jones, R. A., Rounsevell, M. D. A., 2007. The effect of soilErosion on Europe's crop yields. Ecosystems 10, 1209–1219.

[3] Dane, J. H., & Hopmans, J. W. (2002). Water retention and storage. *Methods of soil analysis. Part, 4*, 671-717.

[4] Egboka, B. C. E. (2004). Distress Call and Plea to the Senate Committee for Urgent Actions Against Floods, Soil/Gully Erosion/ Landslides Disasters in the Southeast. Paper Presented to Senate Committee on Environment: Roads/Erosion Senate Delegation to the Southeast. 30p.

[5] Gee GW and Bauder JW1986 Particle size analysis. *In* Methods of Soil Analysis. Part I. Agronomy 9. Ed. A. Klute. pp. 383–411.Am. Soc. Agron., Madison, WI. USA.

[6] Igbokwe, J. I., Akinyede, J. O., Dang, B., Alaga, T., Ono, M. N., Nnodu, V. C., & Anike, L. O. (2008). Mapping and monitoring of the impact of gully erosion in Southeastern Nigeria with satellite remote sensing and Geographic Information System. *The International Archives of the Photogrammetry, Remote Sensing and Spatial Information Sciences*, *37*, B8.

[7] Istvánovics, V. 2009. Eutrophication of lakes and reservoirs, p. 157–165. *In*: G.E. Likens [ed.], Encyclopedia of Inland Waters, vol. 1. Elsevier, Oxford.

[8] Kirkby, M. J., & Bracken, L. J. (2009). Gully processes and gully dynamics. *Earth Surface Processes and Landforms*, *34*(14), 1841-1851.

[9] Middleton, H. E. 1930. Properties of soil which influence soil erosion. U. S. Dept Agronomy Tech. Bull. 178pp.

[10] Mulla, D. J., & McBratney, A. B. (2010). 9 Soil Spatial Variability. *Soil physics companion*, 343.

[11] Nill D., Schwertmann U., Sabel-Koschella U., Berhard M. and Breuer J. 1996. *Soil erosion by water in Africa. Principles, prediction & protection.* GTZ, Germany. 292 pp.

[12] Ofomata, G. E. K. (1985). Soil Erosion in Nigeria: The Views of a Geomorphologist. 7th Inaugural Lecture. *University of Nigeria, Nsukka, Nigeria. pp.*

[13] Ogbonna, J. U. (2012). Understanding Gully Erosion Vulnerability in Old Imo State Using Geographic Information System and Geostatistics. *American Journal of Geographic Information System*, *1*(3), 66-71.

[14] Okoro, B. C., Uzoukwu, R. A., & Chimezie, N. M. (2014). River Basins of Imo State for Sustainable Water Resources Management. *Journal of Civil & Environmental Engineering* 4,130-138.Pimentel, D., Harvey, C., Resosudarmo, P., Sinclair, K., Kurz, D., McNair, M & Blair, R. (1998). Land use, erosion and water resources. *Water resources, environmental planning, management, and development. Tata McGraw-Hill Publishing Limited, New Delhi, India*, 37-71.

[15] Shoghi, H, Ghazavi M, Kazemain, S, Moayedi H (2013) A state of the art review of dispersive soils. Identification methods perspectives. European Journal of Scientific Research107: 322-328.

[16] Thomas, T. 1991. *Aspects of soil degradation and conservation measures in Agucho catchment, West Hararghe.* Soil Conservation Research Project Report 19. University of Bern, Switzerland. 125 pp.

[17] Udo, R. K. (1970). *Geographical regions of Nigeria.* Univ of California Press.

[18] Valentin, C., Poesen, J., & Li, Y. (2005). Gully erosion: impacts, factors and control. *Catena*, *63*(2), 132-153.

[19] Van Oost, K., Govers, G., & Desmet, P. (2000). Evaluating the effects of changes in landscape structure on soil erosion by water and tillage. *Landscape ecology*, *15*(6), 577-589.

[20] Wischmeier, W. H., & Smith, D. D. (1978). Predicting rainfall erosion losses-A guide to conservation planning. *Predicting rainfall erosion losses-A guide to conservation planning*.

[21] Yang, D., Kanae, S., Oki, T., Koike, T., Musiake, K., 2003. Global potential soil erosion with reference to land use and climate changes. Hydrological Processes 17, 2913–2928.

[22] Zuazo, V. H. D., & Pleguezuelo, C. R. R. (2009). Soil-erosion and runoff prevention by plant covers: a review. In *Sustainable Agriculture* (pp. 785-811). Springer Netherlands.

9

# Fluoride Concentration in Ground Water Around an Aluminium Smelter Plant (A Case Study) in Angul–Talcher Industrial Belt, Odisha, India

**Allian Majhi[1, *], Subhra Keshari Biswal[1], Kedar Mohapatra[2]**

[1]Department of Chemistry, Indira Gandhi Institute of Technology, Denkanal, Odisha, India
[2]Department of Chemistry, Gandhi Institute of Technological Advancement, Bhubaneswar, Odisha, India

**Email address:**
allianmajhi@gmail.com (A. Majhi), skbiswal.chemistry1@gmail.com (S. K. Biswal), drkemdar.mohapatra@gmail.com (K. Mohapatra)

**Abstract:** Fluoride is the naturally-occurring stable form of the gaseous element fluorine (F). Fluoride is among the top 15 most abundant components on the Earth's crust [21] and is naturally found in very small amounts in most aquifers in India [23]. An aquifer is an underground unit of saturated earth materials that can provide usable quantities of ground water to a well [5]. This paper gives the result of fluoride measurement in ground water around a major Aluminium smelter plant in India. The concentration of different water quality parameters like pH, electrical conductivity, total alkalinity, bicarbonate, chloride, nitrate, sulfate, total hardness, total dissolved salts, calcium and magnesium are estimated along with fluoride in ground water. Over 216 water samples are taken for analysis for a period of 12 months. The paper also reports the result of correlation coefficient between different water quality parameters.

**Keywords:** Fluoride Concentration, Al-smelter Plant, Ground Water, Physicochemical Parameters, Correlation

## 1. Introduction

Fluorides are normal constituents of water, food and air. Lakes and streams generally contain less than 0.3mg/l of fluoride [16]. Fluorides content in ground water of some areas are higher and depend on the type of the rock through which the water flows [13]. Intake of fluoride in low dose prevents dental caries but fluoride is a cumulative poison under conditions of continuous exposure to sub acute doses [19]. Fluoride is protoplasmic poison [7, 8]. The deleterious effect it has on human and animals are very well known. Fluorosis is a crippling disease caused by the intake of water having higher concentration of fluoride i.e. > 1.5, ppm [6]. The existence of fluoride bearing ground water and fluorosis in India have been reported by many workers. Shortt et. al published the first report on endemic fluorosis in India [18]. The disease was then known to be prevalent in four states in India i.e. Andhra Pradesh, Tamilnadu, Punjab and Uttar Pradesh. Cases of endemic fluorosis have been reported in certain parts of Rajasthan [15, 22, 11, 12, 2, and 3]. No such systematic work has been reported so far regarding the ground water quality of Angul –Talcher industrial belt in the state of Odisha. Besides this literature regarding the fluoride distribution in ground water around an Aluminium smelter plant & related case of fluorosis are scanty [11].

Angul- Talcher Industrial area is in the state of Odisha. This industrial township is reputed due to its vast coal mines, three Thermal power plant (two are owned by NTPC and one Captive Power plant is owned by NALCO), one Aluminum smelter plant (NALCO), some private sector-Ferro-alloy industries & a chemical manufacturing industry. This town is located at about 20°29'N latitude and $34^0$16'E longitude at an elevation of about 73.15 m above mean sea level. Geologically this region is composed of ancient metamorphic rocks like Khondalites, quartzite with charnockite and gneisses forming flanks and valleys. The area under study is mainly covered with Khondalites rock type. Beside coal, the mineral resources comprise of iron, fine clay, chromites, clay, garnet, gold, graphite, kyanite and silimanite, lime stone, manganese ore etc [9]. This industrial belt is flanked with lot of village surroundings and within the periphery of industries. The area under studies is only limited to localities near to aluminium smelter plant producing 650 tons of aluminium per day by Hall-Heroult's process [10].

The effluent carrying drain from the plant is mixed with a natural stream known as sindajhar which passes through number of villages. Water samples are taken from dug well, tube well and ponds of different villages adjacent to the stream and near to the plant. Fluoride distribution with distance from the aluminium smelter plant is taken into consideration.

## 2. Materials and Methods

After studying the nature of effluent emission from the aluminium smelter plant sampling spots in different villages (1-9) have been chosen adjacent to the stream. Beside this another nine sampling spots (10-18) have been selected near to the plant. All the eighteen sampling spot remained within a radius of 6.15km from this smelter plant (Fig-1). All together 216 water samples have been taken for analysis for a period from March-2014 to February-2015.Samples were taken in pre cleaned polyethylene bottle. Fluoride concentration was measured using Orion ion AnalyserEA-940 with chloride ion selective electrode. pH and electrical conductivity were measured using pH –meter and conductivity bridge(both Systronic make). Chloride, total hardness, total alkalinity, bicarbonate, nitrate, and sulfate were estimated titrometrically as per standard procedure as suggested by APHA [20]. Ca and Mg were determined using AAS (Perkin Elmer (303)). The average value of the analysis result for different parameters has been tabulated in Table-1.

***Table 1.*** *Average Hydro chemical data of Ground water around Aluminium smelter plant.*

| Sampling Station | F | pH | TA | $HCO_3^-$ | $Cl^-$ | $NO_3^-$ | $SO_4^{2-}$ | TH | TDS | EC | Ca | Mg |
|---|---|---|---|---|---|---|---|---|---|---|---|---|
| Kulad Pond – 1 | 2.14 | 7.24 | 208 | 239 | 54 | 42.8 | 31.7 | 310 | 360 | 520 | 30 | 18.7 |
| Kulad Pond – 2 | 1.74 | 7.0 | 164 | 189 | 48 | 30.4 | 26.2 | 210 | 322 | 512 | 36 | 22.4 |
| Kulad Borewell | 1.26 | 7.1 | 172 | 212 | 52.4 | 12.6 | 100 | 248 | 580 | 780 | 46 | 22.8 |
| Kulad openwell | 1.82 | 7.3 | 197 | 234 | 62.2 | 22.8 | 62.4 | 308 | 450 | 604 | 40 | 20.4 |
| Bonda Pond | 2.81 | 7.4 | 168 | 197 | 56.2 | 34.6 | 32.6 | 204 | 410 | 590 | 44.6 | 24.1 |
| Bonda well | 0.77 | 7.4 | 184 | 208 | 48 | 20.2 | 25.4 | 178 | 378 | 498 | 41.0 | 18.0 |
| Bondapond II | 3.31 | 7.38 | 276 | 314 | 188 | 12.8 | 44.6 | 140 | 644 | 920 | 35 | 18.9 |
| Bonda tubewell | 1.21 | 7.7 | 278 | 296 | 67.8 | 11.4 | 126 | 152 | 478 | 610 | 40.6 | 30 |
| Chowadhia Pond | 0.76 | 6.8 | 158 | 182 | 42.6 | 26.6 | 22.4 | 190 | 404 | 480 | 48.2 | 36.6 |
| Banarpalpond – I | 0.71 | 7.2 | 164 | 190 | 54.1 | 16.8 | 28.6 | 194 | 510 | 622 | 54.6 | 32 |
| Banarpalpond – II | 0.70 | 7.3 | 178 | 210 | 56.4 | 18.4 | 32.4 | 198 | 540 | 642 | 58.3 | 36.2 |
| Kuruduluwell | 0.73 | 7.7 | 192 | 246 | 64 | 14.1 | 38.7 | 124 | 442 | 640 | 42.4 | 28.8 |
| Gotamarawell | 0.71 | 7.39 | 70 | 85 | 26 | 12.6 | 12.8 | 150 | 135 | 194 | 30 | 18.7 |
| Langulia Bore well | 0.51 | 6.9 | 110 | 92 | 64.6 | 10.4 | 26.6 | 220 | 398 | 460 | 38 | 24 |
| Tulasipal well | 0.64 | 7.22 | 162 | 197 | 24 | 42.6 | 9.4 | 175 | 238 | 340 | 30 | 18.7 |
| Kandasar well | 0.72 | 7.1 | 160 | 184 | 26.4 | 32.4 | 21.8 | 186 | 324 | 440 | 48.2 | 32.4 |
| Samantapur Pond | 0.86 | 7.2 | 170 | 198 | 24.2 | 16.4 | 89 | 146 | 270 | 320 | 45.4 | 25.2 |
| Balaramprasad well | 0.50 | 7.3 | 172 | 186 | 34.4 | 18.2 | 72 | 148 | 470 | 594 | 42 | 24.6 |

*Except pH and conductivity (in ms) the concentration values of the parameters are in mg/l.

***Table 2.*** *Values of correlation Coefficient among the Different water quality parameters of Ground water around aluminium smelter plant.*

| Param- ters | pH | TA | $HCO_3^-$ | $Cl^-$ | $NO_3^-$ | $SO_4^{2-}$ | TH | TDS | EC | Ca | Mg |
|---|---|---|---|---|---|---|---|---|---|---|---|
| $F^-$ | 0.171 | 0.529 | 0.539 | 0.688 | 0.280 | 0.073 | 0.221 | 0.348 | 0.525 | -0.284 | -0.411 |
| pH | | 0.474 | 0.513 | 0.227 | -0.021 | 0.329 | -0.329 | 0.136 | 0.216 | -0.120 | -0.141 |
| TA | | | 0.972 | 0.635 | 0.161 | 0.521 | -0.033 | 0.627 | 0.690 | 0.019 | 0.021 |
| $HCO_3^-$ | | | | 0.594 | 0.159 | 0.470 | -0.027 | 0.604 | 0.695 | 0.044 | 0.014 |
| $Cl^-$ | | | | | -0.012 | 0.104 | -0.091 | 0.675 | 0.753 | -0.117 | -0.179 |
| $NO_3^-$ | | | | | | -0.104 | 0.555 | -0.101 | -0.036 | -0.368 | -0.273 |
| $SO_4^{2-}$ | | | | | | | -0.040 | 0.404 | 0.357 | -0.153 | 0.052 |
| TH. | | | | | | | | 0.085 | 0.098 | -0.086 | -0.022 |
| TDS | | | | | | | | | 0.959 | 0.461 | 0.267 |
| EC | | | | | | | | | | 0.290 | 0.097 |
| Ca | | | | | | | | | | | 0.834 |

Correlation studies were done among various analyzed parameters using standard statistical package. The correlation matrix is given in table-2.

## 3. Results and Discussion

The fluoride concentration in ground water from tube well, dug well, open pond well area is found to be ranging from a minimum of 0.42 to a maximum of 4.3 mg/l near Bonda sampling station with an average value of 0.50 to a maximum of 3.31mg/l. Out of 216 samples analyzed for fluoride concentration, 52 samples were found to be more than 1.5 mg/l as prescribed by ICMR [14]. The highest fluoride concentration was found in the village Bonda adjacent to the stream (i.e sampling point No-5 and 7). Comparing with first nine sampling station adjacent to the effluent carrying drain, other nine sampling station have a fluoride concentration within desirable limit prescribed by Bureau of Indian Standard [1] and WHO [25] for drinking water.

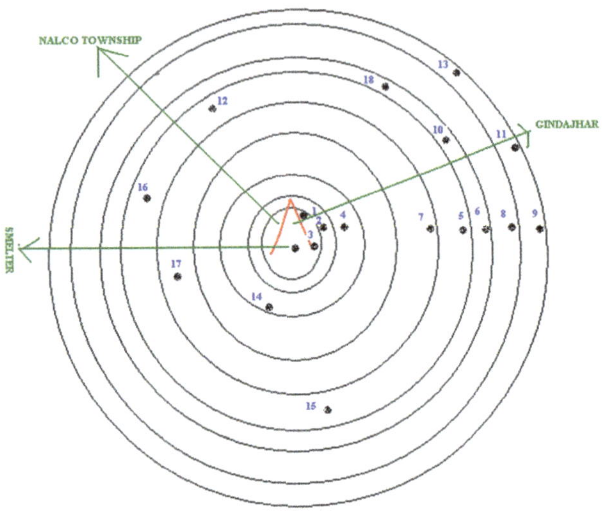


***Figure 1.*** *Sampling site.*

***Table 3.*** *Sampling Points.*

| Sl. No | Sampling point | Sl. No | Sampling point |
|---|---|---|---|
| 1 | Open pond (kulad village) | 10 | Open pond (banarpal village-1) |
| 2 | Open pond (kulad village) | 11 | Open pond (banarpal village-2) |
| 3 | Bore well (kulad village) | 12 | Open well (kurudulu village) |
| 4 | Open well (kulad village) | 13 | Open well (gotamara village) |
| 5 | Open pond(bonda village) | 14 | Open well (langulia beda village) |
| 6 | Open well (bonda village) | 15 | Open well (tulasipal village) |
| 7 | Open pond (Rly.culvert) | 16 | Open well (kandsar village) |
| 8 | Tube well (banda village) | 17 | Open pond (samanthpur village) |
| 9 | Open pond (chowadhia village) | 18 | Open well (balaramprasad) |

To observe the relationship of fluoride with other constituent of water, correlation studies were done and correlation matrix was given in Table-2.

The result revealed that the pH ranged from 6.8 to 7.7. Minimum pH 6.8 was observed in Chowadhiapond and maximum pH 7.7 in Bonda tube well and Kurudulu open well. A positive correlation (r=0.171) was observed between $F^-$ and pH. The relation between fluoride ion and pH is shown in figure 2. According to WHO 1996, the permissible limit of pH is 6.9 to 9.2.A positive correlation as observed between Fluoride and pH (r =0.171).

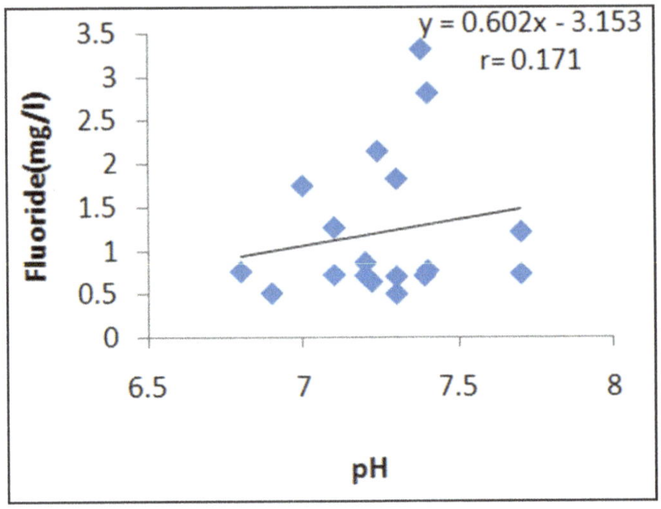


***Figure 2.*** *Relationship between Fluoride and pH.*

Total alkalinity ranged from 70 mg/l to 278 mg/l was observed in the study area. In the present studies, the total alkalinity showed a positive correlation with Fluoride (r = 0.529). This is presented in figure 3.The positive correlation has been reported earlier [24]. The solubility of $CaF_2$ increases with the increase in TA in the ground waters. According to the following reactions:

$$CaF_2 + CO_3^{2-} \longrightarrow CaCO_3 + 2F^-$$

$$CaF_2 + HCO_3^{2-} \longrightarrow CaCO_3 + 2F^- + H_2O + CO_2$$

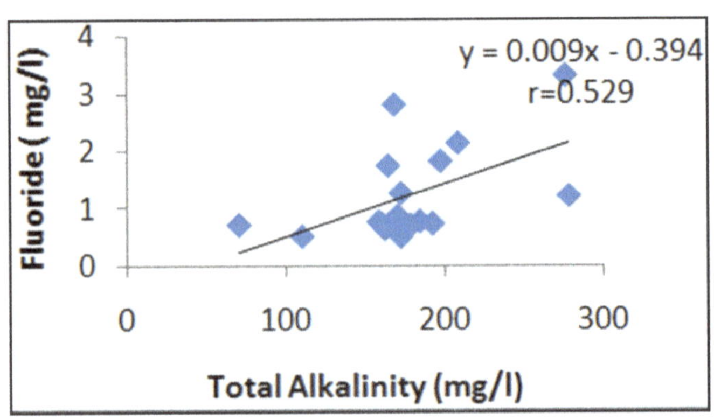


***Figure 3.*** *Relationship between Fluoride and Total alkalinity.*

Total hardness of minimum 124 mg/l and maximum of 310 mg/l were reported in the study area. WHO recommended the safe permissible limit for hardness i.e. 100-500 mg/l. In groundwater, hardness is formed mainly due to carbonate, bicarbonate and chloride of $Ca^{2+}$ and $Mg^{2+}$. All the water samples collected the TH was within the optimum limit. Calcium hardness ranged from 30 mg/l to 58.3 mg/l and the magnesium hardness ranged from 18.0 mg/l to 32.4 mg/l. Chloride content of ground water samples are 24mg/l ranging upto 188mg/l. Fluoride showed a positive correlation with TH (r = 0.221) which is shown in figure 4.

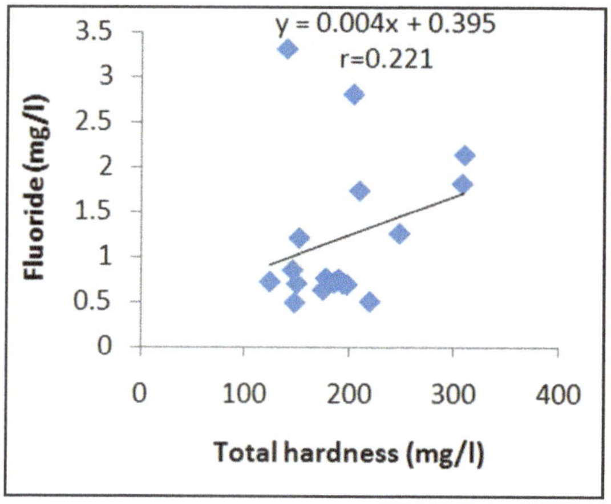


***Figure 4.*** *Relationship between Fluoride and Total hardness.*

A negative correlation was observed between fluoride and calcium (r=-0.284), fluoride and magnesium(r=-0.411). Fluoride was not found related with the depth of water source. It may be due to (i)a negative correlation was observed between fluoride and calcium (ii) fluorine does not occur in free state in nature and it forms fluoride mainly with calcium as fluoropatite [$CaF_2.CaCO_3(PO_4)$], fluespar ($CaF_2$) and apatite($Ca_2F$). The low concentration level of calcium in ground water of this area signifies the above fact that the ground water is not affected by fluoride ore which may be originated from igneous rocks or sedimentary rocks derived from the weathering of igneous rocks [4]. It may be said that the concentration of fluoride in ground water of different villages adjacent to the streams may be due to the effluent of aluminium smelter plant.

An overall correlation analysis is almost a positive correlation between the concentration of fluoride ion with pH, total alkalinity, bicarbonate, chloride, nitrate, sulfate, total hardness, TDS, EC. However $Ca^{2+}$ and $Mg^{2+}$ correlated negatively with fluoride concentration. Out of several parameters only a few were considered for linear regression analysis. The pairs are (a) Fluoride and pH (b) Fluoride and Total Alkalinity (c) Fluoride and Total Hardness. The developed linear relationships fluoride with pH,Total Alkalinity and Total Hardness are given in the equation:

(a) $F = (0.0602*pH) - 3.153$
(b) $F= (0.009*TA) - 0.394$
(c) $F= (0.004* TH) + 0.395$

## 4. Conclusion

The aluminium smelter plant was commissioned in the year 1987. Within the gap of so many years of production time the highest fluoride concentration in the ground water has been reached to a maximum of 4.13mg/l in nearby localities. Moreover many cases of dental and skeletal fluorosis have already been observed among the men and animals, so necessary treatment methods are to be adopted to check high fluoride concentration in the effluent drain of aluminium smelter plant. It may also cause harm to the ecosystem and vegetation, if used for irrigation. So, a ground water management program is suggested for safe and healthy ecosystem.

## Acknowledgements

The authors are grateful Department of Chemistry, Indira Gandhi Institute of Technology, Sarang, Dhenkanal, for extending kind co-ordination in sampling and analysis of water samples.

## References

[1] BIS, Indian Standard Specification for Drinking Water Quality, IS – 10500, 1983.

[2] Choubisa, S. L. et. al., "Fluoride content in domestic water sources of Dungarpur Dist. Of Rajasthan". Indian. J. Environ. Health. 37 (3), 154 – 160, 1995.

[3] Choubisa, S.L. et. al., "Prevalence of flurosis in some villages of Dungarpur Dist. Of Rajasthan". Indian. J. Environ. Health. 38 (2), 119-126, 1995.

[4] Dean, H. T. and Elove. E., "Facts about flourosis", Engineering News Record, 120, April, 1939.

[5] Epa.ohio.gov.gwqcp.fluoride_factsheet, Fluoride in Ohio's Ground water, Fact Sheet 2012-01, Series on Ohio's Ground water Quality, March 2012.

[6] Featherstone J. D., "Dental caries: a dynamic disease process". Aust. Dent. J 5, 2008.

[7] Floyd DeEds, Fluorine in relation to bone and tooth development, The journal of the American Dental Association(JADA), Volume 23, 563, April 1936.

[8] "Fluorides are general protoplasmic poisions" The Journal of the American Medical Association (JAMA), Editorial, Sept 18, 1943.

[9] Geology and mineral Recourses of Odisha-Odisha Minerals, orissaminerals.gov.in.

[10] Gorai Bipra, Jana R.K. and Premchand, Reduction of Emission from Aluminium Indusrtries and Cleaner Technology, Enviornmental & Waste Mangement, 129-138, eprints.nmlindia.org.

[11] Gupta, S. C et. al. "The character of ground water in Nagpur dist, Rajasthan" Indian. J. Environ. Health. 33, 341-348, 1991.

[12] Gupta. S. C. Rathore, G. S. and Doshi, C. S., "Fluoride distribution in ground water of south eastern Rajasthan", Indian. J. Environ. Health. 35 (2) 97 – 109, 1993.

[13] Gupta S, Banerjee S, Saha R, Datta JK, Mondal N Fluoride geochemistry of groundwater in nalhati-1 block of the birbhum district, WEST BENGAL, INDIA, Research report Fluoride 39 (4) 318–320, October-December 2006.

[14] Kumar Manoj and Puri Avinash, A review of permissible limit of drinking water, Indian Journal of Occupational and Environmental Medicine Medknow publications, Jan-Aprl, 16(1), 40-44, 2012.

[15] Kumaran. P, Bhargava G. N and Bhakuni T. S, Fluorides in groundwater and endemic fluorosis in Rajasthan, Indian Journal of Environmental Health, 13, pp. 316-324. 1971.

[16] Meenakshi, Maheshwari RC, Fluoride in Drinking Water and Its Removal, Journal of Hazard Mater, 137(1): 456-63, September 1, 2006.

[17] Mishra. P. C, Meher Kumarmani, Bhosagar Dullav and Pradhan K., Fluoride distribution in different environmental segments at Hirakud Orissa (India), African Journal of Environmental Science and Technology Vol. 3 (9), pp. 260-264, September, 2009.

[18] Shrott, N. E., C. G. and Raghavachari, T. N. S., Endemic flurosis in Nellore district of south India, IndiaMed. Gazatte, 72, 1937.

[19] Simpson, A., Shaw, L. and Smith, A. J. The bio-availability of fluoride from black tea. Journal of Dentistry, 29(1), 15–21. 2001.

[20] Standard methods of examination of water and waste water, American. Public. Health. Association (APHA), Washington, 1991.

[21] Stephen Peckham and Niyi Awofeso, Water Fluoridation: A Critical Review of the Physiological Effects of Ingested Fluoride as a Public Health Intervention The Scientific World Journal, Volume2014, 2014, ArticleID293019, 10pages, http://dx.doi.org/10.1155/2014/293019.

[22] Thergaonkar, v. p. and Bhargava, R. K., "Water quality and incidence of flurosis in Jhunjhunu district of Rajasthan preliminary obs", Indian. J. Environ. Health. 16 (2) 68 – 180, 1974.

[23] UNICEF, 1999. States of the Art Report on the Extent of Fluoride in Drinking Water and the Resulting Endemicity in India. Report by Fluorosis and Rural Development Foundation for UNICEF, New Delhi.

[24] Venkata Chalam Mr, and Jebanesan A, Indian journal of Environmental Protection 8 (10): 734-738, 1998.

[25] WHO, Guideline for Drinking Water Quality, vol. 2, 1984.

# Incident Monitoring of Glyphosate and Aminomethylphosphonic Acid in Natural Waters – Experimental Set-up and Validation

**Britta Jahnke**[1, *], **Bernd Niemeyer**[1, 2]

[1]Institute of Coastal Research / Molecular Recognition and Separation, Helmholtz-Zentrum Geesthacht, Geesthacht, Germany
[2]Institute of Thermodynamics, Helmut-Schmidt-University / University of the Federal Armed Forces Hamburg, Hamburg, Germany

**Email address:**
Britta.Jahnke@hzg.de (B. Jahnke), Bernd.Niemeyer@hsu-hh.de (B. Niemeyer)
[*]Corresponding author

**Abstract:** A measuring system based on sequential injection analysis (SIA) for the detection of the herbicide glyphosate and its main metabolite aminomethylphosphonic acid (AMPA) in natural waters is presented. The system is automated to enable an unattended monitoring of the analytes. Due to its compact manifold it can easily be integrated into existing observing systems or could be used on board of research vessels. Here we show the experimental setup and the results of the system's performance during experimental periods of 20 hours in the laboratory as well as in an observing station situated at the Elbe river in Hamburg, Germany. An incident with elevated glyphosate and AMPA concentrations was simulated by injecting spiked river water samples. The results show a good stability of the system over the experimental period.

**Keywords:** Glyphosate, Roundup, AMPA, Sequential Injection Analysis, SIA, Phosphororganic Analysis

## 1. Introduction

The target compound in this study is glyphosate, the active ingredient of one of the most widely applied herbicides. It was introduced as "Roundup" by Monsanto in the 1970s. 718,600 tons of glyphosate containing pesticides were employed worldwide in 2012 [1] and a continuing increase is to be expected. The herbicides are extensively used to control weeds in agriculture but also in urban areas and railway properties [2]. Glyphosate and its main metabolite aminomethylphosphonic acid (AMPA) are frequently detected in ground and surface waters despite their good adsorption properties in soil. Especially after rainfall events shortly after application glyphosate and AMPA leach into the water phase [3]. An evaluation of measurements in European surface waters between 1993 and 2009 summarized findings of glyphosate in almost one third of the samples. AMPA was even detected in half of the samples [4]. A study investigated the runoff of glyphosate and AMPA from an urban Area in Belgium. Maximum concentrations of 6.1, and 5.8 µg $L^{-1}$, respectively, were reported [5]. Glyphosate is frequently found in groundwater in agricultural areas in Denmark with maximum values of 31 µg $L^{-1}$ [6]. In the Spanish river Llobregat glyphosate concentrations between 20 and 60 µg $L^{-1}$ were found. Three days after application a maximum of 139 µg $L^{-1}$ was reached [7].

In general glyphosate is marketed as relatively nontoxic, because its mode of action is assumed to exclusively target plant organisms [8]. However, lately there have been increasing concerns about its impact on the environment due to the large quantities applied. In May 2015 the IARC (International Agency for Research on Cancer) classified glyphosate as probably carcinogenic to humans [9]. Glyphosate is currently re-evaluated as its approval at the European level expires in June 2016 [10].

Analyses of pesticides in surface waters show that the concentrations are highly dynamic and largely influenced by point sources. The few continuous measurements resulted in high hourly fluctuations [11]. A monitoring of glyphosate could give important information about the extent of its application or events that lead to an increased leaching into the water phase.

Monitoring of contaminants captures temporal events or spatial variations of concentration and thus is an important tool for the assessment of ecosystem condition. Based on this knowledge, particular problems can be addressed and the status of the ecosystem can be improved [12].

Sampling with subsequent analysis at a laboratory is most commonly used, but a trend towards on-site analyses can be observed. The motivation is a higher spatial and temporal resolution of measurement data and the avoidance of typical challenges concerning sampling and transportation of samples to the laboratory, e.g. the risk of contamination or transformation of the analyte of interest between sampling and analysis [13]. Furthermore, incident reaction times can be significantly reduced if the data are constantly reevaluated by e.g. a software.

Existing measuring systems are operated on board of research vessels or integrated in observing systems. Established observing systems operated by research institutes or public authorities offer an infrastructure with power and a central sampling device. Time and effort of analyses are drastically reduced.

In this study an automated measuring system for the detection of glyphosate and its main metabolite AMPA (aminomethylphosphonic acid) in environmental waters is developed. Its principle is based on sequential injection analysis (SIA). The method emanated from flow injection analysis (FIA) and was introduced by Ruzicka and Marshall in 1990 [14].

SIA allows the automation of wet chemical analyses. Various components, e.g. detectors or reactors can easily be integrated into the manifold and several parameters can be adapted by simply modifying the software [15]. Its plain structure and versatility make SIA suitable for use in a compact and mobile measuring system [14]. The method has good green chemical credentials with only small amounts of reagents used. Its portability makes it very suitable for on-site analysis [12].

The working principle is based on the computer-controlled interaction of a syringe pump and a rotary selection valve (Figure 1). The components are connected via flexible tubes (generally made of Teflon).

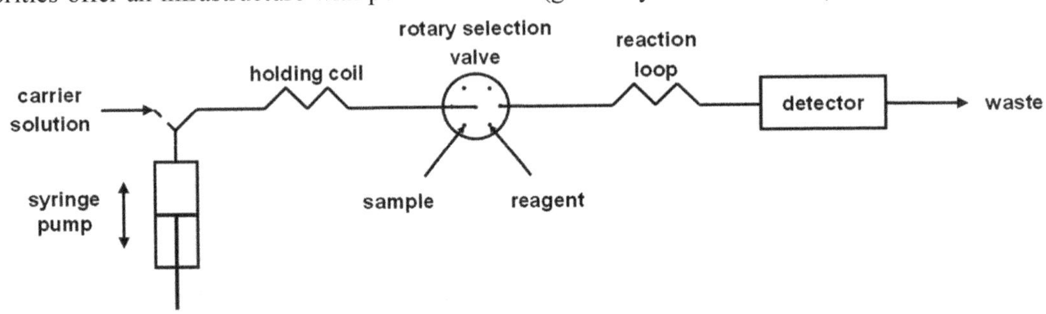


***Figure 1.*** *Scheme of sequential injection analysis (SIA).*

The central port of the rotary selection valve is alternately switched to the other ports so that sample and reagents are drawn into the system and the reaction product generates a detectable measuring signal.

Associated with water monitoring FIA and SIA methods are mainly applied for the observing of industrial sewage discharges and leachates from agriculture and landfills, for nutrient budget studies and for the data acquisition for environmental databases. Since the early 1980s FIA and later SIA systems are often deployed on board of research vessels [16]. Existing systems applied in the freshwater environment mainly concentrate on nutrients (e.g. phosphate, ammonium, nitrate), (heavy) metals (e.g. mercury, cadmium, lead, copper, iron, chromium) or different ions (e.g. chloride, fluoride) [17].

Only a few measuring systems for the monitoring of organic compounds have been described so far. Xu et al. presented a FIA method for the determination of p-aminophenol in industrial waste water with a detection limit of 19 µg/L [18]. Cetylpyridiniumchloride, a cationic tenside, was also detected by a FIA system in pond water, waste water, sediment and soil. In this system, an LOD (limit of detection) of 110 µg/L was reached [19]. Mulchandani et al. developed a FIA method for the analysis of organophosphorus pesticides. The amperometric biosensor detected paraoxon and methyl parathion in distilled water and simulated well water with LODs of 5,5 and 5,3 µg/L, respectively [20]. A flow injection method for the determination of dissolved organic carbon (DOC) in estuarine and coastal waters with an LOD of 800 µg/L was developed by Koshy et al [21].

The analysis of glyphosate and AMPA in this work is based on the o-phthaldialdehyde (OPA) derivatization of primary amines in presence of a thiol as a nucleophile to form a fluorescent product (Figure 2).

AMPA as a primary amine is directly derivatized. Glyphosate is a secondary amine and needs to be oxidized with calciumhypochlorite ($Ca(OCl)_2$) to glycine before the derivatization reaction. This detection principle is generally realized as an HPLC method with post-column derivatization [22, 23]. Its implementation in a SIA system [24, 25] or as a sequential-injection reversed-phase chromatography method [26] is also described. The OPA derivatization of glyphosate and AMPA is very sensitive and the detection is realized using an optical detector, which allows its implementation in a SIA system.

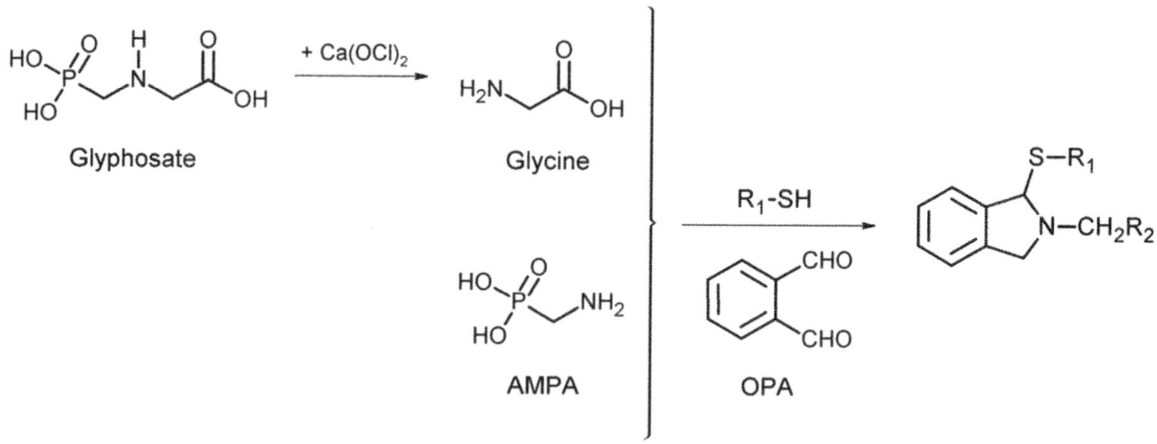


*Figure 2. Derivatization reaction: Glyphosate is oxidized by calciumhypochlorite to glycine which reacts with OPA in the presence of a thiol to form a fluorescent isoindole derivative. AMPA is directly derivatized by OPA with a thiol.*

The proposed method is considered to be a potential incident monitoring tool for the detection of glyphosate and AMPA in areas where leaching of the contaminants into the water phase is likely to occur.

## 2. Experimental

### 2.1. Chemicals and Reagents

All chemicals were of analytical grade and purchased from Sigma-Aldrich (Seelze, Germany) or Merck (Darmstadt, Germany) unless stated otherwise. All reagents were prepared with purified water obtained from a Milli-Q system (Millipore, Schwalbach, Germany) and were filtered (0.2 µm) and degassed. The carrier solution was 0.005 M $KH_2PO_4$. The pH value of the mobile phase utilized for chromatographic separation was adjusted to 2.0 with $H_3PO_4$. A stock solution of $Ca(OCl)_2$ (for synthesis) with a concentration of 2.3 g $L^{-1}$ was prepared. The stock solution was stored at -20°C. The phosphate buffer of the $Ca(OCl)_2$ working solution was prepared by dissolving 17.4 g of $K_2HPO_4$ and 0.5 g of NaCl in 50 mL purified water. The pH was adjusted to 11.0 with NaOH (30%). 11.25 mL of this buffer and 13.75 mL purified water were mixed and 164 µL of the $Ca(OCl)_2$ stock solution was added in order to obtain a final $Ca(OCl)_2$ concentration of 15 mg $L^{-1}$. For the OPA reagent a borate buffer solution was prepared by dissolving 10.25 g of $Na_2CO_3$, 3.25 g of $H_3BO_3$ and 4.75 g of $K_2SO_4$ in ca. 240 mL purified water. The pH was adjusted to 10.0 with NaOH (30%). 400 mg of o-phthaldialdehyde and 500 mg of N-acetyl-L-cysteine was dissolved in 7 mL ethanol and added to the borate buffer. 1.0 mL Brij 35 solution (10%) was added and the volume was adjusted with purified water to 250 mL. This solution was stored at room temperature in a dark glass bottle. Water samples were collected from the Elbe river in Hamburg, Germany. The samples were filtered through 0.45 µm mixed cellulose ester membranes and stored in glass bottles. Filtered river water was spiked by adding 50 µg $L^{-1}$ (laboratory study), or 300 µg $L^{-1}$ (field study) glyphosate and AMPA. The standard solution contained glyphosate and AMPA in mobile phase (pH 2.0) with a concentration of 50 µg $L^{-1}$ (laboratory study) or 500 µg $L^{-1}$ (field study) each.

### 2.2. Instrumentation

The experimental setup is depicted in Figure 2. An HPLC-pump (Knauer Smartline 100, Berlin, Germany) is connected to an ion exchange column (Hamilton PRP-X400, Reno, NV, USA) via a 6-ports/3-channel injection valve (Knauer A1370). This setup is coupled via a T-piece to a 7-port/1-channel valve (Knauer A1374) which is connected via a holding coil (PTFE tubing, 0.8 mm i.d.) to a 2.5 mL syringe pump (Hamilton PSD/4). A heated reactor (45°C) containing 1 m woven PTFE tubing (0.8 mm i.d.) in a heating block is used as reaction coil (Global FIA, Fox Island, WA, USA). All other connections are made of PTFE tubing (0.5 mm i.d.). The syringe pump and the valve are connected via USB to a personal computer and are controlled by scripts written in the python programming language according to Frank [27, 28]. An Agilent fluorescence detector (1200 Series, Boeblingen, Germany) with an 8 µL flow cell is employed in the laboratory setup (ex. 230 nm, em. 445 nm). For the field study this detector is replaced by a compact modular detector: A high power LED (ex. 365 nm, Prizmatix Mic-LED-365, Givat Shmuel, Israel) is coupled to a cuvette holder with a fluorescence cuvette (100 µL, Hellma Analytics, Müllheim, Deutschland). The cuvette holder is connected to a photomultiplier (Counting Unit C8855-01, Hamamatsu Photonics, Hamamatsu, Japan) via an optical fiber (core diameter 1.5 mm) with subsequent collimating lens (Avantes, Apeldoorn, the Netherlands) and bandpass filter (460 nm). The photomultiplier is connected via USB to the personal computer and is also controlled by scripts written in the python programming language.

The setup of the SIA system is shown in Figure 3. The derivatization and analysis of glyphosate and AMPA is carried out according to our recently published method [24]. To allow for a continuous measurement, the setup is completed by a peristaltic pump and a glyphosate standard (500 µg $L^{-1}$ in mobile phase, pH 2.0). The peristaltic pump provides for a constant flow of filtered river water through the sample loop of the injection valve. The glyphosate standard is directly connected to the rotary selection valve.

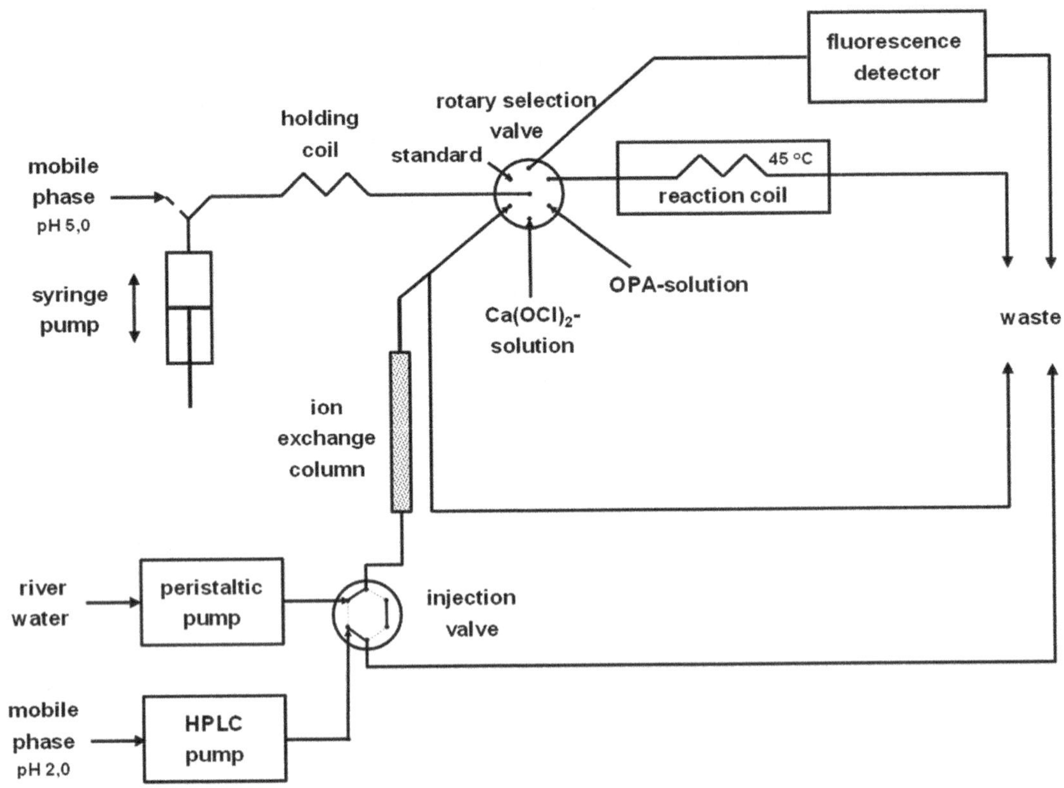


***Figure 3.*** *Scheme of the SIA system utilized for the detection of glyphosate and AMPA.*

For the field study the SIA system was integrated into the observing station Seemannshöft, operated by the Hamburg Institute for Hygiene and Environment and situated at the Elbe river in Hamburg, Germany.

Each measurement cycle consists of four analyses in succession. At first the standard is analyzed by the AMPA-program (addition of OPA-solution) in order to account for a potential drift of the detector's base line. Subsequently the same standard is analyzed by the glyphosate-program (successive addition of $Ca(OCl)_2$- and OPA-solution) to confirm the stability of the calciumhypochlorite solution. Afterwards the injection valve is switched to inject river water into the system and the sample is analyzed by the AMPA program. After completion of the AMPA program, another sample is taken by the injection valve and the glyphosate program starts. The sequence of four measurements is started automatically every 30 minutes over a time range of 20 hours.

In order to simulate an incident, spiked river water is manually introduced into the sample loop of the injection valve with the peristaltic pump temporarily separated from the valve.

# 3. Results and Discussion

### *3.1. Laboratory Conditions*

Over a period of 20 hours the base line (Standard 0) did not notably fluctuate with a standard deviation of only 1.6% (Figure 4). Also the measurement of the glyphosate standard (Standard 50) showed a standard deviation of only 1.2%. Thus both trends proved a very good stability of the system over the experimental period. No loss of activity of both OPA- as well as $Ca(OCl)_2$-solutions could be observed.

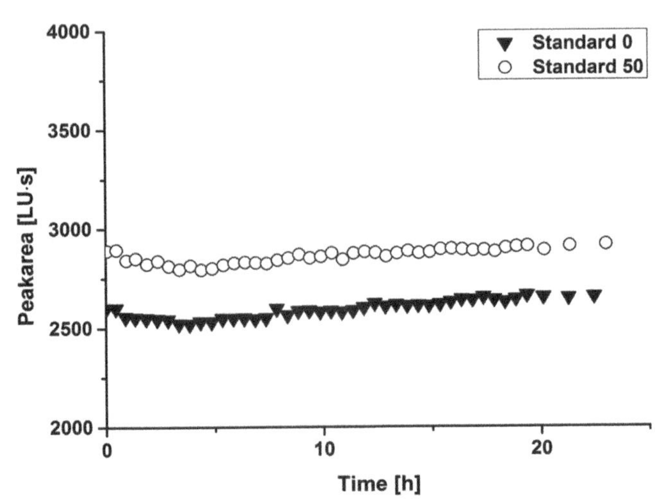


***Figure 4.*** *Analysis of standards over a time range of 20 hours (LU: luminescence units).*

The analysis of the standard was integrated in order to act as an additional parameter for the system's stability. In case of a baseline drift or a deterioration of reagents measured values could be converted based on the results of the standards. According to the data demonstrated, the frequency of the standard's analysis could be reduced in order to further minimize reagent consumption and to decrease the duration of an individual measurement.

Leaking incidents were simulated by introducing spiked glyphosate and AMPA samples into the sample loop of the injection valve. The results of the elevated glyphosate and AMPA concentrations in comparison to samples without the addition of both analytes are depicted in Figure 5.

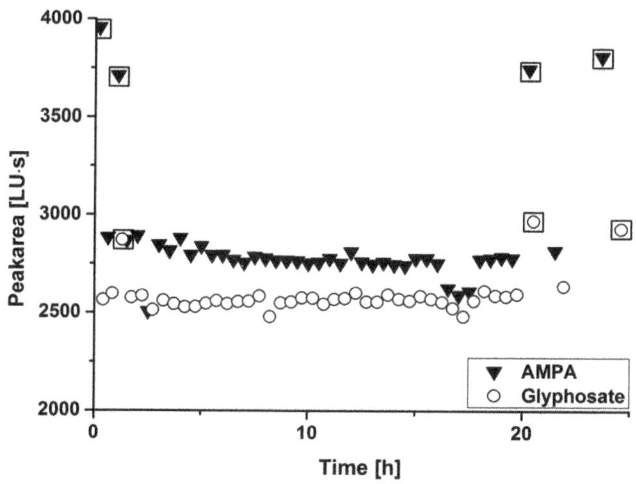


*Figure 5. Results of the laboratory on-line measurement of glyphosate and AMPA (labelled: spiked samples, 50 µg/L).*

Again the results of the on-line measurement of glyphosate and AMPA reflected a good stability of the system over a period of 20 hours with standard deviations of less than ±3.5%. The results are in good agreement with the system's typical sensitivity evaluated from the slopes of the calibration curves of 20.8 $(LU\cdot s)/(\mu g\cdot L^{-1})$ and 7.3 $(LU\cdot s)/(\mu g\cdot L^{-1})$ for AMPA and glyphosate, respectively. The limits of detection are 16 µg/L (glyphosate) and 9 µg/L (AMPA) [24].

### 3.2. Field Study

Within the observation station the SIA system could be easily connected to an available ultrasonic filter from which the peristaltic pump continuously drew filtered river water into the sample loop of the injection valve. After a few measurements the filtered river water blocked the column which was traced back to the filter's pore size of 2.0 µm. Afterwards a river water sample was filtered (0.45 µm) and stored in a glass bottle from where the peristaltic pump drew the water for the time of the further experiment.

Figures 6 and 7 show the respective results from the above described setup under field conditions. The results of the field study were comparable to those obtained in the laboratory. The standard deviations of the repeated measurements of the standard of 3.7% (standard 0) and 3.9% (standard 500) again show a good reliability of the SIA system even integrated into an observing station.

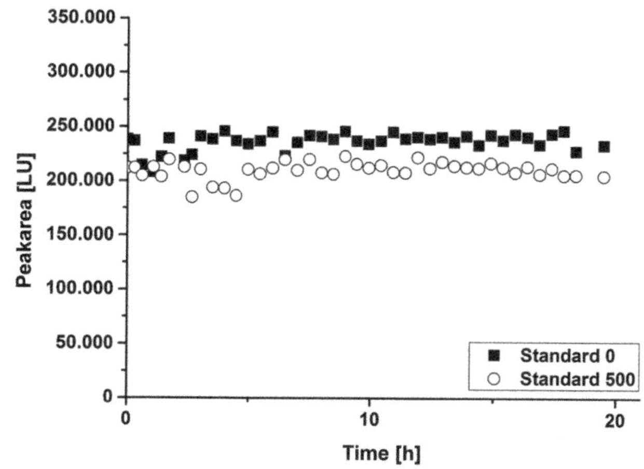


*Figure 6. Analysis of standards over a time range of 20 hours during the field study.*

Actual concentrations of glyphosate and AMPA in the Elbe river are well below the detection limits of the proposed method. However the results of the spiked water samples show that the detection of elevated glyphosate and AMPA concentrations in case of a leaking incident would be possible. The spiked water samples show higher fluorescence signals (labelled values in Figure 7) than river water without the addition of the analytes. In order to detect both analytes in environmentally relevant concentrations a preconcentration would be necessary.

Due to the setup with a constant flow of filtered river water through the sample loop of the injection valve an on-line measurement of glyphosate and AMPA could easily be implemented. The compact size of the system made the temporary integration into an existing observing station feasible.

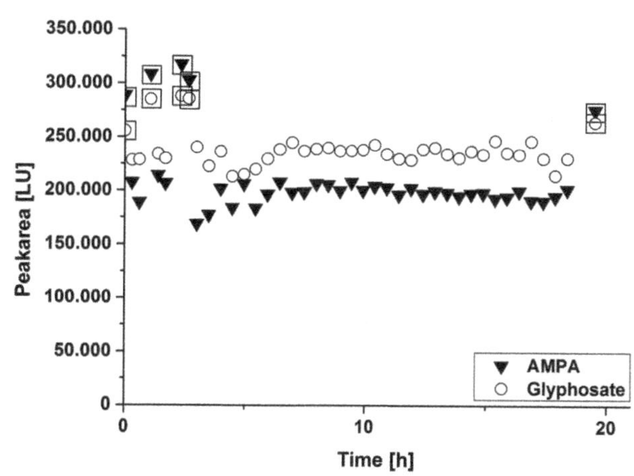


*Figure 7. Results of the on-line measurement of glyphosate and AMPA (labelled: spiked samples, 300 µg/L) during the field study.*

Still the values of the spiked samples taken at the end of the measurement period were lower than those at the beginning of the experiment. This effect could also be observed in preceding experiments in the laboratory, where an occasional manual injection of KOH (0.1 M) in order to regenerate the column resulted in an amelioration.

# 4. Conclusion

The presented SIA system was evaluated with regard to its applicability as an incident monitoring tool for the detection of glyphosate and AMPA in natural waters. The method was successfully tested over a time range of 20 hours. A potential leaking incident of glyphosate and AMPA was simulated by injecting spiked river water samples. The measurement was fully automated and stable during this period.

We showed that this system is applicable for the detection of higher glyphosate and AMPA concentrations in case of a leaking incident. In order to detect lower analyte concentrations a preconcentration of water samples would be necessary. The compact fluorescence detector needs to be optimized for further field studies in order to reach better detection limits. The integration of another syringe pump for the dosing of KOH would be necessary for prolonged experimental periods.

The compact manifold could easily be integrated into an existing observing station. Water continuously flows through the injection valve, which makes the implementation of an on-line measurement very facile. In case of a sufficient system pressure the SIA system could be linked to an existing filter without the peristaltic pump, resulting in an even smaller manifold.

# Acknowledgements

We would like to thank the Hamburg Institute for Hygiene and Environment for the possibility to test our system in their observing station Seemannshöft, Hamburg, Germany.

# References

[1] Glyphosate herbicide sales boom powers global biotech industry. Sustainable pulse. 21. 08. 2014. http://sustainablepulse.com/2014/08/21/glyphosate-sales-boom-powers-global-biotech-industry/#.VYABjmOBiSr (accessed 16.06.2015).

[2] Deutscher Bundestag. Drucksache 16/13993. Einsatz von Pestiziden auf Strecken der deutschen Bahn. 07. 09. 2009.

[3] S. Daouk, L. F. de Alencastro, H. R. Pfeifer. The herbicide glyphosate and its metabolite AMPA in the Lavaux vineyard area, western Switzerland: Proof of widespread export to surface waters. Part II: The role of infiltration and surface runoff. J Environ Sci Heal B. vol. 48, pp. 725-736, 2013.

[4] H. Horth. Monitoring results for surface water and groundwater. European glyphosate environmental information source. 2010.

[5] T. Tang, W. Boenne, N. Desmet, P. Seuntjens, J. Bronders, A. van Griensven. Quantification and characterization of glyphosate use and loss in a residential area. Sci Total Environ. vol. 517, pp. 207-214, 2015.

[6] A. E. Rosenbom, P. Olsen, F. Plauborg, R. Grant, R. K. Juhler, W. Brusch, J. Kjaer. Pesticide leaching through sandy and loamy fields - long-term lessons learnt from the Danish pesticide leaching assessment programme. Environ Pollut. vol. 201, pp. 75-90, 2015.

[7] L. Puertolas, J. Damasio, C. Barata, A. M. V. M. Soares, N. Prat. Evaluation of side-effects of glyphosate mediated control of giant reed (*Arundo donax*) on the structure and function of a nearby Mediterranean river ecosystem. Environ Res. vol. 110, pp. 556-564, 2010.

[8] K. R. Solomon, D. G. Thompson. Ecological risk assessment for aquatic organisms from over-water uses of glyphosate. J Toxicol Env Heal B. vol. 6, pp. 289-324, 2003.

[9] K. Z. Guyton, D. Loomis, Y. Grosse, F. El Ghissassi, L. Benbrahim-Tallaa, N. Guha, C. Scoccianti, H. Mattock, K. Straif. Carcinogenicity of tetrachlorvinphos, parathion, malathion, diazinon, and glyphosate. Lancet Oncol. vol. 16, pp. 490-491, 2015.

[10] Fragen und Antworten zu Glyphosat. Bundesministerium für Ernährung und Landwirtschaft. http://www.bmel.de/DE/Landwirtschaft/Pflanzenbau/Pflanzenschutz/_Texte/GlyphosatFAQ.html (accessed 28. 12. 2015).

[11] K. M. A. Holvoet, P. Seuntjens, P. A. Vanrolleghem. Monitoring and modeling pesticide fate in surface waters at the catchment scale. Ecol Model. vol. 209, pp. 53-64, 2007.

[12] I. D. McKelvie, S. D. Kolev, P. J. Worsfold. More with less: Advances in flow and paper-based monitoring of nutrients in aquatic systems. Pure Appl Chem. vol. 84, pp. 1973-1982, 2012.

[13] D. T. E. Hunt, A. L. Wilson. The chemical analysis of water. 2nd ed. Oxford: The Alden Press, 1986, pp. 92 ff.

[14] J. Ruzicka, G. D. Marshall. Sequential injection - a new concept for chemical sensors, process analysis and laboratory assays. Anal Chim Acta. vol. 237, pp. 329-343, 1990.

[15] N. Dantan. Fließinjektionsanalyse und sequentielle Injektionsanalyse für Prozessmonitoringaufgaben - Einsatzmöglichkeiten in der pharmazeutischen Industrie. Dissertation, Humboldt-Universität zu Berlin, Berlin, Germany, 2002.

[16] K. N. Andrew, N. J. Blundell, D. Price, P. J. Worsfold. Flow-injection techniques for water monitoring. Anal Chem. vol. 66, pp. 916 A-922 A, 1994.

[17] C. E. Lenehan, N. W. Barnett, S. W. Lewis. Sequential injection analysis. Analyst. vol. 127, pp. 997-1020, 2002.

[18] H. Xu, C. F. Duan, Z. F. Zhang, J. Y. Chen, C. Z. Lai, M. Lian, L. J. Liu, H. Cui. Flow injection determination of p-aminophenol at trace level using inhibited luminol-dimethylsulfoxide-NaOH-EDTA chemiluminescence. Water Res. vol. 39, pp. 396-402, 2005.

[19] K. Agrawal, G. Agnihotri, K. Shrivas, G. L. Mundhara, K. S. Patel, P. Hoffmann. Determination of cationic surfactants in environmental samples by flow injection analysis. Microchim Acta. vol. 147, pp. 273-278, 2004.

[20] P. Mulchandani, W. Chen, A. Mulchandani. Flow injection amperometric enzyme biosensor for direct determination of organophosphate nerve agents. Environ Sci Technol. vol. 35, pp. 2562-2565, 2001.

[21] K. Koshy, M. Mataki. Photochemical oxidation and flow injection conductivity determination of dissolved organic carbon in estuarine and coastal waters. Lab Robotics Automat. vol. 12, pp. 157-163, 2000.

[22] DIN 38407-22:2001-10 - Gemeinsam erfassbare Stoffgruppen (Gruppe F) Teil 22: Bestimmung von Glyphosat und Aminomethylphosphonsäure (AMPA) in Wasser durch Hochleistungs-Flüssigkeitschromatographie (HPLC). Deutsches Institut für Normung e. V.: Berlin, Germany, 2001.

[23] Method 547 - Determination of glyphosate in drinking water by direct-aqueous-injection HPLC, post-column derivatization, and fluorescence detection. U.S. Environmental Protection Agency: Cincinnati, OH, USA, 1990.

[24] B. Jahnke, C. Frank, J. F. Fernández, B. Niemeyer. A sequential injection analysis method for the determination of glyphosate and aminomethylphosphonic acid in water samples. Am Chem Sci J. vol. 5, pp. 163-173, 2015.

[25] S. D. Colombo, J. C. Masini. Developing a fluorimetric sequential injection methodology to study adsorption/desorption of glyphosate on soil and sediment samples. Microchem J. vol. 98, pp. 260-266, 2011.

[26] S. D. Colombo, J. C. Masini. A sequential-injection reversed-phase chromatography method for fluorimetric determination of glyphosate and aminomethylphosphonic acid. Anal Methods-UK. vol. 6, pp. 490-496, 2014.

[27] C. Frank, F. Schroeder. Using sequential injection analysis to improve system and data reliability of online methods: Determination of ammonium and phosphate in coastal waters. J Autom Method Manag. vol. 2007, pp. 1-6, 2007.

[28] C. Frank, F. Schroeder, R. Ebinghaus, W. Ruck. A fast sequential injection analysis system for the simultaneous determination of ammonia and phosphate. Microchim Acta. vol. 154, pp. 31-38, 2006.

# 11

# Anaerobic Sediments Decrease the Leaching of Trace Metals to Groundwater

**Amale Mcheik[1, 2, *], Mohamad Fakih[2], Hiba Noureddine[2], Hussein Trabulsi[3], Joumana Toufaily[2], Taysir Hamieh[2], Evelyne Garnier-Zarli[1], Noureddine Bousserrhine[1]**

[1]Laboratory of Biogeochemistry and Ecology of continental regions (IBIOS – BIOEMCO), Department of Biology and Environmental Sciences, Paris-Est University, Creteil, France
[2]Laboratory of Materials, Catalysis, Environment and Analytical Methods (MCEMA), Department of Chemistry, Lebanese University, Beirut, Lebanon
[3]Faculty of Economic Sciences and Business Administration, Lebanese University, Beirut, Lebanon

**Email address:**
Amale.am@hotmail.com (A. Mcheik), fakih.mohamad@gmail.com (M. Fakih), hibanoureddine@yahoo.fr (H. Noureddine), husseintrabulsi@yahoo.fr (H. Trabulsi), joumana.toufaily@ul.edu.lb (J. Toufaily), tayssir.hamieh@ul.edu.lb (H. Taysir), garnier@u-pec.fr (E. Garnier-Zarli), bousserrhine@u-pec.fr (N. Bousserrhine)

**Abstract:** The increased deterioration of water resources in Lebanon from progressive urbanization, agricultural activities and development of industries is, according to the natural authorities, a major critical problem by the year 2010. At our study site, at Al-Ghadir River, aqueous solutions containing heavy metals are extensively released from many industries directly to the river. Sediments and soil, at these sites became contaminated with these elements and their potential mobility is of particular concern since downward leaching of the heavy metals may result in the contamination of the groundwater. The objective of the present work was to investigate the bioleaching of heavy metals from the sediments of Al-Ghadir River to underground water using ex-situ column experiments. In order to conciliate the field conditions and the laboratory constraints, we have chosen to experiment the heavy metal leaching from long-term contaminated and non-destructured sediments. Sediments were incubated under anaerobic conditions and enriched with nutrients to stimulate microbial metabolism. The evolution of carbon metabolism and metals leached from the incubated sediment columns were followed over time and the effect of leaching on the distribution of metals as a function of depth was also studied. Results obtained showed that after a phase of mobilization of the heavy metals and which was enhanced by the bacterial activity, the study of the distribution profile of the heavy metals showed that they were highly readsorbed at the surface of the sediment column and their readsorption was found to decrease with depth.

**Keywords:** Heavy Metals, Bioleaching, Readsorption, Contaminated Sediment, Sediment Columns

## 1. Introduction

Lebanon depends primarily on groundwater that is deteriorating rapidly in quality through possible saline intrusions along its populous coast and pollution from many sources including progressive urbanization, agricultural and industrial activities ([1], [2], [3] and [4]).

These sources are the major contributors to the contamination of the soil and the sediments of Al-Ghadir River with heavy metals hence posing a long-term risk to the groundwater quality in the region. Depending on the environmental conditions, sediments and soils act as a sink for heavy metals, although they may act as a potential site of metal remobilization. The total content of heavy metals at these polluted sites cannot be considered an index for the estimation of the potential risk to other environmental reservoirs because this depends on the soluble fraction of each of the elements.

Historically, as an approximation of potential risk, several extraction procedures have been employed to estimate the mobile, labile or bioavailable pools of heavy metals in soil. The most important methods include: (i) batch extraction and

(ii) column leaching experiments.

The sorption capacity parameter obtained from a batch experiment is useful in providing information about effectiveness of metal-biosorbent system. However, the data obtained under batch conditions are generally not applicable to most treatment system (such as column operations) where contact time is not sufficiently long for the attainment of equilibrium [5]. Hence, there is a need to perform equilibrium studies using columns which are used to provide information about heavy metal release and transport in soil, and are also useful for testing possible soil remediation or stabilization treatments, to assess heavy metal binding and adsorption kinetics, or to study processes such as colloid-facilitated metal transport at the laboratory scale ([6], [7], [8], [9]).

A series of physico-chemical and biological processes are involved in the vertical leaching or mass flow of metals in the soil ([10], [11], and [12]). In turn, these depend on the chemical characteristics of each metal and on the soil parameters, such as pH, the carbonate content, free iron and aluminum oxide contents, the nature and the content of the clay and organic matter fractions [13] and the microbiological activity ([12], [14], [15] and [16]).

Many studies have been carried out aimed at predicting the mobility of results from contaminated soils based on the forms extractable with $CH_3$-COOH and with $CaCl_2$ and on the distribution coefficients , Kd of a given metal in a given soil or based on the forms extractable with $CaCl_2$ and with EDTA but only few studies have been done, which are more realistic talking about the mobility or leaching of heavy metals in the affected soils, such as investigations that involve the leaching with water or with a media of soil columns that simulate field leaching conditions with river water [17]. Thus, the objective of this study was, on the one hand, to determine the retention and the relative leaching of Fe, Mn, Cd, Zn, Cu, Pb and Cr from the contaminated sediments and soils using packed sediment columns and, on the other hand, to determine the efficacy of the sediment themselves at attenuating metal mobility as a function of their physicochemical and biological properties. The mobility of the elements was also studied in the affected sediment columns with an aim to determining the distribution profile of the metals by simulating their leaching in the field by rainwater. These studies can provide data at this time in order to take decisions concerning the need for the immobilization of the metals and for the selection of remediation technologies of the soils.

# 2. Materials and Methods

## *2.1. Study Area and Sampling Methodology*

The first hundred centimeters of sediments were collected from one site of Al-Ghadir River in November 2012 with an auger. This site is located at the end of the river just before its water passes in tunnels near the airport and before its release to the Mediterranean Sea. It was chosen because it was shown that it is the most polluted site chemically and biologically all along the river. Once collected, sediment samples were transferred into a 10 L plastic bottle previously soaked in 10% ultrapure HNO3 for 48h to avoid any metallic contamination then rinsed with milliQ water for 24h.

## *2.2. Experimental Set-Up*

The relative and the effect of leaching on the distribution of the elements at depth were studied under anaerobic conditions on three lab scale columns (10 cm inner diameter and 25 cm height) (Figure 1) using distilled water and 2 different media, Bromfield (amended by [18]) and a synthetic media.

Column flushed with distilled water was used only as a control for comparison purposes. Each of the other columns was operated to simulate the leaching due to the applications of the media (synthetic and Bromfield) continuously to the sediment all along the experiment. These culture media were used to simulate the bacterial growth, accelerate the evolution of sediments and study the impact of the microbial activity on the release of metals. The 2 different media (synthetic and Bromfield) were tested in separate columns. The chemical composition and the ionic strength of the synthetic and Bromfield media were chosen to be comparable to the pore water composition in the river (Synthetic media: $KH_2PO_4$, 0.5g/l; $MgSO_4.7H_2O$, 0.5g/l; $(NH_4)_2SO_4$, 1g/l; Yeast extract, 0.15g/l and Bromfield media: $KH_2PO_4$, 0.5g/l; $MgSO_4.7H_2O$, 0.5g/l; $(NH_4)_2SO_4$, 1g/l; Yeast extract, 0.15g/l; glucose, 10g/l). The pH of the media used in the different columns was 8, this being similar to that of the river water (pH 8.20).

The column consists of a closed system in which the incubation started with a known volume of substrate that the autochthonous microorganisms present in the studied sediments will continuously utilise it for their growth. Each column consists of 2 transparent reservoirs each of 25 cm height and 10 cm inner diameter and connected together by flexible tubes thus allowing a continuous circulation of the effluent along the system. The different effluents were sterilized before use by autoclave. Incubation started when 800 ml of the sterilized effluents were introduced into the column containing 500g of sediments.

Anaerobic conditions were created by degasing the system with N2 and incubations were conducted for 2 months at 28±2℃, at a flow rate of 0.12ml/s and without shaking to avoid resuspension of sediments and consequently to only consider diffusion of metals during the incubation and simulate diffusion of metals in calm and anaerobic areas. Incubations of sediments were realized in duplicates and all manipulations were done in a manner to limit the microbial contamination of the sediments and the reservoirs. These effluents were in continuous circulation between the 2 reservoirs of the column by 2 peristaltic pumps which control the flow at 0.12ml/s to maintain system under non-saturated conditions, the first placed at the end of the 1rst reservoir and the 2nd at the end of the 2nd reservoir. The effluent when entering the soil reservoir crosses a layer of 1 cm of glass beads where their role was to allow the arrival of the effluent on the whole surface of the sediments and avoid the formation of preferential paths.

Under the sediments, a layer of sand with a high permeability and of 1cm thickness was used to allow the cohesion of the base of the column thus permitting the circulation of the leachate. Under the layer of the sand, a plastic grid was used to support the sediments and the sand and prevents their passage and blockage of the pumps hence allowing the continuous circulation of the leachate. Below the grid, another layer of 1 cm of glass beads was used to allow for easier recovery of the leachate and its circulation through the pump till reaching the solution reservoir which contains only the effluent and which was equipped with a rubber cover to allow the sampling of a volume of the effluent by a sterile syringe with a continuous preservation of the anaerobic conditions of the system (Figure 1). Before use, the different parts of the column, the pumps, the glass beads, the containers used and the sand were soaked in 2% (v:v) nitric acid for 24h to eliminate traces of soluble substances and then in ultrapure water for 48h then sterilized by autoclaving and the sand was dried after sterilization.

Samples of 10 ml were taken at different time intervals to achieve chemical and biological analysis of the media. The samples were collected in sterile tubes where the pH and the redox potential (Eh) were measured directly in the leachate. The samples were then centrifuged and filtered through 0.45 μm cellulose acetate filter with 4 ml were kept at 4℃ for organic analysis and 4 ml were acidified by the addition of 2% nitric acid and stored at 4℃ for mineral analysis.

At the end of the experiment, the concentration of the total metals was detected at the different depths of the sediments in the columns. Bacterial counting was also detected at the top and the bottom of the sediments and finally a study of the bacterial diversity was done in the sediments of the different columns to be compared with the results obtained before incubation.

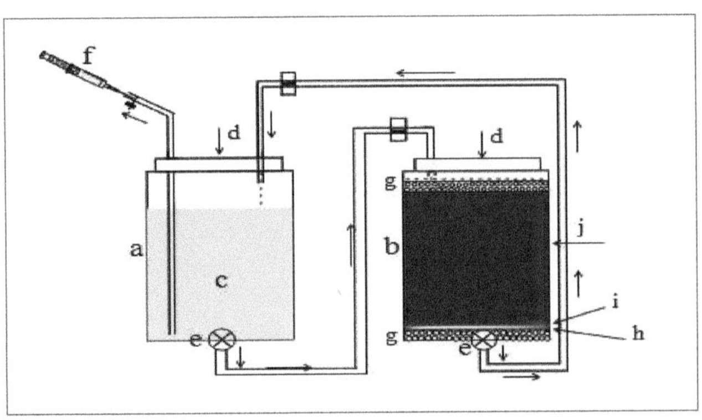


***Figure 1.** Experimental column set-up. a) Solution reservoir, b) sediment reservoir, c) media (effluent), d) $N_2$, e) peristaltic pump, f) sampling syringe, g) glass beads, h) plastic grid, i) layer of sand, j) sediments.*

### *2.3. Study of the Microbial Metabolism and Bacterial Diversity*

#### *2.3.1. Study of the Microbial Metabolism*

During the anaerobic flooding, about 10 mL of soil solution were sampled at each sampling time. The samples were filtered through a 0.45μm (pore diameter) cellulose acetate filter. They were divided into several aliquots for analysis of glucose, organic acids and total metals. pH and Eh were measured in 1 ml of non-filtered solution using a pH meter (PHS-W; Microprocessor pH/mV/temperature Meter).

Glucose was measured enzymatically in 1 ml of the culture medium using a commercial kit (kit D-Glucose/, Roche). Total organic (TOC) and organic carbon (OC) were measured by an elemental analyzer (TOC SKALAR Formacs$^{HT}$). Butyrate and acetate which were the main organic acids produced during glucose fermentation [19] were analyzed in 1 ml of medium solution by HPLC according to [20]. The separation of organic acids was realized by HPLC with a reversed phase column (Thermo ODS hypersil 5 μm, 4.6m*250 mm) and a gradient of $KH_2PO_4$ (20mM) and acetonitrile.

#### *2.3.2. Study of the Microbial Diversity: DNA Extraction, PCR Amplification and DGGE*

DNA extraction was carried out on duplicate of 0.25g of wet sediment samples before their incubation and of the different column sediment samples after their incubation using the PowerSoilTM DNA extraction kit (MoBio laboratories Inc., Ozyme, France) according to the manufacturer's instructions.

16S rRNA amplification was carried out using primers 518r and 338f (Table 1). The PCR was realized on 1μl of DNA added to 14μl of the mixture containing 0.375μl of each of the primer set, 7.5μl of the Taq polymerase (Master Mix) and 5.75μl of sterilized ultrapure water. PCR amplification was performed by using a PTC-200 Peltier-effect Thermal Cycler (BioRad). Amplification conditions were as follows: 94°C for 3 min (initial denaturation), followed by hybridation at 60°C for 1 min, elongation for 3 min at 72°C with the final extension step at 72°C for 15 min. Negative controls (without DNA) were run in all amplification.

DGGE was performed with 9% (wt/vol) acrylamide gel containing a linear chemical gradient ranging from 30 to 60 % denaturant, and run for 15 min at 60°C and 150 V and then for 6h at 60°C and 20V using the BioRad D-CodeTM system (BioRad laboratories, Inc. Ozyme, France). After electrophoresis, the gel was soaked in ethidium bromide nucleic acid gel stain solution for 15 min and then rinsed with milliQ water for 10 min and then photographed under ultraviolet with gel reader Geldoc 2000 (BioRad, Marnes La Coquette, France) and analyzed with the logiciel Quantity one (BioRad, Marnes La Coquette, France).

***Table 1.** Characteristics of the fragments used during the genetic characterization of the total microbial community*

| Fragments | Sequences 5'→3' | Position | Region of 16S rRNA | Reference |
|---|---|---|---|---|
| 338f (F)+ | ACTCCTACGGAGGCAGCAGATTACCGCGGCTGCTG | 338-355 | V3 | Muyzer et al., 1993 [21] |
| 518r (R) | | 518-534 | | |

### *2.4. Metals in Solution and in Sediments*

Sediments were dried at room temperature and sieved with retaining the sediment size fraction of < 250μm. Total metals in sediments were quantified before their incubation as well as at the different depths in the sediments of the different columns after their incubation by flame atomic absorption spectrometry (Ray Leigh Model WFX-210 AAS) using air-acetylene flame. 7 major elements were analyzed: Fe, Mn, Cu, Cd, Zn, Pb and Cr and quantification of metals was based upon calibration curves of standard solutions of respective metals. The precision of the metal analysis was controlled by including duplicate samples in analytical batches.

## 3. Results and Discussion

### *3.1. Microbial Metabolism*

#### *3.1.1. Fermentation Process*

Results presented in Figure 2 showed that in the column which was provided with Bromfield media, the rate of carbon mineralization was high during the first week and was followed by a plateau indicating the consumption of the carbon source. No latency period was observed at the beginning of the experiment meaning a good physiological state of microorganisms and a good adaptation to the experimental conditions. These results have shown that the addition of glucose as a source of carbon which is easily biodegradable has considerably accelerated the processes of bacterial growth. After the 1rst week of incubation, glucose was totally degraded and 2 organic acids, acetate and butyrate, were detected. The maximum of acetate was observed during the 5th day of incubation (250 mg/l) and was followed by a plateau and decreased at the end of the incubation while butyrate increased during the first 5 days to attain a value of 575 mg/l and was followed by a plateau and decreased slightly at the end of the incubation. The production of such organic acids indicates that fermentation metabolism occurred [22]. Furthermore, [23] also demonstrated that in soils enriched with glucose in anaerobic conditions, the microbial metabolism of glucose led to the production of acetate. Acetate decreased between the days 5 and 10 of incubation and then reached a plateau and this is may be due to the bacterial consumption of this organic acid after being produced by them and then the disappearance of this group of bacteria or their adaptation another metabolic pathway after this period hence giving the shape of plateau and the decrease of this acid at the end of incubation may be due to its bacterial consumption as the only source of energy left after this period. Butyrate increased during the first five days when glucose was consumed and then reached a plateau where the bacteria may have adapted to these conditions and then slightly decreased at the end of the incubation, this decrease may be due to the consumption of this source of carbon needed by the bacteria during this period. This column turned black in color also after 10 days of incubation indicating a strong reduction of sulphate as showed by [24].

Most of the organic carbon introduced as glucose was transformed into organic acids and mineralized carbon and the organic carbon not determined can be due to the production of other metabolites such as alcohols, amino acids and other polysaccharides not analyzed and to the increase of the biomass.

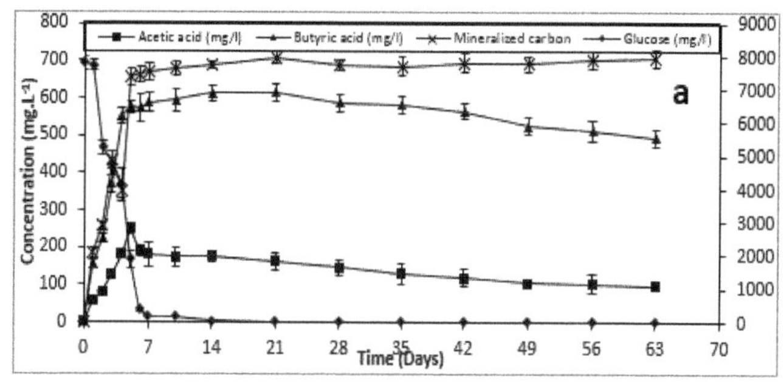


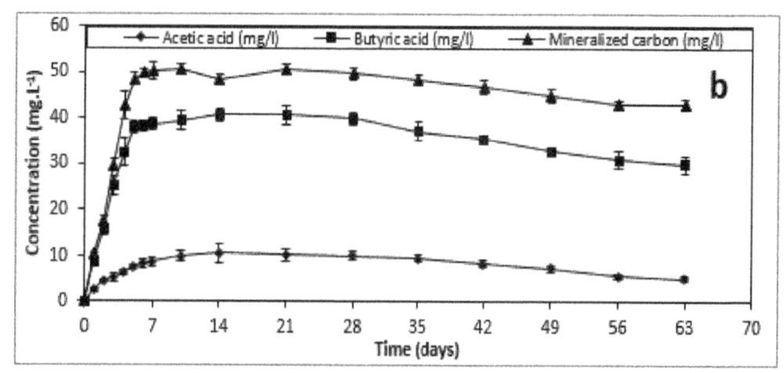


***Figure 2.*** *Evolution of organic acids and mineralized carbon during incubation of the sediments with Bromfield media (a) or synthetic media (b).*

In the column which was supplied with the synthetic media without glucose (Figure 2b), the rate of carbon mineralization was lower than that of Bromfield media and was induced by the bacterial degradation of the endogenous organic matter present in the sediments and which was used as a substrate for these bacteria [19].

In this column, organic acids acetate and butyrate were detected but in concentrations smaller than those obtained in the case of Bromfield media and which reached about 7.5 mg/l and 38.1 mg/l, respectively after 5 days of incubation and then decreased to 5.2 and 30 mg/l, respectively at the end of the incubation and this may be due to a partial consumption by the autochthonous microorganisms [25].

The column which was supplied with distilled water has shown a very low carbon mineralization rate which was mainly produced from the biodegradation of the endogenous organic matter in the sediments and this is due to the complete lack of any other nutritive material for the development of the autochthonous microorganisms. In this column, the production of the organic acids was not observed.

#### *3.1.2. Acidification of the Medium*

The column which was supplied with glucose has shown a strong acidification of the media after 5 days of incubation with a decrease of the pH from 8.00 to 6.7 during the first 5 days of incubation, then after the 1rst week (the day 7), the pH slightly decreased until 6.1 and remained constant till the end of the incubation. This acidification of the medium is due to the fermentation of glucose and to the production of organic acids during the development of the bacteria under anaerobic conditions. In this column, Eh decreased from 175 mV to -24 mV during the first 3 days and after 1 week, Eh reached -243 mV and reached -185 at the end of the incubation.

The sharp decrease in the pH was not observed in the other columns which were supplied with the synthetic media and with the distilled water where it was shown that the column which was supplied with the synthetic media, the pH decreased from 8 to 7.1 during the first 10 days and then increased at the end of the 2$^{nd}$ week to 7.2 and remained constant till the end of the incubation, this decrease in the pH is due to the production of a small concentration of organic acids and their consumption by the bacteria and which may explain the increase of the pH during the 2nd week . In this column, Eh decreased also from 192 to 14 mV in the first 3 days and reached -127 mV after one week and then raised -76 mV at the end of the incubation. The column which was supplied with distilled water, it was shown that the pH slightly decrease from 8.0 to 7.7 and then increased to 7.9 and then remained constant till the end of the incubation, this slight increase may be due to the absorption of protons from the soil components ([18] and [26]), in this column a sharp decrease in the Eh was shown at the beginning of the experiment.

### *3.2. Metals in the Sediments and the Leachates and Their Relationship with the Microbial Metabolism*

#### *3.2.1. Total Metals in Sediments Before Their Incubation*

Table 2 shows the total contents of Cd, Zn, Pb, Cu, Fe, Mn and Cr in the sediments taken from the studied site of Al-Ghadir River (site E) before their incubation. The content of most of these metals is high and based on values considered to be critical by different investigators. Such high previous pollution is not surprising because it is from an area of influence of industries.

***Table 2.*** *Characteristics of the river sediments before incubation.*

| Parameter | | Parameter (μg.g-1 dry sediments) | |
|---|---|---|---|
| pH | 8.2 | Fe | 13570 |
| Sand (%) | 41.2 | Mn | 328 |
| Silt (%) | 17.5 | Cd | 35.4 |
| Clay (%) | 41.3 | Zn | 254 |
| Organic matter (%) | 14.72 | Cu | 146.5 |
| CEC (meq/kg) | 35 | Pb | 125 |
| Carbonate content (%) | 8.78 | Cr | 45.6 |

#### *3.2.2. Metals Leached in the Different Columns*

Figure 3 shows the leaching curves of Fe, Mn, Cd, Zn, Pb, Cr and Cu in packed columns of sediments. In the column which was supplied with Bromfield media, high dissolution of Fe and Mn appeared after one week of incubation (Figure 3). Same trends were observed for the dissolution of Fe and Mn in the column supplied with the synthetic media but dissolved amounts were lower than in the case of Bromfield media and the column which was supplied with distilled water showed traces of leachable metals in the solution. The dissolution of Fe and Mn was maximum during the first days of the experiments and synchronous with the higher decrease of pH. Several processes may be involved in the relationship between lowering of pH and metal release, which has been widely demonstrated for various substrates ([27], [28], [29], [30], [31], [32], [33] and [34]).

However, as shown in the column supplied with the synthetic media, no strict correlation was observed between pH and Fe and Mn release since after the pH decrease during the first week, pH increase slowly while metals were still leached and a reverse relationship (dissolved metals increase with an increasing pH) was found for the final part of the experiment. Same trends as our results are observed in a previous study, in which an increase of the leached metals were observed in soils with a decrease of the pH but both parameters were not obviously correlated [35]. In their study, the only relationship with R2 values above 0.2 was found for Cu and Cr concentration versus pH in addition high correlation was found for Cu and Cr concentration versus pH in addition high correlation was found between C mineralization and the leaching of Fe and Mn, more precisely dissolved iron attained a plateau when glucose is consumed (Figure 2a, 2b) hence suggesting the presence of iron and manganese reducing bacteria. Indeed some fermentative bacteria have been shown to oxidize the organic matter by reducing iron and manganese oxides at the same time to use them as final electron acceptor [19] and the production of butyric acid is characteristic of the microorganisms Clostridium butyricum, known for its iron-reducing activity.

The leaching profile of the other metals was similar to the leaching profile of Fe and Mn in the columns supplied with Bromfield and with the synthetic media, but the leaching profile of the metals in the column supplied with distilled water was very low and nearly negligible and this is due to the complete absence of the nutrients responsible for the growth of the autochthonous bacteria and which showed an important role in their leaching of the heavy metals. In addition, the leaching profile of the metals in the different columns was shown to be high at the beginning of the incubation and decreases at the end. This decrease may be due to the aging of the contamination by the heavy metals which decrease the solubility of the heavy metals with time. In the column supplied with Bromfield media, a high rate of heavy metal leaching was observed beginning from the 3rd day of incubation and then decreases till it reached a plateau where the concentration of most of the measured metals remained constant till the end of the incubation. The shape of the elution curves of the different metals was very similar. Figure 3 showed the leaching curves of Fe, Mn, Zn, Cd, Pb, Cr and Cu leached from the sediments during incubation. The leaching curves of Cd from the sediments (Figure 3) display an initial elution maximum and the maximum amount leached reached about 16% from the column supplied with Bromfield media and 12% for the column supplied with the synthetic media and based on the leaching curves obtained from the studied sediments, all the potentially leachable Cd of the column has passed to the leachates under the experimental conditions.

What is of great importance is that the concentration of the metals leached was important and was affected by bacterial activity in the sediments but also the characteristics of the sediments of the studied region (clay rich sediment with a high carbonate content) (Table 2) have a good capacity to immobilize the soluble fraction of the elements and they showed an important role in the control of the adsorption of the metals in the sediments ([36] and [37]). These characteristics govern the distribution of the elements in the profile of natural soils ([38], [39] and [40]).

The leaching curves of Zn in the sediments (Figure 3) are very similar to those of Cd and the percentage of Zn leached from the different columns is slightly lower than that of Cd. This kind of behavior is in agreement with results showing that Zn is less mobile than Cd ([41] and [42]). Decreased mobilization of this element may be related to the high contents of organic matter and clay in this soil, as these fractions are the dominant controls of Zn adsorption in soils [43].

Considered to be the least mobile heavy elements in the sediment and the soil ([12], [14], [15] and [16]) Pb and Cr are almost completely immobilized in all the columns, the maximum amounts leached was in the column supplied with Bromfield media where the % of Pb and Cr leached reached about 6 and 5%, respectively; and what is important is that the organic matter content and the free oxides and carbonates and their joint action seem to be sufficient for the element to become immobilized in all the columns [44].

Cu showed low mobility in the different columns, about 12% of the total of Cu in the columns was leached. According to the leaching curves, the amounts of Cu leached are higher than those of Pb and Cr which, these later, showed a similar leaching behavior in the calcareous sediments studied and were liberated with great difficulty (Figure 3). As is known Cu, Cr and Pb are the least mobile heavy metals [14] and Cu is strongly adsorbed by all the colloidal components of the sediments such as oxides of iron and manganese, clays and organic matter and the amounts of these components in the studied sediments and soil seem to be sufficient for the immobilization of this element. According to these results, the relative leaching of the elements studied in the polluted sediments calculated from the percentages leached in the soil columns are as follows: Fe>Mn>Cd>Zn>Cu>Pb>Cr. The decrease of Eh in the different columns during the leaching process indicates the installation of the solubilization process. In studies on the leaching of these metals added to packed and undistributed soil columns, [45] found that the mobility of the metals in the soils tended to follow the order of Cd>Zn>Cu>Pb>Cr and the lowest mobility was observed in a fine-loamy soil with a relatively high pH.

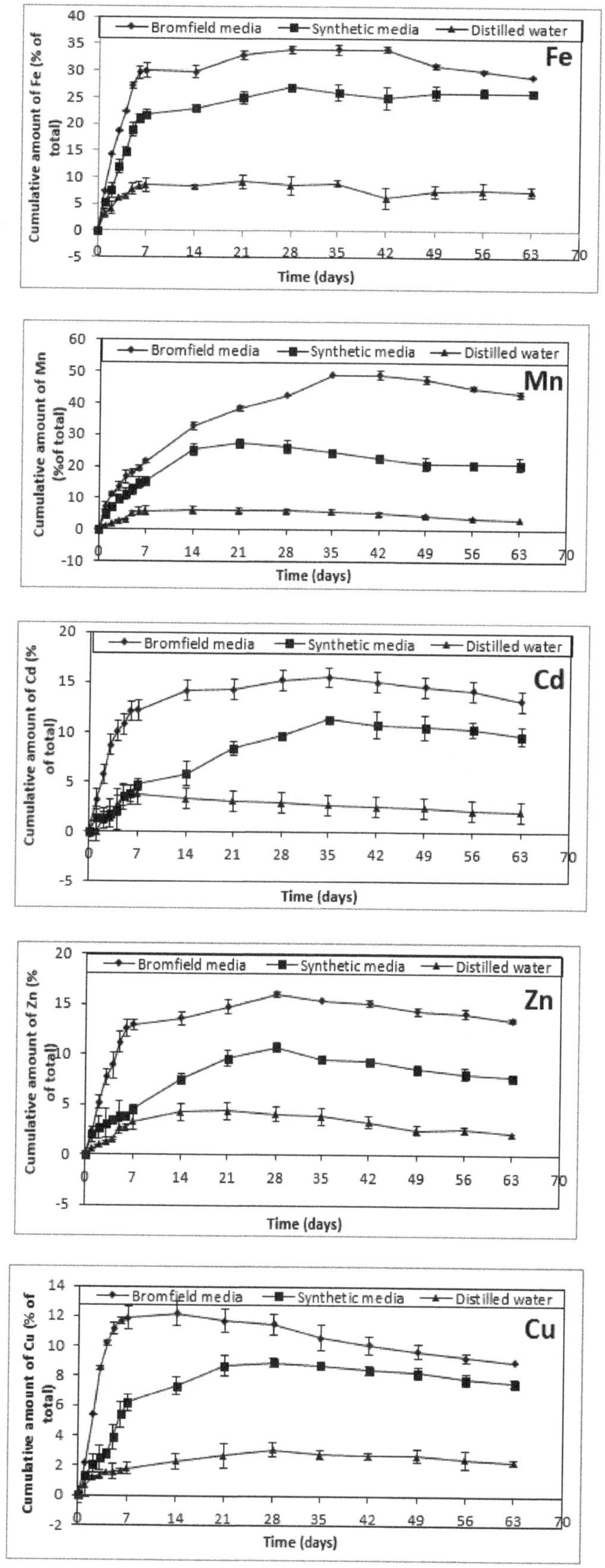

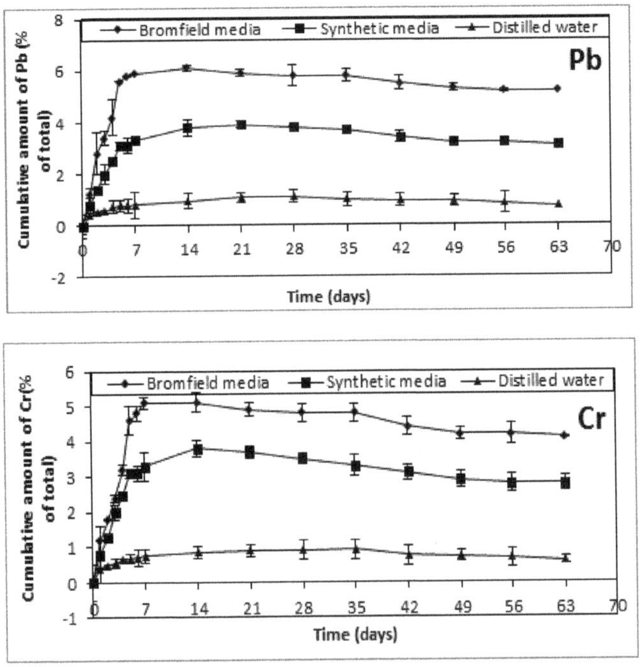


*Figure 3. Cumulative curves of metal leaching in columns of sediments.*

### *3.2.3. Distribution Profiles of the Metals After Leaching*

Figure 4 shows the distribution profiles of Fe, Mn, Cd, Zn, Cu Pb and Cr in the sediment columns after being leached with 800 ml of media. In addition to the results of metal leaching which showed us that the solubilization of the studied heavy metals was homogenous in the different columns (Figure 3), the distribution profile of the studied elements in the different columns which were incubated anaerobically was shown also very similar, an accumulation in the first segments that reaches a greater or lesser depth depending on the metal and the nutritive materials but what was shown and was of great importance is that the concentration of the metals detected in the incubated sediments at the different depths at the end of the incubation, after being leached and after they have been detected in leachates, was shown high at the surface of the sediment columns (depth interval 0-10cm) and decreases with increasing depth (depth interval 10-25cm). This phenomenon can be explained by the fact that the metals after being dissolved and passed with the solution, each time they re-entered with the leachate and reached the sediments, a part of these leached metals was readsorbed to the newly formed phases and which explains the decrease in the concentration of the metals in the leachate at the end of incubation.

In the column, which was supplied with Bromfield media, the accumulation of Fe, Mn, Cd and Zn occurred down to 14 cm and that of Cu, Pb and Cr down to 10 cm, while the column which was supplied with the synthetic media, the accumulation of Fe, Mn, Cd and Zn occurred down to 10 cm and that of Cu, Pb and Cr down to 8 cm. Whereas the column which was supplied with distilled water was considered as a control and showed a very low mobilization of heavy metals.

The addition of the nutritive materials to the incubated sediments was one of the factors responsible for the activation of the autochthonous bacteria and the acidification of the sediments hence the solubilization of the heavy metals but also the composition of the studied sediments (clay sediments with a high contamination of carbonate) was responsible for the readsorption of the solubilized heavy metals to the sediments [40].

These results showed that the leaching of the metals from the sediments in the incubated columns simulates the fate of metals in the real scenery and it gives rise to results on the distribution of the elements that confirmed that the polluted sediments themselves have a good capacity to immobilize the soluble fraction of the elements. However, under conditions of abundant rain and floods of the river in winter, the sediments and the soil in the studied region (with a high level of pollution) do pose a high risk with respect to the contamination of underground waters if one considers the total amount of metals potentially leachable below 25 cm of sediments and soil and what is of great importance is that the acidification of the sediments due to bacterial activity can promote the leaching of certain elements.

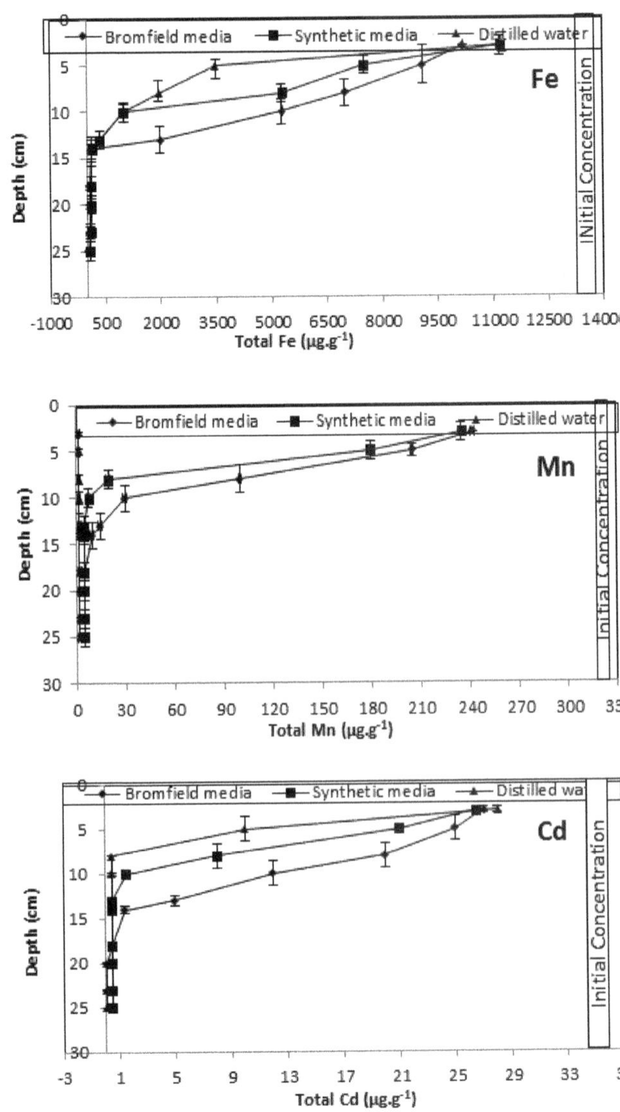

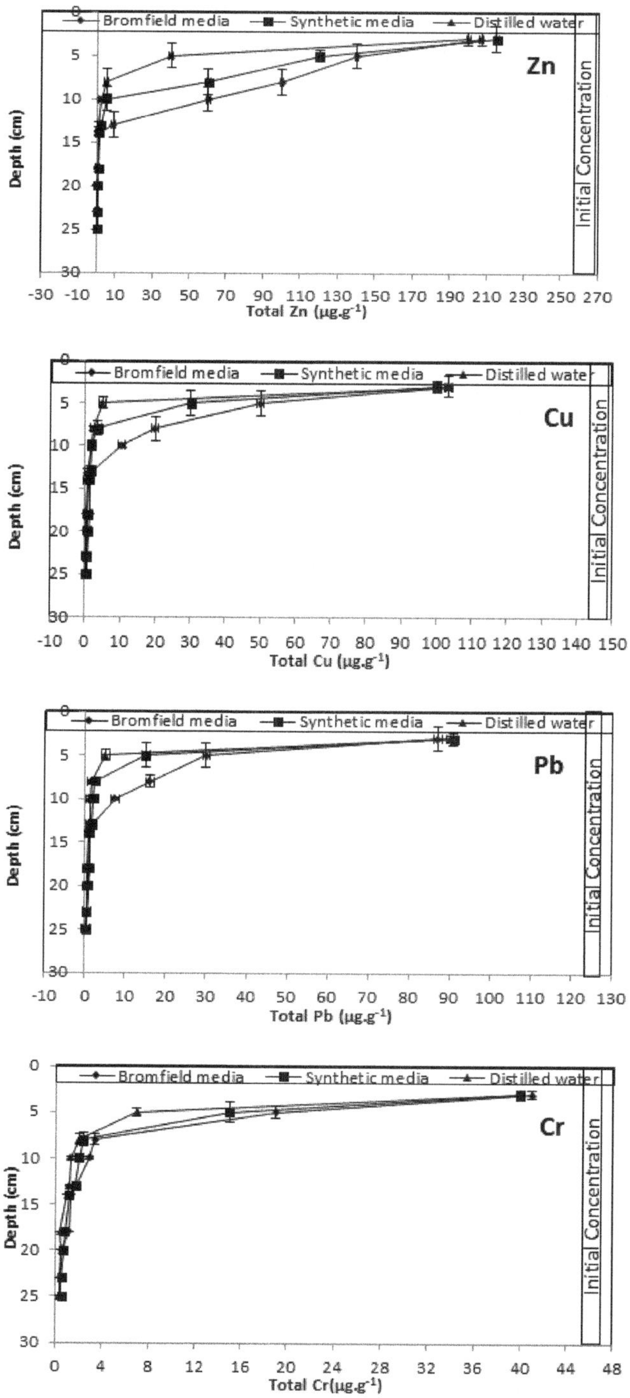


*Figure 4. Vertical distribution profile of metals in the incubated sediments with the different media.*

This latter phenomenon could mainly affect elements that are adsorbed as exchange cations by soil and sediment colloids [46], such as Cd and Zn, which could change from being moderately mobile in soils with pH of 6 to highly mobile if the sediments were to reach pH values below this. The decrease in pH would also affect Cu and Pb [14], which would change from being almost immobile to being moderately mobile.

At the end of the experiment and in order to detect the effect of the heavy metals on the soil microflora which were effective on the solubilization of the heavy metals through the production of the organic acids and the decrease of the pH, bacterial counting was done on the sediments taken from the top and the bottom of the columns at the end of the incubation and compared with the initial diagnosis of the sediment microflora.

Results obtained (Table 3) confirmed a certain toxicity of the metals toward the anaerobic endogenous soil microflora (initial diagnosis of the sediments showed the presence of 1.2*107 CFU/g of dry sediments). This counting putted in evidence a good repartition of this microflora along the whole column where no significant difference was shown between the results obtained at the top and the bottom of the same column but the difference was shown before and after incubation of the sediments in the columns where a slight decrease in the bacterial density was shown at the end of the incubation, this decrease may be due to consumption of the nutritive materials in the column which were important for their survival and which may begin to consume the 2ry metabolites produced by them. A sharp decrease in the bacterial density was shown in the column which was supplied with distilled water and without the addition of external nutritive materials, where the presence of small concentrations of nutritive materials in the sediments was not sufficient to keep the bacterial density the same.

***Table 3.*** *Bacterial counting at the top and the bottom of the columns at the end of the experiment (in CFU).*

| Bacteria (CFU) | Column with Bromfield media | Column with the Synthetic media | Column with the distilled water |
|---|---|---|---|
| Top of the column | 2.1*106 | 7.4*104 | 2.5*103 |
| Bottom of the column | 3.7*106 | 3.5*104 | 1.8*103 |

These results were confirmed with the molecular technique using the PCR-DGGE, which aim was to detect and to compare the temporal variation of the bacterial diversity in

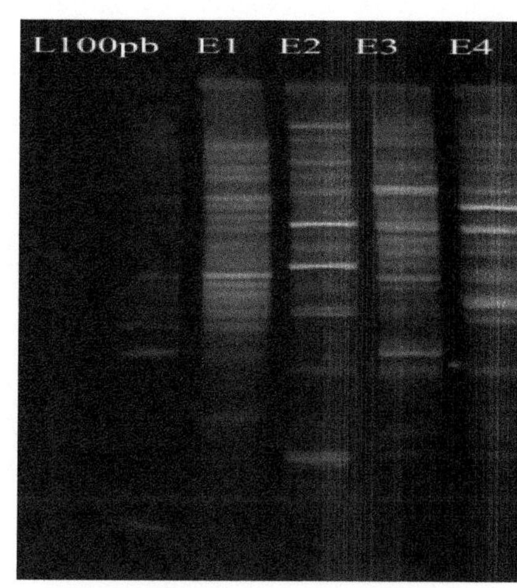


*Figure 5. DGGE profiles of PCR-amplified 16S rDNA gene from sediment samples before and after their incubation with the different media (E1: sediment sample before incubation; E2: sediment sample after incubation with Bromfield media; E3: sediment sample after incubation with the synthetic media; E4: sediment sample after incubation with distilled water).*

the different sediments before and after their incubation to see whether the incubated conditions were responsible for the changes in the microbial community.

Comparison of DGGE profiles shows that the structures of the bacterial communities in the different incubated sediments differ before and after their incubation and also differ from one another before and after their incubation (Figure 5) where many bands appear in common in the different sediments but these bands were present as intense bands in some sediments while as faint bands in other sediments indicating that the incubating conditions were favorable for the growth and the intensification of certain groups of bacteria but also these conditions were responsible for the partial disappearance of many other groups which appeared as faint bands.

## 4. Conclusion

Carefully interpreted column leaching experiments provide valuable information on labile metal pools, metal sorption affinity, the coupling of simultaneous processes such as cation exchanges, carbonate dissolution, readsorption and endogenous bacterial activity as well as on the importance of slow reaction kinetics in metals leaching. Such information should prove useful for the calibration of transport models used to assess the risk of metal leaching at contaminated sites. Results obtained reflect the leaching of the metals at the contaminated sites and showed that their leaching was very low in the calcareous sediment column and varied between metals. This leaching has been accelerated by the stimulation of the endogenous bacterial activity by the addition of complex nutritive media and which mimics the wastewaters released continuously in the river in the studied region.

Cr and Pb showed the greatest difficulty in leaching into the sediment solution, Fe, Mn and Cd reached the greatest concentration in the leachates and Cu and Zn an intermediate behavior. According to this leaching sequence, it seems clear that in sediments and soils carbonated by sorption sites, competition between metals is a major risk for groundwater contamination and raises the probability of absorption by plants especially of the most mobile and toxic metals. This contamination was detected in both the groundwater and plants cultivated and consumed in the region (results not published here) and which threaten the region by the increase in the level of the diseases affecting the people living there. The analysis of the distribution profile of the heavy metals in the incubated sediment columns has showed that the metals have leached from the sediments during incubation in a homogenous manner, but since the system is continuous, the arrival of the leachate rich in solubilized metals has led to the re-adsorption of the metals to the sediments. This repartition depend on the height of the sediment where it was shown that the re-adsorption was maximum at the surface of the column sediments (0-10 cm) and decreases with depth (10-25 cm) convincing the important role of clay and carbonated sediments in the adsorption of heavy metals hence decreasing their leaching to the underground waters and open the door to the thinking about the application of remediation techniques on the studied site to immobilize the heavy metals such as those techniques based on the addition to the polluted soils of limestone, clay minerals, zeolites, hydroxyapatite, Fe oxides and organic materials which are currently being tested in soils from zone with a view to selecting the most appropriate technique.

## References

[1] Jurdi, M. (1992). 'A National Study on the Quality of Potable Water in Lebanon', in Proceedings of the National Workshop of the Status of Water in Lebanon, Beirut, Lebanon: 145–173 (in Arabic).

[2] Khair, K.; Aker, N.; Haddad, F.; Jurdi, M. and Hachach, A. (1994). 'The environmental impacts of humans on ground water in Lebanon', Water, Air Soil Pollut. 78: 37–49.

[3] Jurdi, M. (1998). 'Follow up National Environmental Surveillance, 1996', Unicef Publications, Beirut. (in Arabic).

[4] Sene, K.J.; Marsh, T.J and Hachache, A. (1999), 'An assessment of the difficulties in quantifying the surface water resources of Lebanon', Hydro. Sci. J. 44(1): 79–96.

[5] Low, K.S. and Lee, C.K. (1991). Cadmium uptake by the Moss, Calymperes delessertii, Besch, Biores. Technol. 38: 1–6.

[6] Grolimund, D.; Borkovec, M.; Barmettler, K. and Sticher, H. (1996). Colloid-facilitated transport of strongly sorbing contaminants in natural porous media: A laboratory column study. Environ. Sci.Technol. 30: 3118–3123.

[7] Kretzschmar, R., and H. Sticher. (1997). Transport of humic-coated iron oxide colloids in a sandy soil: Influence of Ca2+ and trace metals. Environ. Sci. Technol. 31:3497–3504.

[8] Temminghoff, E.J.M.; van der Zee, S.E.A.T.M. and de Haan, F.A.M. (1997). Copper mobility in a copper contaminated sandy soil as affected by pH, solid and dissolved organic matter. Environ. Sci. Technol. 31:1109–1115.

[9] Kedziorek, M.A.M.; Dupuy, A.; Bourg, A.C.M. and Compère, F. (1998). Leaching of Cd and Pb from a polluted soil during the percolation of EDTA: Laboratory column experiments modeled with a non-equilibrium solubilization step. Environ. Sci. Technol. 32:1609– 1614.

[10] Ross, S.M. (1994). Retention, Transformation and Mobility of Toxic Metals in Soils. In: Toxic Metal in Soil-Plant Systems, pp. 63–152. (Ross, S.M., Ed.) West Sussex, England, John Wiley & Sons.

[11] Alloway, B.J. (1995). Soil processes and the behavior of heavy metals. In: Heavy Metals in Soils, pp. 11-35. (Alloway, B.J., Ed.) Glasgow, Chapman and Hall.

[12] Andras, P.; Lichy, A.; Kusnierova, M.; Krizani, I; Ladomersky, J.; Ruskova, J. and Hroncova, E. (2009). Heavy metal distribution at dump-field Lubietova - Podlipa and possibilities of clay fraction natural sorbent utilisation. Acta Montanistica Slovaca, 14(2): 127-142.

[13] Stevenson, F.J. (1994). Humus Chemistry. Genesis, Composition and Reactions. Wiley-Intersc. Publ., New-York.

[14] McBride, M.B. (1994). Environmental Chemistry of Soils, New York, Oxford University Press.

[15] Amusan, A.A. and I.F. Adeniyi, I.F. (2005). Genesis, classification and heavy metal retention potential of soils in mangrove forest, Niger Delta, Nigeria. J. Hum. Ecol., 17(4): 255-261.

[16] Onweremadu, E.U. (2008). Physico – chemical characterization of a farmland affected by wastewater in relation to heavy metals. J. Zhejiang Univ. Sci. A., 9(3): 366-372.

[17] Motos, G.D. and Arruda, M.A.Z. (2003). Vermicompost as natural adsorbent for removing metal ions from laboratory effluents, Process Biochem. 39: 81–88.

[18] Bousserrhine, N. (1995). "Etude de paramètres de la réduction bactérienne du fer et application à la defferification de minéraux industriels", Université Henry Poincaré, Nancy I : 331.

[19] Lovley, D.R. (1991). Dissimilatory Fe (III) and Mn (IV) reduction. Microbiological reviews; 55: 259-287.

[20] Tormo, M.A. and Izco, J.M. (2004). "Alternative reversed-phase high-performance liquid chromatography method to analyse organic acids in dairy products", Journal of Chromatography A, 1033(2): 305-310.

[21] Muyzer, G.; de Waal, E.C. et Uitterlinden, A.G. (1993). Profiling of complex microbial populations by denaturing gradient gel electrophoresis analysis of polymerase chain reaction amplified genes coding for 16S rRNA. Appl. Environ. Microbiol; 59: 695-700.

[22] Charlatchka, R. and Cambier, P. (2000). "Influence of Reducing Conditions on Solubility of Trace Metals in Contaminated Soils", Water, Air, & Soil Pollution, 118: 143-168.

[23] Dassonville, F.; Godon, J.J.; Renault, P.; Richaume, A. and Cambier, P. (2004). "Microbial dynamics in an anaerobic soil slurry amended with glucose, and their dependence on geochemical processes", Soil Biology and Biochemistry, 36: 1417-1430.

[24] Fakih, M.; Alphonse, V.; Garnier-Zarli, E. and Bousserrhine, N. (2010). "Mercury mobilization in ferralitic soils under reductive conditions", Science of the Total Environment.

[25] Lovley, D.R. and Phillips, E.JP. (1988). "Novel mode of microbial energy metabolism: organic carbon oxidation coupled to dissimilatory reduction of iron or manganese". Applied and Environmental Microbiology, 54: 1472-1480.

[26] Rajot, J.-L. (1992). Dissolution des oxydes de fer (hématite et goethite) d'un sol ferralitique des Llanos de Colombie par des bactéries ferri-réductrices. Université de Nancy; 185.

[27] Sims, J.L. and Patrick, W.H. (1978). "The distribution of micronutrient cations in soil under conditions of varying redox potential and pH", Soil Science Society of America Journal, 42: 258-262.

[28] Calmano, W.; Hong, J. and Förstner, U. (1993). "Binding and mobilization of heavy metals in contaminated sediments affected by pH and redox potential", Water Science and Technology, 28.

[29] Chuan, M.C.; Shu, G.Y. and Liu, J.C. (1996). "Solubility of heavy metals in a contaminated soil: Effects of redox potential and pH", Water, Air, and Soil Pollution, 90: 543-556.

[30] Stumm, W.; Morgan, J.J. (1996). "Aquatic chemistry, chemical equilibria and rates in natural waters", Trace Metals: Cycling, Regulation, and Biological Role. 3rd ed. John Wiley and Sons, Inc, USA.

[31] Chen, S.Y. and Lin, J.G. (2001). "Bioleaching of heavy metals from sediments: significance of pH", Chemosphere, 44: 1093-1102.

[32] Gundersen, P. and Steinnes, E. (2003). "Influence of pH and TOC concentration on Cu, Zn, Cd and Al speciation in rivers", Water Research, 37: 307-318.

[33] Qureshi, S.; Richards, B.K.; Hay, A.G.; Tsai, C.C.; Mcbride, M.B.; Baveye, P. and Steenhuis, T.S. (2003). "Effect of microbial activity on trace element release from sewage sludge", Environmental Science & Technology; 37: 3361-3366.

[34] Yang, J.Y.; Yang, X.E.; He, Z.L.; Li, T.Q.; Shentu, J.L. and Stoffela, P.J. (2006). "Effects of pH, organic acids, and inorganic ions on lead dissolution from soils", Environmental Pollution, 143: 9-15.

[35] Linde, M.; Öborn, I. & Gustafsson, J.P. (2007). Effects of changed soil conditions on the mobility of trace metals in moderately contaminated urban soils. Water, Air and Soil Pollution, 183: 69-83.

[36] Korte, N.E.; Scopp, J.; Fuller, W.H.; Niebla, E.E. and Alesii, B.A. (1976). Trace element movement in soils: Influence of soil physical and chemical properties. Soil Sci. 122: 350–359.

[37] Sánchez-Camazano, M. and Sánchez-Martín, M.J. (1993). Mobility of cadmium as influenced by soil properties, studied by soil thin layer chromatography. J. Chromatograph.: 357-362.

[38] Sánchez-Camazano, M.; Sánchez-Martín, M.J. and Lorenzo, L.F. (1994). Content and distribution of cadmium in natural soils as influenced by soil properties. Sci. Total Environ. 156: 183–190.

[39] Hargital, L. (1995). Some aspects of the mobility and distribution of toxic heavy metals contaminants in soil profiles and river sediments. Int. J. Environ. Anal. Chem. 59: 317–325.

[40] Sánchez-Camazano, M.; Sánchez-Martín, M.J. and Lorenzo, L.F. (1998). Significance of soil properties for content and distribution of cadmium and lead in natural calcareous soils. Sci. Total Environ. 218: 217–226.

[41] Tyler, L.D. and McBride, M.B. (1982). Mobility and extractability of cadmium, copper, nickel, and zinc in inorganic and mineral soil columns. Soil Sci. 134: 189–205.

[42] Hickey, M.G. and Kittrick, J.A. (1984). Chemical partitioning of cadmium, copper, nickel and zinc in soils and sediments containing high levels of heavy metals. J. Environ. Qual. 13: 373–376.

[43] Missana, T.; Garcia-Guttierez, U. and Alonso, U. (2008). Sorption of strontium onto illite/smectite mixed clays. Phys. Chem. Earth, 33(Supl. 1): 156-162.

[44] Amer, M.W.; Khalili F.I. and Awwad, A.M. (2010). Adsorption of lead, zinc and cadmium ions on polyphosphate-modified kaolinite clay. J. Environ. Chem. Ecotoxicol., 2(1): 001-008.

[45] Camobreco, V.J.; Richards, B.K.; Steenhuis, T.S.; Peverly, J.H. and McBride, M.B. (1996). Movement of heavy metals through undisturbed and homogenized soil columns. Soil Sci. 161: 740–750.

[46] Kabata-Pendias, A. et Pendias, H. (2001). Trace elements in soils and plants. 3rd CRC Press, BocaRaton, London, New-York, Washington, DC: 403.

# 12

# Risk Assessment on Trace Metals in Groundwater and Springs in Urban Environment of Zanzibar Island

**Abdul Ali Juma Mohamed*, Sara Abdalla Khamis, Haji Mwevura, Miza Ali Kombo, Said Suleiman Bakari**

Department of Natural Sciences, State University of Zanzibar (SUZA), Zanzibar, Tanzania

**Email address:**
jumabdull@yahoo.com (A. A. J. Mohamed)
*Corresponding author

**Abstract:** Urban-west region is the most highly populated region of Zanzibar. More than half of the Zanzibar populations reside at this region. Groundwater is the main resource of drinking water in this area, and currently there is no drinking water treatment plant; hence, the suitability and appropriateness of groundwater resources for drinking and other domestic uses is of public and scientific concern. In this study, groundwater samples were collected from 30 sites in urban areas of Zanzibar Island in December 2012. Measurements, and investigation of the levels of trace metals in water including Cd, Co, Cr(III), Cu, Fe, Li, Mn, Mo, Sb, Se, Ti, and Zn were measured in thirty locations in Zanzibar urban west region using Inductively Coupled Plasma Emission Spectroscopy (ICP-OES). The paper also highlights and depicts the estimate of the degree of metal contamination ($C_d$), hazard quotient which yields hazard index (HI), heavy metal evaluation index (HEI), and heavy metal pollution index (HPI). Using the four heavy metals (Cd, Zn, Cu, and Fe), the degree of contamination ($C_d$) in all water sources were less than 1, thus they fall in to the category of low degree of contamination. The general trend of the average daily dose (ADD) of Zn, Cu, Fe, and Cd in water sources was Zn > Cu > Fe > Cd. As the severity of metal toxicity is governed by several factors, such as dose, nutrition, age, and even life style. Therefore, this trend might not guarantee the absence of human health risk. Generally, from risk assessment on trace metals using risk indices ($C_d$, HPI, HI and HEI), the analyzed water sources might not be at the alarming health risk. However, due to an increasing level of environmental pollution, water sources might be potential sink of contaminants; this is significant reason that makes the treatment of drinking water and monitoring implementation inevitable.

**Keywords:** Zanzibar, Degree of Contamination, Risk Assessment, Health Risk, Environmental Pollution

## 1. Introduction

Water is an essential compound for all forms of life in the Earth planet. Springs and groundwater are the two main sources of water used for domestic and other social purposes in Zanzibar. The exploitation of water from groundwater aquifers has become a common practice in different areas of Zanzibar Island. Moreover, during rainy seasons, rainwater is also a potential source of fresh water for domestic and other social purposes and activities.

The issue of water quality is very poignant in many countries, but the vulnerability of water to contaminants due to natural and anthropogenic sources has become a common problem worldwide [1].Water is an inevitable need to man and his environment; it has existed throughout the history of the earth crust even before the existence of man. However, the water qualityis persistently pollutedin many countries of the world [2].

The importance of water quality for the health of human being and welfare is of a great interest and attention in the world. Access to sufficient, clean, and safe water is essential for human survival, and is one among the fundamental human rights. Regarding public health, limited access to safe water obviously undermines other public sectors such as public health, which can lead to communicable diseases. The provision of potable drinking water has been highly assembled in national and international documents [3].

The issue of water quality is one of the poignant agenda in many countries. The exploitation and utilization of water from groundwater aquifers has become a common practice in different areas of Zanzibar Island. Nevertheless, the vulnerability of water to contaminants due to natural and anthropogenic sources seems to be a common phenomenon [1]. Moreover, both the qualitative and quantitative aspects of water resources might be vulnerable to deterioration due to environmental degradation. Very recently, about 2.7 million liters of water are anticipitated to be lost at Chunga water supplies located at west region of Zanzibar owing poor utilization of the environment.

Fresh water supply is an important concern worldwide, about 90% of fresh water in the world originated from groundwater aquifers [4]. An aquifer is an underground component of saturated earth materials that can provide usable quantities of ground water to a well [5].

In Zanzibar, rapid rate of urbanization has increased demand of water, which necessitated the usage of groundwater to supplement the existing water supply systems, in most areas of the Island. Accordingly, the trend of developing privately owned water wells has increased to an amount that has outnumbered the public and community water source [1].

Water resources evaluation typifies a major concern of the present world due to the importance of water for human being and society in general [6]. Good water quality management is fundamental for the production of clean water for the human consumption and safety. Qualitative as well as quantitative determination of the contaminants, along with the predictions of their possible source(s) are vital in an attempt of tackling contamination problems in groundwater and surface water [1]. Therefore, an evaluation of drinking water quality is a crucial concern for implementing sustainable water-use strategies, plan, and policy.

Moreover, information about water quality analysis is always used as a basis to discuss utilizing groundwater in order to avoid associated water illnesses and health problems. [7] Groundwater quality depends, to some extents, on its chemical composition. Monitory of the levels of cations and anions play important and indicative role in assessing degree of groundwater contamination [8]. Nevertheless, an access to good water quality supply and sanitation services has become a hot issue in most of developing countries, and still these services do not reach a significant proportion of the world's population. It has become a more challenging agenda due to factors such as rapid rate of urbanization and environmental pollution.

Harmful impurities from various sources such as, agricultural and domestic effluents, municipal sewerage are among the key factors, which deteriorate the quality portable water [9]. Trace metals are among a wide range of contaminants, which receive health concernsdue to their potential toxicities even at very low concentrations. Trace metals gain entrance into human systems via contaminated drinking water, air, and food. In the body, the metals can compete with, and displace essential minerals such as zinc, copper, magnesium, and calcium; and interfere with body biological organs, systems and functions [10] Over a certain period, metals have tendency to bio-accumulate in tissues of living organisms [11]. However, some trace metals (such as zinc, copper, iron, and manganese) are the micronutrients; the body in small amounts for metabolic activities requires them. At higher concentrations, these same elements, can cause adverse health effects or illness [12].

Toxic metals such as cadmium (Cd), lead (Pb), arsenic (As), chromium (Cr), and thallium (Tl) do not have beneficial effects in humans. Long-term exposures to these metals may cause severe disruptions in the normal dysfunctioning of the organs and systems where the metals are accumulated. Pb, for example, is linked with a broad range of negative pregnancy outcomes, including cardiovascular diseases, early membrane rupture and spontaneous abortion, and erectile dysfunction [13]. Cadmium is very toxic and is linked to several cases of poisoning through food. Small quantities of cadmium cause adverse changes in the arteries of human kidney causing kidney damage. Moreover, cadmium replaces zinc biochemically and causes high blood pressures [14].

Metals such as Cd, Cr, V, As, Mn, Ti, Co, Cu, Fe, Pb, Ni, Zn, and their compounds have been suggested to be initiators or promoters of carcinogenic activity in animals. Furthermore, Be, Sb, Al, Hg, Ni, Cd and Co can cause undesirable reproductive/fertility problems [15].

Zinc toxicity can leads to diarrhea [16], manganese may slow down the intellectual development of the child [17]. Iron has been associated with repeated blood transfusions, genetic and metabolic diseases [18], and copper toxicity is related to several health concerns, including kidney disease, stomach cramps, nausea, vomiting, diarrhea, cancer, and liver damage [19].

The presence of chemical species in fresh water sources, such as groundwater aquifers and springs can be of natural or anthropogenic origins. The constituents present in water may pose many health risks to human, and other life when exceed the maximum contamination level. Because of the absence of drinking water treatment plants in Zanzibar, it is mandatory to assess, and to keep regular drinking water monitoring programs. Thus the objective of this study is to examine and highlighting the levels of twelve metals in groundwater and spring sources in urban west region of Zanzibar. The paper also reveals the correlation coefficients between the analyzed parameters, the estimate of the degree of metal contamination ($C_d$). Other hazard and pollution indices were also studied.

## 2. Materials and Methods

December 2012, samples of water were collected from twenty-eight ground water sources and two springs located at Urban-west region and its territoriesof Zanzibar Island. Geographic coordinates of all sampling points were recorded using the Global Positioning System (GPS). Sampling sites are shown in Figure 1). Water samples from 30 locations were collected in sterile bottles (1000 mL) preserved, and analyzed by using standard methods for the examination of water and wastewater [20]. These water sources were closed hand dug well (CHDW), open hand dug well (OHDW), public bore well (PBW), spring water (SW), and bore well owned by private individual or community owned well (BWP). Prior to sampling process, all the bottles were cleaned and rinsed thoroughly with water to be analyzed. All reagents used were of analytical grade. The water samples were stored at 4°C prior to laboratory chemical analysis. Metal analyzed are Cd, Co, Cr (III), Cu, Fe, Li, Mn, Mo, Sb, Se, Ti, and Zn. The concentrations of these metals were determined using ICP-OES (Thermo Scientific iCAP 6000). The ICP-OES is robust and can undergo multi elemental analysis [1, 3].

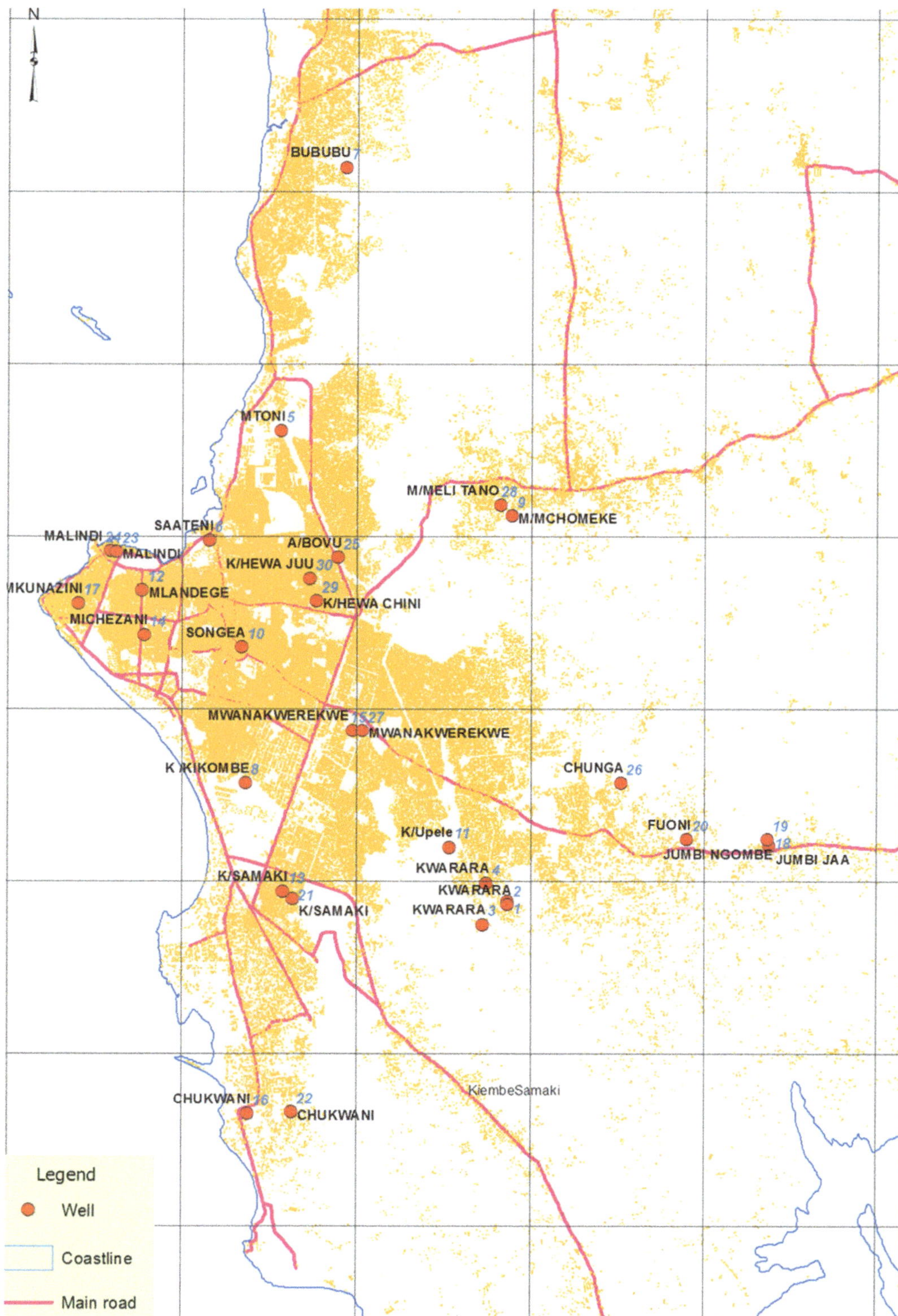


*Figure 1. Map of sampling sites in Zanzibar Urban-West Region and territories.*

### 2.1. Data Analysis

IBM Statistical Package for the Social Sciences (SPSS) version 16 was used to give descriptive statistics from the data analysis. Concentrations of the analyzed trace metal within and between the water sources were not significant different (Table-1).

*Table 1. Summary of the descriptive statistics of the analyzed parameters.*

| Parameters | df | F | Sig. | Mean | Std. Deviation | Minimum | Maximum | N |
|---|---|---|---|---|---|---|---|---|
| Cd | 4 | 0.946 | 0.454 | 0.0006 | 0.00031 | 0.0001 | 0.0014 | 30 |
| Co | 4 | 1.912 | 0.140 | 0.0002 | 0.00026 | BDL | 0.0009 | 30 |
| Cr(III) | 4 | 1.262 | 0.311 | 0.0124 | 0.01809 | BDL | 0.0465 | 30 |
| Cu | 4 | 0.419 | 0.794 | 0.0069 | 0.02028 | BDL | 0.111 | 30 |
| Fe | 4 | 0.543 | 0.706 | 0.0031 | 0.00794 | BDL | 0.035 | 30 |
| Li | 4 | 1.196 | 0.337 | 0.0075 | 0.01758 | 0.0002 | 0.0985 | 30 |
| Mn | 4 | 0.329 | 0.856 | 0.0009 | 0.00233 | BDL | 0.0089 | 30 |
| Mo | 4 | 0.582 | 0.678 | 0.0002 | 0.00055 | BDL | 0.0028 | 30 |
| Sb | 4 | 0.710 | 0.593 | 0.0021 | 0.00173 | BDL | 0.0074 | 30 |
| Se | 4 | 0.867 | 0.497 | 0.0027 | 0.00413 | BDL | 0.0135 | 30 |
| Ti | 4 | 1.531 | 0.224 | 0.0038 | 0.00183 | 0.00031 | 0.0071 | 30 |
| Zn | 4 | 0.413 | 0.798 | 0.0158 | 0.04358 | 0.00013 | 0.2339 | 30 |

### 2.2. The Estimate of the Pollution Indices and Degree of Metal Contamination

The present study evaluated heavy metal pollution index (HPI) proposed by [21], the degree of contamination ($C_d$) developed by [22]. $C_d$ is also gives an indication for the extent of metal pollution [23]. Degree of contamination in water resources can fall in to one of three categories, namely, low ($C_d < 1$), medium ($C_d$ =1-3), and high ($C_d > 3$) [23]. Another pollution index is "Heavy metal evaluation index" (HEI), it gives an overall quality of the water with respect to heavy metals [24]. Figure A2 shows the value of the calculated indices. Toxicity response due to Fe, Cu, Cd, and Zn exposures are shown in Table-2.

*Table 2. Toxicity response due to Fe, Cu, Cd, and Zn exposures.*

| Toxicity Response [25] | | Exposure Parameters | |
|---|---|---|---|
| Parameter | Oral RfD (mg/Kg/day) | IR | 2.2 L/day |
| Fe | 0.7 | EF | 365 Days/year |
| Cu | 0.04 | ED | 70 years |
| Cd | 0.0005 | BW | 70 Kg |
| Zn | 0.3 | AT | 25550 days |

The pollution indices and other hazard indices were calculate using the follow shown below:

#### 2.2.1. Degree of Contamination ($C_d$)

$$C_d = \sum_{i=1}^{n} C_{fi} \quad (1)$$

$C_{fi}$ is a sub-index showing contamination factor, and is calculated according to Eq. (2)

$$C_{fi} = \frac{CM_i}{CS_i} - 1 \quad (2)$$

$C_{fi}$, $CM_i$, and $CS_i$, are the contamination factor, analytical value, and upper permissible concentration for $i^{th}$ component respectively, while n denotes the 'normative value'.

#### 2.2.2. Heavy Metal Pollution Index (HPI)

$$HPI = \sum_{i=1}^{n} WiQi / \sum_{i=1}^{n} Wi \quad (3)$$

Where: Wi and Qi are unit weight and the sub-index of the $i^{th}$ parameter, respectively, and n is the number of parameters considered. The sub-index (Qi) is calculated according to Eq. (4).

$$Qi = \sum_{i=1}^{n} \frac{|Mi - Ii|}{Si - Ii} \times 100 \quad (4)$$

#### 2.2.3. Heavy Metal Evaluation Index (HEI)

$$HEI = \sum_{i=1}^{n} \frac{H_c}{H_{mac}} \quad (5)$$

Where:
$H_c$ = Analyzed concentration of the respective parameter.
$H_{mac}$ = Maximum admissible concentration of the respective parameter.

#### 2.2.4. Hazard Quotient (HQ)

$$HQ = \frac{ADD}{RfD} \quad (6)$$

Where ADD is the average daily dose, which is calculated using equation 7.

$$ADD_i = \frac{Ci \times IR \times EF \times ED}{BW \times AT} \quad (7)$$

Where:
$C_i$ = Concentration of the monitored parameter
IR = Ingestion rate
EF = Exposure frequency
ED = Exposure Duration
BW = Body weight
AT = Average time

### *2.2.5. Hazard Index (HI)*

$$HI = \sum HQ_i \quad (8)$$

### *2.2.6. Chronic Daily Intake CDI*

Chronic daily intake is calculated using modified equation 9 [26].

$$CDI = Ci \times \frac{DI}{BW} \quad (9)$$

## 3. Results and Discussion

### *3.1. Correlation of Trace Metals in the Water Samples*

As shown in Table 3, the correlation among the analyzed trace metals in water samples are expressed by Pearson coefficient. Copper, selenium, and manganese had strong correlation with zinc (r = 0.933), chromium III (r = 0.827), and manganese (r = 0.63), respectively. Negative correlations were also observed, for instance, iron had negative correlations with cadmium, cobalt, and chromium (III). The correlation between same parameters is not necessarily uniform, and it can remarkably differ from one study area to another. In the context of analytical chemistry, the correlation matrix is both fundamental and pervasive concept, and it does not necessarily mean the presence of causality among the analyzed parameters.

***Table 3.*** *The Correlation between Trace metals in the water samples.*

| Correlations | | | | | | | | | | | | |
|---|---|---|---|---|---|---|---|---|---|---|---|---|
| | **Cd** | **Co** | **Cr(III)** | **Cu** | **Fe** | **Li** | **Mn** | **Mo** | **Sb** | **Se** | **Ti** | **Zn** |
| Cd | 1 | -0.13 | 0.292 | 0.172 | -0.16 | 0.55 | 0.068 | 0.227 | -0.17 | 0.274 | 0.191 | 0.199 |
| Co | | 1 | -0.43 | -0.219 | -0.18 | -0.22 | -0.25 | -0.209 | 0.115 | -0.47 | -0.374 | -0.231 |
| Cr(III) | | | 1 | 0.06 | -0.24 | 0.398 | -0.17 | 0.396 | -0.24 | 0.827 | 0.705 | 0.007 |
| Cu | | | | 1 | 0.553 | 0.023 | 0.63 | 0.032 | 0.043 | 0.137 | 0.175 | 0.933 |
| Fe | | | | | 1 | -0.07 | 0.406 | -0.106 | -0.04 | -0.15 | 0.26 | 0.616 |
| Li | | | | | | 1 | 0.138 | 0.339 | -0.25 | 0.325 | 0.353 | 0.091 |
| Mn | | | | | | | 1 | -0.035 | -0.04 | 0.089 | -0.094 | 0.65 |
| Mo | | | | | | | | 1 | -0.34 | 0.382 | 0.31 | 0.046 |
| Sb | | | | | | | | | 1 | 0.019 | -0.247 | -0.088 |
| Se | | | | | | | | | | 1 | 0.628 | 0.011 |
| Ti | | | | | | | | | | | 1 | 0.153 |
| Zn | | | | | | | | | | | | 1 |

### *3.2. Cadmium, Cobalt, Cr(III) and Copper in Different Water Sources*

The concentrations of cadmium, cobalt, chromium III, and copper were in the ranges of 0.0001 – 0.0014, BDL – 0.0009, BDL – 0.0465, BDL – 0.111 $mgL^{-1}$ respectively (Table-1 and Figure A1). In decreasing order of the concentrations, the general trend of the analyzed trace metals in different water sources were: chromium (III) > copper > cadmium > cobalt (Figure 2). Cadmium was detected in all 30 sampling areas followed by copper, chromium (III), and cobalt, which were detected in 18, 14, and 14 of the sampling sites respectively (Figure A1). While the natural sources of cadmium in the analyzed water samples cannot be directly excluded, the detected concentration of cadmium are associated with activities in which cadmium is used as important component in different application, such as pigments, plastics, and battery.

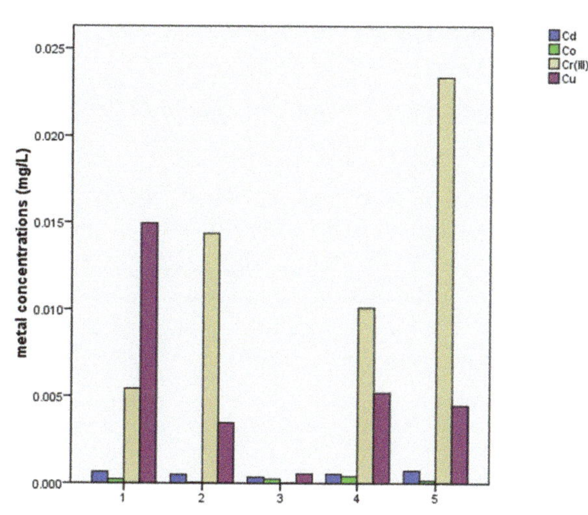


***Figure 2.*** *Cadmium, Cobalt, Cr(III) and Copper in Different Water Sources.*

### 3.3. Iron, Lithium, Manganese and Molybnenumin Different Water Sources

The concentrations of iron, lithium, manganese and molybdenum were in the ranges of BDL – 0.035; 0.0002 – 0.0985; BDL – 0.0089, and BDL – 0.0028 $mgL^{-1}$ respectively (Table-1 and Figure A1).The general trend of the analyzed trace metals in different water sources in decreasing order of their concentrations were: lithium > iron > manganese > molybdenum (Figure 3). Lithium was detected in all 30 sampling areas followed by iron, manganese, and molybdenum, which were detected in 24, 14, and 3 of the sampling sites respectively (Figure A1). As in the case of cadmium, the detected concentration of lithium are associated with anthropogenic activities in which lithium-containing materials such as batteries are improperly disposed in the environment. Because the soil is the major sink of all the metals in Earth crust, therefore the occurrence and the detected concentrations of the trace metals in water sources is not an anomalous event.

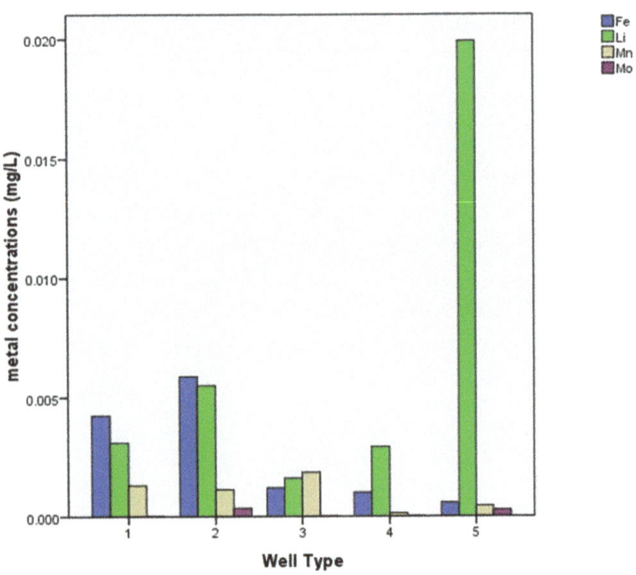


***Figure 3.*** *Iron, Lithium, Manganese and Molybnenum in Different Water Sources.*

### 3.4. Antimony, Selenium, Titanium and Zinc in Different Water Sources

The concentrations of antimony, selenium, titanium and zinc were in the ranges of BDL – 0.0074; BDL – 0.0135; 0.00031 – 0.0071, and 0.00031 – 0.0071 $mgL^{-1}$ respectively (Table-1 and Figure A1).The general trend of the analyzed trace metals in different water sources in decreasing order of their concentrations were: zinc > titanium > selenium > antimony (Figure 4). Zinc and titanium were detected in all 30 sampling areas followed by antimony, and selenium, which were detected in 24, and 16 of the sampling sites respectively (Figure A1). These trace metals are naturally present in the earth crust; however, anthropogenic pressure may enhance their levels in a given environment. The leachates of domestic effluents with residues of titanium containing shampoos and soaps may penetrate the soil and conduit into groundwater resources.

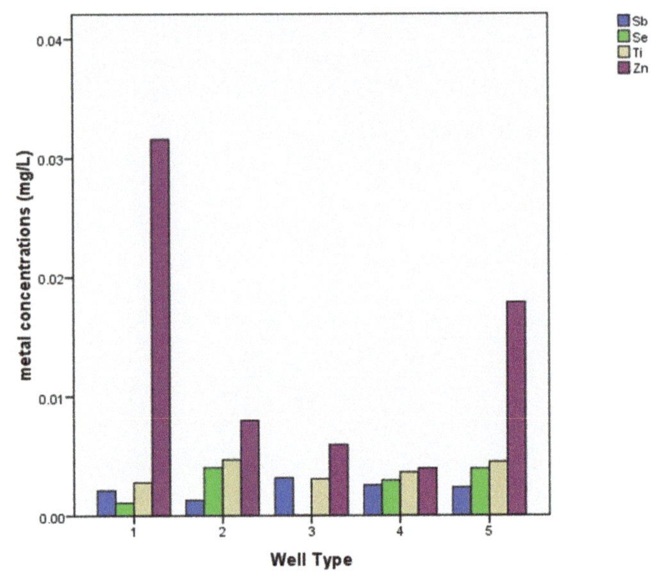


***Figure 4.*** *Antimony, Selenium, Titanium, and Zinc in Different Water Sources.*

### 3.5. Occurrence of Metals in Water Sources

Table-4 gives a summary of the occurrence of metals in five different sources (CHDW, OHDW, SW, PBW, and BWP) from which water samples were collected.Cadmium, zinc, and lithium prevailed in all water sources. Titanium also existed in more than 80% of the water sources. Molybnenum abundance was the least inthe prevailing metals.

***Table 4.*** *Percentage occurrence of the analyzed parameters in water sources.*

| Water source | (%) Cd | % Co | % Cr III | % Cu | % Fe | % Li | % Mn | % Mo | % Sb | % Se | % Ti | % Zn |
|---|---|---|---|---|---|---|---|---|---|---|---|---|
| CHDW | 100 | 12.5 | 25 | 50 | 87.5 | 100 | 50 | 0 | 87.5 | 50 | 100 | 100 |
| OHDW | 100 | 25 | 75 | 75 | 87.5 | 100 | 25 | 12.5 | 62.5 | 75 | 100 | 100 |
| SW | 100 | 50 | 0 | 50 | 100 | 100 | 100 | 0 | 100 | 0 | 100 | 100 |
| PBW | 100 | 80 | 60 | 40 | 80 | 100 | 60 | 0 | 100 | 40 | 100 | 100 |
| BWP | 100 | 37.5 | 50 | 75 | 50 | 100 | 37.5 | 25 | 25 | 50 | 87.5 | 100 |

### *3.6. Human Health Risk Assessment Due to Toxic Metal*

Heavy metals are toxic elements, and they are in the list of the noxious pollutants in the context of environmental and health sciences. Ingestion of considerable amount of heavy metals via different exposure pathways may result into several detrimental ill effects. Using the four heavy metals (Cd, Zn, Cu, and Fe), the degree of contamination ($C_d$) in all water sources were less than 1, thus they all fall in to the category of low degree of contamination. Similar trend was found in the study conducted in Tehran water sources [27].

The general trend of the average daily dose (ADD) of Zn, Cu, Fe, and Cd in water sources is:

$$Zn > Cu > Fe > Cd.$$

With regard to severity of the toxicity, cadmium is highly toxic, so this trend likely does not show alarming situation in the context of cadmium risk in the water sources. As the severity of toxicity is governed by several factors such as dose, nutrition, age and even life style. Therefore, this trend might not guarantee the absence of human health risk (Figure 5).

For the case of $C_d$, there is a remarkable spatial variation in different water sources. While for HPI the spatial variations is less remarkable. HPI for all water samples were well below the critical value (Figure 6).

All water samples (100%) to some extent showed some levels of HEI. Moreover, all water samples showed some ADD levels with respect to Cd, Zn, and copper metal, while iron metal was over 60%. Nevertheless, the spatial variations of HI are more apparent as compared to HEI (Figure 7 & Figure A2).

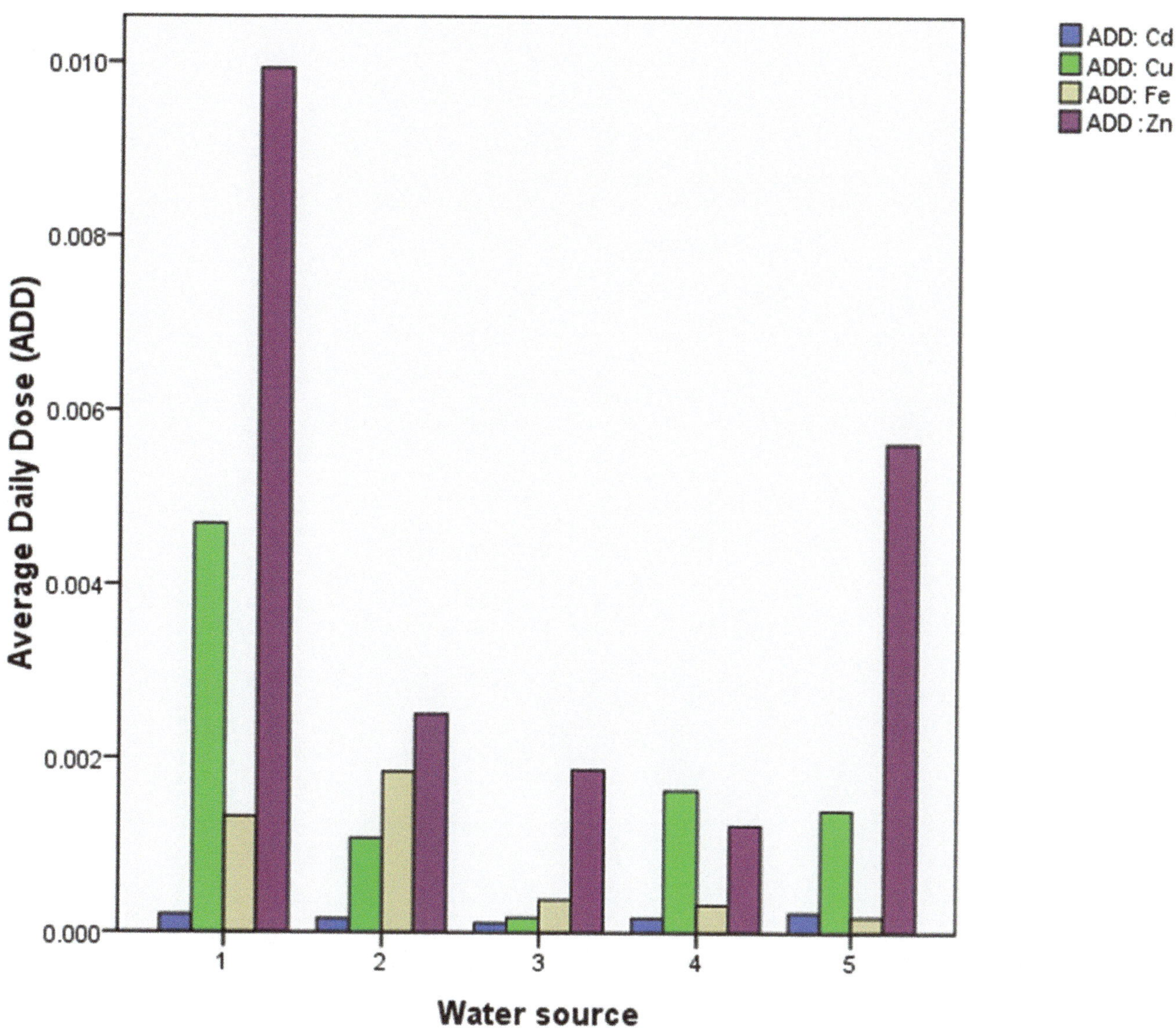


***Figure 5.*** *ADD of Cd. Cu, Fe and Zn.*

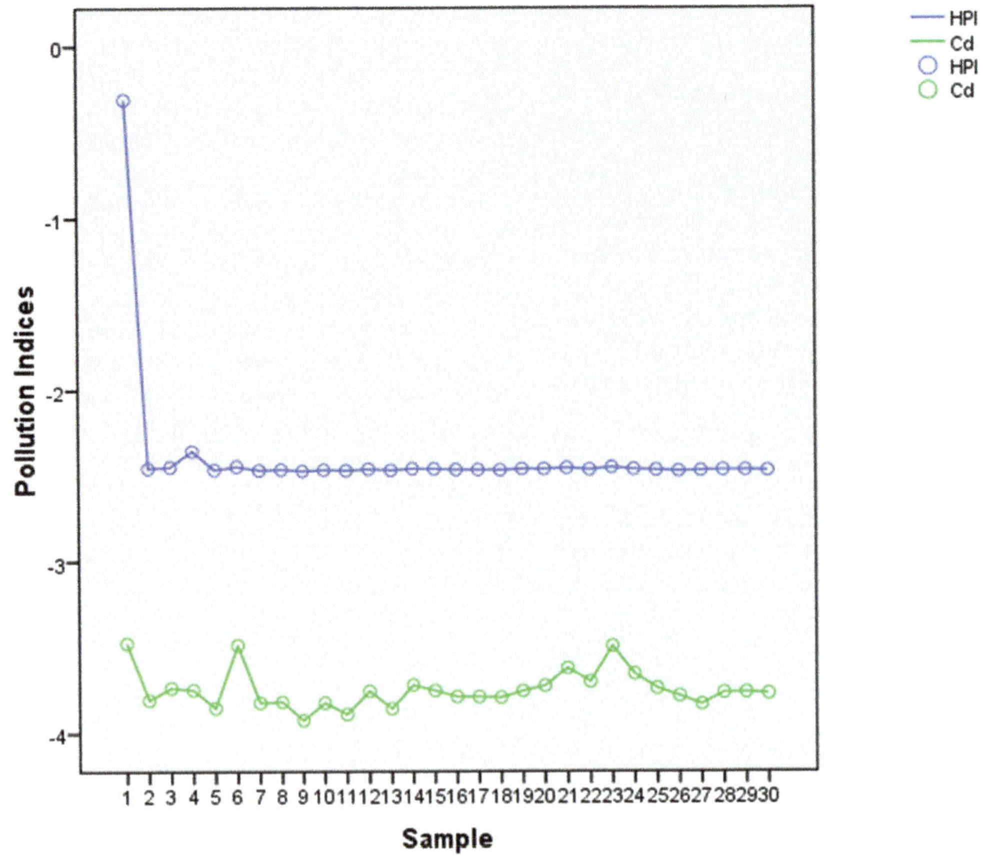


***Figure 6.*** *HPI and $C_d$ in water samples.*

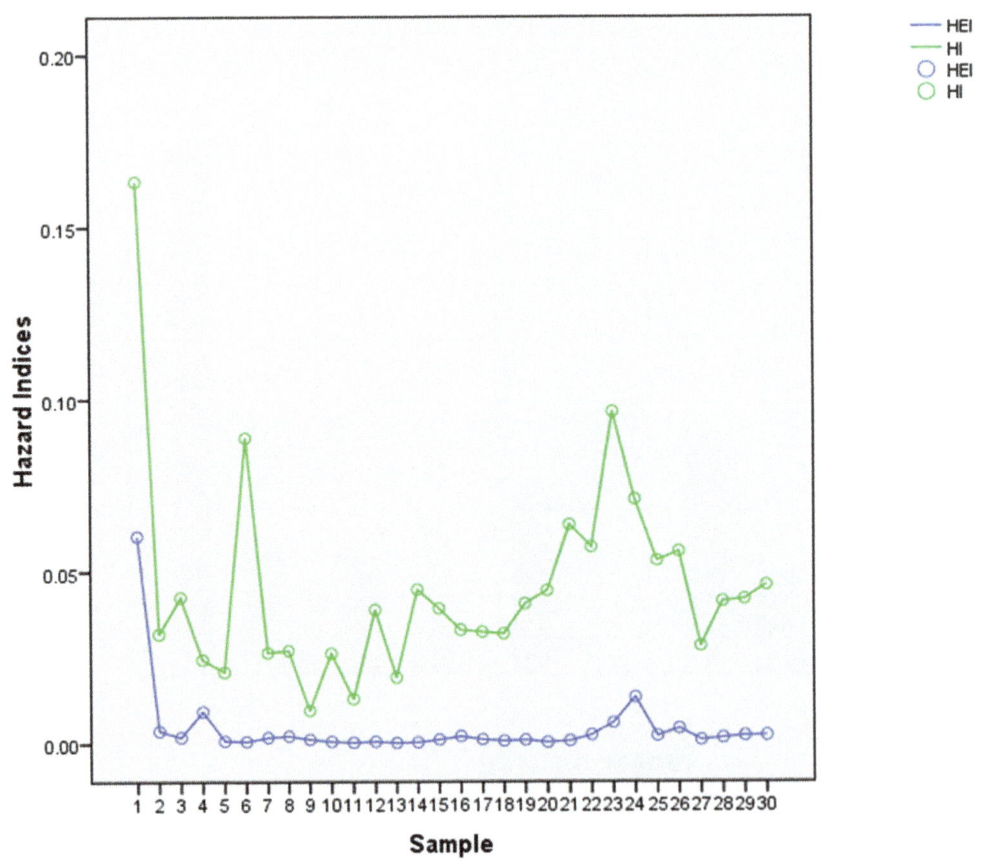


***Figure 7.*** *Hazard index in water samples.*

## 4. Conclusion

From the twelve analyzed metals, cadmium, zinc, and lithium existed at varying concentrations in all water sources. Other metals were also detected though at relatively low abundances. Generally, from risk assessment on trace metals using risk indices ($C_d$, HPI, HI and HEI), the analyzed water sources are not at the alarming health risk, however, the increasing level of environmental pollution, water sources might be potential sink of contaminants, this is significant reason that make the treatment of drinking water and monitoring implementation inevitable.

## Appendix

| Sampling site | Cd | Co | Cr III | Cu | Fe | Li | Mn | Mo | Sb | Se | Ti | Zn | Well Type |
|---|---|---|---|---|---|---|---|---|---|---|---|---|---|
| 1 | 0.8 | nd | 0.2 | 111 | 28.5 | 3.8 | 8.9 | nd | 2.2 | 1 | 4.2 | 234 | 1 |
| 2 | 0.4 | nd | 1.4 | 4.1 | 4.5 | 4.7 | 8.9 | nd | 1.6 | 7.5 | 3.2 | 13.9 | 2 |
| 3 | 0.6 | nd | 3.5 | 2.6 | 5.4 | 7.5 | nd | nd | 3.3 | 0.2 | 4 | 2.8 | 2 |
| 4 | 0.1 | nd | 2.3 | nd | 35 | 4.2 | nd | nd | 1.4 | 1 | 7.1 | 23.1 | 2 |
| 5 | 0.3 | 0.5 | nd | 1.1 | 1.9 | 3 | 0.1 | nd | 3.3 | nd | 5.8 | 1.6 | 3 |
| 6 | 1.4 | 0.5 | 0.5 | nd | 1.7 | 2.2 | nd | nd | 2.1 | 1.1 | 4.5 | 0.7 | 4 |
| 7 | 0.4 | nd | nd | nd | 0.5 | 0.2 | 3.6 | nd | 3 | nd | 0.3 | 10.3 | 3 |
| 8 | 0.4. | 0.1 | nd | nd | 1.6 | 0.5 | 0.39 | nd | 0.7 | nd | 3 | 11.3 | 4 |
| 9 | 0.1 | 0.9 | 11.7 | 3.4 | 0.9 | 0.5 | nd | nd | 2.2 | nd | 3.1 | 3 | 4 |
| 10 | 0.4 | 0.6 | nd | 0.6 | 0.9 | 1.3 | 0.2 | nd | 1.2 | nd | 2.6 | 1.5 | 1 |
| 11 | 0.2 | 0.5 | nd | nd | 0.8 | 0.9 | 0.2 | nd | 3 | nd | 1.7 | 0.5 | 4 |
| 12 | 0.6 | 0.4 | nd | 0.7 | 1 | 0.8 | 0.1 | nd | 3.9 | nd | 2.4 | 1 | 5 |
| 13 | 0.3 | nd | nd | nd | 0.7 | 0.8 | 0.9 | nd | 2.7 | 0.1 | 2.5 | 0.19 | 1 |
| 14 | 0.7 | 0.6 | nd | nd | 1.5 | 0.6 | 0.4 | nd | 3.6 | nd | 0.8 | 0.13 | 1 |
| 15 | 0.6 | 0.6 | nd | 0.4 | 1.6 | 7.6 | 0.1 | nd | 1.2 | nd | 3.7 | 4.5 | 5 |
| 16 | 0.5 | 0.1 | nd | nd | 0.7 | 4.4 | nd | nd | 3.4 | nd | 1.9 | 11.7 | 5 |
| 17 | 0.5 | nd | nd | nd | 0.7 | 3.1 | nd | nd | 3.8 | nd | 2.3 | 6.1 | 1 |
| 18 | 0.5 | 0.1 | nd | nd | 0.1 | 7.7 | nd | nd | 2.8 | nd | 3.8 | 4.3 | 2 |
| 19 | 0.6 | 0.1 | nd | 2.6 | 1.7 | 3.5 | nd | nd | 1.1 | nd | 2.7 | 1.3 | 2 |
| 20 | 0.7 | 0.3 | nd | 0.1 | 0.6 | 0.5 | nd | nd | 2.6 | 0.4 | 1.7 | 0.9 | 1 |
| 21 | 1 | 0.3 | nd | nd | 1 | 4.1 | nd | nd | 0.6 | nd | 2 | 3 | 1 |
| 22 | 0.8 | nd | 43.2 | 7.8 | nd | 10.6 | nd | nd | nd | 6.9 | 6 | 7 | 1 |
| 23 | 1.4 | nd | 38 | 6.5 | 0.7 | 98.5 | 2.9 | 1 | nd | 7.1 | 6 | 29.2 | 5 |
| 24 | 0.9 | nd | 46.5 | 8.4 | nd | 16.5 | nd | 0.9 | 0.4 | 6.9 | 6 | 74.6 | 5 |
| 25 | 0.7 | nd | 42.6 | 11.4 | nd | 7.6 | nd | nd | 7.4 | 12.1 | 6 | 1.9 | 5 |
| 26 | 0.6 | nd | 38.3 | 22.6 | nd | 10.5 | 0.1 | nd | 4.6 | 13.5 | 5.7 | 4.1 | 4 |
| 27 | 0.4 | nd | 36.4 | 3.9 | nd | 4 | nd | nd | nd | 1.2 | 5.3 | 2.3 | 5 |
| 28 | 0.6 | nd | 36.4 | 3.9 | nd | 3.6 | nd | nd | Nd | 6.4 | 5.1 | 5.8 | 2 |
| 29 | 0.6 | nd | 37.8 | 4.2 | 0.2 | 6.4 | 0.1 | nd | Nd | 8 | 5.8 | 9 | 2 |
| 30 | 0.6 | nd | 33.5 | 10.2 | 0.1 | 6.4 | nd | 2.8 | nd | 8.9 | 5.6 | 3.7 | 2 |

***Figure A1.*** *Analyzed parameters all in* $\mu gL^{-1}$*; nd: not detected; 1 = CHDW; 2 = OHDW; 3 = SW; 4 = PBW; 5 = BWP.*

| Sample | $C_d$ | HPI | HEI | ADD: Cd | ADD: Cu | ADD: Fe | ADD: Zn | HI | Water Source |
|---|---|---|---|---|---|---|---|---|---|
| 1 | -3.473 | -0.3080 | 0.06033 | 2.51E-05 | 0.003489 | 0.00090 | 0.007351 | 0.1633 | 1 |
| 2 | -3.805 | -2.4553 | 0.00369 | 1.26E-05 | 0.000129 | 0.00141 | 0.000437 | 0.0318 | 2 |
| 3 | -3.735 | -2.4482 | 0.00184 | 1.89E-05 | 8.17E-05 | 0.00170 | 0.000088 | 0.0425 | 2 |
| 4 | -3.747 | -2.3572 | 0.00938 | 3.14E-06 | 0 | 0.01100 | 0.000726 | 0.0244 | 2 |
| 5 | -3.851 | -2.4646 | 0.00079 | 9.43E-06 | 3.46E-05 | 0.00060 | 5.03E-05 | 0.0207 | 3 |
| 6 | -3.485 | -2.4459 | 0.00061 | 0.000044 | 0 | 0.00053 | 0.000022 | 0.0888 | 1 |
| 7 | -3.822 | -2.4674 | 0.00181 | 1.26E-05 | 0 | 0.00016 | 0.000324 | 0.0264 | 3 |
| 8 | -3.816 | -2.4637 | 0.00214 | 1.26E-05 | 0 | 0.00050 | 0.000355 | 0.0270 | 4 |
| 9 | -3.925 | -2.4722 | 0.00119 | 3.14E-06 | 0.000107 | 0.00028 | 9.43E-05 | 0.0097 | 4 |
| 10 | -3.822 | -2.4661 | 0.00055 | 1.26E-05 | 1.89E-05 | 0.00028 | 4.71E-05 | 0.0262 | 1 |
| 11 | -3.889 | -2.4697 | 0.00024 | 6.29E-06 | 0 | 0.00025 | 1.57E-05 | 0.0130 | 4 |
| 12 | -3.756 | -2.4623 | 0.00053 | 1.89E-05 | 0.000022 | 0.00031 | 3.14E-05 | 0.0388 | 5 |
| 13 | -3.856 | -2.4683 | 0.00019 | 9.43E-06 | 0 | 0.00022 | 5.97E-06 | 0.0192 | 1 |
| 14 | -3.719 | -2.4587 | 0.00038 | 0.000022 | 0 | 0.00047 | 4.09E-06 | 0.0447 | 1 |
| 15 | -3.752 | -2.4603 | 0.00114 | 1.89E-05 | 1.26E-05 | 0.00050 | 0.000141 | 0.0392 | 5 |
| 16 | -3.787 | -2.4650 | 0.00208 | 1.57E-05 | 0 | 0.00022 | 0.000368 | 0.0330 | 5 |
| 17 | -3.789 | -2.4649 | 0.00118 | 1.57E-05 | 0 | 0.00022 | 0.000192 | 0.0324 | 1 |
| 18 | -3.792 | -2.4669 | 0.00079 | 1.57E-05 | 0 | 0.00003 | 0.000135 | 0.0319 | 2 |
| 19 | -3.754 | -2.4606 | 0.00100 | 1.89E-05 | 8.17E-05 | 0.00053 | 4.09E-05 | 0.0407 | 2 |
| 20 | -3.724 | -2.4618 | 0.00037 | 0.000022 | 3.14E-06 | 0.00019 | 2.83E-05 | 0.0444 | 1 |
| 21 | -3.621 | -2.4552 | 0.00081 | 3.14E-05 | 0 | 0.00031 | 9.43E-05 | 0.0636 | 1 |
| 22 | -3.700 | -2.4645 | 0.00251 | 2.51E-05 | 0.000245 | 0.00000 | 0.00022 | 0.0571 | 1 |
| 23 | -3.490 | -2.4516 | 0.00609 | 0.000044 | 0.000204 | 0.00022 | 0.000918 | 0.0965 | 5 |
| 24 | -3.653 | -2.4627 | 0.01353 | 2.83E-05 | 0.000264 | 0.00000 | 0.002345 | 0.0710 | 5 |
| 25 | -3.738 | -2.4672 | 0.00226 | 0.000022 | 0.000358 | 0.00000 | 5.97E-05 | 0.0532 | 5 |
| 26 | -3.782 | -2.4724 | 0.00440 | 1.89E-05 | 0.00071 | 0.00000 | 0.000129 | 0.0559 | 4 |
| 27 | -3.830 | -2.4701 | 0.00106 | 1.26E-05 | 0.000123 | 0.00000 | 7.23E-05 | 0.0284 | 5 |
| 28 | -3.763 | -2.4667 | 0.00166 | 1.89E-05 | 0.000123 | 0.00000 | 0.000182 | 0.0414 | 2 |
| 29 | -3.761 | -2.4662 | 0.00226 | 1.89E-05 | 0.000132 | 0.00006 | 0.000283 | 0.0420 | 2 |
| 30 | -3.769 | -2.4683 | 0.00235 | 1.89E-05 | 0.000321 | 0.00003 | 0.000116 | 0.0462 | 2 |

***Figure A2.*** *Pollution indices in water samples.*


## Acknowledgement

We highly acknowledge to His Majesty Government of Brunei Darussalam for funding this project. State University of Zanzibar for supporting the study. We also thank Hjh Faridah, Azzlin, Najib and other UBD chemistry laboratory staffs for their tireless assistance.


## References


[1] Mohamed AAJ, Ibrahim AR, Lee HL, Hanna M, Mohamed M, Fadilah T, Nur AR (2016) Assessment of Groundwater Quality Parameters in Northern Region of Zanzibar Island, J Chem Eng Bioanalyticalchem, 1: 1-8.

[2] Shittu A, Chifu EN, Abdussalam BS, Hafeez YH (2016) Evaluation of Radioactivity Concentration in Drinking Water Collected from Local Wells and Boreholes of Dutse Town, North West, Nigeria Int J Envi Monit Analy, 4: 6-11.

[3] Mohamed AAJ, Ibrahim AR, Lee HL, Khamis SA, Mwevura H, Mbwana KH. (2016) An Index Approach to Metallic Pollution in Groundwater Sources of South Region of Pemba Island, Scie J Analyt Chem, 4: 12-21.

[4] Armon R, Kitty J (1994) The Health Dimension of Groundwater Contamination. In: Groundwater Contamination and Control, Holler (Ed.), Marcel Dekker, Inc, New York, USA.

[5] Epa.ohio.gov.gwqcp.fluoride_factsheet, Fluoride in Ohio's Ground water, Fact Sheet 2012-01, Series on Ohio's Ground water Quality, March 2012.

[6] Oiste AM (2014) Groundwater quality assessment in urban environment, Int J EnvirSci Tech, 11:2095–2102.

[7] Nagarajan R, Rajmohan N, Mahendran U, Senthamilkumar S(2010) Evaluation of groundwater quality and its suitability for drinking and agriculture use in Thanjavur city, Tamil nadu, India, Envir Monit Asse, 171: 289-308.

[8] Mattiesen H (2008) Detailed chemical analyses of groundwater as a tool for monitoring urban archaeological deposits: results from Bryggen in Gergen, J Archaeological Sci, 35: 1378–1388.

[9] Milenkovic N, Damjanovic M, Ristic M (2005) Study of heavy metal pollution in sediments from the iron gate (danuae river), serbia and Montenegro, Polish Journal of Environmental Studies, 14: 781-787.

[10] Wu JL, Zeng, HA, Yu H, Ma L, Xu, LS, Qin BQ (2012) Water and sediment quality in lakes along the middle and lower reaches of the Yangtze River, China Water Resour. Manage, 12:3601–3618.

[11] Ikejimba CC, Sakpa S (2014) Comparative study of some heavy metals' concentrations in water and Tympanotonusfuscatusvar radula samples of Egbokodo River, Warri, Nigeria, Int J Modern Biol Re, 2: 7-15.

[12] Valavanidis A, Vlachogianni T (2010) Metal pollution in ecosystem, ecotoxicology studies & risk assessment in the marine environment. Dept. of Chemistry, University of Athens University Campus Zografou, 15784 Athens, Greece.

[13] EPA (1986) Air quality criteria document for lead. Environmental Criteria and Assessment Office, US Environmental Protection Agency, Research Triangle Park, North Carolina.

[14] Rajappa B, Manjappa S, Puttaiah ET (2010) Monitoring of heavy metal concentration in groundwater of Hakinaka Taluk, India, Contemporary Eng Sci, 3: 183-190.

[15] Nriagu JO (1988) A silent epidemic of environmental metal poisoning? Environmental Pollution, 50: 139-161.

[16] Osibanjo O, Majolagbe AO (2012) Physicochemical quality assessment of groundwater based on land use in Lagos city, Southwest, Nigeria, Chem J, 2: 79-86.

[17] Buschmann J, Berg M, Stengel C, Winkel L, Sampson, ML, Trang, PTK, Viet PH (2008) Contamination of drinking water resources in the Mekong Delta floodplains: Arsenic and other trace metals pose serious health risks to population, EnvirInt, 34, 756–764.

[18] Fraga CG, Oteiza PI (2002) Iron toxicity and antioxidant nutrients, Toxicology, 180: 23-32.

[19] EPA (2013) Consumer Factsheet on: COPPER. Retrieved on February 26, 2016 from http://water.epa.gov/drink/contaminants/index.cfm

[20] APHA (2005) Standard methods for the examination of water and waste water, 21st edition, American Public Health Association, Washington, DC, USA.

[21] Prasad B, Bose JM (2001) Evaluation of heavy metal pollution index for surface and spring water near a limestone mining area of the lower Himalayas, Envir Geology, 41: 183–188.

[22] Backman B, Bodis D, Lahermo P, Rapant S (1997) Application of a groundwater contamination index in Finland and Slovakia, Envir Geology, 36: 55–64.

[23] Rubio B, Nombela MA, Vilas F ( 2000) Geochemistry of Major and Trace Elements in Sediments of the Ria de Vigo (NW Spain): an Assessment of Metal Pollution, Marine Pollution Bulletin, 40: 968-980.

[24] Edet AE, Offiong OE (2002) Evaluation of water quality pollution indices for heavy metal contamination monitoring. A study case from Akpabuyo - Odukpani area, Lower Cross River Basin, (southeastern Nigeria), Geo J, 57: 295–304.

[25] Wongsasuluk P, Chotpantarat S, Siriwong W, Robson M (2014) Heavy metal contamination and human health risk assessment in drinking water from shallow groundwater wells in an agricultural area in Ubon Ratchathani province, Thailand, Environ Geochem Health, 36: 169–182.

[26] Muhammed S, Shari MT, Khan S (2011) Health Risk Assessment of Heavy metals and their source apportionment in drinking water of Kohitan region, northern Pakistan. Microchem J, 98: 334-343.

[27] Nasrabadi T (2015) An Index Approach to Metallic Pollution in River Waters Int. J. Environ. Res., 9(1): 385-394.

# 13

# Monitoring Soil and Underground Water EC in a Rice Field Affected by the Great East Japan Earthquake

**Yoshiko Muto, Eiichi Kurashima*, Kiyohito Yamamoto, Koji Harashina**

Faculty of Agriculture, Iwate University, Morioka, Japan

**Email address:**
mutoy@iwate-u.ac.jp (Y. Muto), kurasima@iwate-u.ac.jp (E. Kurashima), yama3@iware-u.ac.jp (K. Yamamoto), hkoji@iwate-u.ac.jp (K. Harashina)

**Abstract:** The Otomo rice field zone in Rikuzentakata City, Iwate Prefecture was catastrophically damaged by large-scale subsidence and the tsunami that followed the Great East Japan earthquake, which occurred on March 11th, 2011. Electrical conductivity of underground water, bulk electrical conductivity of the soil, and various meteorological elements were observed at a fixed point for eight months. The electrical conductivity of underground water fluctuated irregularly at the beginning of the observation period, temporarily reaching up to 5 $Sm^{-1}$ at sea level. After some time, an overall decreasing trend prevailed, and when the observations ended the conductivity of the water had dropped to 0.55 $Sm^{-1}$. The bulk electrical conductivity of the soil also decreased gradually, from 0.4 to 0.3 $Sm^{-1}$, over the eight months, which is likely linked to the interactions between rainfall and seawater intrusions. The decrease in soil conductivity has been more gradual here than in the regions affected by the tsunami following the earthquake in the Indian Ocean off Sumatra on December 26th, 2004, and in our study area it has not yet decreased to a level that would allow the resumption of rice farming. It is proposed that this difference is a result of the subsidence in Iwate Prefecture.

**Keywords:** East Japan Great Earthquake, Subsidence, Rice Field, Hydrological Elements, Electrical Conductivity

## 1. Introduction

The fourth largest earthquake in recorded history occurred in the Tohoku region, on the Pacific shore of Japan, on March 11th, 2011. The magnitude of this earthquake was 9.0, and its hypocenter was located 70 km off the Oshika peninsula, Miyagi Prefecture [1]. The tsunami caused by this earthquake exceeded 10 m across 530 km of the Tohoku coastline, and exceeded 20 m over 200 km. The maximum tsunami height in the region was measured as 40.1 m, at Ryori point in Ofunato City, Iwate Prefecture; this was also the maximum height across all of Japan [2]. The desalinization effect at farmland was monitored after the tunami in a lot of studies [e.g., 3, 4, 5].

In this study, we investigate a rice field zone in Rikuzentakata City, which received considerable damage during the 2011 tsunamis. The Otomo rice field zone in Rikuzentakata City, Iwate Prefecture is located at around 39°N and 142°E, and has a total area of 120 ha, developed behind the reclamation dyke constructed along the Hirota Bay coastline, which faces the Pacific Ocean.

The reclamation dyke in this area was constructed with a view to preventing tsunami disasters; it was begun in 1960, immediately following the Chile earthquake tsunami, and completed in 1967. The rice field area it encloses is the largest along the entire coastline of Iwate Prefecture. During the 2011 event, the reclamation dyke was completely destroyed, and the tsunami flowing from Kadonohama Bay to the east also penetrated the rice field zone, resulting in catastrophic damage from large-scale subsidence, erosion, and the large amounts of debris that the seawater carried. Following this, a temporary dyke was completed in October 2011, the chronic flooding of the area by seawater was eased, and the removal of debris was begun.

Restoration of the area has been ongoing rapidly since June 2013, following the removal of the debris. Land improvement has almost been completed in the affected rice field zones through large-scale backfilling behind the temporary sand bag dyke, and wet rice culture recommenced in 2014.

This study focuses on the collection and analysis of data

over eight months spanning the period of debris removal and the beginning of land restoration. During this measurement period, electrical conductivity is used as an index of salinity in the soil and subsurface water, and is considered along with various other meteorological factors. In 2015, as the reconstruction of a permanent reclamation dyke is begun, this research provides basic information for the management and prevention of salt damage in farmland, which can also be compared with the saltwater damage following the tsunami of the 2004 Sumatran earthquake [e.g., 6].

## 2. Materials and Method

### *2.1. Observation Station and Subsidence of the Surrounding Area*

The single stationary measurement point for this study is located around 1 km southeast of the Hirota Bay coast, at 38.99392°N, 141.6933°E. As the breakage of the water supply facilities differs across the affected rice fields and the agricultural roads around the rice fields are severed in many places, the fixed-point observation station was chosen with a consideration of the ease of communication with land owners, leaseholders, and improvement organizations, and of physical access to the point. Figure 1 shows the location of the Otomo rice field zone and the approximate site of the observation station, with a panoramic photograph taken from a nearby observatory.

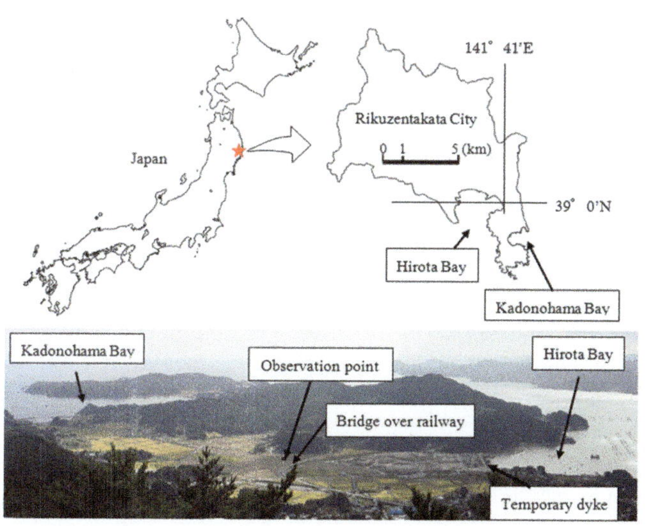


***Figure 1.*** *Location of the study area, and panoramic photograph of the Otomo rice field zone taken from a nearby observatory on October 2nd, 2014.*

Figure 2(a) shows a panoramic view of the study area one month after the tsunami struck, when the rice field was flooded with seawater and was covered with a large amount of debris. Figure 2(b) shows the rice field zone two years after the tsunami, with the specific location of the observation point shown.

The horizontal distribution of subsidence around the observation point is estimated using a digital elevation model (DEM). Altitude information before the earthquake is provided by a numeric map with 50 m mesh, constructed from the contour lines of 1: 25,000 topographical maps, while altitudes following the earthquake are given as basic map information on a 5 m mesh acquired during aerial laser surveys [7]. A precise comparison of altitudes is problematic because the two datasets differ in their measurement methods and spatial resolutions, however the difference in 5 m mesh altitude after the earthquake that was the nearest 50 m mesh altitude point before the earthquake was assumed to represent the ground subsidence level.

(a) April 13th, 2011

(b) April 11th, 2013

***Figure 2.*** *Photographs taken from the railway bridge that passes through the rice field zone.*

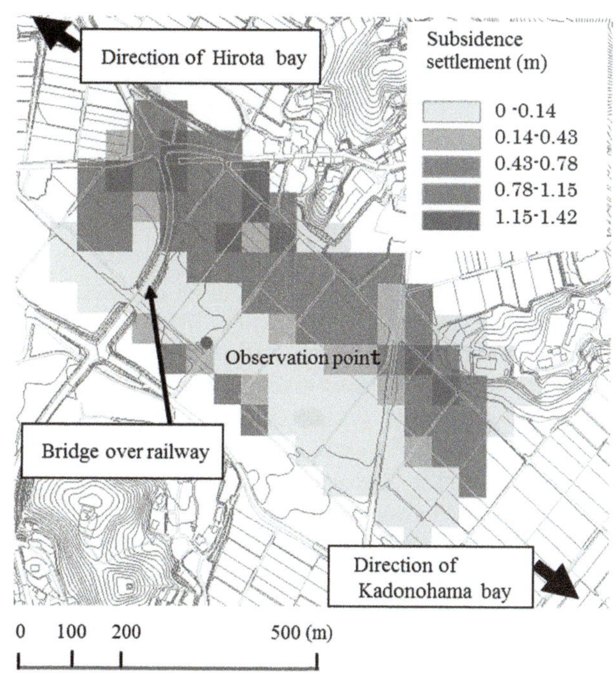


***Figure 3.*** *Approximate horizontal distribution of subsidence around the observation point following the 2011 earthquake.*

Figure 3 shows the horizontal distribution of subsidence around the observation point. From this, it can be seen that the ground subsidence level is greatest to the northeast of the observation point. In support of this, the annual average difference in tide levels at Ofunato tidal observatory, 6.5 km from the observation point, was found to be 1.4 m in 2012 [8], and the maximum ground subsidence in this region is in accordance with the daily tidal variation.

### *2.2. Observation of Electrical Conductivity and Meteorological Elements*

Figure 4 shows the sensor arrangement at the observation point. The structure in the upper part of the image is the above-mentioned bridge over the railway. A pipe was laid at a depth of 1.28 m, in order to sample from the lower groundwater levels, and a pressure-type water gauge and electrical conductivity sensors were set up at the bottom of the pipe. Observations of the soil by relative permittivity sensors (Hydra probes) may be influenced by residual water and irregularities in the Earth's surface, and so the sensors were installed in two comparatively smooth two locations, at a separation of 20 m.

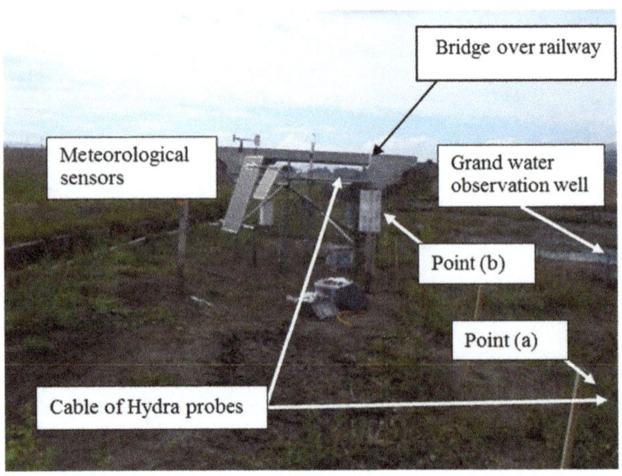


***Figure 4.*** *Arrangement of sensors at the observation station.*

The ground altitudes of the two installation points, (a) and (b), are 0.51 m and 0.46 m, respectively. In consideration of this difference in elevation, the Hydra probes were set up 10 cm and 20 cm below the ground level at point (a), and 5 cm and 15 cm below ground level at point (b), allowing the sensors' installation altitudes to be unified between the two points. Unfortunately, the sensors installed at depths of 20 cm at point (a) and 5 cm at point (b) failed during the early days of the observation period. Physical damage to the sensors is likely to have resulted in aberrant values, and thus their data were excluded from subsequent analyses. The particle size distributions of soils were measured. At observation (a) the soil at 10 cm depth comprised 66% sand, 4% clay, and 30% silt; they can be classified as sandy loam. Similarly, the soil around 15 cm below the surface at observation point (b) was classified as loam, with 55% sand, 4% clay, and 41% silt. Very little vegetation cover was observed around the observation point, although some graminaceous weeds were scattered across the tsunami-stricken rice field.

In parallel with these observations, direct and reflected solar radiations, and downward and upward long wave radiations, precipitation, atmospheric pressure, and soil heat flow were measured. Additionally, in order to estimate the sensible heat flux, three-component wind velocity and sonic temperature were observed using an ultrasonic anemometer. These sensors, with the exception of the rain gauge and the aerotonometer, were arranged on a frame 2 m in height and 2×2 m in length and breadth. The rain gauge was installed in a lower position, and the aerotonometer was set up alongside the data logger. The soil heat flux sensor was installed 1 cm below the ground surface. The ultrasonic anemometer was connected to a CR1000 data logger (Cambell), and the measurement interval was set to 0.1 s. Other sensors were connected to a CR3000 data logger (Cambell), with measurement intervals of 20 s, restricted by the measurement time of the Hydra probe.

A fixed-point camera was set up on the frame with the sensors, in order to visually capture the changing conditions around the observation point. Shooting began on September 12th, but the battery of the camera set up was installed in a low position and was submerged beneath the high tide waters during Typhoon No.17 on September 30th, 2012, becoming inoperational until November 7th. After this, measurements were made every 30 min during the daytime until April 30th; a total of 3552 scenes were captured.

Details of the observed elements and sensor models with their accuracies are shown in Table 1, although air temperature, relative humidity, wind direction and wind velocity data were not used in the following analysis. This study utilized the above-mentioned data from 236 days between September 7th, 2012 and April 30th, 2013. During the beginning of the observation period, construction of the temporary dyke was ongoing as an alternative to the older, destroyed dyke, and associated debris removal was taking place. At the end of the period, it was not possible to continue the observations, due to the beginning of land restoration.

***Table 1.*** *Models and measurement accuracy of sensors used for the observation.*

| Observed element | Manufacture | Model | Measurement accuracy |
|---|---|---|---|
| Air temperature<br>Relative humidity | Vaisala | HMP155 | ±0.13°C<br>±1% (0–90%)<br>±1.7% (90–100%) |
| Wind direction<br>Wind speed | Young | YG5103 | ±3° 0.3 m/s |
| Atmospheric pressure | Vaisala | PTB110 | ±0.3 hPa 20°C |
| 4-Components net radiation | Hukseflux | NR01 | ±10% 12 h totals |
| Heat flow in soil | REBS | PHF-01 | ±5% |
| 3-Component Wind speed Sonic temperature | Young | YG81000 | ±0.05 m/s ±2°C |
| Precipitation | IKEDA | RT-5 | ±3% at 20–80 mm/h |
| Grand water level | STS | ATM/N | ±0.1% FS 5 m |
| Electric conductivity<br>Temperature (Grand water) | TOA DKK | CT-27112B | ±0.5% FS 20 S/m<br>±0.2°C |
| Electric conductivity<br>Temperature (Soil) | Stevens | Hydra probe | ±2% or 0.005 S/m<br>±0.6°C |
| Fix point shooting | Kadec | Eye II | |

### 2.3. History of Hydrology Factors in the Affected Rice Fields

The amount of evaporation is one of the hydrological factors that can be estimated from the observed meteorological elements. The energy equation for the ground surface is given as

$$\left(I_{s}\downarrow - I_{s}\uparrow\right) + \left(I_{ir}\downarrow - I_{ir}\right)\uparrow - I_{g} = I_{sen} + I_{lat} \quad (1)$$

where $I_s\downarrow$ (Wm-2) is the downwelling short wave radiation flux, $I_s\uparrow$ (Wm-2) is the upwelling short wave radiation flux, $I_{ir}\downarrow$ (Wm-2) is the downwelling long wave radiation flux, $I_{ir}\uparrow$ (Wm-2) is the upwelling long wave radiation flux, $I_{sen}$ (Wm-2) is the sensible heat flux, $I_{lat}$ (Wm-2) is the latent heat flux, and $I_g$ (Wm-2) is the soil heat flux. Then, turbulent heat fluxes $I_{sen}$ can be expressed by the eddy covariance method as:

$$I_{sen} = C_{p}\rho_{a}\overline{\theta' w'} \quad (2)$$

where $C_p$ (J kg-1 K-1) is the specific heat of air at a constant pressure, $\rho_a$ (kg m-3) is the air density, $\theta$ is the air temperature (°C), $w$ (m s-1) is the vertical velocity, bar donates the average, and prime marks denote random instantaneous departures from the mean. In using Equation (2), a change in the coordinates based on wind direction, a correction of the attack-dependent error and an internal crosswind correction were applied [9, 10]. The imbalance of the heat budget between the sum of sensible heat and latent heat obtained by the eddy covariance method and other observed elements has been noted, and it has been shown that the ratio of right-hand values to left-hand values of Equation (1) is 0.84 over an average of 50 samples [11]. Therefore, there is a possibility that this represents a margin of error in the amount of evaporation estimated by this method. However, it is difficult to unambiguously identify the causes and patterns of this error from the limited observations of this study.

On the basis of the fixed-point photography, it was estimated that the study area was covered by snow for 11 days, from January 9th to February 26th, 2013. During this time, it is necessary to consider the changes in temperature and phase within the snow cover. However, it was assumed that the amount of evaporation over these 11 days was negligible, on the basis of previous reports that found that evaporation from snow surfaces during the melt season was very small [12].

### 2.4. Corrections of Electrical Conductivity of Underground Water and Bulk Electrical Conductivity of Soil

The measurements of electric conductivity of underground water are based on the 2-AC bipolar method. Owing to the fact that the electrical conductivity of solutions shows considerable temperature dependence (2%/1°C), the electrical conductivity of underground water was converted to its corresponding value based on a temperature of 25°C. The temperature-corrected electrical conductivity at 25°C, $EC_{w25}$ (S m-1) can therefore be expressed as:

$$EC_{w25} = \frac{EC_{wt}}{\left[1 + 0.02\left(t - 25\right)\right]} \quad (3)$$

where $EC_{wt}$ (S m-1) is the electrical conductivity of underground water at $t$ (°C).

The Hydra Probe is a sensor that measures the permittivity and the temperature of the soil, and electrical conductivity EC (S m-1) is expressed as

$$EC = 2\pi f \varepsilon_{0}\varepsilon_{r}'' \quad (4)$$

where $\varepsilon_r''$ is the imaginary component of the relative complex permittivity, $f$ is the frequency (50 MHz for the Hydra Probe), and $\varepsilon_0$ is the dielectric constant in a vacuum (8.854 × 10-12 F m-1). As the relative permittivity depends on the temperature, the relative permittivity was converted to its corresponding value based on a temperature of 25°C. The temperature-corrected dielectric permittivity $\varepsilon_{r25}''$ is therefore expressed as [13]:

$$\varepsilon_{r25}'' = \frac{\varepsilon_{r}'' \times 1.0755}{0.693 + 0.0153 \times T} \quad (5)$$

where $T$ (°C) is the soil temperature. In this study, we obtain the electrical conductivity of bulk soil $EC_a$ by substituting $\varepsilon_{r25}''$ into equation (4).

## 3. Results and Discussion

### 3.1. Hydrological Factors

Figure 5 shows the daily precipitation and estimated daily evaporation. The total precipitation over the study period was 692 mm, while evaporation was 336 mm. Observations of antecedent precipitation at the AMeDAS (Automated Meteorological Data Acquisition System) point [14], nearly 8 km from the observation point of this study, totaled 1070 mm between October 2011, when the temporary dyke was completed, to August 2012 when our observations began (during the debris removal work). Flooding from the inflow of rain and seawater that originated around the demolished reclamation dyke and subsided areas is frequently seen, representing a particular hydrological phenomenon of the area around the observation point. It is difficult to quantify the inflow of such water, and therefore its extent is demonstrated by the scenes captured by fixed- point photography. Figure 6 shows the flooded area compared with dry conditions (Figure 6a). Figure 6b illustrates the flooding associated with Typhoon No.17, which was linked to a barometric depression of 992 hPa and an associated sea level rise. Figure 6c shows a flooding event believed to be derived from rainfall, as it is correlated with a continuous rainfall event of 43.5 mm. In total, 181 of the 3552 pictures captured during the fixed-point photography reveal flooding during the observation period.

Figure 7 illustrates the hourly groundwater level. The water table rose from 0.25 to 0.53 m between 28th and 29th October 2012, in combination with a total precipitation of 79 mm during this period. Additionally, it showed a gradual decrease during the low-rainfall period from the middle of December to the beginning of April. A periodic change in groundwater level was also seen under the influence of rainfall events, as noted above. Figure 8 shows a periodic analysis of 512 h ($512=2^9$) from 1st to 22nd of October and November 2012, using the fast Fourier transform algorithm. Twelve-hour and 24-hour cycles, as well as variations in sea level, were identified as amplitudes although the signals derived from rainfall events were also confirmed.

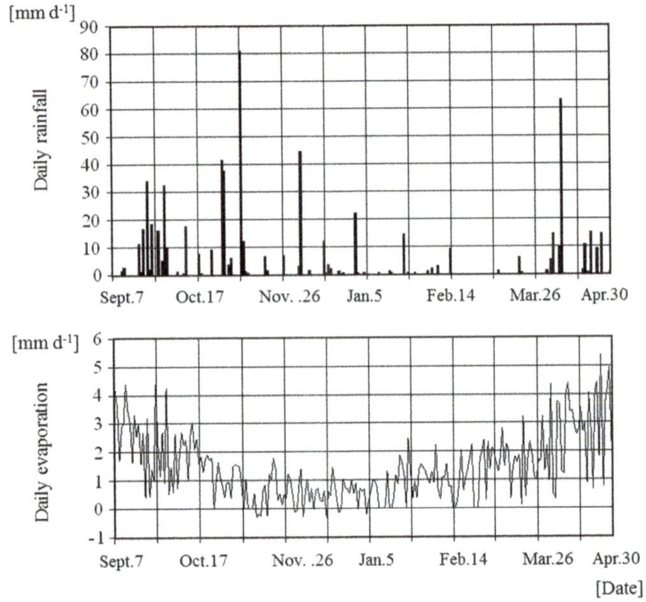


***Figure 5.*** *Observed daily rainfall and estimated daily evaporation during the study period.*

Figure 9 shows the time varying correlation coefficient between the tidal level at the Ofunato tidal observatory [8] and the groundwater levels during a period of seven days with no rainfall from 19th to 25th November. This demonstrates that the temporal difference between tidal level and groundwater level variations was almost three hours, with an average correlation coefficient of 0.54, implying that changes in groundwater level are closely synchronized with tidal levels. Hilton et al. studied the distributions of salt water and fresh water in the groundwater of the Florida peninsula in the United States, and found that the diffusion range of saltwater inland from the coast was around 0.6 km [15]. It should also be noted that

Rikuzentakata was not one of the 21 saltwater intrusion regions identified across the Japanese Islands by Hamaguchi et al. [16]. Although these studies represent only circumstantial evidence, it seems that the saltwater intrusion in our study region, which can be attributed to the subsidence and the fully-destroyed reclamation dyke, caused a periodic change in the groundwater level at our observation station, around 1 km away from the coastline.

(a) Dry condition(12:00 Sept. 14th, 2012)

(b) During the approach of Typhoon No.17 (16:30 Sept. 30th, 2012)

(c) Immediately following rainfall (12:00 Dec. 4th, 2012)

***Figure 6.*** *Flooding at the observation station.*

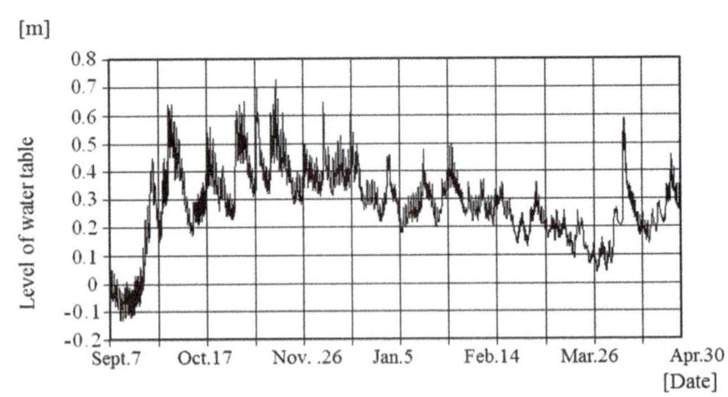


***Figure 7.*** *Hourly water table levels.*

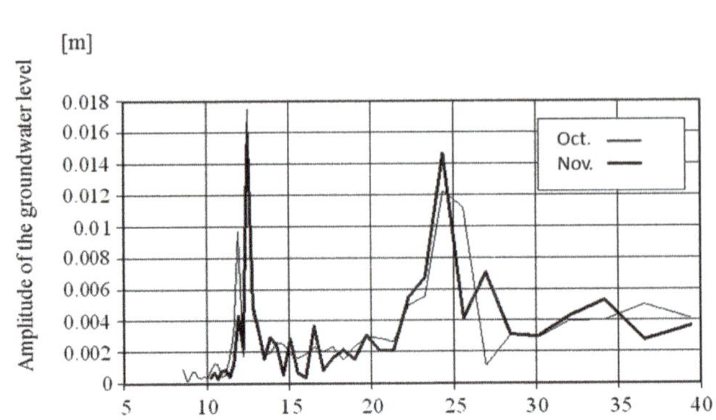


***Figure 8.*** *Periodic analyses of groundwater level during October and November 2012, by fast Fourier transformation.*

### *3.2. Changes in Electrical Conductivity*

Kume et *al.* [17] suggested that the salinity of the underground water of permeable farmland in the Nagapattinam district, India, which was affected by the tsunami caused by the earthquake in the Indian.

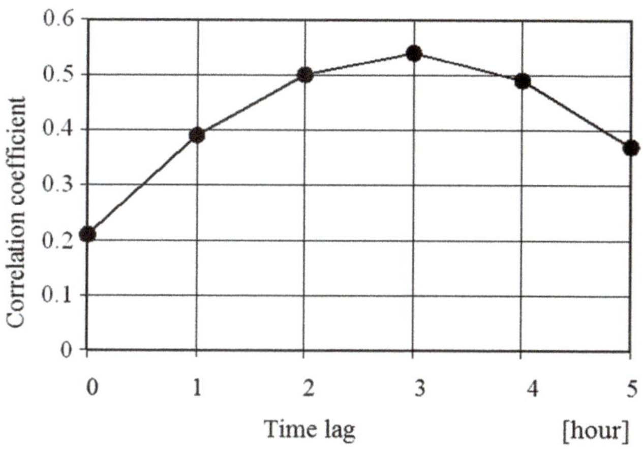


***Figure 9.*** *Correlation coefficients of sea level and groundwater level during a no-rainfall period.*

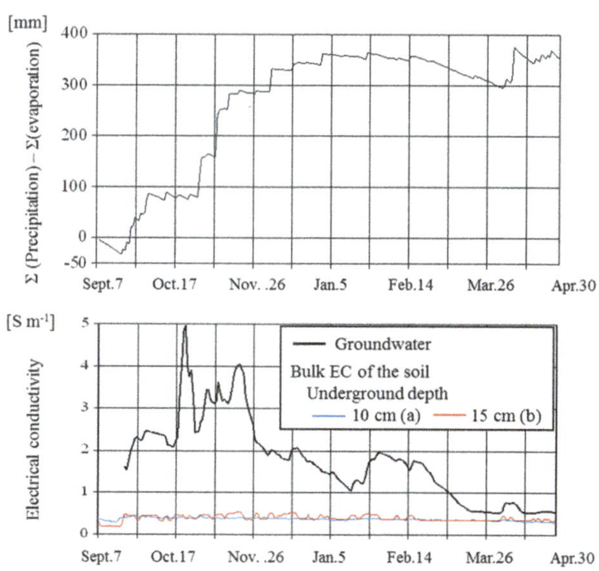


***Figure 10.*** *Daily mean electrical conductivity of soil and underground water, and daily change in accumulated input (precipitation) minus output (evaporation).*

Ocean off Sumatra on December 26, 2004, had returned to pre-tsunami levels by the monsoon rainfall of 2006. Illangasekare, et *al.* [18] estimated that the salinity of coastal aquifers in Sri Lanka, where seawater mixed with ground water via open wells during the tsunami of the same earthquake, would be decreased by the monsoon climate. They also suggested that evapotranspiration during the dry season was a major factor contributing to the increase in salinity of underground water in the same region.

Figure 10 shows the daily mean electrical conductivity of underground water, in addition to the integrated value of precipitation minus evapotranspiration. No clear relationship can be identified between the changes in electrical conductivity and the relative water supply, as both have varied irregularly since the beginning of the observation period. However, it can be suggested that an increase in precipitation will increase the groundwater level, during which salt water mixing may be temporarily facilitated. It can be seen that electric conductivity reached 5 Sm-1 on October 21. This high value was possibly induced by a concentration of seawater in the region. Subsequently, after some irregular variations, the conductivity began to gradually decrease after November 19th. The change in the electrical conductivity of underground water at a point at which seawater often infiltrated indicates that dilution by fresh water was generally superior to the supply of salt water, although the variations are not typically temporally distinct. Finally, the conductivity of the underground water seemed to converge on the bulk electrical conductivity of the soil, as described below. The daily mean bulk soil electrical conductivity $EC_a$ remained around 0.4 Sm-1 at both points (a, 10 cm below the ground, and b, 15 cm below the ground ) from the beginning of the observation period to the end of December, after which it decreased, to around 0.3 Sm-1 when the observation ended (Figure 10). The electrical conductivity of a saturation extract ( $EC_e$ ) is the best index for the salinity resistance of plants [e.g.19]. Rhoades et al. [20] presented the regression expressions between $EC_e$ ($S\ m^{-1}$) and $EC_a$ ($S\ m^{-1}$) for different soils at their field-capacity water contents, and $EC_e$ of the loam can be expressed as

$$EC_e = 5.47 \times EC_a - 0.047 \quad (6)$$

$EC_e$ of the studied area is estimated to be at least 1.6 S m-1 on the basis of this equation. Additionally, the approximate relationship between the relative yield of rice plants, $Y_r$ (%), and $EC_e$ can be expressed as [21]:

$$Y_r = 100 - \left[12 \times (10 \times EC_e - 3)\right] \quad (7)$$

Thus, a rice harvest cannot be expected according to the calculations using these equations, even for $EC_a = 0.3$.

Two monsoon seasons have passed since the Asian tsunami hit the region around the Indian Ocean, and it has been shown that the electrical conductivity of the soil in these areas has decreased dramatically, even in the farmland of the Andaman Islands where sea water sometimes intrudes at high tides, especially during new moons and full moons [6]. It has also been shown that the electrical conductivity of farmland in the Nagapattinam District in southern India, affected by the same tsunami, has decreased to pre-tsunami levels as a result of the accumulation of rainfall [22]. It is clear that the decrease in the electrical conductivity of soil at the observation point is slower than that found in previous studies [6, 17, 18, 22], and it is suggested that the difference is due to the relative presence or absence of ground subsidence.

It can be concluded that land restoration by a large-scale filling is an appropriate method, based on the present electrical conductivity of the soil. Furthermore, it will be necessary to continue monitoring the salinity of underground water and soil in this area in order to fully track and manage the regeneration of this region.

## 4. Conclusions

This study focused on the Otomo rice field zone in Rikuzentakada City, Iwate Prefecture, which experienced considerable damage as a result of the subsidence and the tsunami following the Great East Japan Earthquake in 2011. A fixed-point observation station was set up to enable long-term observation of meteorological elements, as well as the electrical conductivity of underground water and soil. These observations were begun in September 2012 following the removal of the tsunami debris, and continued until April 2013 when restoration construction was begun.

The observation point (38.99392°N, 141.6933°E) is located 1 km from the coast, and was chosen on the basis of ease of access and communication with the landowners. It was shown that non-uniform subsidence of up to 1.4 m had occurred around the observation point. This subsidence represented a remote cause of flooding, and floods due to rainfall events were also frequently observed around the observation point. It was also shown that the groundwater level covaried with the tide, and exhibited periodic changes.

Rainfall and evaporation were compared with the electrical conductivity of underground water over the entire eight-month observation period. Although a distinct relationship cannot be identified between these factors, the electrical conductivity of underground water showed a gradual decrease after undergoing a period of irregular change.

The bulk electrical conductivity of the soil did not show any major changes, and a gradual decrease was observed over the entire observation period. It was calculated that a harvest of rice could not be expected, even though the bulk electrical conductivity had decreased to 0.3 $Sm^{-1}$. These results contrasted with previous reports of lands affected by the 2004 Asian tsunami, in which the electrical conductivity was significantly decreased by rain infiltration. It is therefore supposed that the difference between these situations is due to the subsidence in our study area.

On the basis of these soil conditions, land restoration through large-scale earth filling appears to be an appropriate approach.

## Acknowledgements

Part of this study was funded by the Japanese River Improvement Fund in 2013 (code: 25-1215-005).

## References

[1] US Geological Survey (2011) Magnitude 9.0 - NEAR THE EAST COAST OF HONSHU, JAPAN. http://earthquake.usgs.gov/earthquakes/eqinthenews/2011/usc0001xgp/. Accessed 3 May 2013.

[2] The 2011 Tohoku Earthquake Tsunami Joint Survey (TTJS) Group (2011) The 2011 off the Pacific coast of Tohoku Earthquake Tsunami Information.

[3] Chibai K, Kanmuri H, Kati T (2012) Desalinization technique for tunami-hit farmland by infiltration of rain and discharge from underdrain. J. Jpn. Soc. Soil Phys. 121: 29-33 (Japanese).

[4] Yamamoto K, Kobayashi A, Harashina K, Kurashima E, Muto Y, Tukada Y (2015) Change in salinity concentration measured by electromagnetic survey in the paddy fields damaged from tsunami. Water, Land and Environ. Eng. 83: 671-676 (Japanese).

[5] Yamamoto K, Kobayashi A, Muto Y, Kurashima E, Harashina K, Tukada Y (2015) Change in salinity concentration for irrigation tank embankment submerged in sea water due to the 2011 tohoku tunami. IDRE Jornal 299: I_165-I_175 (Japanese with English summary).

[6] Raja R, Chaudhuri SG., Ravisankar N, Swarnam TP, Jayakumar V, Srivastava RC (2009) Salinity status of tsunami-affected soil and water resources of South Andaman, India. Curr Sci 96: 152-156.

[7] Geospatial Information Authority of Japan (2012) Maps & Geospatial Information. http://www.gsi.go.jp/ENGLISH/page_e30031.html. Accessed 8 Oct. 2012.

[8] Japan Meteorological Agency (2014a) Tidal observation data set. http://www.data.jma.go.jp/gmd/kaiyou/db/tide/genbo/index.php. Accessed 30 May 2014.

[9] Gash J. H. C. Dolman A. J. (2003) Sonic anemometer (co)sine response and flux measurement: I. The potential for (co)sine error to affect sonic anemometer-based flux measurements. Agric For Meteorol 119: 195-207. doi: 10.1016/S0168-1923(03)00137-0.

[10] Schotanus P, Nieuwstadt FTM, deBruin HAR (1983) Temperature measurement with a sonic anemometer and its application to heat and moisture fluxes. Boundary Layer Meteorol 26: 81-93.

[11] Wilson K, Goldstein A, Falge E, Aubinet E, Baldocchi D, Berbigier P, Bernhofer C, Ceulemans R, Dolman H, Field C, Grelle A, Ibrom A, Law BE, Kowalski A, Meyers T, Moncrieff J, Monson R, Oechel W, Tenhunen J, Valentini R, Verma S (2002) Energy balance closure at FLUXNET sites. Agric For Meteorol 113: 223-243, doi: 10.1016/S0168-1923(02)00109-0.

[12] Kojima K (1967) A field experiment on the influence of evaporation of snow upon snow melt. Low temperature science Series A, Physical sciences 25: 119-126 (Japanese with English summary, 125).

[13] Bellingham K (2007) The Stevens Hydra Probe inorganic soil calibrations. *Stevens Water Monitoring Systems Inc*: 1-5. http://www.stevenswater.com/catalog/products/soil_sensors/datasheet/The%20Stevens%20Hydra%20Probe%20Inorganic%20Soil%20Calibrations.pdf. Accessed 6 May 2015.

[14] Japan Meteorological Agency (2014b) Climate of Japan. http://www.data.jma.go.jp/obd/stats/data/en/index.html. Accessed 30 May 2014.

[15] Hilton H, Cooper J, Francis AK, Harold RH, Robert EC (1964) Relation of salt water to fresh ground water. Sea water in coastal aquifers. USA government printing Office. Washington DC USA: 1-11. http://pubs.er.usgs.gov/. Accessed 10 June 2015.

[16] Hamaguchi T, Kamal A, Sumi T (2013) Fundamental study on development of horizontal seawater intrusion model in coastal aquifer. Kyoto University Disaster Prevention Research Institute Annuals 56B: 585-589 (Japanese with English summary, 585).

[17] Kume T, Umetsu C, Palanisami K (2009) Impact of the December 2004 tsunami on soil, groundwater and vegetation in the Nagapattinam district, India. J Environ Manage 90: 3147–3154. doi: 10.1016/j.jenvman.2009.05.027.

[18] Illangasekare T, Scott WT, Clement TP, Villholth KG, Perera APGR. L, Obeysekera J, Gunatilaka A, Panabokke C, Hyndman DW, Cunningham K, Kaluarachchi JJ, Yeh WW-G, van Genuchten MT, Jensen K (2006) Impacts of the 2004 tsunami on groundwater resources in Sri Lanka. Water Resour Res 42: W05201. doi: 10.1029/2006WR004876.

[19] US Salinity Laboratory Staff (1954) Diagnosis and improvement of saline and alkali soils, Agriculture. Handbook 60; U. S. Government Printing Office: Washington D. C., USA, pp 55-68.

[20] Rhoades JD, Chanduvi F, Lesch S (1999) Soil salinity assessment method and interpretation of electrical conductivity measurements. FAO Irrigation and Drainage Papers 57: 15-62.

[21] Dobermann A, Fairhurst T (2000) Rice: Nutrient disorders & nutrient management. Handbook series. Potash & Phosphate Institute (PPI), Potash & Phosphate Institute of Canada (PPIC) and International Rice Research Institute. 191p.

[22] Chandrasekharan H, Sarangi A, Nagarajan M, Singh VP, Rao DMP, Stalin P, Natarajan K, Chandrasekaran B, Anbazhagan S (2008) Variability of soil-water quality due to Tsunami-2004 in the coastal belt of Nagapattinam district. J Environ Manage 89: 63-72. doi: 10.1016/j.jenvman.2007.01.051.

# Facade Greening: A Way to Attain Sustainable Built Environment

**Muhammad Shamsuddeen Abdullahi, Halil Zafer Alibaba***

Department of Architecture, Eastern Mediterranean University (E.M.U.), Gizamagusa, North Cyprus via Mersin 10, Turkey

**Email address:**
halil.alibaba@gmail.com (H. Z. Alibaba), Muhammadshamsudeen@yahoo.com (M. S. Abdullahi)

**Abstract:** In urban areas today, natural vegetation are been replaced with concrete pavements, buildings and other structures, which bring with it several consequences such as; noticeable increase in the amount of energy used in heating and cooling buildings which led to urban heat island (UHI) effect, degrading of air quality, increasing the amount of hard surfaces in cities which led to rise in temperature and also increase the volume of storm water collected. The technology of integration of plants on building façade became a trend in the last decade. This paper outlined different types of green façade and their benefits to the built environment. The paper through review of related literature is aimed at discussing how façade greening positively affect the built environment through the moderation of air quality; mitigate urban heat stress through transpiration cooling and shading. The paper also shows that façade greening will help in attaining a healthier living and a pollution free work place in the urban areas and any other place. The research concluded that the use of greenery on building façade provides a way to attaining a sustainable built environment.

**Keywords:** Green Façade, Sustainability, Urban Heat Island, Air Quality

## 1. Introduction

The integration of vegetation into building and construction of cities has been in use for many centuries, one of the famous examples of such is the Hanging Gardens of Babylon ordered by the ancient civilization of Tigris and Euphrates river valleys in Mesopotamia (Dunnett et al., 2010). Olive trees and grape vines were cultivated purposely to climb building walls in the Greek and Roman Empire cities (Newton et al., 2007). Before the period of industrial revolution there were no dedicated spaces for parks or neighborhood garden in urban planning. But during the period of industrial revolution, when industrial cities were first seen, town planning practice took a new dimension, urban parks, rows of trees along the sides of streets and neighborhood gardens were introduced as part of urban planning.

Nowadays we are living in a world where little is untouched by man. And the urban areas are becoming unhealthier to live in due to increase in air temperature, which led to an increase in the amount of energy required in heating and cooling buildings. The architects and the engineers are now faced with the daunting task of designing and constructing buildings that are environmental friendly as much as possible, while urban planners are looking for where to plant their greenery. This is what gave green facade more popularity (Wong et al., 2009).

"The ficus and climbing plants had developed in an unvaryingly perpendicular direction, imposed by the density of the element which had produced them. Motionless, even after parted them with my hands, the plants immediately returned to their original position. This is the realm of verticality" (Dunnett et al., 2008). The use of vegetation on building facades in form of living walls, green facades and vertical gardens offers an outstanding number of public and private benefits such as: aesthetic, social, ecological and environmental, and it fits in the principle of ecological engineering (Odum, 1995).

The use of plants on building façade gives an essential contribution to improving sustainability of the built environment. Using them as an architectural feature that upgrades façade is ecologically and aesthetically accepted. Plants when properly used on building façade in our design, would led to an energy conscious design approach that would prevent the deterioration of natural environment in the densely populated urban areas (Krushe et al., 1982).

The advantage of green façade is not only limited to environment and nature but, also help in lowering the operating cost for buildings in long term. It also helps in moderation of the micro climate of a given area when use in larger area, because of the additional layer of vegetation on the building facades of the area. In this case, the vegetation serves as dust filter, as humidifier and also generates cold air, therefore, promoting human health (Sheweka et al.)

## 2. Literature Review

### 2.1. Typologies of Green Façade

Green façade can simply be described the vertical integration of planting system on the façade of building or any other structure, the plants may be rooted in the ground or in planter boxes at different levels of the façade as shown in figure 1 (Suleiman, 2013). There are two different typology of green façade that are available until now namely; the one with plants rooted into the soil and that with plants rooted in an artificial substrate at grade as shown in figure 2. (Mir, M. A. 2011)

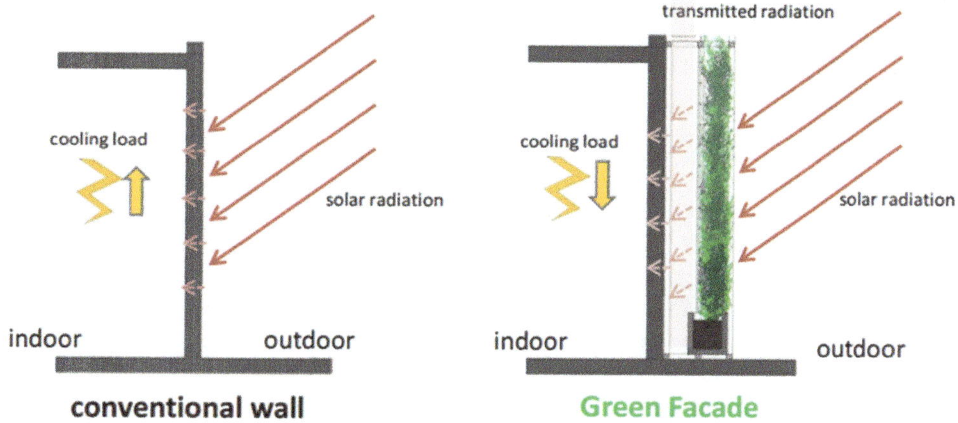


*Figure 1. Showing the difference between a conventional wall facade and a green façade (Suleiman, 2013).*

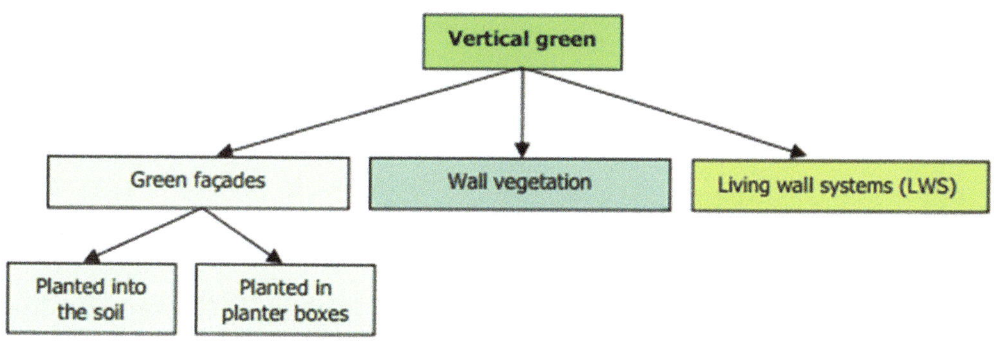


*Figure 2. Showing two different typologies of green façade (Mir, M. A. 2011).*

### 2.2. Plants Planted into the Soil

In this typology the plants are planted on the ground, their roots are in natural ground, and they grow from the ground against the wall. In this category watering system is not required because; they tap their water from the natural source i.e. rain water and ground water, it take a long time for this type of green to cover the surface of the wall. This category can also be divided into two; self-climbing (directly on wall) and the one with supporting structure (indirectly on the wall), as shown in figure 3 (Mir, M. A. 2011).

The sticky root structure of some plants enables them to attach directly on the façade, covering the whole surface. The type of plants species used will determine how efficiently façade would be covered and how long will it take, and how many plants should be used at what intervals. Some plants species doesn't have the sticky root structure to attach them to the façade directly, thus, they require supportive structure to grow and cover the façade as shown in figure 4 (Mir, M. A. 2011).

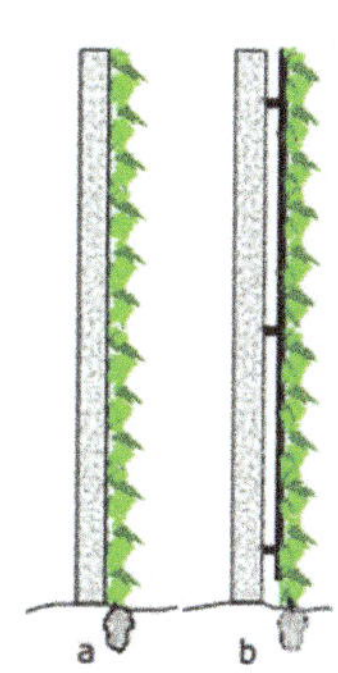


*Figure 3. Shows the type of green façade in which the plants are rooted into the ground; use of self-climbing plants, a) directly on wall, b) with supporting structure (Mir, M. A. 2011).*

*Figure 4. Showing a typical example of supportive structure for plants with non-sticky root structures (greenscreen.com).*

## *2.3. Plants Planted in Planter Box*

In this typology, as the name implies, the plants are planted in planter boxes. The planter boxes are placed either at the bottom of the façade (figure 5a and b) or on roof top, or at the bottom of each floor (figure. 5c). This category requires a continuous watering system because; the roots of the plants are not directly on natural ground. It also requires a long period to cover the surface of the wall, unless in the case of figure 4c where the planter boxes are placed at the bottom of each floor. This category can also be sub divided into two; plants directly on the wall and indirectly on the walls (on supporting structure) figure 5b and c.

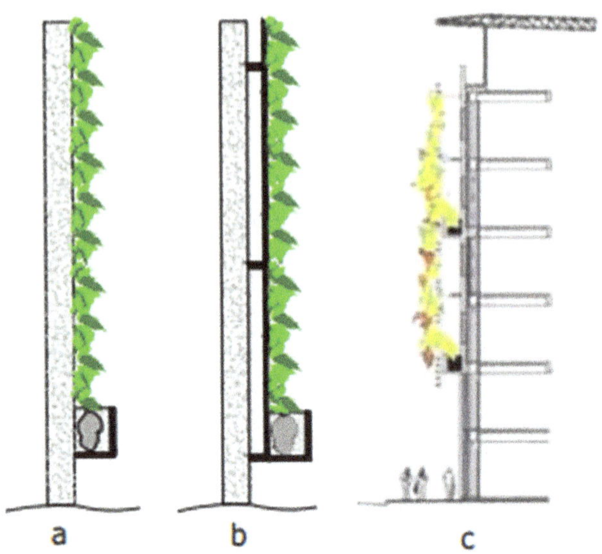


*Figure 5. a) Showing planter box at the bottom with plants directly on the wall, b) planter box at the bottom with plants on supporting structure, c) planter box at the bottom of floors with plants of supporting structure (Mir, M. A. 2011).*

## *2.4. Importance of Greenery on Built Environment*

The rate at which urban areas are developing leaves less available space for greening. Major councils in the city of Sydney accept façade greening as space as part of planning process (Mark, 2011). This shows the importance of green areas in urban areas where urbanization is growing fast.

In developing countries, urbanization is growing rapidly and the cities are becoming denser in terms of population. Urbanization is causing the replacement of green areas with buildings and other structures, which brings more hard surface development and less green areas. The use of greenery on building façade in urban environment can solve the above mentioned problems by creating a natural green environment. In advanced and developed countries like Europe, Australia and New Zealand, this has become a successful trend as part of their urban development planning practice (Ling, et al., 2012).

## *2.5. Effect of Green Façade on Urban Heat Island*

According to Encyclopedia of Earth, "heat island refers to urban air and surface temperatures that are higher than nearby rural areas" (Agency, E. 2012). The difference of air temperature between the cities and their surrounding rural areas is ranging from 1°C to 7°C. All this is as a result of the replacement of green areas with paved surfaces and other structures. By replacing trees and shrubs with asphalt, concrete and other impermeable surfaces which only absorb rather than reflect the heat, causes rise in the overall surface ambient temperature of the area. Urbanization is a factor that lead to increase in urban heat island phenomenon, due to the fact that nature is been replaced with manmade structures, which are good in absorbing and retaining heat than normal in nature (Al-musaed et. at., 2007).

According to Al-musaed (2007), cited by sheweka et. al., (2012), both urban and rural areas are been affected by heat island which causes a high demand of energy during the summer for cooling, air pollution and greenhouse gas emittion and other health related illnesses. In the figure 6 below, we can clearly see the effect of urban heat island.

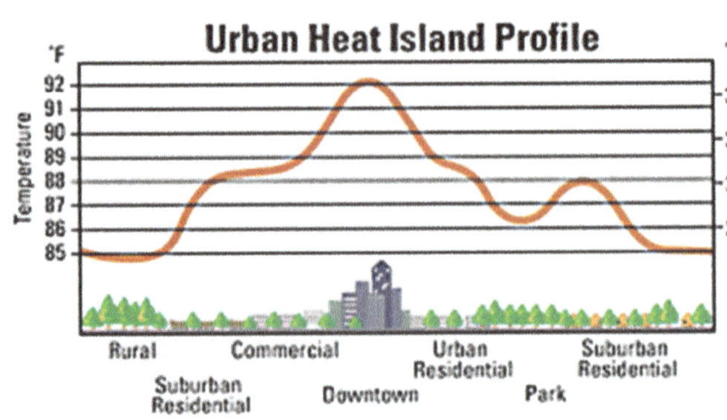


*Figure 6. Urban heat island profile (Agency, E. 2012).*

The micro climate of a particular area can be affected by facades and the streets around it due to their impermeable nature. This causes discomfort and increase in the amount of energy required for cooling (Oke, 1987). In this case, use of greenery on concrete roofs and façade whereby the heat will be consumed through the process of evapotranspiration can be a possible solution a (Figure 7).

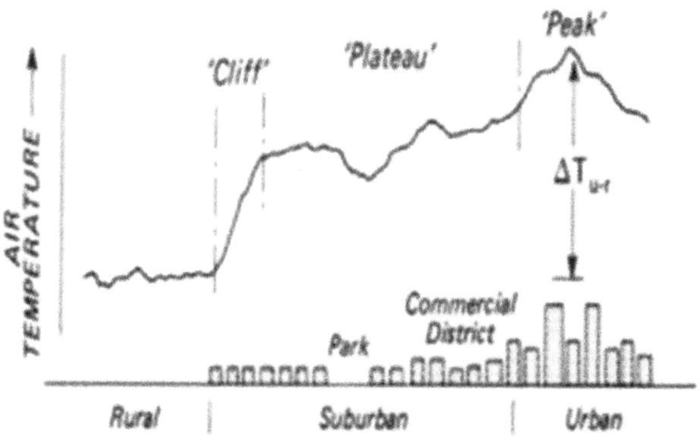


*Figure 7. Longitudinal section of urban island effect (sheweka et. al., 2012).*

In addition, use of greenery on building roofs and facades to cover the impermeable surfaces can contribute to vertical mixing of air; therefore the temperature over the surface may likely be lower than that of the surrounding built up areas. Fresh air is always denser than warm air, so, here the warm air on the heated surfaces is replaced by the fresh air produced by the vegetation around, therefore reducing heat effect (Jonhston et. al., 2004).

Plants has the ability to absorb "smog" and producing oxygen, therefore helps in improving air quality. The planted vegetation on the façade or roof can reduce the "smog" by either absorbing to some extend the suspended particles in air or by moderating the temperature (Dinsdale et. al., 2006).

According to climates studies the use of greenery on building roofs and facades have great effect in moderating temperature in urban canyon, it will be more effective if both the two are combined i.e. green roofs and green facades, to cover most of the urban surfaces (Alexandri et. al,. 2007).

### *2.6. Façade Greening to Improve Aesthetical Value of Urban Living*

According to Frank Lloyd Wright "A doctor can bury his mistakes, but an architect can only advise his clients to plant vines." (Wright, 1931). This brought a good idea for those who use their roof tops for gardening; instead they can hang them on the walls to make it more visible for people (Alter, 2008).

The master of green façade Edouard Francois was quoted saying "Man can live solely within architecture. He needs a complex building which must be decorated. Only in this way can he be happy". Indeed, in Francois' view, working with nature offers a welcome complexity: 'Watch a tree. It has a thousand branches, it moves, grows, changes color' (Alter, 2008).

Sir William Dobell once said "A sincere artist is not one who makes a faithful attempt to put on to canvas what is in front of him, but one who tries to create something which is, in itself, a living thing." (Mike, 2011)

Now that it is a known problem, that green spaces in urban areas are disappearing day by day due to urbanization. Integrating greenery on building façade as part of city planning will have an impact urban areas. If carefully designed and constructed, it can also be applied to other structures like bridges and can also serve as a cover for ugly and old buildings. Green façade will definitely improve the environment aesthetically (Michelle, 2011).

If properly designed and maintained, green façade can enhance the appearance of building and any other structure. Another thing is the careful selection of plants to use which varies in their looks and texture; if properly done it can also enhance the building appearance. In parts of the world where graffiti is not accepted, façade greening can serve as protection which makes it almost impossible (Mir, M. A. 2011).

High degree of maintenance is the weakness of façade greening, it requires a lot of maintenance work such as trimming, replacing the death once and watering to keep it natural and in healthy condition (GhaffarianHoseini, 2012) (Ghaffarian Hoseini, 2012).

### *2.7. Green Façade to Improve Urban Climate*

Green façade have a direct positive impact on both the cities and their inhabitants. Those living in the cities can see the plants used on green façade as hanging gardens. Walls that are directly exposed to sun and UV will conduct heat and transfer it into the building, which will lead to a significant rise of temperature in the building. This shows the Urban Heat Island effect. The plants used on green façade lose water via their leaves in a process called evapotranspiration which helps in moderating the temperature around the buildings. Green façade in urban areas helps in moderating the microclimate therefore, improving the living condition in such cities (Peter, 2013).

Now that green areas are disappearing in urban areas, the excessive heat built up due to the presence of more hard surfaces which doesn't absorb but only reflect the heat, combined with other pollutants distorts the weather of a particular area and its surrounding environment. This is what led to urban heat island effect. The urban heat island effect is an area of the city that is extremely warmer than its neighboring rural area. Integration of plants on building façade can drastically bring down the urban heat island effect and make urban life more healthier (Patrick, 2011).

Proper design and construction of green façade will promote the cooling process in buildings, by making the problem of rain water management less severe; the soil either on the ground or in the planter boxes trap and retain the collected rain water, which is absorbed by the plants through their roots. This process helps in cooling the building structure. Even the heat absorbed by the plants through direct expose to sun and UV will be evaporated through the process called evapotranspiration (Paul, 2010)

### *2.8. Green Façade to Reduce Carbon Footprint and Improve Air Quality*

Biologically, plants absorb carbon dioxide in order to synthesize food from sun light; this process is generally known as photosynthesis. Oxygen is been released as a bi product of this process. This shows that the more plants we have in urban areas the more clean air we get (Patrick, 2011).

Plants growing on building façade absorb carbon dioxide and release oxygen. The plants also serve as filter to other pollutants like; cadmium and lead. They also act as carbon filters by absorbing the greenhouse gases emitted by the atmosphere and store them in their tissues. The arrangement of plants on living walls are thick which gives room for planting more on a small area. The carbon absorbed by 20m$^2$ of a living wall is equivalent to that absorbed by a medium sized tree (Patrick, 2011).

Integration of greenery to building façade can improve the air quality in cities. Air quality can also be improved through the process called bio filtration. Bio filtration can simply be described as a technique used in controlling pollution using living materials; here the living materials serve as filter for the air and dust that passes through it. The microbes contained by the plants absorb must of the pollutants. In this process the plants absorb the carbon dioxide produced, therefore releasing oxygen to the environment (Appropedia, 2012).

It is proved that; green façade have the capabilities of saving energy in buildings. They also help in preventing the formation of dust by absorbing dangerous atmospheric particles (Marie, 2012).

According to Patrick (2011), the moisture level of green wall needs to be monitored and controlled, due to too much amount of water retention which may contribute to high moisture index leading to mold growth effect.

## *2.9. Thermal Impact of Green Façade*

### *2.9.1. Temperature Reduction*

According to research, in humid climate the use of greenery on building façade in an urban canyon can help in achieving a substantial benefits of maximum temperature reduction of up to 8.4°C (Alexandri et. al,. 2008). This is important because, energy consumption of buildings is influenced by the distribution of ambient air in a canyon as higher temperatures in canyon increase heat convection to a building and similarly increases the cooling load (Santamouris, 2001).

By directly shading the heat absorbing surfaces of a building with vegetation, it will help in alleviating UHI through evapotranspiration cooling (McPherson, 1994). The use of greenery on building façade can drastically reduce the temperature of a building to as much as 50% by shading walls from direct sunlight. The process of evapotranspiration helps in converting large amount solar radiation into latent heat which does not cause any rise in temperature. Furthermore, an efficient green façade have the ability to protect the building from intense solar radiation during the summer period and can use its leaf cover to reflect and absorb between 40% and 80% of the received radiation, which depend on the amount and the type of vegetation used (sheweka et. al., 2012).

In another studies, at the University of Toronto since 1992, surface temperature of vertical greenery have been observed in different settings (Bass, et. al., 2003). The results consistently shows that areas with vertical greenery are cooler than areas with light coloured bricks, walls and other surfaces found in the urban areas. Another experiment in japan shows that climbing plants can reduce the temperature of a veranda with south-western exposure (Hoyano, 1988). While in Africa, the temperature reduction of 2.6°C behind panels covered with vines (Holm, 1989). Thus, temperature fluctuations on the wall surface can be moderated from between 10°C and 60°C to between 5°C and 30°C (Peck, et. al., 1999)

### *2.9.2. Shading and Insulation*

The use of greenery on building façade as shades to solar radiation is obvious, with the advantage that the traditional concrete or brick facades which radiate the heat in and around the building, while greenery does not. This depends crucially on the density of the foliage (sheweka et. al., 2012).

In a double skin façade, the temperature is generally lower if greenery is used in between the skins. In a case where plant is used instead of slat in double skin façade the temperature never exceed 35°C, while with slats, temperature can exceed 55°C. The amount of energy required for air conditioning can be reduced by up to 20% with the use of plants in an internal double skin façade (Stec, et. al., 2004).

The physiology of plant implies that some portion of sunlight received by the vegetation is used for the process of photosynthesis, while the remaining is used for evapotranspiration, hence, making them mechanism for temperature moderation. This is what makes vegetation effective in blocking solar radiation without increasing it temperature. The transmittance factor of a leaf is 0.2, while it absorbance is 0.5 (sheweka et. al., 2012).

According to an experiment "Bioshader" conducted in University of Brighton, in this experiment an office space was used as a case study, the windows of the office were covered with vegetation. The temperature recorded was between 3.5°C to 5.6°C lower than when the windows are covered with other materials. The solar transmittance of the foliage was also measured, which ranges from 0.43, with a single layer leaves, and up to 0.14 with five layers of leaves, and also a reduction in solar radiation of up to 37% in a single layer leaf cover and 86% in five layer leaf cover as shown in figures (7, 8 and 9). (Miller, et. al., 2007).

In another experiment, it suggested that more thermal energy flows into the non-shaded walls due to their direct exposure to sun light, which results in higher wall surface temperature. The heat absorbed by the non-shaded surface will later advance to the inner wall surfaces making the temperature interior rise, which lead to more energy consumption for cooling (Neila, 2004).

***Figure 8.*** *Bioshader experiment (Miller, et. al., 2007).*

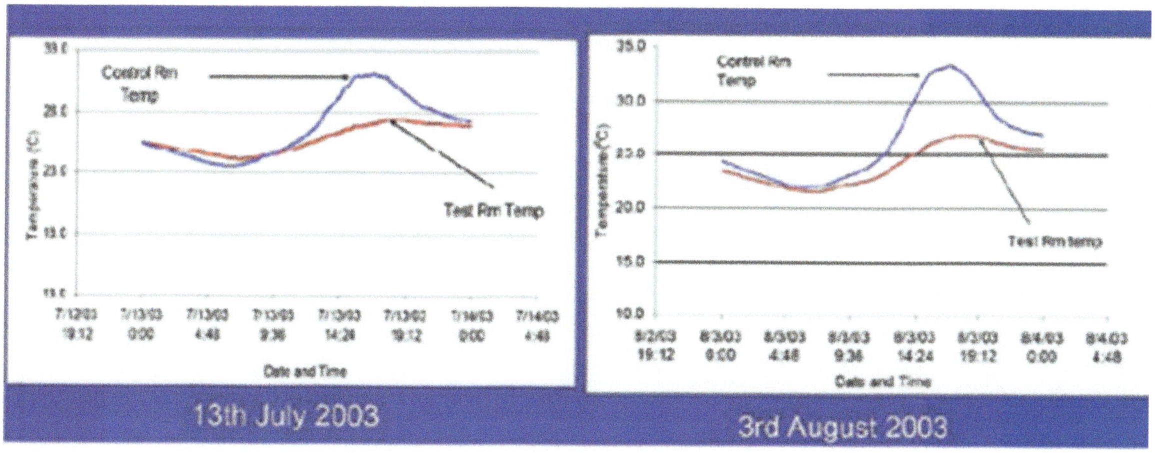


***Figure 9.*** *Bioshader experiment; showing temperature distribution graph (Miller, et. al., 2007).*

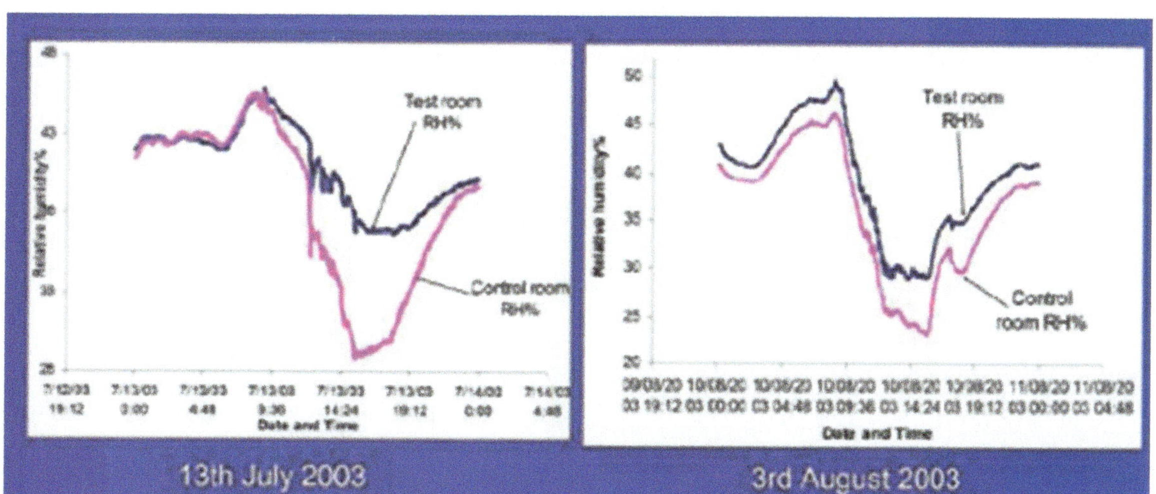


***Figure 10.*** *Bioshader experiment showing relative humidity distribution (Miller, et. al., 2007).*

### *2.9.3. Evaporative Cooling*

According to a study, in tropical climate, it was confirmed that vegetation has cooling effect. The study was done by measuring the temperature atmosphere at different altitudes. The maximum difference in temperature recorded was 4.2°C. Based on this study, it shows that the cooling effect is limited with the height (Wong, et. al., 2003).

In the project of the institute of physics of the Humbolt University Berlin, Adolershof, combines rain water management and energy savings with natural conditioning through vegetated walls. The shadow casted on the façade by the vegetation affect the cooling of the building, which influence the energy consumption of the building Figure 11 (www.gebaeudekuehlung.de).

*Figure 11.* Institute of Physics, University of Humboldt (www.gebaeudekuehlung.de).

Evapotranspiration is the most significant environmental benefit of green roofs and facades in the urban areas. This process has effect on urban hydrology, in reducing the temperature of urban surfaces and improving the management of rainwater run off (Stec, et. al., 2004).

More so, experiment conducted at University of Brighton "Bioshader" which compared an office space with window covered with vegetation and with other covers, verifies that humidity level in the office with bioshade is higher than that without. This shows that, the use of greenery on building façade provides additional moisture to the dry indoor environment (Miller, et. al., 2007).

Lastly, the cooling effect of greenery due to evapotranspiration, helps in moderate the temperature around the building, and the evaporated water by the plants increase the humidity up to 2 kg of water per $m^3$ of dry air (Papadakis, et, al., 2001).

## 3. Findings and Discussion

Integration of greenery to façade system of buildings and other structures is not only limited to new projects; it can also be applied to existing once in order to bring them back to life, either from sustainability point of view or aesthetic point of view. From sustainability point of view, among the properties of plant is making their own food through the process called photosynthesis, the process requires carbon dioxide to be done and the byproduct of this process is oxygen; therefore plants absorb the carbon dioxide which pollute the environment and produce oxygen which is good for humans and their environment. And from the aesthetic point of view green façade can help in hiding the mistakes of designers and engineers and it can also make the appearance of a building and other structures more pleasing to the eye.

The major factor that is causing the urban heat island (UHI) effect is high degree of urban development; for example, how many square meters of green area are in order to construct houses and streets for the growing population in the cities, and most of the materials used in these constructions are concrete and asphalt, which are all impervious materials; they absorb heat in the day time and realize it at night. This effect can be reduced through the process called evapotranspiration; the process through which liquid water is converted to water vapor and it is removed from the evaporating surfaces, such as pavements, soil and wet vegetation.

Using greenery of façade systems can give aesthetic difference in the environment where people carry out their daily activities. Studies show that, the presence of plant within an environment helps in improving mental health and the wellbeing of the inhabitants of that environment. Apart from being essential to life on this planet, they add beauty and attraction with their unique colors and textures. Nowadays, improving aesthetics is among the primary objectives of designing green façade projects. If we look at the large areas we have on façade of buildings and other structures in our cities today, we can utilize them as hanging gardens; that will also improve the aesthetics of urban streets.

Surfaces in urban area that are not used such as roof and façade could be used in order to moderate the urban temperature; this can be achieved by covering such surfaces with greenery. The thermal performance of a building can drastically be affected by the influence of greenery on the microclimate. The climbing plants used in greening facades serves as shading devices to the façade, which prevents it from direct solar exposure, therefore reducing the amount of energy require in cooling the interior.

# 4. Conclusion

The integration of vegetation on buildings and other structures façade can of great benefit to urban environment; it can also be used as a tool for passive thermal control of buildings. If carefully design and maintained, green façade have the ability to moderate urban temperature.

In conclusion, the benefits of façade greening can be categorize into two; private benefits and public benefits

Private Benefits:

- Green facade helps in improving indoor air quality: when used in the interior of building, the plants on the façade have the ability to filter the noxious gases from the carpets and other furniture. The greenery also absorbs the airborne pollutants like dust and other unwanted gases.
- Protecting the building structure: Generally building facades are exposed sun, water and other weathering elements and with time most of the construction materials start to break down, due to heating and cooling which causes contraction and expansion of the materials. Green façade can protect the building façade from direct exposure to sun light.
- Energy efficiency in building: Green façade can serve as a wind breaker during the winter; it can also reduce the ambient temperature through evapotranspiration process. If applied internally helps in reducing the amount of energy required in cooling the outdoor air for indoor use.

Public Benefits:

- Green façade helps in reducing urban heat island effect: this effect is caused by the replacement of green areas with hard surfaces, which lead to conversion of sunlight to heat. It is a well-known fact that greenery cools the building by shading it from direct exposure to sunlight which leads to the reduction of reflected heat and also through evapotranspiration, therefore with this we can say façade greening helps in natural cooling processes of buildings and environment and moderating ambient temperature in urban areas.
- Green façade helps in improving outdoor air quality: there is a lot of pollution in urban area due to presence of high number of industries, vehicles and other things that emit a lot of unwanted gases. One of the benefits of green façade is that the plants have the ability to absorb the airborne pollutants and other deposits on their leaf.
- Green façade helps in improving the appearance of a building or a street, due to the nature of the plants used. They may have different appearance and texture which will give variation in the appearance.

Finally, the integration of greenery on building façade shows great potentials in attaining a sustainable built environment through reduction of urban heat island effect, improving both indoor and outdoor air quality, making the environment more aesthetically pleasing and also reducing the carbon foot prints in urban areas.

# References

[1] Odum, H. T., (1995). Scales of ecological engineering. Ecol. Eng. 6 (1996), 7 19. Pekkanen J, Timonen KL, Tiittanen P, Vallius M, Lanki T, Sinkko H., 2000. Exposure and Risk Assessment for Fine and Ultrafine Particles in Ambient Air. National Public Health Institute.

[2] Newton J, Gedge D, Early P, Wilson S. Building greener guidance on the use of green roofs, green walls and complementary features on buildings. London, UK: CIRIA; 2007 & [LBNL website http://eetd.lbl.gov/HeatIsland/].

[3] Badrulzaman J, Ismail S, Mohd N, Moh'd H. (2013) Impact of vertical greenery system on internal building corridors in the tropics. *Procedia- Social and Behavioral Sciences (105 (2013) 558-568).*

[4] Shewake M S, Moh'd N M. (2012) Green façade as a new sustainable approach towards climate change. *Energy Procedia (18 (2012) 507-520).*

[5] Ling, C. Z, Hoseini, A. G. (2012) Greenscaping Buildings: Amplification of vertical greening towards approaching sustainable urban structures. Journal of creative sustainable Architecture and Built Environment.

[6] Mir, M. A. (2011). Green façade and Building Structures Delft University of Technology, Delft.

[7] Mark, P., (2011) "Going out on a limb", www.greenwall.com.au, Horticulture, "First Green Roofs…Now Green Walls, Sustainable Horticulture", (2009).

[8] Alter, L. (2008) "11 Buildings Wrapped in Gorgeous Green and Living Walls", treehungger.com.

[9] Mike, (2011). "Green Wall Benefits – Aesthetics", www.Greenovergrey.com.

[10] Michelle, (2011). "The Green City Guidelines", thegreencity.com.

[11] MMA Architectural Systems Ltd. Green Wall Benefits, "Green Wall Systems", (2010).

[12] Wright, F. L. (1931). *To the young man in architecture.* Chicago: The Lakeside Press, R.R. Donnelley & Sons Company.

[13] (Verne, (i.e.) quoted in Lambertini and Leenhardt, 2007:9), (Dunnett, N. & Kingsbury, N. (2008). Planting Green Roofs and Living, Walls. London: Timber Press.).

[14] Al-musaed A (2007a) Heat island effects upon the human life on the city of Basrah, building low energy cooling and advanced ventilation technologies the 21st century. In: PALENC 2007, the 28th AIVC Conference, Crete Island, Greece.

[15] OKE, T. R. (1987): Boundary Layer Climates. Wiley and Sons, 372 pp).

[16] JOHNSTON, J.; NEWTON, J. (2004): Building Green A guide to using plants on roofs, walls and pavements. Major of London 2004.

[17] ALEXANDRI, E.; JONES, P. (2007): Temperature decreases in an urban canyon due to green walls and green roofs in diverse climates. Building and Environment, In Press, Corrected Proof, Available online 4 September 2007.

[18] SAILOR, D (2008): Energy and urban climate benefits of green roofs. World Green Roof Congress London, 16 17 September 2008.

[19] Agency, E. (2012). Heat island. Retrieved from http://www.eoearth.org/view/article/153461

[20] www.camelclimatechange.org

[21] Gail, H. and Erin, A. (2013): landscape design: Aesthetic characteristics of plants. http://www.ufl.edu/

[22] (Krushe P, Krushe M, Althaus D, Gabriel I. Okologisches Bauen Herausgegeben vom Umweltbundesamt. Wiesbaden und Berlin: Bauverlag; 1982.).

[23] Alexandri E, Jones P. (2008) Temperature decreases in an urban canyon due to green walls and green roofs in diverse climates. Building and Environment; 43:480 93.

[24] Santamouris M. The canyon effect. In: Santamouris M, editor. Energy and climate in the urban built environment. London, UK: James and James Publishers; 2001. p. 69 96.

[25] McPherson EG. Preserving and restoring urban biodiversity: cooling urban heat islands with sustainable landscapes. In: Platt RH, Rowntree RA, Muick PC, editors. The ecological city. Amherst, US: University of Massachusetts Press; 1994. p. 151 72.

[26] Climate booklet for urban development. Ministry of Economy Baden-Wurttemberg in Cooperation with Environmental Protection Department of Stuttgart; 2008.

[27] Bass B, Baskaran B. Evaluating rooftop and vertical gardens as an adaptation strategy for urban areas. Institute for Research and Construction. NRCC-46737, Project number A020, CCAF report B1046. Ottawa, Canada: National Research Council; 2003.

[28] Hoyano A. Climatologically uses of plants for solar control on the effects on the thermal environment of a building. Energy and Buildings 1988; 11: 181 9.

[29] Holm D. Thermal improvement by means of leaf cover on external walls a simulation model. Energy and Buildings 1989; 14(1): 19 30.

[30] STEC, W.J.; VAN PAASSEN, A.H.C.; MAZIARZ, A. (2004): Modelling the double skin façade with plants. Energy and Buildings, 37 (2005) 419-427.

[31] NEILA, FJ. (2004): Bioclimatic architecture in a sustainable environment. Editorial Munilla-Lería.

[32] WONG, NH.; CHEN, Y; ONG, CL.; SIA, A. (2003): Investigation of thermal benefits of rooftop garden in the tropical environment. Building and Environment, Volume 38, Issue 2, February 2003, Pages 261 270.

[33] http://www.gebaeudekuehlung.de/en_regenwasser.html

[34] Papadakis G, Tsamis P, Kyritsis S. An experimental investigation of the effect of shading with plants for solar control of buildings. Energy and Buildings 2001; 33: 831 6.

15

# Study on Bioagents/Bioaerosols Standoff Detection by Lidar

**Yang Hui[1], Sun Yanfei[1], Wang Tiedong[1], Zhao Xuesong[2]**

[1]New Star Application Technology Institute, Hefei, China

[2]Key Lab. of Environmental Optics & Technology, CAS, Hefei, China

**Email address:**

sanpedroman@163.com (Yang Hui), hope32@163.com (Sun Yanfei), 18905607537@189.com (Wang Tiedong), xszhao@aiofm.ac.cn (Zhao Xuesong)

**Abstract:** Scattering, fluorescence and polarization are the important data source for bioagent or bioaerosol time-space observation and identification. This paper stated and discussed the theory and data inversion principles for Mie scattering, laser induced fluorescence and polarization sensing Lidar. The sensing and data inversion results for bioagent/bioaerosol extinction coefficient, horizontal linear depolarization ratio were also demonstrated. The signal and *SNR* simulation of fluorescence lidar were also demonstrated. The sensing results revealed that the three kinds of detecting technology approaches are reasonable and potential for bioagent/bioaerosol characterization and recognition.

**Keywords:** Bioagents/Bioaerosols, Lidar, Laser Induced Fluorescence, Depolarization Ratio, Mie Scattering

## 1. Introduction

Biological warfare agents (BWAs) or bioagents are live pathogenic microorganisms and biotoxin used in biological terrorism attack or in the battlefield, which are capable to cause mass infections or poisoning of people, animals and/or plants at reasonable economical cost. Bioagents can be sorted according to the causal agent as bacteria, viruses, rickettsiae and toxins [1, 2, 3]. Bioagents are relatively inexpensive to produce and can yield a significant impact as a terrorist weapon. Up to now, there are more than 1200 kinds of bioagents that can be produced to biological weapon. The infection of bioagents is mainly caused through absorption from respiratory tract, alimentary tract, or direct contact of skin, the deterrence will be much more awful enhanced by the recombinant DNA technique, and the death rate of un-immunized people will be nearly 100%. The World Health Organization estimated that attacking a large city with 50 kg of anthrax spores would produce 95,000 deaths and an additional 125,000 sicknesses [4]. And for viral hemorrhagic fevers, less than 10 organisms can cause decease and many of the most dangerous biological warfare agents are infectious when less than 10,000 organisms or spores are inhaled [5]. On the other hand, because of the common intrinsic fluorescence characteristics of fluorophore, such as tryptophan, tyrosine, etc., it's difficult to distinguish harmless bacteria from fatal bacteria, even genetically in some cases. Bioagents can be disseminated through at least three major ways: by vector (i.e., insects), by contaminated water or food supplies, and the most effective way of dispersion is the form of aerosol particles suspending in the air. As for the biological aerosol (bioaerosol), it includes plant and insect debris, fungal and plant spores, pollen, cells, viruses and bacteria. Abundance in the atmosphere, bioaerosol affects the cloud microphysical processes, helps to the biodiversity and disease transmission through long range transport.

Typical point detection and discrimination system based sample technique can detect bioagents/bioaerosols with a low time resolution and a relatively high false alarm rate, the bioactivity can be deteriorated during the sample procedure, and the exterior factors such as dehydration, variation of temperature will also affect the recognition result.

Early warning of a biological attack such as lidar standoff detection offers many potential capabilities not easily provided by point detection systems such as sensitivity over a wide area, easy to deploy, capability of mapping, allowing for early warning to downwind assets and population, etc.

Assisted by plume modeling technique, mapping and tracking of the hazard cloud by lidar can help to determine origin, evolution of a biological attack, thus guide treatment and decontamination efforts. For example, at a condition of 20km/h wind velocity, the population on the ground can get a early warning time of about 6 minutes if the windward [6]. For these reasons the need to develop methods to remotely detect and discriminate biological aerosols from background aerosols and ultimately to discriminate biological warfare agents from naturally occurring aerosols, is paramount.

Elastic Mie scattering, laser induced fluorescence (LIF) and polarization scattering are three major bioagents/bioaerosols remote sensing techniques. In this paper, the theory, technology, ability of detection and other factors affecting detection are discussed individually.

## 2. Principle of Bioagents/Bioaerosols Remote Detection by Lidar

Bioagents/bioaerosols remote detection is based on the interaction physical process between the laser beam and particles on the path, such as rayleigh scattering, raman scattering, Mie scattering, and the fluorescence process induced by incident ultraviolet laser beam. The fluorophores in the biological material, such as trypophan, tyrosine, Phenylalanine, and NADH (nicotinamide adenine dinucleotide) absorb the incident radiation at a ultraviolet wavelength and re-emit fluorescence at one or more longer wavelengths, and riboflavin absorbs ultraviolet and incident radiation and emits at visible wavelength.

Bioagents/bioaerosols might be non- spherical particle, the backscattering emission is not linear depolarized when illuminated by a polarized light. Through the wave plates and polarization analyzer and double-channel PMT, the received parallel and perpendicular signal can be collected and recorded individually, and finally, the bioagents/bioaerosols can be discriminated with depolarization ratio. Using depolarization ratio database for bioagents/bioaerosols and interferents, the bioagents, such as anthrax, plague, ricin, etc. and bioaerosol, such as dust, smoke, pollen, etc. can be differentiated.

So, Mie scattering lidar is used mainly for searching, positioning and space mapping of the bioagents, fluorescence lidar is used mainly to determine the fluorescence character of bioagents, and polarization lidar can be used to determine the type of bioagents cloud and provide exact early warning.

First, confirm that you have the correct template for your paper size. This template has been tailored for output on the A4 paper size. If you are using US letter-sized paper, please close this file and download the file "MSW_USltr_format".

### *2.1. Lidar Configeration*

The overall schematic diagram of lidar is illustrated in Figure 1. The lidar is consisted of laser emitting sub-system, receiving sub-system and data acquisition sub-system. The returned fluorescence signal is rather weak, so the signal can be collected by photon counter, intensified CCD or PMT array. The doubled and tripled output pulse with 30mJ energy from a flash-pump Nd:YAG laser was used for Mie scattering, polarization scattering and fluorescence measurements. Beam expanders were employed to obtain outgoing laser beams with a 2-cm diameter and 0.5mrad divergence (80% of the total energy). Linear polarization purity of 1/500 for the laser beam was obtained with a polarization cube. A 300mm Cassegrain telescope with an adjustable FOV collects elastically scattered 532nm Mie, 355nm parallel and perpendicular power polarization and 355nm fluorescence signal. The scattering, polarization and fluorescence signal were accumulated for at least 60 seconds, and raman signal was accumulated for at least 30 minutes.

The main specifications of lidar are summarized in table 1.

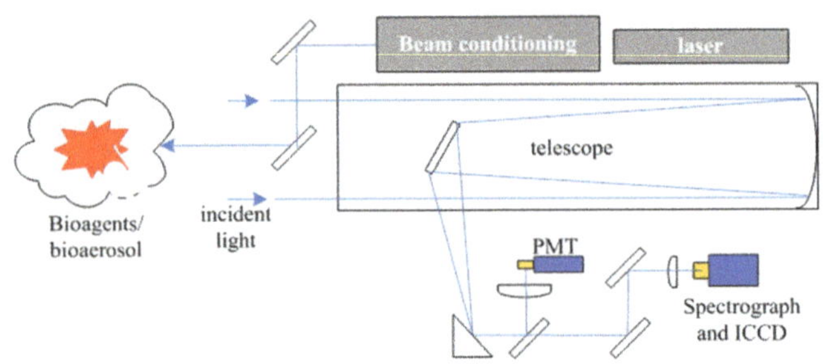


***Fig. 1.*** *Sketch of bioagent/bioaerosol sensing lidar.*

***Table 1.*** *Main specifications of fluorescence sensing lidar.*

| Laser | Nd:YAG |
|---|---|
| Wavelength(nm) | 532, 355, 387(Raman) |
| Laser pulse energy(mJ) | >40 |
| Repetition rate(Hz) | 20 |
| Pulse width(ns) | 20 |
| Accumulated pulse counts | 2000 |
| Telescope | Schmidt-cassegrainian |
| Diameter(mm) | Φ300 |
| Field of view(mrad) | 1-2 adjustable |
| Receiver | |
| Filter central wavelength(nm) | 465 |
| Filter width(nm) | 60 |
| Total optical transmittance | 0.16 |
| PMT quantum yield | 0.2 |
| Range resolution(m) | 15 |
| Fluorescence central wavelength(nm) | 465 |
| NADH Quantum yield | 0.1 |
| NADH fluorescence cross section | $5e^{-12}$ |

### *2.2. Theory of Mie Scattering and Fluorescencel Lidar*

The elastic Mie lidar equation assuming single scattering and is describing the received optical power as [7]

$$P(\lambda_L, z) = P_L \cdot \frac{A_0}{z^2} \cdot \xi(\lambda_L) \cdot \beta(\lambda_L, z) \cdot \xi(z) \frac{c \cdot \tau_L}{2} \cdot T(z) \qquad (1)$$

where $P_L$ is the outgoing effect of the laser and is calculated as $P_L = E_L / \tau_L$, $A_0$ is the aperture area, z is the range in meter, $\xi(\lambda_L)$ is the optical efficiency of the system, $\beta(\lambda_L, z)$

is the volume backscatter coefficient for the elastic backscattering, $\xi(z)$ is the overlap function between the transmitter and the receiver field-of-view. The speed of light is c and $\tau_L$ is the pulse length of the laser, T(z) is the total transmission. The laser wavelength is denoted as $\lambda_L$. The receiver noise is simulated by the detector and optical background induced noise. The total transmission at range z is in general calculated as

$$T(z) = \exp[-2\int_0^z \alpha(\lambda_L, z)dz] \quad (2)$$

The extinction coefficient $\alpha(\lambda_L, z)$ can be divided into two terms, one for the atmospheric background and one for the bioagents/bioaerosol cloud,

$$\alpha(\lambda_L, z) = \alpha_{atm}(\lambda_L, z) + \alpha_c(\lambda_L, z) \quad (3)$$

The backscattering coefficient extinction coefficient $\beta(\lambda_L, z)$ can be also divided into two terms, one for the atmospheric background and one for the bioagents/bioaerosol cloud,

$$\beta(\lambda_L, z) = \beta_{atm}(\lambda_L, z) + \beta_c(\lambda_L, z) \quad (4)$$

The lidar ratio of atmosphere can be calculated by atmospheric mode and for the natural aerosols is set normally as $R_{atm} = \alpha / \beta = 50$.

The fluorescence equation is similar to the lidar equation and is written as

$$P(\lambda, z) = P_L \cdot \frac{A_0}{z^2} \cdot K_0(\lambda_L) \cdot T(z) \cdot \xi(z) \cdot N(z) \cdot \frac{\sigma^F(\lambda_L) \cdot L^F(\lambda)}{4\pi} \cdot \frac{c_0 \cdot \tau_L}{2} \quad (5)$$

where $K_0(\lambda_L)$ is the optical filter function including the optical efficiency, N(z) is the particle concentration, $\sigma^F(\lambda_L)$ is the fluorescence cross section for the present substance and wavelength and $L^F(\lambda)$ is the fingerprint function for the present substance.

### *2.3. Ppolarization Lidar Equation*

According to equation (1), the vertical and perpendicular backscattering signal of 355nm at distance z can be given as

$$P_\perp(z) = P_L \cdot \frac{A_0}{z^2} \cdot \xi(\lambda_L) \cdot \beta(\lambda_L, z) \cdot \xi(z) \frac{c_0 \cdot \tau_L}{2} \cdot T(z) \cdot p_\perp(z) \quad (6)$$

$$P_\parallel(z) = P_L \cdot \frac{A_0}{z^2} \cdot \xi(\lambda_L) \cdot \beta(\lambda_L, z) \cdot \xi(z) \frac{c_0 \cdot \tau_L}{2} \cdot T(z) \cdot p_\parallel(z) \quad (7)$$

Where, $p_\perp(z)$ and $p_\parallel(z)$ are the constant of two channel individually and they depend on the quantum yield of PMT and optical receiving, so the depolarization ration is given as

$$\delta(z) = \frac{P_\perp(z) / p_\perp(z)}{P_\parallel(z) / p_\parallel(z)} = \frac{P_\perp(z)}{P_\parallel(z)} \cdot \frac{p_\parallel(z)}{p_\perp(z)} \quad (8)$$

Where, $K = \frac{p_\parallel(z)}{p_\perp(z)}$ can be obtained by experimental data.

## 3. Data Retrieval for Mie Lidar

In the hypothesis of single scattering, The Fernald method [8] was used to invert the following distributions of the aerosol extinction coefficients, the former is backward solution and the latter is forward solution.

$$\alpha_c(z) = -S\alpha_{atm}(z) + \frac{X(z) \cdot \exp[2(S-1)\int_z^{z_c} \alpha_{atm}(z')dz']}{\frac{X(z_c)}{\alpha_c(z_c) + S\alpha_{atm}(z_c)} + 2\int_z^{z_c} X(z)\exp[2(S-1)\int_z^{z_c} \alpha_{atm}(z'')dz'']dz'} \quad (9)$$

$$\alpha_c(z) = -S\alpha_{atm}(z) + \frac{X(z) \cdot \exp[-2(S-1)\int_z^{z_c} \alpha_{atm}(z')dz']}{\frac{X(z_c)}{\alpha_c(z_c) + S\alpha_{atm}(z_c)} - 2\int_z^{z_c} X(z)\exp[-2(S-1)\int_z^{z_c} \alpha_{atm}(z'')dz'']dz'} \quad (10)$$

Where, $X(z) = P(z) \cdot z^2$, $S = S_c / S_{atm}$, $S_c$ is the aerosol extinction-to-backscattering ratio which differs with the aerosol concentration, size distribution and chemical component, its value is chosen 40~50 suggested by Sasano and Browell [9], $S_{atm}$ is the extinction-to-backscattering ratio of atmospheric molecular valued by $8\pi/3$, $S_{atm} = 8\pi/3$, $Z_c$ is calibration height which is selected by the calculation program automatically within the height range of 4.0~6.0km at daytime or 6.0~8.0 km at nighttime. At the calibration height the content of aerosol drops to its minimized value and close to zero, so $R(Z_c) = X(z_c) / \beta_{atm}(z_c)$ comes to its minimized value. So $\alpha_c(z_c)$ can be deduced by the aerosol and atmospheric molecular scattering ratio

$$R(Z_c) = ((\beta_c(z_c) + \beta_{atm}(z_c) / \beta_{atm}(z_c)) = 1.01 \quad (11)$$

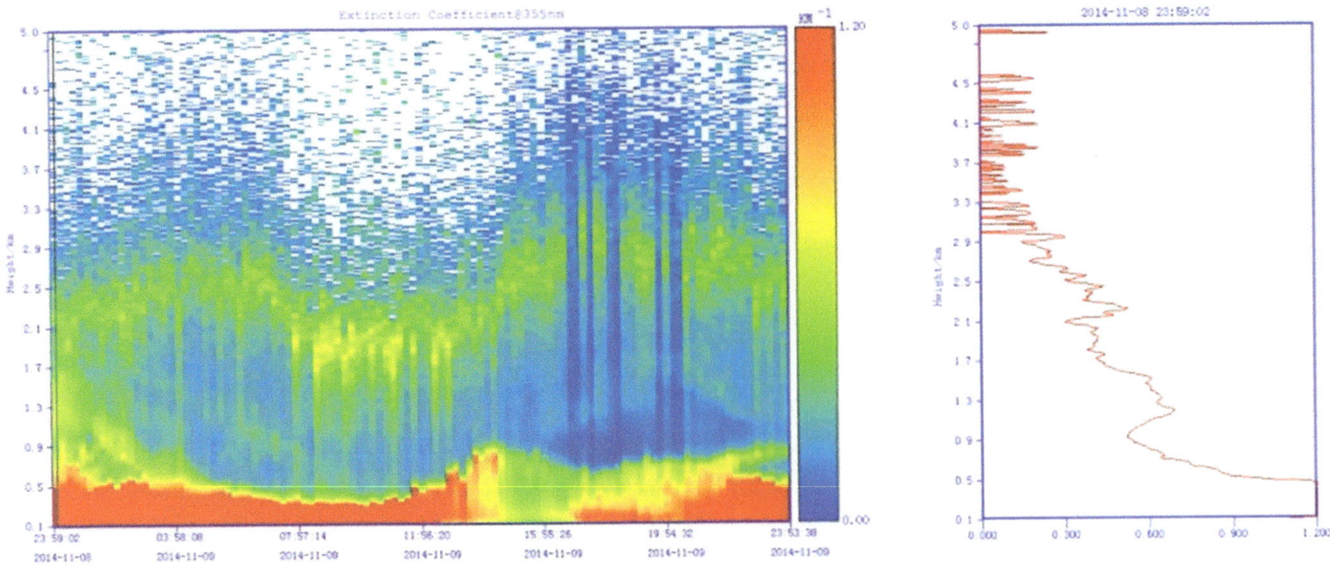

(a)

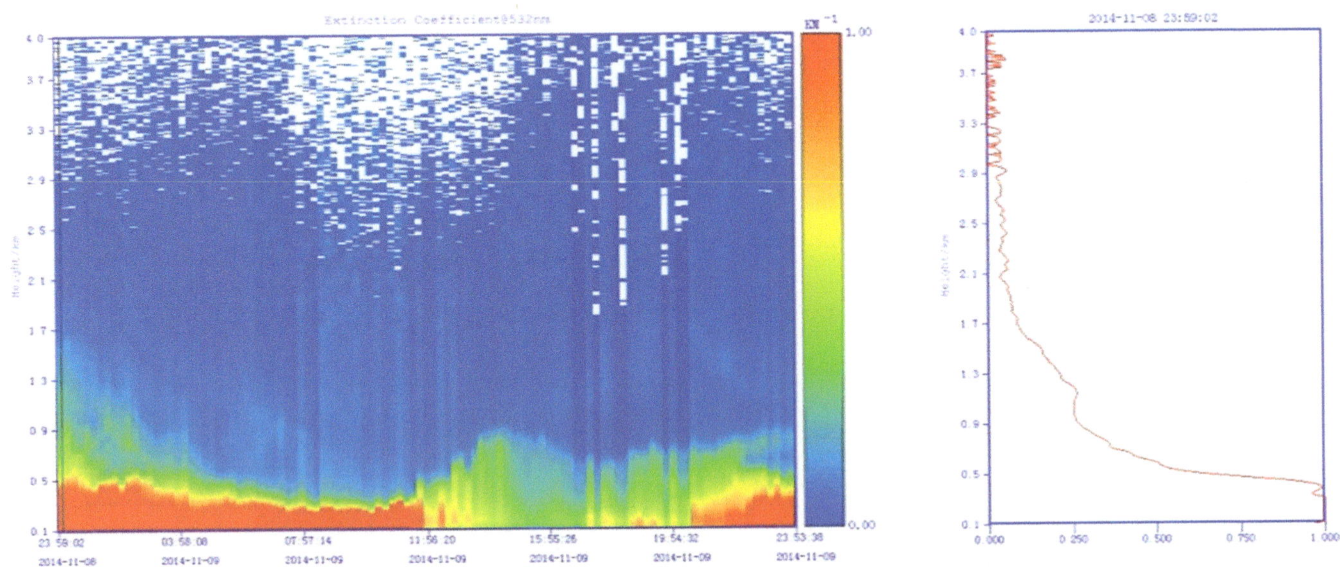

(b)

***Fig. 2.*** *Time and space distributions of bioaerosol, (a)355nm, (b)532nm.*

Figure 2 shows the time and space distributions of aerosol extinction coefficient measured by our multi-channel Lidar and retrieved with the Ferald method. To the right part of each diagram shows the distribution of aerosol extinction coefficient of each laser shot. The measurements were made at hefei city, china, the lidar was laid verticallly. From figure (a) and (b), it can be easily concluded that:

At the bottom of boundary layer, the extinction coefficient distributions remained almost the same trend for the two wavelengths, the maximal value was up to 1.0km-1.

The extra aerosol cloud layer between heights from 1.7km to 2.9km were retrieved through the 355nm channel, the transmittance of 532nm channel is obviously higher, but 355nm channel is more suitable for the fine aerosol particles monitoring with a relatively high laser energy.

# 4. *SNR* of Fluorescence Lidar

The SNR of fluorescence signal can be described as

$$\mathrm{SNR}(z)=\frac{\mathrm{N_s}(z)}{\sqrt{\mathrm{N_s}(z)+2(\mathrm{N_b}+\mathrm{N_d})}}\times\sqrt{\mathrm{n}} \tag{12}$$

Where, *Ns* (*z*) is photo number received at height *z*, $N_b$ (*z*) and $N_d$ are the photo number of background and detector dark current respectively. The more photo number received, and less background and detector noise, the *SNR* (*z*) will be higher. And also, the increased laser pulses can enhance the *SNR*. The simulated fluorescence signal and *SNR* profiles are shown in figure 3.

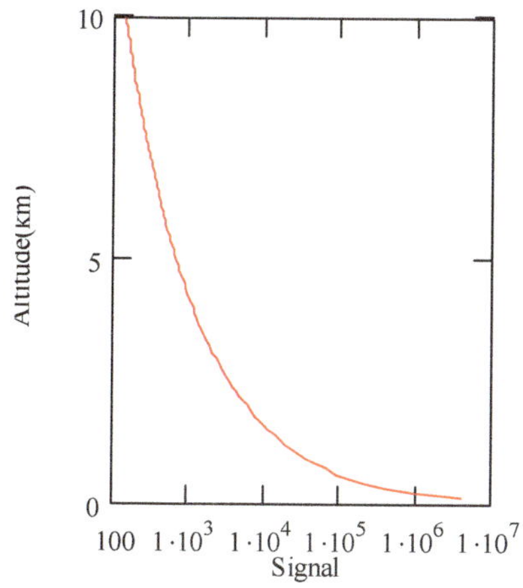


(a)

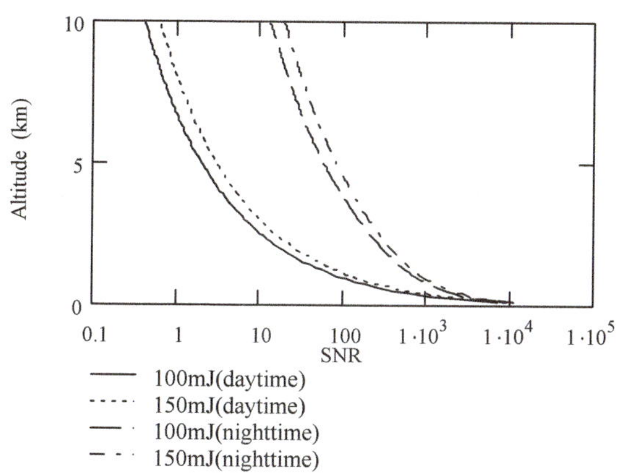


(b)

*Fig. 3. Profiles of simulated fluorescence signal (a) and SNR (b).*

The fluorescence signal profile was simulated with standard atmospheric mode. The number of laser pulse is set as n = 2000, the night *SNR* is much better than that of daytime. The requirement of *SNR* for real lidar system is set to *SNR*>10.

## 5. Results of Depolarization Detection

The time and space distributions of depolarization ratio at 355nm and extinction coefficient at 532nm are shown in figure 4. The profiles were obtained by the aerosol fine particle lidar on November 27, 2014 at Hefei city. The lidar was manufactured and adjusted by AIOFM. Similarly, figure 4(a) shows the extinction coefficient profiles, and the profile at an exact single spot time is shown on the right. From figure 4, it can be shown that:

In the boundary layer, the height of 500m close-ground layer decreased with the time;

In the boundary layer, the complex multi-layer structured extinction coefficient distributions were observed, it means that the aerosol concentration along the laser beam path was affected seriously by the wind, humidity, press and other atmospheric condition;

The two separated vertical distributions of depolarization ratio layer were obvious, the bottom layer is ground layer and the upper layer is surface layer and boundary layer. In the ground layer of 500m height, the value of depolarization ratio remains ness than 0.01, it means that, in this layer, the aerosol particle is in the shape of fine spherical.

After 04:00pm, in the upper surface layer and boundary layer, the zigzag structured distribution of depolarization ratio was observed clearly. At some points along the path, the maximized value of depolarization ratio reached 0.3, it means that the height, the aerosol particle is not spherical caused by some complex physical and photochemistry processes, such as, coagulation, condensation, cohesion, and chemical reaction with the pollution materials released by the industrial and living activities, etc.

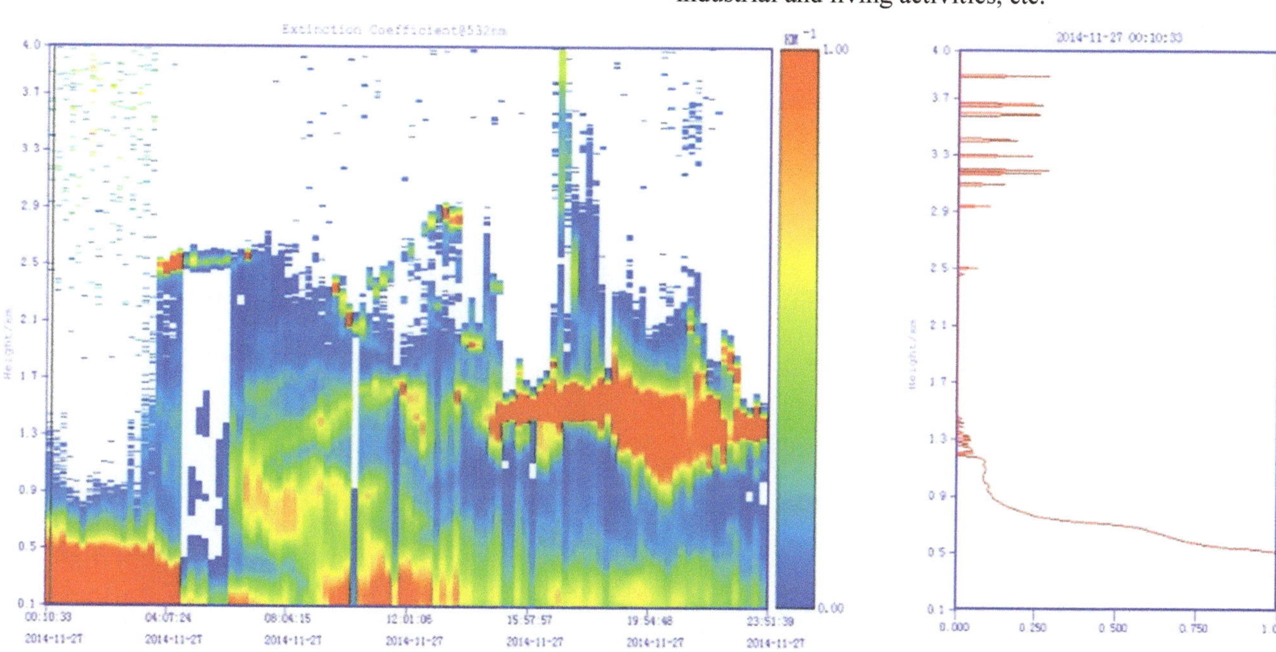


(a)

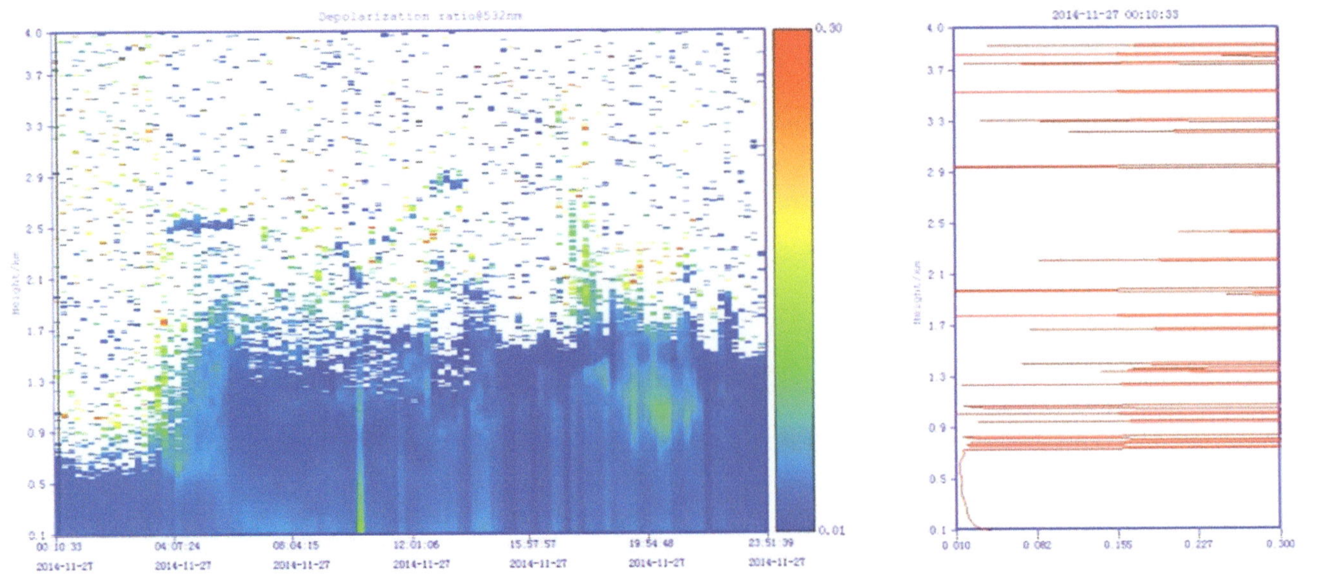


(b)

*Fig. 4. Vertical and time distributions of extinction coefficient and linear depolarization ratio.*

# 6. Discussions and Conclusions

The remote sensing and recognition of bioagents/bioaserosol is the hot spot in the field of photochemistry, biochemistry, etc. Scattering, fluorescence and polarization data are the effective data source of bioagents/bioaserosol time-space distribution monitoring and characteristics recognition. In this paper, the principle, data retrieve algorithms of Mie scattering, ultraviolet laser induced fluorescence lidar are introduced, and the results of vertical time-space evolvement, distributions of extinction coefficient, depolarization ratio observation were discussed.

With the combined multi-dimension observation data, the bioagents/bioaserosol recognition and discrimination appear feasible in principle. In this paper, the polarization data of 355nm wavelength is discussed, and for the following next generation lidar system, polarization characteristics on more wavelengths will be measured and tested, at the same time, the PMT array will be integrated into the system for fluorescence measurement of bioaerosol or bioagents stimulants. So, future work is required to make it reality, and the following questions must be addressed: Except the time-space distribution, how to discriminate exactly bioagents/bioaserosol with observed fluorescence data? Except 355nm wavelength, what wavelength is more effective for depolarization ration measurement, the shorter one or longer one? Does the depolarization ration is wavelength dependent? And what is effect of atmospheric factors, such as relative humidity, background aerosol fluctuation?

# Acknowledgment

The work was supported by the National Natural Science Foundation of China named "Study on Technology of Ultraviolet Laser-induced Fluorescence LIDAR for Bioagent Remote Sensing" No. 41375026.

# References

[1] Matoušek, J.; Bendík J. & Linhart P. CBRN: biologické zbraně. CBRN "Boological weapons" Sdružení požárního a bezpečnostního inženýrství, 2007. (in Czech).

[2] Klietmann, W. F. & Ruoff, K. "Bioterrorism: Implications for the Clinical Microbiologist", Clinical Mikrobiology Reviews, 2001, 14(2), pp. 364-81.

[3] Burger, H. A.; Solomon, W. R. "Sampling and analysis of biological aerosols". Atmospheric Enviroment, 1987, 12 (2), pp. 451-456.

[4] Health Aspects of Chemical and Biological Weapons, World Health Organization, Geneva, Switzerland, 1970.

[5] D. R. Walt and D. R. Franz, "Biological Warfare Detection", Analytical Cemistry, pp. 378 A-746 A, 2000.

[6] G. M. Dougherty, D. S. Clague, and R. R. Miles, B. Field-capable biodetection devices for homeland security missions. SPIE Opt. Photon. Global Homeland Security III, vol. 6540, 2007.

[7] Measures, R. M., Laser Remote Sensing: Fundamentals and Applications, John Wiley & Sons, New York, 1984.

[8] Fernald, F. G., Analysis of atmospheric lidar observation: some comments, Appl. Opt., 1984; 23, pp. 652~655.

[9] Y. Sasano and E. V. Browell, Light scattering characteristics of various aerosol types derived from multiple wavelength lidar observations, Appl. Opt. 1989, 28, pp. 1670~1679.

[10] Xiaoying Cao, Gilles Roy and Robert Bernier, Lidar polarization discrimination of bioaerosols, Optical Engineering, 2010, 49(11), pp. 116201-1~116201-12.

[11] Nianwen Cao, Cunxiong Zhu, Shaobo Yang, Yinhai Xie, Fengkai Yang, and Feng Zhang, Identification of Aerosol Species Using Polarization Lidar, Optics and Spectroscopy, 2014, 117( 3), pp. 486-491.

# Investigation and Evaluation on Heavy Metal Contaminations of Green Salads and Potato Fried in Different Restaurants and Fresh Vegetables in Some Egyptian Governorates

**Hanaa S. M. Abd El-Rahman, Abeer M. N. H. El-Dakak, Zein H.**

Special Food and Nutrition Dept., Food Technology Research Institute, Agricultural Research Center, Giza, Egypt

**Email address:**

Hanaasayed97@yahoo.com (H. S. M. A. El-Rahman), Abeereldakak@Gmail.com (A. M. N. H. El-Dakak), Hosamzin@hotmail.com (H. Zein)

**Abstract:** The content of Cd, Pb, Al, and As in fresh vegetables, green salads and potato fried purchased from the Egyptian market and restaurant were determined using atomic absorption spectrometry (AAS).The results of this study showed that there was wide variation in the concentration of these metals in vegetables collected from different sites. The highest Cd level in watercress (8.33 mg/Kg) was found in Cairo. Vegetables grown in Cairo and Alexandria had cadmium many folds higher than those of other towns. Lead concentration in potato of Alexandria is more than permitted level. Aluminium (Al) content was high concentration in all vegetable samples and arsenic (As) concentration mostly was appeared in potato. Analytical results indicated that the concentration of Cd, Pb, and As in green salad samples were so few. The highest concentration of Al was detected in salad collected from popular restaurant 8. The results show that the heavy metal content in unprocessed potato appears to be much higher than that in processed potato. The highest levels of Cd, Pb, and Al were detected in popular restaurant 9. Cd and As contents of international chain restaurant 1 were found in high level (2.958 and 0.95mg/Kg).

**Keywords:** Heavy Metals; Food Contaminants, Cadmium, Lead, Aluminium, Arsenic, Fresh Vegetables, Green Salad, Potato Fried

## 1. Introduction

Food safety is a major public concern worldwide. During the last decades, the increasing demand for food safety has stimulated research regarding the risk associated with consumption of foodstuffs contaminated by pesticides, heavy metals and/or toxins (1).Vegetables constitute an important part of the human diet since they contain carbohydrates, proteins, as well as vitamins, minerals and trace elements (2, 3).Vegetables also act as buffering agents for acidic substances obtained during the digestion process. However, these plants may contain both essential and toxic elements, such as heavy metals, at a wide range of concentrations (4).Publicity regarding the high level of heavy metals in the environment has created apprehension and fear in the public as to the presence of heavy metal residues in their daily food. The public is confused and alarmed about their food safety. Keeping in mind the potential toxicity and persistent nature of heavy metals, and the frequent consumption of vegetables and fruits, it is necessary to analyze these food items to ensure the levels of these contaminants meet agreed international requirements (5).

Heavy metals are easily accumulated in human vital organs and threaten human health. Vegetables are part of human diet to take up a lot of essential nutrients and certain trace elements in a short period. In this situation, safety of vegetables is very important (6, 7, 8, and 9). Heavy metals are one of a range of important types of contaminants that can be found on the surface and in the tissue of fresh vegetables. Heavy metals, such as cadmium, copper, lead; chromium and mercury are important environmental pollutants, particularly in areas under irrigated with waste water. Several investigations of water, soil and vegetables pollution by waste water are available (10, 11, 12, 13, 14, 15, and 16).

Strictly speaking, heavy metals are defined as those with higher density than 5 mg $mL^{-1}$ (17) but the collective term now includes arsenic, cadmium, chromium, copper, lead, nickel, molybdenum, vanadium and zinc. Some interest also exists in aluminium, cobalt, strontium and other rare metals. Physiologic roles are known for iron (haemmoeties of heamoglobin and cytochromes), copper (amine oxidases, dopamine hydrolase and collagen synthesis), manganese (superoxide dismutase), zinc (protein synthesis, stabilization of DNA and RNA) with low requirements of chromium (glucose homeostasis). Other heavy metal ions are not believed to be essential to health even in trace amounts. Vegetables, especially those of leafy vegetables grown in heavy metals contaminated soils, accumulate higher amounts of metals than those grown in uncontaminated soils because of the fact that they absorb these metals through their leaves (18).Research findings show that at least 20 million hectares of land in North and South Africa, South America, Middle East, Southern Europe, South West America, Mexico and a significant part of Central and East Asia is irrigated by raw sewage, mainly for cultivation of vegetables. Consequently, this usage ends to soil contamination and heavy metals accumulation both in soil and crops (19). Heavy metal accumulation in soils is of concern in agricultural production due to the adverse effects on food quality (safety and marketability), crop growth (due to phytotoxicity) (20, and 21).

Metals such as lead, mercury, cadmium and copper are cumulative poisons. These metals cause environmental hazards and are reported to be exceptionally toxic. On consumption of food in the diet, the trace metal contents of food are directly taken into the body (22, and 23).These metals enter the human body mainly through two routes namely: inhalation and ingestion, and with ingestion being the main route of exposure to these elements in human population. Heavy metals may enter the human body through inhalation of dust, direct ingestion of soil and consumption of food plants grown in metal contaminated soil (24, and 25).Heavy metals intake by human populations through the food chain has been reported in many countries with this problem receiving increasing attention from the public as well as governmental agencies, particularly in developing countries(26, and 27).The allowable Limit of Heavy Metals, as safe values for copper, lead, and cadmium in fruit and vegetables recommended by the WHO / FAO are 40, 0.3, and 0.2 mg/kg, respectively. Vegetables take up metals by absorbing them from contaminated soils, as well as from deposits on different parts of the vegetables exposed to the air from polluted environments (28). It has been reported that nearly half of the mean ingestion of lead, cadmium and mercury through food is due to plant origin (fruit, vegetables and cereals).

Metal contamination soils may be widespread in urban areas due to past industrial activity and the use of fossil fuels (29).Additional sources of heavy metals for plants are: rainfall in atmospheric polluted areas, traffic density, use of oil or fossil fuels, for heating, atmospheric dusts, plant protection agents and fertilizers which could be adsorbed through leaf blades (30, 31, 32, 33).The content of essential elements in plants is conditional, the content being affected by the characteristics of the soil and the ability of plants to selectively accumulate some metals (34).

Prolong consumption of unsafe concentrations of heavy metals through foodstuffs may lead to the chronic accumulation of heavy metals in the kidney and liver of humans causing disruption of numerous biochemical processes, leading to cardiovascular, nervous, kidney and bone diseases (35).The main source of human exposure to Pb and Cd is food, which is believed to provide about 80-90% of daily doses (36, 37). Pb and Cd toxicity is well documented and is recognized as a major environmental health risk throughout the world. Pb affects humans and animals of all ages, but the effects of lead are most serious in young children. Cadmium is a toxic and carcinogenic element. The International Agency for Research on Cancer has identified Cd as a known human carcinogen. Pb and Cd poisoning results from the interaction of the metal with biological electron-donor groups, such as the sulfhydryl groups, which interferes with a multitude of enzymatic processes. Clinical manifestations of Pb toxicity include symptoms referable to the central nervous system, the peripheral nervous system, the hematopoietic system, the renal system, and the gastrointestinal systems. Cd is a cumulative nephrotoxicant that is absorbed into the body from dietary sources and cigarette smoking (37, 36).Potentially toxic metals are also present in commercially produced foodstuffs. Results of (38) indicated that the cadmium concentrations in shoots and roots varied both with different Cd levels and type of vegetable. Generally Cd accumulation in various plant parts in vegetable crops was increased with the increasing cadmium concentrations in the growth medium. Cd increased more sharply in roots than shoots. Celery contained higher Cd in the edible parts than other vegetable species. Permissible level of consumption of Cadmium, for human is 70$\mu day^{-1}$.

The effect of environmental pollution on contamination of foods and on their safety for human consumption is a serious global public issue and widely addressed (39, 40). Some of these elements are toxic to humans even at a very low level. Excessive content of Pb and Cd metals in food is associated with etiology of a number of diseases especially with cardiovascular, kidney, nervous as well as bone diseases (41, 42, 43,17). They have also been implicated in causing carcinogenesis, mutagenesis and teratogenesis (44, 45).Lead is a toxic element that can be harmful to plants, although plants usually show ability to accumulate large amounts of lead without visible changes in their appearance or yield. In many plants, Pb accumulation can exceed several hundred times the threshold of maximum level permissible for human(46). The introduction of Pb into the food chain may affect human health, and thus, studies concerning Pb accumulation in vegetables have increasing importance (47).

Although a maximum Pb limit for human health has been established for edible parts of crops (0.2 mg/kg) (48), soil Pb thresholds for producing safe vegetables are not available. Further investigations Zhang and Zhou (49) showed that the Al-based coagulants at the tested concentrations had a poisonous effect on the germination of vegetable seeds. While aluminum can be toxic at higher levels, it is considerably less toxic than either mercury or lead. In fact, aluminum is found at easily measurable levels in various biological fluids and tissues. However, at high levels aluminum has the potential to cause a number of health problems such as anaemia and other blood disorders, colic, fatigue, dental caries, dementia dialactica, kidney and liver dysfunctions, neuromuscular disorders, osteomalacia and Parkinson's disease (50). Arsenic is regarded as human carcinogen from extremely low levels of exposure, having no possible beneficial metabolic functions for humans. Its low level exposure cause nausea and vomiting decreased production of RBCs and WBCs, abdominal pain and its long term exposure causes darkening of skin and appearance of small corns on palm soles. Other affect includes anorexia, fever, fluid loss, goiter, hair loss, headache, herpes, impaired healing, jaundice, keratosis, kidney and liver damage, muscle spasms, pallor, peripheral neuritis, sore throat, weakness and interferes with the uptake of folic acid (50).

There were positive curvilinear or linear relationships between the inhibitory rate of seed germination and the concentration of Al in the acidic and neutral conditions except for the toxic effects of PAC (polyaluminum-chloride) on *Brassica chinensis*in the neutral condition (49). Moreover there were obvious differences in root elongation of *Brassica chinensis* exposed to $AlCl_3$ in various pH conditions.

The aim of the present study was to evaluate the safety of selected fresh vegetables in different regions of Egypt, green salads, and potatoes fried available in Egyptian restaurants in the regions of Cairo and Giza in 2014, as determined by the content ofPb, As, Al, and Cd.

# 2. Materials and Methods

## 2.1. Materials

Various kinds of fresh vegetables (33 samples) were randomly collected from the main farmlands around the city (Cairo, Giza, Alexandria, menoufia, and kalyopia). Green salads (12 samples) and potato fried were collected from International chain, local, and popular restaurants in February and March during the year 2014 submitted to a laboratory for analysis. All reagents which used in these methods were obtained from Merck-Darmstadt.

## 2.2. Determination of Mineral Contents

The content of metals in vegetables and green salads samples was determined according to a procedure advised by the Regional Center for Food and Feed. The samples were brought in plastic bags to Laboratory. Vegetables samples were cleaned peeled (if necessary) and washed to obtain edible parts prior to analysis. Green salads and potato fried samples were homogenized. All samples were weighed (1g±0.01) in Pyrex tubes, add 10 ml of concentrated nitric acid, and allow standing overnight. Heat carefully on a hot plate unit the production of red $NO_2$ fames has ceased. Cool the beaker and a small amount (2-4 mL) of 70 % $HClO_4$. Heat again and allow evaporating to a small volume transfer the sample to a 50 mL flask and diluting to volume with distilled water (51).The samples were analyzed by Contraa 700-P, atomic absorption spectrophotometer- Analytical Jena.

## 2.3. Statistical Analyses

Analysis using the Dunks test was carried out to examine the statistical significance of differences in the mean concentration of metals between groups of vegetables using SPSS, version 10. A probability level of $p<0.05$ was considered statistically significant.

# 3. Results

Heavy metal content obtained on tomatoes, cucumber, lettuce, watercress, and potatoes collected from different agricultural locations (Cairo, north and south Giza, Alexandria, Menoufia, Kalopia, and Sohag) in Egypt are given as mean values and standard Error in Table 1. Results of monitoring studies carried out in Egypt in 2014 showed that the heavy metal content of vegetable samples were regionally dependent. The highest Cd level in watercress (8.33 mg/Kg) was found in Cairo and higher in potatoes (4.41 mg/Kg) from Kalyopia while the lowest was determined in the cucumber (0.0000198 mg/Kg) from north Giza. Cd content of potatoes were found partly high levels in Cairo, north Giza, and Alexandria (1.3, 0.33, and 0.95 mg/Kg) compared with other agricultural location. The content of Cd in lettuce from Cairo, Alexandria, Menoufia, and south Giza was 1.06, 1.69, 0.81, and 1.31 mg/Kg, respectively.

The highest Pb content was found in potatoes (61.1mg/Kg) from Alexandria while the lowest were found in Tomatoes (0.0000062, 0.000006, 0.000004, and 0.000069) from north Giza, Alexandria, Menoufia, and Kalyopia, respectively. On the other hand, the Pb content of lettuce, potatoes were high (0.57, 0.51, and 0.543 mg/Kg) in Cairo, Menoufia, and Kalyopia independent of region.

Al contents varied in agricultural locations among the vegetable samples. The highest levels were 4178.67 and 1820.85 mg/Kg in potatoes from north Giza, and lettuce from Alexandria while the lowest Al levels were found in cucumber from Cairo, Alexandria, and tomatoes from Kalypia (3.369, 3.547, 3.34 mg/Kg, respectively).

As content of vegetable samples were found at low levels. As content of vegetable samples were changed between 0.00042 (in cucumber from North Giza) to 0.319 (in potato from North Giza) mg/Kg.

Concentration of some heavy elements found in green salad collected from international chain restaurant, local restaurant, and popular restaurants were summarized in Table 2. Cd, Pb, and As contents of green salad collected from different restaurants were found at low levels. Al content of salad samples in all restaurant categories ranged from 1.603 to 914.11mg/Kg. The highest concentration of Al was detected in salad collected from popular restaurant 8 (914.11 mg/Kg) while international chain restaurant (2), local restaurants (3,6), and popular restaurant (10,11) had the lowest concentration (1.603,3.998, 4.03, 3.93, and 4.32mg/Kg).

The concentrations of heavy metals (Cd, Pb, Al, and As) in potatoes fried from three level restaurants (international chain, local, and popular) analyzed are reported in Table 3. The highest levels of Cd, Pb, and Al were detected in popular restaurant 9. Cd and As contents of international chain restaurant 1 were found in high level (2.958 and 0.95mg/Kg) compared with other restaurants, While Pb contents was found in low level (0.033 mg/Kg). As content in potatoes fried from popular restaurant 7 and 8 were non-detectable amounts. At the same form, potatoes fried contained Al at the low concentration. Al values showed a notable increase from 56.88 mg/Kg in popular restaurant 7 to 2305.87 mg/Kg in popular restaurant 9. Pb was detected at lowest levels in potatoes fried from local restaurant 6 (0.0019 mg/Kg) and Cd was lowest in potatoes fried from local restaurant 4 (0.419 mg/Kg).

***Table 1.*** *Levels of Cd, Pb, Al, and As (mg/kg) in some vegetables*

| Name of location | Vegetables | Cd | Pb | Al | As |
|---|---|---|---|---|---|
| Cairo | Tomatoes | $0.00043\pm0.000018^{a}$ | $0.000035\pm0.000022^{a}$ | $340.92\pm16.38^{i}$ | $0.0013\pm0.00025^{ab}$ |
| | Cucumber | $0.00025\pm0.00034^{a}$ | $0.000011\pm0.0000035^{a}$ | $3.369\pm1.57^{a}$ | $0.0014\pm0.00017^{ab}$ |
| | Lettuce | $1.06\pm0.10^{e}$ | $0.572\pm0.045^{h}$ | $544.76\pm40.7937^{k}$ | $0.020\pm0.004^{c}$ |
| | Watercress | $8.33\pm0.197^{i}$ | $0.38\pm0.01086^{c}$ | $104.76\pm3.9^{bcdefg}$ | $0.038\pm0.0017^{e}$ |
| | Potatoes | $1.303\pm0.14^{f}$ | $0.0123\pm0.00038^{a}$ | $138.66\pm28.61^{defgh}$ | $0.02\pm0.004^{cd}$ |
| North Giza | Tomatoes | $0.000034\pm0.000017^{a}$ | $0.0000062\pm0.0000025^{a}$ | $439.60\pm21.24^{j}$ | $0.0011\pm0.000056^{ab}$ |
| | Cucumber | $0.0000198\pm0.000016^{a}$ | $0.00019\pm0.000076^{a}$ | $605.62\pm18.25^{k}$ | $0.00042\pm0.000061^{a}$ |
| | Lettuce | $0.00036\pm0.00003^{a}$ | $0.000016\pm0.000003^{a}$ | $6.30\pm2.05^{a}$ | $0.0064\pm0.00053^{b}$ |
| | Potatoes | $0.33\pm0.0042^{b}$ | $0.0066\pm0.000032^{a}$ | $4178.67\pm168.59^{n}$ | $0.319\pm0.011^{k}$ |
| South Giza | Tomatoes | $0.00024\pm0.00003^{a}$ | $0.00074\pm0.00059^{a}$ | $74.31\pm3.95^{abcde}$ | $0.0018\pm0.00021^{ab}$ |
| | Cucumber | $0.000023\pm0.000053^{a}$ | $0.0000053\pm0.0000025^{a}$ | $148.74\pm16.60^{efgh}$ | $0.0014\pm0.00017^{ab}$ |
| | Lettuce | $1.31\pm0.052^{f}$ | $0.44\pm0.022^{d}$ | $87.895\pm6.53^{bcdef}$ | $0.0899\pm0.0038^{g}$ |
| | Watercress | $0.0012\pm0.00025^{a}$ | $0.447\pm0.007^{d}$ | $844.39\pm18.47^{l}$ | $0.023\pm0.00151^{cd}$ |
| | Potatoes | $0.00087\pm0.0001^{a}$ | $0.000006\pm0.000002^{a}$ | $207.57\pm7.27^{h}$ | $0.0032\pm0.00024^{ab}$ |
| Menoufia | Tomatoes | 0.000364±0.0051a | $0.000004\pm0.000001^{a}$ | $12.16\pm1.56^{a}$ | $0.0022\pm0.00028^{ab}$ |
| | Cucumber | $0.000046\pm0.00001^{a}$ | $0.000049\pm0.000023^{a}$ | $102.37\pm3.0065^{bcdefg}$ | $0.00086\pm0.000093^{a}$ |
| | Lettuce | $0.81\pm0.045^{d}$ | $0.44\pm0.0236^{d}$ | $172.0023\pm12.57^{gh}$ | $0.023\pm0.0025^{cd}$ |
| | Watercress | $0.003\pm0.000011^{a}$ | $0.0011\pm0.00039^{a}$ | $66.17\pm7.99^{abcd}$ | $0.0019\pm0.00034^{ab}$ |
| | Potatoes | $0.12\pm0.268^{a}$ | $0.51\pm0.0185^{f}$ | $133.11\pm13.08^{defg}$ | $0.045\pm0.001^{f}$ |
| Kalyopia | Tomatoes | $0.00006\pm0.00007^{a}$ | $0.000069\pm0.000102^{a}$ | $3.34\pm0.22^{a}$ | $0.0018\pm0.0002^{ab}$ |
| | Cucumber | $0.00081\pm0.000023^{a}$ | $0.000007\pm0.000002^{a}$ | $842.005\pm9.51^{l}$ | $0.0021\pm0.000095^{ab}$ |
| | Lettuce | $0.00023\pm0.0003^{a}$ | $0.00006967\pm0.000096^{a}$ | $4.06\pm1.\ 1^{a}$ | $0.002\pm0.00015^{ab}$ |
| | Watercress | $0.00075\pm0.000041^{a}$ | $0.0000095\pm0.0000028^{a}$ | $931.85\pm26.05^{m}$ | $0.0013\pm0.00021^{ab}$ |
| | Potatoes | $4.41\pm0.29^{h}$ | $0.54\pm0.0156^{g}$ | $554.01\pm3.43^{k}$ | $0.21\pm0.0032^{j}$ |
| Sohag | Tomatoes | $0.00045\pm0.000056^{a}$ | $0.0048\pm0.000196^{a}$ | $69.42\pm1.25^{abcd}$ | $0.0012\pm0.00021^{ab}$ |
| | Cucumber | $0.00046\pm0.000055^{a}$ | $0.0056\pm0.00058^{a}$ | $36.23\pm2.07^{ab}$ | $0.0014\pm0.00043^{ab}$ |
| | Lettuce | $0.00021\pm0.000023^{a}$ | $0.0049\pm0.00032^{a}$ | $161.20\pm2.16f^{gh}$ | $0.0024\pm0.00028^{ab}$ |
| | Potatoes | $0.00021\pm0.000022^{a}$ | $0.0033\pm0.00024^{a}$ | $52.97\pm2.87^{abc}$ | $0.0014\pm0.00014^{ab}$ |
| Alexandria | Tomatoes | $0.000138\pm0.00002^{a}$ | $0.000006\pm0.00003^{a}$ | $965.59\pm6.7^{m}$ | $0.00055\pm0.000053^{a}$ |
| | Cucumber | $0.00097\pm0.00067^{a}$ | $0.0020\pm0.000299^{a}$ | $3.547\pm0.53^{a}$ | $0.00057\pm0.000012^{a}$ |
| | Lettuce | $1.69\pm0.079^{g}$ | $0.442\pm0.025^{d}$ | $1820.85\pm138.21^{n}$ | $0.16\pm0.0035^{i}$ |
| | Watercress | $0.61\pm0.025^{c}$ | $0.34\pm0.025^{b}$ | $121.80\pm8.04^{cdefg}$ | $0.025\pm0.0035^{d}$ |
| | Potatoes | $0.95\pm0.228^{e}$ | $61.1\pm0.0044^{e}$ | $106.07\pm4.78^{bcdefg}$ | $0.12\pm0.003^{h}$ |

Each value represents the mean ± SE; the mean values with different superscript alphabets indicate significant differences ($P<0.05$).

***Table 2.*** *Heavy metal concentrations in green salad (mg/kg).*

| Name of location level | Cd | Pb | Al | As |
|---|---|---|---|---|
| International chain restaurant(1) | $0.0004\pm0.000014^{de}$ | $0.000014\pm0.000056^{a}$ | $5.31\pm0.17^{ab}$ | $0.00046\pm0.00005^{a}$ |
| International chain restaurant(2) | $0.0001\pm0.00002^{a}$ | $0.000025\pm0.000035^{ab}$ | $1.603\pm0.34^{a}$ | $0.0033\pm0.0003^{g}$ |
| Local restaurant(3) | $0.0003\pm0.00004^{bcd}$ | $0.00023\pm0.000016^{e}$ | $3.998\pm0.18^{a}$ | $0.00015\pm0.00004^{a}$ |
| Local restaurant(4) | $0.00066\pm0.00012^{g}$ | $0.000038\pm0.000017^{abc}$ | $12.52\pm0.47^{b}$ | $0.0029\pm0.0004^{f}$ |
| Local restaurant(5) | $0.00049\pm0.00017^{e}$ | $0.000013\pm0.000044^{a}$ | $293.95\pm5.7^{c}$ | $0.00054\pm0.0004^{ab}$ |
| Local restaurant(6) | $0.00023\pm0.0003^{bc}$ | $0.000016\pm0.000041^{de}$ | $4.03\pm0.21^{a}$ | $0.0016\pm0.0004^{d}$ |
| Popular restaurant(7) | $0.00031\pm0.00005^{cd}$ | $0.000015\pm0.000003^{a}$ | $290.59\pm5.58^{c}$ | $0.0011\pm0.00005^{c}$ |
| Popular restaurant(8) | $0.00015\pm0.000015^{ab}$ | $0.00011\pm0.0000028^{bcd}$ | $914.11\pm1.28^{d}$ | $0.00035\pm0.00003^{a}$ |
| Popular restaurant(9) | $0.00044\pm0.00007^{ef}$ | $0.000006\pm0.000002^{a}$ | $5.79\pm0.15^{ab}$ | $0.0021\pm0.0001^{e}$ |
| Popular restaurant(10) | $0.00014\pm0.00002^{ab}$ | $0.0018\pm0.00015^{f}$ | $3.93\pm0.33^{a}$ | $0.00087\pm0.0001^{bc}$ |
| Popular restaurant(11) | $0.00018\pm0.00003^{ab}$ | $0.0000049\pm0.0000007^{a}$ | $4.32\pm0.12^{a}$ | $0.00051\pm0.00006^{ab}$ |
| Popular restaurant(12) | $0.00082\pm0.00022^{h}$ | $0.00011\pm0.00004^{cd}$ | $6.51\pm0.15^{ab}$ | $0.0011\pm0.00013^{c}$ |

Each value represents the mean ± SE; the mean values with different superscript alphabets indicate significant differences ($P<0.05$).

*Table 3. Concentration of heavy metal in potatoes fried(mg/kg).*

| Name of location level | Cd | Pb | Al | As |
|---|---|---|---|---|
| International restaurant(1) | 2.958±0.034$^{d}$ | 0.033±0.003$^{b}$ | 88.55±7.44$^{ab}$ | 0.95±0.0028$^{f}$ |
| International restaurant(2) | 0.779±0.18$^{ab}$ | 0.044±0.003$^{c}$ | 78.597±6.83$^{a}$ | 0.126±0.0011$^{d}$ |
| Local restaurant(3) | 0.503±0.047$^{a}$ | 0.08±0.004$^{d}$ | 537.57±8.13$^{d}$ | 0.037±0.0034$^{a}$ |
| Local restaurant(4) | 0.419±0.085$^{a}$ | 0.031±0.003$^{b}$ | 190.44±5.27$^{c}$ | 0.121±0.0014$^{c}$ |
| Local restaurant(5) | 0.52±0.073$^{a}$ | 0.36±0.003$^{g}$ | 93.21±2.62$^{ab}$ | 0.054±0.0039$^{b}$ |
| Local restaurant(6) | 2.961±0.297$^{d}$ | 0.0019±0.0003$^{a}$ | 163.53±1.34$^{bc}$ | 0.127±0.0014$^{d}$ |
| Popular restaurant(7) | 1.052±0.264$^{b}$ | 0.28±0.004$^{e}$ | 56.88±4.051$^{a}$ | Nd |
| Popular restaurant(8) | 2.09±0.115$^{c}$ | 0.32±0.0022$^{f}$ | 78.30±5.003$^{a}$ | Nd |
| Popular restaurant(9) | 2.9997±0.22$^{d}$ | 0.73±0.0045$^{h}$ | 2305.87±1.25$^{e}$ | 0.199±0.305$^{e}$ |

Each value represents the mean ± SE; the mean values with different superscript alphabets indicate significant differences (P<0.05). ND = Not detected. Levels were below the detection limit

# 4. Discussion

The toxicity of the metals in agricultural products depends upon relative level of exposure of crops to the contaminated soils as well as the deposition of toxic elements in the polluted air by sedimentation. Different metals show the toxicity at different concentrations and can be potentially toxic at sufficiently high concentrations. However, certain metals exhibit toxic effects even at relatively low concentrations. Metal toxicity depends on the element, its chemical form and its oxidation state. The results in Table 1 show a high degree of contamination in vegetables, when compared with the permissible levels given by the FAO and WHO (36). Our data showed that the overall toxic metal accumulation was greater in leafy vegetables lettuce, watercress and tubers such as potato. Heavy metal content has been reported in various common vegetables of many towns in Egypt. Our data showed that large concentrations of cadmium (toxic trace element) in lettuce (Cairo, Alexandria, menofia, and South Giza), watercress (Cairo and Alexandria) and potatoes (Cairo, Alexandria, and Kalyopia). The above data reveal that cadmium content in same vegetable differs from town to town. According to above data, cadmium content in vegetables of Cairo and Alexandria are many folds higher than those of other towns. It might be because the metal uptake in vegetables is influenced by several factors such as metal concentrations in agricultural soils, soil pH, physico-chemical characteristics of the soil, soil classification, etc. Moreover, vegetable consumption (per person per day) is not same for the residents of different regions. Abdel-Shafy*et al.*, (52) reported the levels of the metals in the surface layer of soil irrigated with the Nile water and ground water in Qatta village Egypt were arranged in the following ascending orders Fe > Zn >Mn>Pb> Cu > Cd and Fe >Mn> Zn = Cu >Pb> Cd, respectively. The level of heavy metals in different parts of potato plants irrigated with Nile and ground water for over four successive seasons in Qatta village. In the period of continuous soil irrigation with both the Nile and ground water, most of metals, namely cadmium and lead, are accumulated in plants. The level of heavy metals in soil does not depend essentially on the sources of irrigation, but first of all on the metal constituents of the fungicide applied. Okwulehie and Ogoke (53) reported that cadmium in the range of 0.7-0.94 ppm for certain mushrooms in Nigeria. Banerjee *et al.*,(54) recorded cadmium levels in some fruits and vegetables in India to be in the range of 0.03 to 7.32 µg/g dry weight. Cadmium content of tomato and cucumber analyzed in Egypt was below the WHO standards limit (0.2 mg/kg) (35).The high contamination levels found in some vegetables may be related to pollutants in irrigation water, farm soil or pollution from highway traffic (55).In a study Bahemuka and Mubofu (4) reported that the cadmium content in vegetables of Tanzania ranged between 0.1 and 0.6 mg/kg dry weight. Studies have shown that metals such as iron, copper, cadmium, chromium, lead, mercury and nickel have the ability to produce reactive oxygen species. The result of this is lipid peroxidation, DNA damage and altered calcium homeostasis (56, 57). Elsokkary and Sharaf (58) reported that the bioaccumulation ratio of Cd in the plants followed the order: Chard>Spinach> Lettuce> Parsley>Rocket> Coriander.

The main cause for concern in terms of contamination of vegetables in Alexandria and Menoufia by heavy metals relates to Lead (Pb). Although a maximum Pb limit for human health has been established for edible parts of crops in China is 0.2 mg/kg (48) but this limit by WHO standards is 0.3 mg/kg (35). Data showed that in potato of Alexandria, lead concentration is more than permitted level, so they are not suitable for consumption. Lead is a toxic element that can be harmful to plants, although plants usually show ability to accumulate large amounts of lead without visible changes in their appearance or yield. In many plants, Pb accumulation can exceed several hundred times the threshold of maximum level permissible for human (46). The introduction of Pb into the food chain may affect human health and thus, studies concerning Pb accumulation in vegetables have increasing importance (47, 3). On the whole, some vegetables that were studied in this study were contaminated by lead and they were toxic to consumer. Zhuang*et al.*, (59) have found higher contents of Cd and Pb concentrations than the maximum permissible levels in vegetables collected from six sampling sites around Dabaoshan mine located at Shaoguan city, Guangdong, southern China. It was found that most of the vegetable types marketed in Kathmandu were grown along the bank of rivers, roadsides and highways irrigated with wastewater as well as from vegetable farms with possible use of fertilizers and pesticides in excessive amount (source: personal communication with vegetable grosser). Continuous irrigation of agricultural land with sewage and wastewater may cause heavy metal accumulation in the soil and vegetables (60).Besides, pesticides and fertilizers are known to be the main sources of heavy metal pollution in agricultural areas (61).Moreover, the transportation and marketing systems of vegetables play a significant role in elevating the contaminant levels of heavy metals which may pose a threat to the quality of the vegetables with consequences for the health of the consumers of locally produced foodstuffs (62, 63).It is concluded that atmospheric deposition from urban and agricultural areas may play an important role in the enrichment of agricultural produce from Cd and/ or Pb (64). Abdel-Sabour and Rabie (65) also studied ten sample of vegetable plant species, plants grown on soils (irrigated with industrial wastewater of Shebin El-Qanater and Mostorod collectors) and (irrigated with polluted water of El-Shaboura canal), whereas its soils have been subjected to prolonged domestic and industrial wastewater irrigation (about 35 years). Results revealed that vegetable plants species varied in their affinity to accumulate metals in their edible parts. Irrigation with different wastewater significantly increased the concentration of Pb and Cd in vegetable plants especially the leafy species. Rashad and Shalaby, (66) measured the levels of heavy metals at different sites with different distances and directions from two dumpsites in Alexandria for vegetation. The concentrations of metals in leaves and roots of tomatoes, carrots and potatoes plants were higher in plants grown at the site close to Abis (the municipal solid waste) dumpsite and decreased with increasing distance. The only difference between Abis and El-Montaza is that the recorded levels of heavy metals at Abis area were higher than obtained at El-Montaza area. This could be due to the other pollution sources found at Abis area, including the cement company and other petroleum companies near Abis area which is a serious source of their metals.

A market basket survey showed that Aluminium (Al) content was high concentration in all vegetable samples. Al content in some common vegetables in North Giza which show that potato had the highest amount while the lettuce had lowest (Table 1). In Alexandria the highest variation in Al level was observed in lettuce and tomato and lowest in cucumber. However the Kalyopia vegetables (cucumber and watercress) contain much higher amount of Al compared to Sohag. It can be seen that total Al in vegetable samples ranged from 3.34 to 4178.67 mg/kg, in green salad was 1.603 to 914.11 mg/kg, and in processed potato ranged from 56.88to 2305.87 mg/kg. Similar high concentrations of Al in black tea leaves were from 685 to 1200 mg/kg. In the other vegetable samples the concentration of Al is much higher, ranging from 70 to 400 mg/kg. The highest concentration of about 400 mg/kg of Al was found in lamb's lettuce and in garden lettuce was reported in the literature (67).Müller *et al.*,(68) also found higher Al concentrations in various sorts of lettuce (200–1000 mg/kg). Since Al-tolerant plants can accumulate Al from soil solution, higher concentrations of Al may be found in vegetables (up to 80 mg/kg) (68, 69), in different kinds of lettuce (up to 1000 mg/kg), in herbs and spices(up to 300 and 1000 mg/kg Al, respectively) and in black tea leaves (from 600 to 1200 mg/kg of Al) (68). Al enters drinking waters mainly due to the use of $Al_2(SO_4)_3$that is applied as a coagulant to clarify turbidity (70).Another possible source of the entry of Al into foodstuffs is the use of Al containing food additives, which in cheese processing may increase Al concentrations up to 70 mg/kg(68). Lokeshappa*et al.*, (71) analyzed the Aluminium in selected vegetables available in Powai area, Mumbai, India. The content of Aluminium in spinach and coriander was in high concentration and in moderate concentrations in tulsi, potato and ladyfinger. Cadmium was present in small concentrations in all samples except coriander and carrot. It was found that the overall toxic metal accumulation was greater in leafy vegetables viz. spinach, coriander and tubers such as potato.

Data in Table 1 showed that arsenic (As) concentration mostly was appeared in potato and in fallow order by town: North Giza >Kalyopia> Alexandria. Several countries, including the UK and Australia, currently use a 1 ppm limit for arsenic in food and this is often cited as a "safe" level for rice. Most of the values exceeded the maximal permissible limit of food As standard 1.0 mg/kg dry weight (72), but this limit by WHO standards is 0.43 mg/kg (35). As reported in the literatures, the total arsenic contents in vegetable products were < 0.004 to 0.303 mg/kg fresh weight (73, 74, 75, 76). The average arsenic concentration in the vegetables collected from some arsenic prone areas of Bangladesh was 0.28 mg/kg fresh weight, which was higher than that of the United Kingdom, 0.003 mg/kg fresh weight (77), and Croatia, 0.0004 mg/kg fresh weight (78). Concerning plants growing on contaminated soils of Egypt, several investigators studied the accumulation of heavy metals in these plants. ElSabbagh (79) found that Cd, Cu, Pb and Zn were accumulated in vegetable plants growing on contaminated soils in industrial area. Vegetables Cd ranged from 0.03-0.34 mg/100g. The cultivated vegetables revealed that it absorbs soilPb with varying concentrations from < 0.4-2.5 mg/100g, green onion ranks the first order as it contains 2.5 mg/100g followed by Radish. Abdel-Maksoud (80) reported that using wastewater for irrigation lead to the accumulation of Fe, Cd and Pb in vegetable crops such as Spinach, Potato, Tomato and Celery growing on polluted soil at Giza Governorate. Heavy metal contents of leaves *Ficusretusa* showed that their concentrations increased with the increase in the concentration of these metals in the soil: plants grown at AbouQuir had the highest concentrations (81).

According to the results of this study, the heavy metal (Cd, Pb, and As) contents of green salad in washing samples are so small that one must draw the conclusion that migration of heavy metal from the washing into the green salad can be ignored. However Al was high in some restaurant categories as well as in vegetable samples (Table 2). Washing treatments significantly ($P \leq 0.05$) reduced concentrations of heavy metal contents in the leaves of *Ficusretusa* (81).El-Desoky and Ghallab (82) found that Contents of Pb in the unwashed plant samples (6.00-40.18 ppm) are quite high, compared with the washed samples (0.34 to 29.53 ppm) which were collected from the area around the superphosphate factory. Amounts of cadmium accumulate on plant surface from dusts carried from the factory chemnies.

The results show (Table 3) that heavy metal is widely contained in processed potato (potatoes fried). Furthermore, the heavy metal content in unprocessed potato appears to be much higher than that in processed potato. These results suggest that the presence of heavy metal from processed potato due to the containers used in the manufacturing, processes of heat, type of oil, and oil contamination. These differences in restaurant categories may be probably due to raw material, process conditions and analytical procedures. Harmankaya*et al.*, (83) found that Cd and Pb contents of all chips (from several markets in Konya in Turkey) were at the low levels. Fried potatoes from the Egyptian market content of the studied metals were found to be ranged from 0.054 to 0.10 of Cd, 0.065 to 0.159 of Pb (64). Heavy metal contents in fried potatoes were found to be ranged from 0.054 to 0.10 of Cd and 0.065 to 0.159 mg/kg of Pb. The levels of Cd in all samples are higher than the permissible limits (84).

In conclusion, it is well known that environmental pollution is a product of urbanization and technology and other attendant factors of population density, industrialization and mechanization that serve to provide the necessities of the population. Aluminium (Al) content was high concentration in all vegetable samples. Data showed that potato of Alexandria, lead concentration is more than permitted level. Among the restaurant categories Cd, Pb, and As contents of green salad were found at low levels. The highest levels of Cd, Pb, and Al were detected in popular restaurant 9. Cd and As contents of international chain restaurant 1 were found in high level. If these vegetables are to be used for a short period of time or in small doses, washing can reduce the surface contamination to a great extant. Hence, it is suggested that consumers wash the vegetable prior to use. This result will therefore help the government, individuals and communities to take necessary measure in controlling heavy metal pollution and to minimize exposure of people living town of Egypt. Next study will be on the Egyptian soil and water.

## Acknowledgements

The authors are thankful to Prof. Dr. Hanaa Sidky Special Food and Nutrition Dept., Food Technology Research Institute, Agricultural Research Center, for valuable helping in this study.

## References

[1] D'Mello, J.P.F. (2003). Food safety: Contaminants and toxins. Cambridge: CABI Publishing.

[2] Abdola, M. and Chmtelnicka, J. (1990).New Aspects on the Distribution and Metabolism of Essential Trace Elements after Dietary Exposure to Toxic Metals. Biol. Trace Element Res., 23:25-53.

[3] Bigdeli, M. and Seilsepour, M. (2008). Investigation of Metals Accumulation in Some Vegetables Irrigated with Waste Water in Shahre Rey-Iran and Toxicological Implications.American-Eurasian J. Agric. & Environ. Sci., 4 (1): 86-92.

[4] Bahemuka, T.E. and Mubofu, E.B. (1999).Heavy metals in edible green vegetables grown along the sites of the Sinza and Msimbazi rivers in Dar es Salaam, Tanzania.Food Chem. 66: 63-6.

[5] Bahemuka, T.E. and Mubofu, E.B. (1999).Heavy metals in edible green vegetables grown along the sites of the Sinza and Msimbazi rivers in Dar es Salaam, Tanzania. Food Chemistry 66, 63-66.

[6] Radwan, M.A. and Salama, A.K. (2006).Market basket survey for some heavy metals in Egyptian fruits and vegetables.Food ChemToxicol. 44: 1273-8.

[7] Liu, W.X.; Li, H.H.; Li, S.R. and Wang, Y.W. (2006). Heavy metal accumulation of edible vegetables cultivated in agricultural soil in the suburb of Zhengzhou City, People's Republic of China.Bulletin of Environmental Contamination and Toxicology 76 163–170.

[8] Zhou DM, Hao XZ, Wang YJ, Dong YH and Cang L (2005). Copper and Zn uptake by radish and pakchoi as affected by application of livestock and poultry manures. Chemosphere 59 167–175.

[9] Kulkarni, D.K.; Delbari, A. S. and MahajanD.M. (2014). Bio concentration factor (bcf) for heavy metals detection and selection of hyper-accumulator plantscase study of pune-india and tehran –iran.Indian Journal of Fundamental and Applied Life Sciences ISSN: 2231-6345

[10] Damek-Poprawa, M. and Sawicka-Kapusta, K. (2003).Damage to liver, kidney and teats with reference to burden of heavy metals in yellow-necked mice from areas around steelworks and zinc smelters in Poland.Toxicol., 186(1-2): 147-158.

[11] Mench, M.; Vangronsveld, J.; Didier, V. and Clijsters, H.(1994).Evaluation of metal mobility, plant availability and immobilization by chemical agents in a limed-silty soil.Environ. Pollut., 86: 279-286.

[12] Wegelin, T.; Jenka, B. and Etter, R. (1995). Heavy metals in soils and plants of allotment gardens in the city of Zurich (Switzerland).Soil Environ., 5(1): 269-270.

[13] Gimmler, H.; Degenhardt, B.; Lang, S. and Track, C. (1998). Uptake of sodium, boron and heavy metals from an alkaline top layer and translocation to leaves and lower roots of Vitisviniferaduring a long time duplex study. J. Appl. Bot., 72: 191-202.

[14] McLaughlin, M.J.; Maier, N.A.; Corell, R.L.; Smart M.K. and Grant, C.D. (1998). In-situ immobilization techniques to remediate cadmiumcontaminated agricultural soils.pp: 253-460. In Proc. 6th Int. Conf. on Contaminated Soils, Edinburgh, UK. 17-21 May 1998. Telford, London.

[15] Meuser, H.; Schleuss, U.; Taubner, H. and Wu, Q. (1998). Soil properties of coal, iron and steel industrial sites in Essen, Germany. Z. P. flanzenernaehr. Bodenkd. 161: 197-203.

[16] Trautmannsheimer, M.; Schramel, P.; Winkler, R. and Bunzl, K. (1998). Chemical fractionation of some natural radionuclides in a soil contaminated by slags. Environ. Sci. Technol., 32: 238-243.

[17] Tossavainen, M. and Frossberg, E. (1999).The potential leachability from natural road construction materials. Sci. Total Environ., 239: 31-47.

[18] Jarup, L. (2003). Hazards of heavy metal contamination.Br. Med. Bull., 68: 167-182.

[19] Jassir, M.S.; Shaker, A. andKhaliq, M.A. (2005). Deposition of heavy metals on green leafy vegetables sold on roadsides of Riyadh city, Saudi Arabia. Bulletin Environmental Contaminants Toxicology. Vol.75, pp.1020 – 1027.

[20] Carr, R. (2005). WHO guidelines for safe waste water use: More than just numbers. J. Iris. Drain, 54: 103-111.

[21] Msaky, J.J. and Calvert, R. (1990). Adsorption behavior of copper and zinc in soils: influence of pH on adsorption characteristics. Soil Sci., 150(2): 513-522.

[22] Fergusson, J.E. (1990). The Heavy Elements: Chemistry, Environmental Impact and Health Effects. Oxford Press, pp: 382-399.

[23] Ellen, G.; Loon, J.W. and Tolsma, K. (1990).Heavy metals in vegetables grown in the Netherlands and in domestic and imported fruits.Z LebensmUntersForsch, 190(1): 34-39.

[24] Yargholi, B. and Azimi, A.A. (2008).Investigation of Cadmium absorbtion and accumulation in different parts of some vegetables.American Eurasian J. Agric and Environ. Sci., 3(3): 357-364.

[25] Cambra, K.; Martínez, T.; Urzelai, A. and Alonso, E. (1999). Risk analysis of a farm area near a lead- and cadmium-contaminated industrial site.J. Soil Contam., 8(5): 527-540.

[26] Dudka, S. and Miller, W.P. (1999). Permissible concentrations of arsenic and lead in soils based on risk assessment.Water Air Soil Poll., 113(1/4): 127-132.

[27] Nriagu, J.O. (1989). A global assessment of natural sources of atmospheric trace metals.Nature, 338(6210): 47-49.

[28] Islam, E.; Xiao,Y.; Zhen, H. and Qaisar, M. (2007). Assessing potential dietary toxicity of heavy metals in selected vegetables and food crops.J Zhejiang UnivSciB 8(1):1-13.

[29] Zurera-Cosano, G.; Moreno-Rojas, R.; Salmeron- Egea, J. and Pozo Lora, R.(1984). Heavy metal uptake from greenhouse border soils for edible vegetables. J. Sci. Food Agric., 49(3): 307-314.

[30] Wong, J.W.C. (1996). Heavy metal contents in vegetables and market garden soils in Hong Kong. Environ Technol., 17(4): 407-414.

[31] Kovacheva, P.;Djingova, R. andKuleff, I. (2000).On the representative sampling of plants for multielement analysis. Phyto. Bal. 6: 91-102.

[32] Lozak, A.;Soltyk, K.;Ostapczuk, P. andFijalek, Z. (2002).Determination of selected trace elements in herbs and their infusions. Sci. TotalEnviron. 289: 33-40.

[33] Atrouse, OM.; Oran, S.A. and Al-Abbadi, S.Y. (2004). Chemical analysis and identification of pollen grains from different Jordanian honey samples. Inter. J. Food Sci. Tech. 39: 413-417.

[34] Sobukola, O.P.;Awonorin, S.O.;Idowu, M.A. andBamiro, F.O. (2008). Chemical and physical hazard profile of 'robo' processing – a street vended melon snack. Inter. J. Food Sci. Technol. 43(2): 237-242.

[35] Divrikli, U.; Horzum, N.; Soylak, M. and Elci, L. (2006). Trace heavy metal contents of some spices and herbal plants from western Anatolia, Turkey. Int. J. Food Sci. Technol. 41: 712-716.

[36] FAO/WHO Codex Alimentarius Commission (2001). Food Additives And Contaminants. Joint FAO/WHO Food Standards Program, 01/12a, pp. 1-289.

[37] Krejpcio,Z.; Sionkowski, S. and Bartela, J. (2005).Safety of Fresh Fruits and Juices Available on thePolish Market as Determined by Heavy Metal Residues. Polish Journal of Environmental Studies Vol. 14, No 6, 877-881.

[38] Satarug, S. and Moore, M. R. (2004). Adverse health effects of chronic exposure to low-level cadmium in foodstuffs and cigarette smoke. Environ.Health Perspect.112, (10), 1099.

[39] Ni, W.Z. and Long, X.X.( 2002).Studies on the criteria of cadmium pollution in growth media of vegetable crops based on the hygienic limit of cadmium in food.J. Plant.

[40] Soylak, M.; Özcan, B. and Elci, L. (2004). Application of the HMDTC-HMA coprecipitationmethod for the enrichment and atomic absorption spectrometric determination of some trace metal ions in various samples. Kuwait Journal of Science Engineering31:47-59.

[41] Harmankaya, M.; MusaÖzcan,M.; Duman, E. and Dursun, E. (2012). Mineral and heavy metal contents of ice-cream wafer, biscuit and gofret wafers.Journal of Agroalimentary Processes and Technologies 18 (4): 259-265.

[42] WHO (1992). Cadmium, Environmental Health Criteria, Geneva.Vol.134.

[43] WHO (1995). Lead. Environmental Health Criteria, Geneva, vol. 165.

[44] Steenland, K. andBoffeta, P. (2000). Lead and cancer in humans: where are we now? Am. J. Ind. Med. 38: 295-299.

[45] IARC (1993).Cadmium and cadmium compounds. In: Beryllium, Cadmium, Mercury and exposure in Glass manufacturing Industry. IARC Monographs on the evaluation of carcinogenic risks to humans, International Agency for Research on Cancer, Lyon 58: 119-237.

[46] Pitot, C.H. andDragan, P.Y. (1996). Chemical carcinogenesis, 5th edition: In: Casarett, D. (ed.), Toxicology Inter. Edi., McGraw Hill, New York, pp. 210-260.

[47] Wierzbicka, M. (1995). How lead loses its toxicity to plants. Acta Soc. Bot. Pol., 64:81-90.

[48] Coutate, T.P. (1992).Food, the Chemistry of Its Component, 2nd Ed. Royal Society of Chemistry, Cambridge, p.265.

[49] CDPM (Chinese Department of Preventive Medicine) (1994).Threshold for Food Hygiene.Beijing: China StandardPress (In Chinese).

[50] Zhang, K.S. and Zhou, Q.X. (2005). Toxic effects of Al-based coagulants on Brassica chinensisand Raphanussativus growing in acid and neutral conditions.Environ. Toxicol., 20(2):179-187.

[51] NAS/NRC (National Academy of Sciences/National Research Council)(1999). Arsenic in drinking water. Washington, DC. pp. 251-257.

[52] Dokiya,Y.; Yamamoto,Y.; Toda, S. and Fuwa, K. (1975).Comparison of Mineral Acids in Wet Digestion of Plant Materials for Flame and Flameless Atomic Absorption Measurement of Metals.Spectroscopy Letters: An International Journal for Rapid Communication 8(10): 759-769

[53] Abdel-Shafy, H. A.;Aly,R. andSahab, A. F.(2005). Fate of metalloid fungicide in water, soil and plant in Egypt. Environment Protect ion Engineering 31(1): 69-80.

[54] Okwulehie,I. C. and Ogoke,J. A. (2013). Bioactive, nutritional and heavy metal constituents of some edible mushrooms found in Abia State of Nigeria. Int. J. Appl. Microbiol. Biotechnol. Res.1:7-15.

[55] Banerjee, D.; Bairagi, H.; Mukhopadhyay, S.; Pal, A.;Bera, D. and Ray, L. (2010).Heavy metal contamination in fruits and vegetables in two districts of west Bengal, India.Electronic Journal of Environment and Agriculture and Food Chemistry 9:1423-1432.

[56] Igwegbe, A.O.; Belhaj, H.; Hassan, T.M..andGibali, A.S. (1992). Effect of a highways traffic on the level of lead and cadmium in fruits and vegetables grown along the roadsides. J Food Safety 13: 7-18.

[57] Stohs, S.J. and Bagchi, D. (1994).Oxidative mechanisms in the toxicity of metal ions.Free Radical Biol. Med. 18(2): 321-336.

[58] Otitoju, O. and Onwurah, I.N.E. (2005).Superoxisedismutase (SOD) activity and serum calcium level in Rats exposed to a locally produced insect Powder "Rambo". Anim. Res. J.2(1): 261-266.

[59] El-Sokkary, LH. and Sharaf, A.1. (1993) Enrichment of soils and plant crops irrigated with wastewater by Zn and Cd. 4th Nat. Cong 011 Pollution Control of Agric. Environ, Egypt. Soc. Soil Sci. Cairo, pp. 262 – 268.

[60] Zhuang, P.; Zou, B.; Li, N.Y. and Li, Z.A. (2009). Heavy metal contamination in soils and food crops around Dabaoshan mine in Guangdong, China: implication for human health. Environnemental Geochemistry and Health.6: 707-715.

[61] Singh, K.P.; Mohon, D.; Sinha, S. and Dalwani, R. (2004). Impact Assessment of Treated/Untreated Waste Water Toxicants Discharge by Sewage Treatment Plants on Health, Agricultural and Environmental Quality in Waste Water Disposal Area.Chemosphere.55: 227–255.

[62] Kabata-Pendias, A. and Pendias, H. (2001).Trace Elements in Soils and Plants.3rd Edn. Boca Raton: CRC Press.

[63] Smarghandi, M.; Karimpoor, M. and Sadri, Gh.H. (2000). Determination of heavy metals in Irrigated vegetables with polluted water in hamadan city. Journal of Man Environment.5: 24-31.

[64] Shakya, P. R. and Khwaounjoo, N. M. (2013). Heavy metal contamination in green leafy vegetables collected from different market sites of kathmandu and their associated health risks. Scientific World: 11(11):37-42.

[65] Salama, A.K. and Radwan, M.A. (2005). Heavy metals (Cd, Pb) and trace elements (Cu, Zn) contents in some foodstuffs from the Egyptian market.Emir. J. Agric. Sci. 17 (1): 34-42.

[66] Abdel-Sabour, M.F. and. Rabie, F. H. (2003).Accumulation of Heavy Metals in VegetablePlants Grown in Mostorod Area. Egypt. 1. Soil Sci .43, No. I, pp. 63-76.

[67] Rashad, M. and Shalaby, E. A. (2007).Dispersion and Deposition of Heavy Metals around Two Municipal Solid Waste (MSW) Dumpsites, Alexandria, Egypt. American-Eurasian J. Agric. & Environ. Sci., 2 (3): 204-212.

[68] Ščančar, J.; Stibilj, V. and Milačic, R.(2004).Determination of aluminium in Slovenian foodstuffs and its leachability from aluminium-cookware.Food Chemistry 85:151–157.

[69] Müller, M.; Anke, M. and IllingGünther, H. (1998). Aluminium in food stuffs. Food Chem.61, 419–428.

[70] Oniawa, P. C.; Ikadeh, G. C. and Nweze, S. E. (1997). Aluminium contents of some raw and processed Nigerian foods. Food Chem.58, 351–353.

[71] Martin, R. B. (1986). The chemistry of aluminium as related to biology and medicine. Clin. Chem.32, 1797–1806.

[72] Lokeshappa. B.; Shivpuri, K.; Tripathi.V. and Dikshit, A. K. (2012).Assessment of Toxic Metals in Agricultural Produce. Food and Public Health 2(1): 24-29.

[73] NFA (National Food Authority) (1993). Australian Food Standard Code.Canberra' Australian Government Publication Service; March.

[74] Dabeka, R.W.; McKenzie, A.D.; Lacroix, G.M.A.; Cleroux, C.; Bowe, S.; Graham, R.A.;Conacher, H.B.S. andVerdier, P. (1993).Survey of arsenic in Total Diet food composites and estimation of the dietary intake of arsenic by Canadian adults and children. Journal of AOAC International 76, 14-25.

[75] Schoof, R.A.; Yost, LJ.; Eickhoff, J.; Crecelius, E.A.; Cragin, D.W.; Meacher, D.M. andMenzel, D.B. A. (1999). Market Basket Survey of Inorganic Arsenic in Food. Food and Chemical Toxicology 37, 839-846. 758

[76] Urieta, I.; Jalon, M. andEguileor, I. (1996). Food surveillance in the Basque country (Spain). II. Estimation of the dietary intake of organochlorine pesticides, heavy metals, arsenic, aflatoxin M1, ironand zinc through the Total Diet Study, 1990/91. Food Additives and Contaminants 13, 29-52.

[77] Ysart, G.; Miller, P.; Crews, H.; Robb, P.; Baxter, M.; De L'Argy, C.; Lofthouse, S.; Sargent, C. and Harrison, N.(1999). Dietary exposure estimates of 30 elements from the UK Total Diet Study. Food Additives and Contaminants 16, 391-403.

[78] MAFF. (1997).Total Diet Study: aluminium, arsenic, cadmium, chromium, copper, lead, mercury, nickel, selenium, tin and zinc. Food surveillance information sheet, No. 191. Ministry of Agriculture, Fisheries and Food (MAFF): HMSO, London, U.K.

[79] Sapunar-Postruznik, J.; Bazulic, D. andKubala, H. (1996).Estimation of dietary intake of arsenic in the general population of the Republic of Croatia. The Science of The Total Environment191, 119-123.

[80] EI-Sabbagh, T.A. (1991) Environmental impact of industrial area- Mostorod- on the concentrations of some elements in water. soil and plant, M.Sc. Thesis. Institute of Environ. Studies and Res., Ain Shams Univeristy, Cairo, Egypt.

[81] Abdel-Maksoud, M.M. (1993) Presence of selected contaminant irrigation and drainagewater in Giza Govern. M.Sc. Thesis. Cairo Univ., Egypt.

[82] Abou El Saadat, E.M.; Hassan, M.R.; Hassan, I.A. and Weheda, B.M. (2011). Heavy Metal Content in Leaves of Ficusretusa Collected from Contaminated and Uncontaminated Sites in Northern Egypt: Mitigation of Toxicity by Washing Treatments. Universal Journal of Environmental Research and Technology 1(4): 408-415.

[83] El-Desoky, M.A. and A. Ghallab (2000). Pollution Impact On Soils And Plants In An Industrial Area Near Assiut City. Ass. Univ. Bull. Environ. Res. 3(1) 1-19.

[84] Harmankaya,M.; Özcan, M. and Endes, Z. (2013).Mineral contents of several corn and potato chips. Journal of Agroalimentary Processes and Technologies 19(2), 222-227.

[85] Walker, J. M. (1988). Regulation by other countries in foods and the human environment, Proceeding, No.2 "Cadmium Accumulation in Australian Agriculture". National Symposium, Canberra, 1-2 March 1988, Australian Government Publishing Service, Canberra, 176- 85.

17

# Environmental Risk Assessment Methodology (ERAM) for Oil Pollution

**Huseyin Murat Cekirge**

Department of Mechanical Engineering, Prince Mohamed Bin Fahd University, Al Khobar, KSA

**Email address:**
hmcekirge@usa.net

**Abstract:** Coastal areas have economical, archaeological, cultural, economic and environmental significance. During the last decades, a rapid development of recreation and other tourism activities has been noticed. However, these coastal areas are often visited by oil tankers for transportation purposes; this paper considers risk analysis by using index. Each index and methodology are presented in detail. This analysis provides a mathematical tool to determine oil pollution risks that lead to handle with marine environment and oil pollution prevention.

**Keywords:** Environmental Risk Assessment, ERA, Index Method, Coastal Safety, Oil Spill Risk Analysis

## 1. Introduction

Ecological risk assessment from the oil spill has become the most common practice on the areas where the ship transportation, oil terminals, pipelines drilling rigs and oil plat form are active. Oil spill is a result of accidents in the above mentioned activities which pollutes the water and the surrounding environments which directly affects the ecological system of marine life and may have the adverse effects on human health.

In the past few decades many researches, risk reduction techniques, risk assessment methodology, models and reports have been developed and some of the recent researches are given below:

Trends in tanker/terminal are examined oil spills in the Greek Sea port and the main concern in this study was coastal zone. The paper uses a model for ship incidents to identify the sequence of events in a network and proposes a systematic approach for analysis of incidents which will help in taking preventive measures during operation and planning. It is discussed that human errors and operational failure can also be the cause of big incidents, [1] and [2].

The recent methodologies and advances are studied in the methodology using maritime transportation system (MTS) with incident/accident probability model and consequences model, and also presented improved graphical format to represent maritime transportation risk in geographic manner. All of these enhancements and improvement were made in the context of two years oil transportation risk study conducted from 2006-2008 in the Puget Sound and surrounding waters, they also presented the comparison of risk relation analysis between a one way zone, an escorting and double hull requirement, [3], [4], [5] and [6].

The Statmap model for risk analysis is described in detail with example and explains all the input and output facilities of the model, [7].

The risk assessment of oil spill pollution in China waters from Chinese shipping industry is presented. In the paper, all the traffic data and the risk from the vessel traffic are presented; and also showed that with increase in oil import there is a greater risk than those who exports, but in China oil spill is relatively low, [8].

The assessments of environment risks created by increasing oil transportation in the Gulf of Finland for assessment are explained. The risk assessment tool was implemented as Perl Model and they concluded that the initial impression of the methodology and the tool are effective and useful, [9].

A dynamic risk assessment model for drift grounding is proposed, this model may function as a decision support tool. This model estimates the environmental risk of drift grounding accidents for oil tankers in real time and in forecast model, [10].

An oil spill risk assessment model developed for the US water is presented, [11]. The whole methodology was discussed in detail and risk was calculated as the product of spill probability and import indices, in this study they presented the impact indices incorporated the total risk, The potential impact was characterized by the amount of oil spill, value, and type of oil, time of oil spill and place of spill.

The environmental risk assessment (ERA) is studies for soil pollution from the accident of oil spill at industrial site for drilling, [12]. Also, it is presented the methodology for qualitative and quantitative assessment of environmental risk and also the methodology for risk assessment of soil pollution based on modules and steps which require all the hazard analysis and all the technical data.

The most recent studies for risk analysis by using the available data for oil spill in the Crete over 1995-1999 are reviewed, [13], all the incident and accident of oil spill is thoroughly analyzed and presented, In the study, it is proposed a formula for combining the various risks It was found that all the Northern coast of the Heraklion and East Heraklion have high degree of risk whereas the Southern Crete have low risk.

The geographic information based risk assessment of oil spill in the coastal areas of Akwalbom state of Nigeria is presented [14]. A combination of hazard and vulnerability data layers were used; and the resulting risk layer were formed and classified into risk zones of very high, high, moderate and marginal. From the study, it was found that the environs in the study are in the category of very high risk because of increasing investment in the oil section.

This concept paper is intended to present a background for development of a common approach (methodology) for ecological risk assessment Environmental Risk Assessment Methodology (ERAM) from oil pollution with application for all sea regions. The potential developer is expected to follow the main principles described here and to use the latest achievements in this field. The methodology/model should include risk assessment from both accidental and operational oil pollution from ships, oil terminals, pipelines, drilling rigs and oil platforms. It is expected that a software must be developed for practical implementations of the ERAM.

In the all factors for risk analysis are presented in detail and computational methodologies are explained and an example is given with coastal environmental sensitivity indexes.

## 2. Principles

The following main principles are expected to be followed during development of ERAM:

- Effectiveness; risk assessment should accurately measure the risks to the extent necessary to achieve an appropriate level of protection,
- Transparency, the reasons and evidences supporting the action recommended by risk assessment, and areas of uncertainty (and their possible consequences to those recommendations), are clearly documented and made available to decision-makers,
- Consistency, risk assessment should achieve a uniform high level of performance, using a common process and methodology,
- Comprehensiveness, all aspects of influence of oil pollution should be considered when assessing risks and making recommendations - economic, environmental, social and cultural,
- Precautionary, risk assessment should include a level of precaution when making assumptions, and making recommendations, to account for uncertainty, unreliability, and inadequacy of information. The absence of, or uncertainty in, any information should therefore be considered as an indicator of potential risk.

## 3. Methodology

The following should be taken in mind during development of the ERAM for oil pollution:

1. The methodology should be standardized, Generally the ERAM is to be developed in respect to:
   - shipping traffic, oil terminals, pipelines, oil platforms and drilling rigs,
   - hydrological and meteorological conditions,
   - ecological sensitive/protected areas,
2. Assessment of response options available in case of oil spill for the region (effectiveness, logistics and limitations),
3. Estimation of the possible impact of the most probable and the worst case scenario of oil spills on resources and habitats in regard of the response techniques used,
4. Risk assessment should include procedures related to analyses, evaluation and acceptance of the risk,
5. The ERAM should be applicable for both, ports and coastal areas.

## 4. Software Application

All factors are effecting risk against oil pollution will be to be calculated and explained in the report. The respective section for description of the application should.

The ERAM should be made as a software application which will give the possibility risk consist of at least the following:

1. Diagram/flowchart for the logic of the programme,
2. Description of the list of variables and required input data/databases with data format,
3. Description of the output data – the output data should be presented both in a table format and in GIS format for visualisation,
4. User guide for work with the software.

## 5. Example: A Port with an Oil Terminal

A real case for application of the software is to be presented for one port (with an oil terminal).

### *5.1. Risk Factors*

When considering the probabilities of accidents, the risk

factors caused by the facility are discussed, for this purpose; operation and constitution of the facility, particularities and the amount of the materials handled in the facility are used.

Other than this, when considering the environmental risks, geological, biological, ecological and socio-economical structures that might be damaged during or afterwards the accident, are carefully studied.

Headlines of the risk factors are as below:

1. Risks caused by the coastal facility, operation risks (staff, administration, organization etc,) and risks caused by the handled material (amount and particularities of the material etc,)
2. Environmental risks caused by the coastal facility (geological, biological, ecological and socio-economical structures in vicinity of the facility).

### *5.1.1. Risks Caused by the Coastal Facility*

Risks caused by the coastal facility are studied under two headlines: risks concerning the operation of the facility and risks concerning the handled material.

a. Risks caused by the operation deal with risks concerning the staff of the facility, the constitution of the organization and the operational administration,
b. Risks caused by the material, furthermore include the risks caused by the type and the amount of the material.

Depending on the location of the facility, as the destroying made by the spills caused by both of the risks mentioned above will differ, the risk factors must be additionally depending on the location of the facility.

The risk factors and the weight coefficients dependent on location, are given in Table 1.

***Table 1.*** *Risk Factors Caused by Facility and Weight Coefficients Dependent on Location.*

| Row | Risk Factor. $F_i$ | Weight Coefficient Dependent on Location ($WL_i$) | | |
|---|---|---|---|---|
| | | Open Sea | Settlement Region | Sensitive Region |
| Factors Considering Operational Risk of the Facility | | | | |
| 1 | Experience of crew at the port and of operators, $A_1$ | 0 | 2 | 2 |
| 2 | Crew sufficiency, $A_2$ | 2 | 2 | 2 |
| 3 | Condition of communication systems, $A_3$ | 0 | 2 | 2 |
| 4 | Condition of urgent intervention plans, $A_4$ | 2 | 1 | 1 |
| 5 | Condition of intervention tools to oil spill, $A_5$ | 2 | 2 | 3 |
| 6 | Accessibility to instalation, $A_6$ | 1 | 2 | 3 |
| 7 | Arriving frequency of ships, $A_7$ | 1 | 1 | 1 |
| 8 | Education condition of crew, $A_8$ | 2 | 2 | 2 |
| 9 | Ship traffic simulation, $A_9$ | 2 | 1 | 3 |
| 10 | Attitude of company at a dangerous situation, operational limits, $A_{10}$ | 2 | 2 | 2 |
| 11 | Maintenance period of the system, $A_{11}$ | 2 | 2 | 2 |
| 12 | Installation age, $A_{12}$ | 1 | 1 | 1 |
| 13 | Intervention beginning time, $A_{13}$ | 2 | 2 | 2 |
| Factors Considering Type and Amount of the Material | | | | |
| 14 | Black oil transport, $P_1$ | 2 | 2 | 3 |
| 15 | White oil transport, $P_2$ | 1 | 1 | 1 |
| 16 | Metallic oil transport, $P_3$ | 1 | 1 | 1 |
| 17 | Number of oil tankers arriving to dock, $P_4$ | 0 | 2 | 2 |
| 18 | Maximum pumping capacity, $P_5$ | 0 | 1 | 1 |
| 19 | Tanker traffic in vicinity of the facility, $P_6$ | 2 | 1 | 2 |
| 20 | Oil generation plants in vicinity, $P_7$ | 1 | 1 | 1 |
| 21 | Length of oil pipe line belonging to facility, $P_8$ | 1 | 1 | 1 |
| 22 | Number of oil storage tanks in facility, $P_9$ | 1 | 1 | 1 |
| 23 | Automatic on off valves, $P_{10}$ | 2 | 2 | 2 |
| 24 | Accidents occurred in facility, $P_{11}$ | 2 | 2 | 2 |

Values that risk factors caused by the facility will take are classed from 0 to 5 as below:

- 0: No risk for the corresponding factor,
- 1: Insignificant risk for the corresponding factor,
- 2: Poor risk for the corresponding factor,
- 3: Significant risk for the corresponding factor,
- 4: Heavy risk for the corresponding factor,
- 5: Maximum risk for the corresponding factor.

Explanation and values of risk factors are given in the tables below:

#### 5.1.1.1. Experience of Crew and of Operators

| Evaluation Criteria | Risk Factor Value. $A_1$ |
|---|---|
| Good and experienced | 1 |
| Good and inexperienced | 2 |
| Poor and experienced | 3 |
| Poor and inexperienced | 4 |

#### 5.1.1.2. Crew and Equipment Sufficiency

| Evaluation Criteria | Risk Factor Value. $A_2$ |
|---|---|
| Sufficient | 1 |
| Insufficient | 2 |
| Unknown | 3 |

#### 5.1.1.3. Condition of Communication Systems

| Evaluation Criteria | Risk Factor Value. $A_3$ |
|---|---|
| 3 types of communication | 1 |
| 2 types of communication | 2 |
| 1 type of communication | 3 |

Communication systems can be studied as wireless, mobile phone and telephone.

#### 5.1.1.4. Condition of Urgent Intervention Plans

| Evaluation Criteria | Risk Factor Value. $A_4$ |
|---|---|
| A+B+C+D | 1 |
| A+B+C, A+B+D | 2 |
| A+C. A+D | 3 |
| A | 4 |
| E | 5 |

A = Port has urgent intervention plan,
B = It has regional urgent intervention plan,
C = Usage probability of local opportunities,
D = Usage probability of international opportunities,
E = No urgent intervention plan.

#### 5.1.1.5. Condition of Intervention to Oil Spill Tools

| Evaluation Criteria | Risk Factor Value, $A_5$ |
|---|---|
| A+B, sufficient | 1 |
| A, sufficient | 2 |
| B, sufficient | 3 |
| A or B, insufficient | 4 |
| A + B, insufficient | 5 |

A = Outfit for intervention to oil spill and sufficient staff member are available,
B = some outfit for intervention and staff as subcontractor are available.

#### 5.1.1.6. Accessibility to Installation

| Evaluation Criteria | Risk Factor Value, $A_6$ |
|---|---|
| Reachable from land by heavy vehicles | 1 |
| Reachable from land by light vehicles | 2 |
| Reachable only from sea | 3 |
| Reachable only from air | 4 |

It is evaluated in terms of accessibility to the situation of the probable spill source.

#### 5.1.1.7. Number of Arriving of Ships

AF = the number of arriving of ships.

| Evaluation Criteria | Risk Factor Value, $A_7$ |
|---|---|
| $AF \le 25$ | 1 |
| $25 < AF \le 50$ | 2 |
| $50 < AF \le 100$ | 3 |
| $100 < AF \le 200$ | 4 |
| $AF > 200$ | 5 |

#### 5.1.1.8. Education Condition of Staff

| Evaluation Criteria | Risk Factor Value, $A_8$ |
|---|---|
| Good T1 + Good T2 | 1 |
| Acceptable T2 + Good T1 | 2 |
| Poor T2 + Good T1 | 2 |
| Acceptable T1 + Acceptable T2 | 3 |
| Poor T2 + Acceptable T1 | 4 |
| Poor T1 + Poor T2 | 5 |

T1 = Intervention education to first level spill,
T2 = Intervention education to second level spill.

#### 5.1.1.9. Ship Traffic Simulation

| Evaluation Criteria | Risk Factor Value, $A_9$ |
|---|---|
| Iron/Windlass breakdown risk | 1 |
| Rope split risk | 2 |
| Dock/buoy damage | 2 |
| Grounding risk | 3 |
| Ramming risk | 4 |
| Collision risk | 5 |

#### 5.1.1.10. Attitude of the Company at Dangerous Situations, Operational Limits

| Ev Evaluation Criteria | Risk Factor Value, $A_{10}$ |
|---|---|
| Company stops working at dangerous situations | 0 |
| Company continues working at dangerous situations | 1 |

Dangerous situation is defined as the situation which might cause any spill;

- air conditions (strong wind and flow),
- busy sea traffic,
- terrorist threats.

#### 5.1.1.11. Maintenance Period of the System

MP = the maintenance period of system.

| Evaluation Criteria | Risk Factor Value, $A_{11}$ |
|---|---|
| $MP \le 3$ months | 1 |
| 3 months $< MP \le 6$ months | 2 |
| 6 months $< MP \le 1$ year | 3 |
| 1 year $< MP \le 2$ years | 4 |
| $MP > 2$ years | 5 |

#### *5.1.1.12. Instillation Age*

AI = the age of installation.

| Evaluation Criteria | Risk Factor Value, $A_{12}$ |
|---|---|
| AI ≤ 5 years | 0 |
| 5 years < AI ≤ 10 years | 1 |
| 10 years < AI ≤ 15 years | 2 |
| 15 years < AI ≤ 20 years | 3 |
| 20 years < AI ≤ 25 years | 4 |
| AI > 25 years | 5 |

#### *5.1.1.13. Intervention Beginning Time*

IT = beginning time of intervention.

| Evaluation Criteria | Risk Factor Value, $A_{13}$ |
|---|---|
| IT ≤ ½ hour | 0 |
| ½ hour < IT ≤ 1 hour | 1 |
| 1 hour < IT ≤ 2 hours | 2 |
| 2 hours < IT ≤ 3 hours | 3 |
| IT > 3 hours | 5 |

#### *5.1.1.14. Black Oil Transportation*

AO = the amount of handling black oil.

| Evaluation Criteria | Risk Factor Value, $P_1$ |
|---|---|
| AO = 0 | 0 |
| AO ≤ 500 tones | 1 |
| 500 tones < AO ≤ 1 000 tones | 2 |
| 1 000 tones < AO ≤100 000 tones | 3 |
| 100 000 tones < AO ≤ 500 000 tones | 4 |
| AO > 500 000 tones | 5 |

#### *5.1.1.15. White Oil Transportation*

AW = the amount of handling white oil.

| Evaluation Criteria | Risk Factor Value, $P_2$ |
|---|---|
| AW = 0 | 0 |
| AW ≤ 500 tones | 1 |
| 500 tones < AW ≤ 1 000 tones | 2 |
| 1 000 tones < AW ≤ 100 000 tones | 3 |
| 100 000 tones < AW ≤ 100 000 tones | 4 |
| AW > 500 000 tones | 5 |

#### *5.1.1.16. Metallic Oil Transport*

MM = the amount of handling metallic oil.

| Evaluation Criteria | Risk Factor Value, $P_3$ |
|---|---|
| MM = 0 | 0 |
| MM ≤ 500 tones | 1 |
| 500 tones < MM ≤ 1 000 tones | 2 |
| 1 000 tones < MM ≤ 100 000 tones | 3 |
| 100 000 tones < MM ≤ 500 000 tones | 4 |
| MM > 500 000 tones | 5 |

#### *5.1.1.17. Dock Traffic*

TS = the number of oil tankers arriving to dock.

| Evaluation Criteria | Risk Factor Value, P5 |
|---|---|
| TS ≤ 10 | 1 |
| 10 < TS ≤ 100 | 2 |
| 100 < TS ≤ 500 | 3 |
| 500 < TS ≤ 1 000 | 4 |
| TS > 1 000 | 5 |

#### *5.1.1.18. Pumping Capacity*

PK = the total pumping capacity in the facility.

| Evaluation Criteria | Risk Factor Value, $P_5$ |
|---|---|
| PK ≤ 50 $m^3$/hour | 1 |
| 50 $m^3$/hour < PK ≤ 250 $m^3$/hour | 2 |
| 250 $m^3$/hour < PK ≤ 1 000 $m^3$/hour | 3 |
| 1000 $m^3$/hour < PK ≤ 4 000 $m^3$/hour | 4 |
| PK > 4 000 $m^3$/hour | 5 |

#### *5.1.19. Tanker Traffic in Vicinity of The Facility*

| Evaluation Criteria | Risk Factor Value, $P_6$ |
|---|---|
| No tanker traffic in vicinity the facility | 0 |
| Tanker traffic in vicinity of the facility | 1 |

#### *5.1.1.20. Oil Generation Plants in Vicinity*

OP = the number of oil generation plants in vicinity.

| Evaluation Criteria | Risk Factor Value, $P_7$ |
|---|---|
| OP = 0 | 0 |
| 0 < OP ≤ 5 | 1 |
| 5 < OP ≤ 10 | 2 |
| 10 < OP ≤ 15 | 3 |
| 15 < OP ≤ 25 | 4 |
| OP > 25 | 5 |

#### *5.1.1.21. Length of Oil Pipe Line Belonging to Facility*

$PL_{Land}$ and $PL_{Sea}$ = length of the oil pipe line at land and at open sea.

| Evaluation Criteria | Risk Factor Value, $P_8$ |
|---|---|
| $PL_{Land}$ = 0 m | 0 |
| $PL_{Sea}$ = 0 m | 0 |
| $PL_{Land}$ < 250 m | 1 |
| $PL_{Sea}$ < 250 m | 2 |
| $PL_{Land}$ < 250 m and $PL_{Sea}$ < 250 m | 2 |
| $PL_{Land}$ > 250 m | 2 |
| $PL_{Land}$ > 250 m and $PL_{Sea}$ < 250 m | 3 |
| $PL_{Sea}$ > 250 m | 4 |
| $PL_{Land}$ < 250 m and $PL_{Sea}$ > 250 m | 4 |
| $PL_{Land}$ > 250 m and $PL_{Sea}$ > 250 m | 5 |

#### *5.1.1.22. Number of Oil Stocking Tanks in the Facility*

OF = the number of oil tanks at shore.

| Evaluation Criteria | Risk Factor Value, $P_9$ |
|---|---|
| OF = 0 | 0 |
| OF = 1 | 1 |
| OF = 2 | 2 |
| OF = 3 | 3 |
| OF = 4 | 4 |
| OF ≥ 5 | 5 |

#### *5.1.1.23. Presence of Automatic on-off Valves*

| Evaluation Criteria | Risk Factor Value, $P_{10}$ |
|---|---|
| A+B+C | 1 |
| A+B, A+C, B+C | 2 |
| A | 3 |
| B, C | 4 |
| None of them | 5 |

A = Automatic on-off valve at loading point,
B = Automatic on-off valve in the tank,
C = Pipeline control on-off valve.

### *5.1.1.24. Accidents Occurred in Facility*

Accident Intensity (AI) = Number of Accidents / Operation Time of Facility.

| Evaluation Criteria | Risk Factor Value, $P_{11}$ |
|---|---|
| $0 \leq AI < 0.001$ | 1 |
| $0.001 \leq AI < 0.01$ | 2 |
| $0.01 < AI \leq 0.1$ | 3 |
| $\leq AI$ | 4 |

### *5.1.2. Environmental Risks*

In order to calculate the environmental risks, it is necessary to study the geological, ecological and socio-economical structure of the region. In this paper, it is assumed that shore facility is on the centre, each region of 20 kilometers of diameter is divided into circular strips of 1000 meters and risk assessments are carried out for coastlines and for sea areas in these circular strips. This distance is chosen by considering that oil spills move off during the day generally 15-20 kilometers. It is assumed that spills move by moving along the direction of the blowing wind with a velocity approximately at 3 % of the wind velocity and at 100 % of the surface flow. The distance of 20 kilometers is used. Maximum wind velocity in this area is assumed to be 10-15 knots.

The sum of the risk values in a strip cannot exceed maximum value of the corresponding risk factor, if it does; maximum value of the risk factor will be taken as the risk value.

Factors and their explanations belonging to environmental risks are given in Table 2 as below:

***Table 2.*** *Risk Factors and Comparative Weight Coefficients.*

| Row | Risk Factor, Fi | Weight Coefficient (WLi) |
|---|---|---|
| Risk Factor of Coastline | | |
| 1 | Coastal risk factors, $RK_1$ | 1.00 |
| Socio-Economical Risk Factors | | |
| 2 | Settling, $RED_1$ | 0.60 |
| 3 | Tourism Areas, $RED_2$ | 1.00 |
| 4 | Archeological Areas, $RED_3$ | 0.40 |
| 5 | Industrial Plants, $RED_4$ | 0.70 |
| 6 | Fishing, $RED_5$ | 1.00 |
| 7 | Art Buildings, $RED_6$ | 0.60 |
| 8 | Accessibility to Coastline, $RED_7$ | 1.00 |
| Biological and Ecological Risk Factors | | |
| 9 | Naval Biological Variety, $RED_8$ | 1.00 |
| 10 | Birds, $RED_9$ | 1.00 |
| 11 | Declared Areas of Private Status, $RED_{10}$ | 1.00 |

#### *5.1.2.1. Coastal Risk Factors*

At indicating the risk factors of coastline, Environmental Sensitivity Index (ESI) [15] is taken as essential. Coastal risk factors are expressed with the values between 1 and 11 with respect to the structure of coastline; these values are given in Table 3 as below:

***Table 3.*** *Coastal Risk Factors.*

| Row | Shore Composition | Risk Factor Value $RK_i$ |
|---|---|---|
| 1 | Cliff and steep rocks | 1 |
| 2a | Rocks on surface | 2 |
| 2b | Dock on a pile | 3 |
| 3 | Shore with sand of mid-fine thickness | 4 |
| 4 | Shore with sand of thick thickness | 5 |
| 5 | Art buildings | 6 |
| 6 | Shore with a mixture of sand and grit | 7 |
| 7 | Shore with a mixture of grit and stone | 8 |
| 8 | Shore with big stones | 9 |
| 9 | Tide area | 10 |
| 10 | Reed beds | 11 |

#### *5.1.2.2. Settlements*

PS = the population of settlements;

| Evaluation Criteria | Risk Factor Value, $RED_1$ |
|---|---|
| $O < PS < 100$ | 0 |
| $100 \leq PS < 500$ | 1 |
| $500 \leq PS < 2\,000$ | 2 |
| $2\,000 \leq PS < 5\,000$ | 3 |
| $5\,000 \leq PS < 10\,000$ | 4 |
| $PS \geq 10\,000$ | 5 |

#### *5.1.2.3. Tourism Areas*

| Evaluation Criteria | Risk Factor Value, $RED_2$ |
|---|---|
| Not Available | 0 |
| A+C+E, G | 1 |
| A+C+F, B+C+E | 2 |
| A+D+E, B+C+F | 3 |
| A+D+F, B+D+E | 4 |
| B+D+F | 5 |

A: The facility is in the Marmara or in the Black Sea region (less desired sea areas),

B: The facility is in the Aegean or in the Mediterranean region (more sea area),

C: Has 3 stars or less,

D: Has 3 stars or more,

E: Has 100 beds or less,

F: Has 100 beds or more,

G: Beaches, parks and recreation areas.

#### *5.1.2.4. Archeological Areas*

| Evaluation Criteria | Risk Factor Value, $RED_3$ |
|---|---|
| None of them | 0 |
| A | 1 |
| A+C | 2 |
| B | 3 |
| B+C | 4 |
| A+B+C | 5 |

A: Archeological areas are on the coast

B: Archeological areas are on the sea

C: Museum area

#### *5.1.2.5. Industrial Plants*

| Evaluation Criteria | Risk Factor Value, $RED_4$ |
|---|---|
| Not Available | 0 |
| A | 1 |
| B | 2 |
| C, A+B | 3 |
| A+C, B+C | 4 |
| A+B+C | 5 |

A: Industrial plants on the coast that do not use sea as direct or indirect source,

B: Industrial plants that use sea as indirect source: Factories that make transportation by their docks, depots, production and martial plants,

C: Industrial plants that use sea as direct source: Shipyards, electric generation plants etc.

#### *5.1.2.6. Fishing*

| Evaluation Criteria | Risk Factor Value, $RED_5$ |
|---|---|
| Not Available | 0 |
| A | 1 |
| B | 2 |
| C | 3 |
| A+C, B+C | 4 |
| A+B+C | 5 |

A: Small fishing activities (fishhook, sportive fishing, small boats),

B: Hunting ground, bigger fishing activities (trawl, larger grounds),

C: Fish farms.

#### *5.1.2.7. Art Buildings*

| Evaluation Criteria | Risk Factor Value, $RED_6$ |
|---|---|
| Not Available | 0 |
| A | 1 |
| B | 2 |
| C | 3 |
| D | 4 |
| E | 5 |

A: Slipway ground; coastal structure which allows fishing boats and small tonnage boats to be grounded for their maintenance and repair jobs, which has right equipment and which has a sandy or concreted inclined area for repair jobs,

B: Emergency port; coastal structure without any major substructure and superstructure where fishing boats at any length and at any waterline can take shelter in, in order to avoid bad weather conditions,

C: Fisher port; coastal structure protected by breakwaters to serve fishing boats at any length and at any waterline; which possesses a dry dock and a backyard for local fishers' needs; loading - unloading areas and tying docks; water, electricity, net drying field, market place, administration; fuel pumps, preliminary cooling system and a slipway for demands of the fishing boats,

D: Yacht dock; tourism certificated coastal structure protected from wind and sea effects, which provides a secure tying and a direct walking way onto the yachts; which also offers technical and social substructure, administration, supporting, maintenance and repair services,

E: Dock; coastal structure naturally or artificially protected from wind and sea effects, and also convenient for the ships' boarding – disembarkation, loading – unloading, tying and waiting.

#### *5.1.2.8. Accessibility to Coastline*

| Evaluation Criteria | Risk Factor Value, $RED_7$ |
|---|---|
| Reachable from land by heavy vehicles | 0 |
| Reachable from land by light vehicles | 1 |
| Reachable only from sea | 2 |
| Reachable only from air | 3 |

#### *5.1.2.9. Naval Biological Variation*

| Evaluation Criteria | Risk Factor Value, $RED_8$ |
|---|---|
| Not available | 0 |
| A, B, C, D, E, F, G, ID | 1 |
| Presence of any two classes | 2 |
| Presence of any three classes | 3 |
| Presence of any four classes | 4 |
| Presence of any five classes | 5 |
| Presence of any six classes | 6 |
| Presence of all classes together, H | 7 |

A: Turtle, B: Dolphin,

C: Seal, D: Fish,

E: Mollusks, F: Shrimp,

G: Sponge, H: Endemic species,

ID=Insufficient data, it is assumed that only one class is present.

#### *5.1.2.10. Bird Species*

| Evaluation Criteria | Risk Factor Value, $RED_9$ |
|---|---|
| Not Available | 0 |
| A | 1 |
| B | 2 |
| C, A+B | 3 |
| B+C, A+C | 4 |
| A+B+C, D | 5 |

A: Local species

B: Migrant species

C: Extinct species

D: Endemic species

#### *5.1.2.11. Declared Areas of Private Status*

| Evaluation Criteria | Risk Factor Value, $RED_{10}$ |
|---|---|
| Not available | 0 |
| A, B, C, D, E | 1 |
| Presence of any two classes | 2 |
| Presence of any three classes | 3 |
| Presence of any four classes | 4 |
| Presence of any five classes | 5 |

A = RAMSAR [15] / Wet Lands,

B = Private Environmental Protective Areas,

C = National Parks,

D = Hunting, Wild Life Fields,

E = Other Areas of Private Status.

### *5.2. Method of Calculating Risk Value*

#### *5.2.1. Calculation of Risk Caused by Coastal Facility*

Coastal facility risk caused by materials ($R_{material}$) is worked out by the product of the risk factor value and the weight coefficients depending on location of the facility,

$A_i$ = Risk factor value caused by material

$WL_i$ = Weight coefficient depending on location

$$R_{material} = \sum_{i=1}^{13} A_i \cdot WL_i \quad (5.1)$$

Coastal facility risk caused by enterprise ($R_{enterprise}$), is worked out by the product of the risk factor value and the weight coefficients depending on location of the facility,

$P_j$ = Risk factor value caused by enterprise

$WL_j$ = Weight coefficient depending on location

$$R_{enterprise} = \sum_{j=1}^{9} P_j \cdot WL_j \quad (5.2)$$

Maximum risk constant value ($N_F$), is worked out by the sum of the products of the weight coefficients depending on maximum location and maximum value of each risk factor;

$A_{i\ max}$ = Maximum value of the risk factor caused by material,

$P_{j\ max}$ = Maximum value of the risk factor caused by enterprise,

$N_F$ = Maximum risk constant;

$$N_F = \sum_{i=1}^{13} A_{i\ max} \cdot WL_i + \sum_{j=1}^{9} P_{j\ max} \cdot WL_j = 167 \quad (5.3)$$

On the other hand, total risk or probability caused by coastal facility ($R_{facility}$) is equivalent to the normalization of the sum of material risks ($R_{material}$) and maximum risk constant ($N_F$),

$$R_{facility} = (R_{enterprise} + R_{material}) / N_F \quad (5.4)$$

#### *5.2.2. Calculation of Environmental Risk*

Assuming that shore facility is on the centre, each region with 20 kilometers of diameter is divided into circular strips of 1 000 meters and risk assessments are carried out for coastlines and for sea areas in these circular strips.

When working out environmental risk values, each strip is firstly evaluated on itself and then one risk value that represents the entire coastline is calculated.

Environmental risk factors are differed in every strip in terms of length as well as classification of coast. For this reason when studying the coastal risk factors, the risk value is worked out by considering the weight averages of the coast class distances in the strip,

$RK_{k,I}$ = value of $k^{th}$ coastal risk factor of $i^{th}$ strip

$LK_k$ = coastline length of $k^{th}$ coastal risk factor in the strip

$LD_{total,\ I}$ = total length of coastline on $i^{th}$ strip

M = total number of coastal risk value on $i^{th}$ strip

$RKV_i$, = coastal risk factor value of the corresponding strip;

$$RKV_i = (\sum_{k=1}^{m} RK_{k,i} \cdot LK_k) / LD_{total,\ i} \quad (5.5)$$

At this point, environmental risk values except coastal risk value ($REV_i$) have to be picked from corresponding tables.

According to the importance rate of each environmental risk factor, a weight coefficient

($EA_i$) is determined,

Maximum environmental and coastal risk factor at any $i^{th}$ strip, $EA_{total,i}$ is worked out as below;

$$EA_{total,\ i} = EA_{coastal} \cdot RKV_{maximum} + \sum_{i=1}^{n} EA_i \cdot REV_{maximum,\ i} \quad (5.6)$$

n = total number of environmental risk factors at $i^{th}$ strip,

Total risk value at $i^{th}$ strip $RV_{total,i}$is worked out as below;

$$RV_{total,i} = (RKV_i \cdot EA_i + \sum_{i=1}^{n} REV_i \cdot EA_i) / EA_{total,i} \quad (5.7)$$

Total coastline length of each strip ($LD_{total,\ i}$) will be different from another, Strips with longer coastlines are riskier than the ones with short coastlines. For this reason, total coastline length of $i^{th}$ strip ($LD_{total,\ i}$) must be included into risk evaluation.

Because probable source of contamination is assumed to be the center, more distant from this point more risk value would reduce. Therefore, to continue the evaluation, strip influence weight coefficient ($A_{strip\ influence,\ i}$) must be used. Strip influence weight coefficients are indicated according to Table 4.

***Table 4.*** *Strip Numbers and Strip Influences.*

| Strip Number | ±10 | ±9 | ±8 | ±7 | ±6 | ±5 | ±4 | ±3 | ±2 | ±1 |
|---|---|---|---|---|---|---|---|---|---|---|
| $A_{strip\ influence}$ | 0.3 | 0.3 | 0.3 | 0.5 | 0.5 | 0.5 | 0.7 | 0.7 | 1 | 1 |

Environmental risk ($R_{environmental}$) value or probability is worked out as below,

$$R_{environmental} = (\sum_{i=1}^{p} RV_{total,\ i} \cdot LD_{total,\ i} \cdot A_{strip\ influence,\ i}) / (\sum_{i=1}^{p} LD_{total,\ i} \cdot A_{strip\ influence,\ i}) \quad (5.8)$$

#### *5.2.3. Influence of Total Coastline Length of Facility Effect Area on Risk*

Coastline length between circular strips is different for each facility with respect to geographic factors. While coastline length and distance inside strip are approximately equivalent for rough coastlines, coastline length increases for the opposing situations, If the corresponding facility is situated on a geography which has less than 5 kilometers between its two coasts, total coastline length to evaluate may increase up to 100 kilometers as there is the risk that spill might reach the opposite coast.

Therefore, influence of total coastline length of the facility on risk is fairly high. Larger effect area of a facility, more the risk value will increase.

Minimum value of total coastline length ($LD_{minimum}$), is 20 kilometers which is the total length of a smooth coastline. This situation has no effect on risk.

Maximum value of total coastline length ($LD_{maximum}$) is assumed to be 100 kilometers. Thus, total risk is predicted to increase of 30% and this value is at the same time equal to 130% of maximum value of total coastline influence coefficient ($A_{LKmaximum}$).

Even if total coastline length is upper than maximum value, increase of the risk value will be equivalent to maximum increase.

Increase of coastline influence coefficient ($A_{LK}$) is predicted to be linear at the time when coastline lengths are between minimum and maximum values, Provided that total coastline length of facility effect area ($LD_{total}$) is in kilometers, coastline influence coefficient ($A_{LK}$) is worked out as below;

$$A_{LK} = 0.00375 \cdot LD_{total} + 0.925 \quad (5.9)$$

### *5.2.4. Calculation of Spill Risk*

Risk factors caused by facility and environmental risk factors are the values which would not vary in short periods, However, spill risk is rather more variable because of tanker traffic of the facility, Maximum waste amount that would spill is proportional to transferring load, Under these circumstances, high capacity tankers have higher spill risk. If total product arriving to the facility between intervals of time is studied, a curve of oscillation would be observed, then, spill risk varies as well, and spill risk can be determined as three values: maximum, average and minimum,

$L_i$ = transferring load by $i_{th}$ ship (tones)

N = total tanker number;

Average transferring load ($L_{average}$) is worked out as below:

$$L_{average} = \sum_{i=1}^{N} L_i / N \quad (5.10)$$

From the column of "Transferring Load" in Table 5 which gives probable waste spill and risk factor value, amount which comes as against calculated $L_{average}$ is picked, From hence, value in the column of "Risk Factor Value (RV)" is determined. Average spill risk value ($R_{spill,\ average}$) is equivalent to the risk factor value (RV),

$$R_{spill,\ average} = RV \quad (5.11)$$

*Table 5. Probable Waste Spills and Risk Factor Values.*

| Transferring Amount, $M_{tonnage}$ (tonnes) | Designation of Spill | Risk Factor Value, RV |
|---|---|---|
| $M_{tonnage}$> 100 000 | Very severe catastrophe | 12 |
| 70 000 <$M_{tonnage} \leq$ 100 000 | Severe catastrophe | 11 |
| 50 000 <$M_{tonnage} \leq$ 70 000 | Serious catastrophe | 10 |
| 30 000 <$M_{tonnage} \leq$ 50 000 | Catastrophe | 9 |
| 20 000 <$M_{tonnage} \leq$ 30 000 | Very severe disaster | 8 |
| 10 000 <$M_{tonnage} \leq$ 20 000 | Severe disaster | 7 |
| 5 000 <$M_{tonnage} \leq$ 10 000 | Serious disaster | 6 |
| 3 000 <$M_{tonnage} \leq$ 5 000 | Disaster | 5 |
| 2 000 <$M_{tonnage} \leq$ 3 000 | Very severe spill | 4 |
| 1 000 <$M_{tonnage} \leq$ 2 000 | Severe spill | 3 |
| 500 <$M_{tonnage} \leq$ 1 000 | Serious spill | 2 |
| 0 <$M_{tonnage} \leq$ 500 | Minor spill | 1 |
| 0 | Insignificant spill | 0 |

Values above being valid for open sea, risk is increased one level up at settling areas and at sensitive zones.

After having determined average value of the transferring load, standard deviation of the transferring load (σ) would be defined in terms of transferring load ($L_i$), average transferring load ($L_{average}$) and tanker number (N) as:

$$\sigma = \sqrt{[\sum_{i=1}^{N} (L_i - L_{average})^2]/N} \quad (5.12)$$

If standard deviation of transferring load (σ) and average transferring load ($L_{average}$) are summed up, maximum transferring load ($L_{maximum}$); if these two are substracted, minimum transferring load (Lminimum) would be obtained,

$$M_{maximum} = M_{average} + \sigma. \quad (5.13)$$

$$M_{minimum} = M_{average} - \sigma. \quad (5.14)$$

Afterwards, risk factor values which come up against $M_{maximum}$ and $M_{minimum}$ are picked from Table 5. In this way, maximum and minimum spill risks ($R_{spill,\ maximum}$ and $R_{spill,\ minimum}$) are worked out;

$$R_{spill,\ maximum} = RL_{maximum} \quad (5.15)$$

$$R_{spill,\ minimum} = RL_{minimum} \quad (5.16)$$

Under the case that $RL_{maximum}$ and $RL_{minimum}$ are equivalent, average value (RL) is used in calculations.

### *5.2.5. Final Risk Calculation*

Final risk value concerning a facility (Risk) is obtained by the multiplication of risk value caused by shore facility ($R_{facility}$), environmental risk value ($R_{environmental}$) and total coastline influence coefficient ($A_{LK}$). This risk value determines risk or probability which comes up against unit spill risk.

At this condition, probability or risk value coming up against unit spill force becomes in the table with 12 steps;

$$Risk = (0,9 \cdot R_{facility} + 0.1 \cdot R_{environmen\ tal}) \cdot 12 \quad (5.17)$$

Under the condition that Risk exceeds maximum possible value which is 12, risk will be assumed to be equivalent to the maximum value. Designation and probabilities of these values are given in Table 6.

Since spill risk value possesses maximum, average and minimum value, final risk value concerning the shore facility will be between an interval depending on spill risk value. As spill risk value approaches to maximum, final risk value will proportionally increase. Minimum spill risk will similarly decrease the final risk value.

At this condition, final risk value ($Risk_{final}$) is worked out in the table with 144 steps as below;

$$Risk_{final,\ maximum} = Risk\ R_{spill,\ maximum} \tag{5.18}$$

$$Risk_{final,\ average} = Risk\ R_{spill,\ average} \tag{5.19}$$

$$Risk_{final,\ minimum} = Risk\ R_{spill,\ minimum} \tag{5.20}$$

Final risk value for the facility is indicated in the interval of;

$$Risk_{final,\ minimum} < Risk_{final,\ average} < Risk_{final,\ maximum} \tag{5.21}$$

***Table 6.*** *Designation of Risk and Probability Relation.*

| Risk Level or Probability Level (Risk) | Designations | Probability Range, Pr |
|---|---|---|
| 1 | Extremely remote | $Pr \leq 10^{-6}$/ years |
| 2 | Remote | $10^{-6}$/ years < Pr ≤ $10^{-5}$/ years |
| 3 | Very rare | $10^{-5}$/ years < Pr ≤ $10^{-4}$/ years |
| 4 | Rare | $10^{-4}$/ years < Pr ≤ $10^{-3}$/ years |
| 5 | Low | $10^{-3}$/ years < Pr ≤ $10^{-2}$/ years |
| 6 | Possible | $10^{-2}$/ years < Pr ≤ 0.02 / years |
| 7 | Moderate | 0,02/ years < Pr ≤ $10^{-1}$/ years |
| 8 | Reasonably probable | $10^{-1}$/ years < Pr ≤ 0.2 / years |
| 9 | Quite frequent | 0.2 / years < Pr ≤ 1 / years |
| 10 | Frequent | 1 / years < Pr ≤ 2 / years |
| 11 | Regular | 2 / years < Pr ≤ 10 / years |
| 12 | Very frequent | Pr > 10 / years |

For this calculation; the risk becomes,

$$0 \leq Risk_{final} \leq 144. \tag{5.22}$$

If 20% of this value is evaluated; it becomes,

$$0 \leq Risk_{final} < 24\ . \tag{5.23}$$

Its probability can be worked out from Table 8.

Another situation is either between 20% and 35% or

$$24 \leq Risk_{final} < 42\ . \tag{5.24}$$

In this situation, it is necessary to be ready for an unwanted incident; to revise the intervention equipment, to follow the incoming ships and to arrange the staff education.

Unacceptable risk situation is;

$$42 \leq Risk_{final} \tag{5.25}$$

In this situation, handling material and equipment condition need to be studied in detail.

### *5.3. Risk Analysis for an Oil Company Terminal Terminal for Sea Spills*

An oil company is located close to shore in the borders of a coast line. Since oil products are handled in the facility, there is a high risk of oil contamination. As indicated in the previous sections of this supplemental, environmental risks are studied such as operational risks of the facility, type of the transferring material, risks caused by its amount and risks which appear after having studided the geological, ecological and socio-economical situation of the region. Risk values caused by the facility and by the material are as given in Table 7. The values about environmental risk values pointed are summarized in Table 8.

All risk values belonging to an oil company terminal given in Table 9. According to this table, total risk caused by the coast facility is calculated as $R_{facility} = 0.39$.

***Table 7.*** *Risk Value Caused by Operation and Material.*

| Row | Risk Factor, $F_i$ | Designations | Risk Value |
|---|---|---|---|
| Factors Considering Operational Risk of the Facility | | | |
| 1 | Experience of crew at the port and of operators, $A_1$ | Good and experienced | 2 |
| 2 | Crew sufficiency, $A_2$ | Sufficient | 2 |
| 3 | Condition of communication systems, $A_3$ | 3 kinds of communication | 2 |
| 4 | Condition of urgent intervention plans, $A_4$ | A+C, A+D | 3 |
| 5 | Condition of intervention tools to oil spill, $A_5$ | A or B, insufficient | 8 |
| 6 | Accessibility to installation, $A_6$ | Accessible from land by heavy vehicles | 2 |
| 7 | Arriving frequency of ships, $A_7$ | 2 < AF ≤ 4 | 3 |
| 8 | Education condition of crew, $A_8$ | Acceptable T2 + Good T1 | 4 |

| Row | Risk Factor, $F_i$ | Designations | Risk Value |
|---|---|---|---|
| 9 | Ship traffic simulation, $A_9$ | Grounding risk | 6 |
| 10 | Attitude of the company at dangerous situations, operational limits, $A_{10}$ | The company stops working at dangerous situations | 0 |
| 11 | Maintenance period of the system, $A_{11}$ | 6 months < MP ≤ 1 year | 6 |
| 12 | Instalation age, $A_{12}$ | 10 years < AI ≤ 20 years | 2 |
| 13 | Intervention beginning time, $A_{13}$ | 1 hour < IT ≤ 2 hours | 4 |
| Factors Considering Type and Amount of the Material | | | |
| 14 | Black oil transport, $P_1$ | AO = 0 | 0 |
| 15 | White oil transport, $P_2$ | 100000 tones < AO ≤ 500000 tones | 4 |
| 16 | Metallic oil transport, $P_3$ | MM = 0 | 0 |
| 17 | Number of oil tankers arriving to dock, $P_4$ | 10 < TS ≤100 | 4 |
| 18 | Maximum pumping capacity, $P_5$ | 250 $m^3$/hour < PK ≤1000 $m^3$/hour | 3 |
| 19 | Tanker traffic in vicinity of the facility, $P_6$ | Tanker traffic in the vicinity of the facility | 2 |
| 20 | Oil generation plants in vicinity, $P_7$ | 0 < OP ≤ 5 | 1 |
| 21 | Length of oil pipe line belonging to facility, $P_8$ | $PL_{Land}$ < 250 m and $PL_{Sea}$> 250 m | 4 |
| 22 | Number of oil storage tanks in the facility, $P_9$ | 5 ≤ TS | 5 |
| 23 | Automatic on off valves, $P_{10}$ | A + B, A + C, B + C | 4 |
| 24 | Accidents occurred in the facility, $P_{11}$ | 0,001 ≤ KY< 0,01 | 4 |

***Table 8.*** *Environmental Risk Values.*

| Risk Factors | Maximum Risk Factor Value | Weight Coefficients, WLi | Strip Number | | | | | | | | | | | | | | | | | | | |
|---|---|---|---|---|---|---|---|---|---|---|---|---|---|---|---|---|---|---|---|---|---|---|
| | | | 10 | 9 | 8 | 7 | 6 | 5 | 4 | 3 | 2 | 1 | -1 | -2 | -3 | -4 | -5 | -5 | -7 | -8 | -9 | -10 |
| Risk Factor of the Coastline | | | | | | | | | | | | | | | | | | | | | | |
| Coastal Risk Factors. RK1 | 11.00 | 1.00 | 5.39 | 8.98 | 6.43 | 5.96 | 4.14 | 7.58 | 8.00 | 4.41 | 4.33 | 7.26 | 7.48 | 7.84 | 7.57 | 8.64 | 7.76 | 7.57 | 6.27 | 7.21 | 7.38 | 6.27 |
| Socio-Economical Risk Factors | | | | | | | | | | | | | | | | | | | | | | |
| Settling. RED1 | 5.00 | 0.60 | 0 | 0 | 0 | 3 | 3 | 5 | 5 | 0 | 0 | 0 | 0 | 0 | 5 | 5 | 5 | 5 | 5 | 0 | 0 | 0 |
| Touristic Areas. RED2 | 5.00 | 1.00 | 1 | 0 | 0 | 0 | 0 | 0 | 0 | 0 | 0 | 0 | 0 | 0 | 0 | 2 | 0 | 0 | 0 | 0 | 1 | 1 |
| Archeological Areas, RED3 | 5.00 | 4.00 | 0 | 0 | 0 | 0 | 0 | 0 | 0 | 0 | 0 | 0 | 0 | 0 | 0 | 0 | 0 | 0 | 0 | 0 | 0 | 0 |
| Industrial Plants, RED4 | 5.00 | 0.70 | 0 | 2 | 2 | 2 | 2 | 0 | 0 | 0 | 2 | 2 | 2 | 2 | 2 | 0 | 0 | 0 | 0 | 0 | 0 | 0 |
| Fishing, RED5 | 5.00 | 1.00 | 2 | 2 | 1 | 0 | 0 | 0 | 0 | 0 | 0 | 0 | 0 | 0 | 0 | 0 | 0 | 0 | 0 | 0 | 0 | 0 |
| Art Buildings, RED6 | 5.00 | 0.60 | 0 | 0 | 0 | 0 | 0 | 0 | 0 | 0 | 0 | 0 | 0 | 0 | 3 | 0 | 0 | 0 | 0 | 0 | 0 | 0 |
| Accessibility to the Coastline, RED7 | 3.00 | 1.00 | 0 | 0 | 0 | 0 | 0 | 0 | 0 | 0 | 0 | 0 | 0 | 0 | 0 | 0 | 0 | 0 | 0 | 0 | 0 | 0 |
| Biological and Ecological Risk Factors | | | | | | | | | | | | | | | | | | | | | | |
| Naval Biological Variety. RED8 | 8.00 | 1.00 | 3 | 3 | 3 | 3 | 2 | 2 | 2 | 1 | 1 | 1 | 1 | 1 | 1 | 2 | 2 | 4 | 3 | 3 | 3 | 3 |
| Bird, RED9 | 5.00 | 1.00 | 0 | 0 | 0 | 0 | 0 | 0 | 0 | 0 | 0 | 0 | 0 | 0 | 0 | 0 | 2 | 0 | 2 | 0 | 0 | 0 |
| Declared Areas of Private Status RFD10 | 5.00 | 1.00 | 0 | 0 | 0 | 0 | 0 | 0 | 0 | 0 | 0 | 0 | 0 | 0 | 0 | 0 | 0 | 0 | 0 | 0 | 0 | 0 |
| Risk Value in the Strip. RVtotal.i | | | 0.22 | 0.29 | 0.23 | 0.23 | 3.22 | 0.24 | 0.25 | 0.10 | 0.13 | 0.18 | 3.19 | 0.19 | 0.28 | 0.30 | 0.28 | 3.28 | 0.27 | 0.19 | 0.22 | 0.20 |
| Strip Influence Weight Coefficient (Astrip influence,i) | | | 0.3 | 0.3 | 0.3 | 0.5 | 0.5 | 0.5 | 0.7 | 0.7 | 1 | 1 | 1 | 1 | 0.7 | 0.7 | 0.5 | 05 | 0.5 | 0.3 | 0.3 | 0.3 |
| Strip Coastline Length (LKi) | | | 1.13 | 2.67 | 2.76 | 2.58 | 2.75 | 1.07 | 1.12 | 1.3 | 1.42 | 1.53 | 1.09 | 3.58 | 5.8 | 1.85 | 1.35 | 1.37 | 1.42 | 1.3 | 1.63 | 1.2 |
| Environmental Risk (Renvironmental) | | | 0.23 | | | | | | | | | | | | | | | | | | | |

Finally, $R_{environmental} = 0.23$ and $A_{LK} = 1.07$ are obtained.

*Table 9. Risk Values.*

| Row | Risk | Risk Value |
|---|---|---|
| 1 | Total risk caused by coast facility ($R_{facility}$) | 0.40 |
| 2 | Environmental risk ($R_{environmental}$) | 0.23 |
| 3 | Coastline influence coefficient ($A_{LK}$) | 1.07 |
| 4a | Maximum spill risk ($R_{spill, maximum}$) | 5.00 |
| 4b | Average spill risk ($R_{spill, average}$) | 5.00 |
| 4c | Minimum spill risk ($R_{spill, minimum}$) | 4.00 |
| 5 | $(0.9\ R_{facility} + 0.1\ R_{environmental})\ A_{LK}\ 12$ | 4.87 |
| 6a | Maximum final risk ($Risk_{final, maximum}$) | 24.37 |
| 6b | Average final risk ($Risk_{final, average}$) | 24.37 |
| 6c | Minimum final risk ($Risk_{final, minimum}$) | 19.50 |

Using the above values;

Probability which comes up against unit spill force or risk value (Risk) is worked out in the table with 12 steps as

Risk = 4.79 .

This is the probability of a spill, Table 8.

Final risk varies from 19.59 to 24.37. This is an acceptable risk value. However;

- It is necessary to take particular cautions for big tankers,
- It is necessary to develop the staff education,
- It is necessary to keep the intervention equipments in good condition and sufficient,
- It is necessary to take essential cautions and to make the appropriate agreements for Tier 2.

### *5.4. Coast Classification Under the Influence Area of an Oil Terminal*

As shore areas vary between themselves, they are all influenced in several different ways in the contaminations caused by oil and oil products. For this reason, shore areas are classified according to their influence from contamination. This classification is called "Coastline Sensitivity Determination". While making this determination; geomorphologic composition of the coast, coast ecology, settling, industry and location of tourism facilities are taken into consideration.

In order to determine the coastline sensitivity, defined risk factors are listed in Table 10 as below:

*Table 10. Risk Factors in Coastline Sensitivity.*

| Coastal Risk Factors | | Socio – Economical Risk Factors | Biological and Ecological Risk Factors |
|---|---|---|---|
| Shore Index | Shore Type | | |
| 1 | Cliff and steep rocks | Settling | Fish |
| 2a | Rocks on the surface | Tourism Areas | |
| 2b | Dock on a pile | Archeological Areas | Birds |
| 3 | Shore with sand of mid-fine thickness | Industrial Plants | |
| 4 | Shore with sand of thick thickness | | |
| 5 | Art buildings | Fishing | Other Biological Living (turtle, seal, dolphin) |
| 6 | Shore with a mixture of sand and grit | | |
| 7 | Shore with a mixture of grit and stone | Art Buildings | Declared Areas of Private Status (RAMSAR area, private environmental protective area, national parks, hunting-wild life protective areas, other areas of private status) |
| 8 | Shore with big stones | | |

The maps according to the shore classification studies as well as the details about the studies which remain inside the influence area of an oil terminal installation and which are obtained according to the Environmental Sensitivity Index (ESI) [16] are given in the tables below:

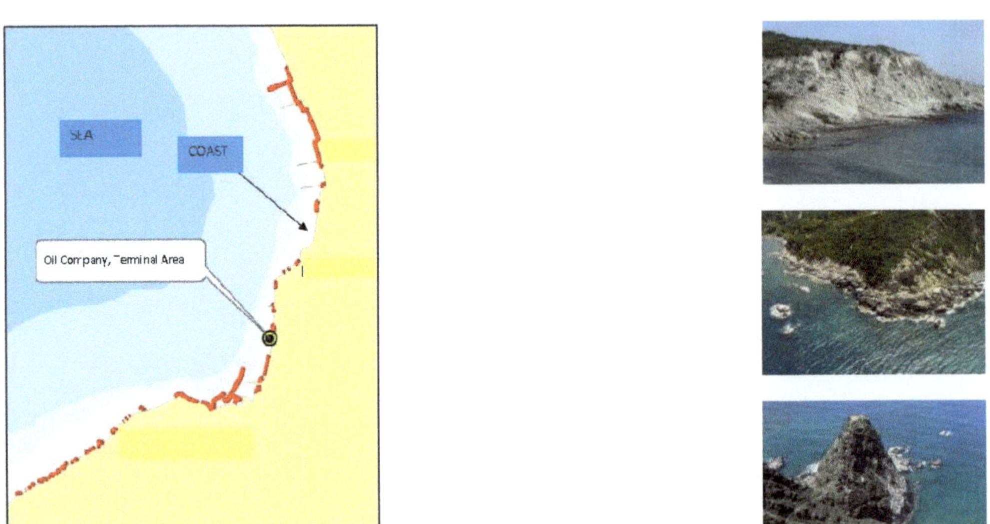


*ESI 1. Cliff and Steep Rocks.*

In general, there is not any possibility of access to the coastline from cliff and steep rocks. Beneath the steep rocks, in general there lie mixed sediment and a shore of grit and stone. There is a large variety and density of genres, although typical organisms such as mussels, whelks, sea algaes and lichens live on the rocky shores; the number of genres on the shore is less. As rock surfaces are not smooth, small puddles resulting from tides can be seen. Typical organisms such as crustaceans, whelks, lichens and even sea algues live.

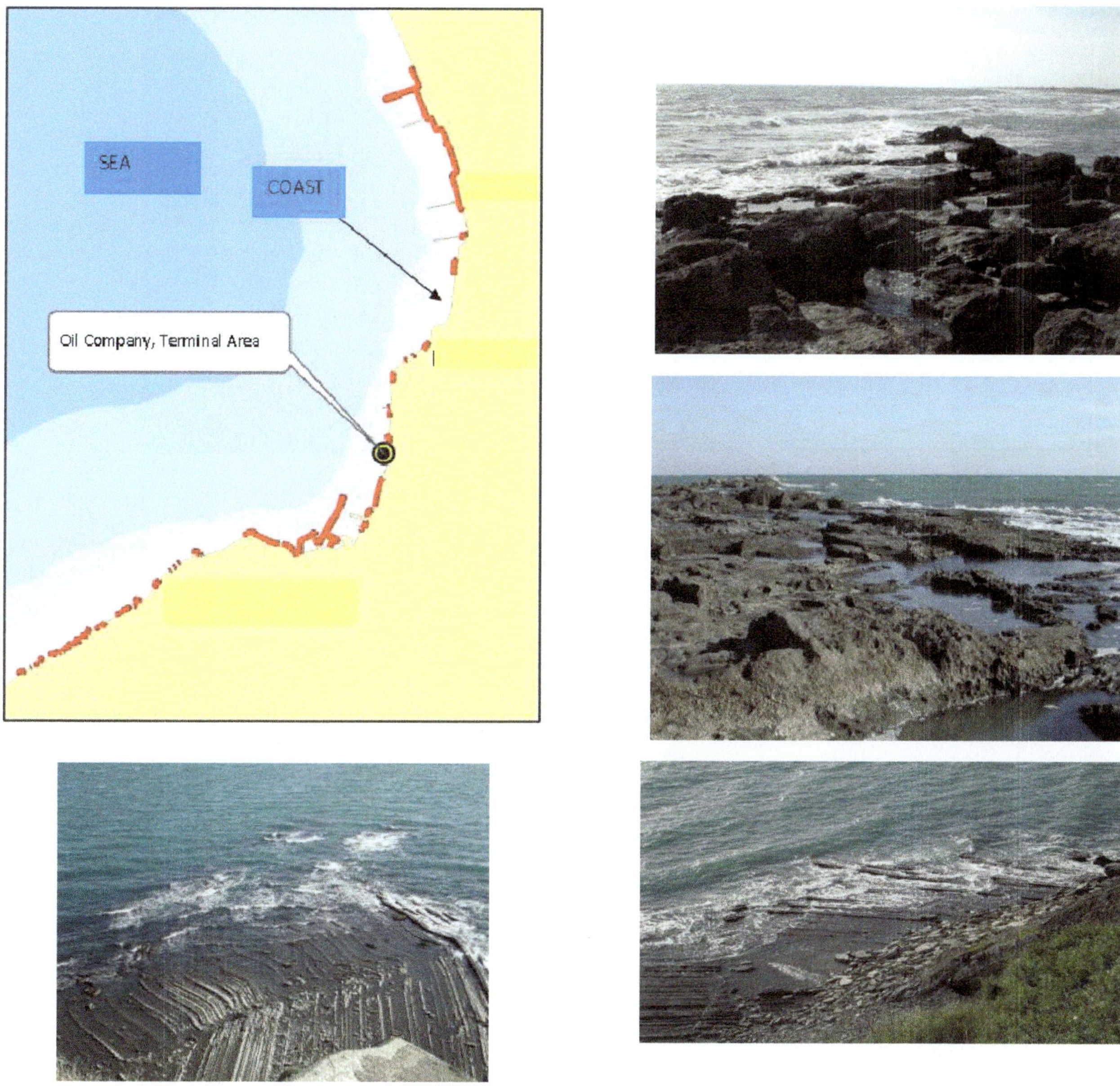


***ESI 2a.*** *Rocks on the Surface.*

Most of the coastlines of this kind are only accessible from land. In the tide area, local rocks on the surface are dominant. This type of coastline also includes volcanic blackstones and big stone pieces are seen on the shore. At high from the coastline, sandy coasts or of sand and grit can be observed. As rock surfaces are not smooth, small puddles resulting from tides can be seen. Typical organisms such as crustaceans, whelks, lichens and even sea algues live.

*ESI 2b.* *Dock on a Pile.*

These are the staging structures built on conrete or steel piles. Piles largely exposed to oil are placed perpendicularly or close to perpendicular. Organisms such as algaes which hold on piles are observed, however these plants and organisms are rare where the sea is wavier. Docks are rather located in the industrial areas.

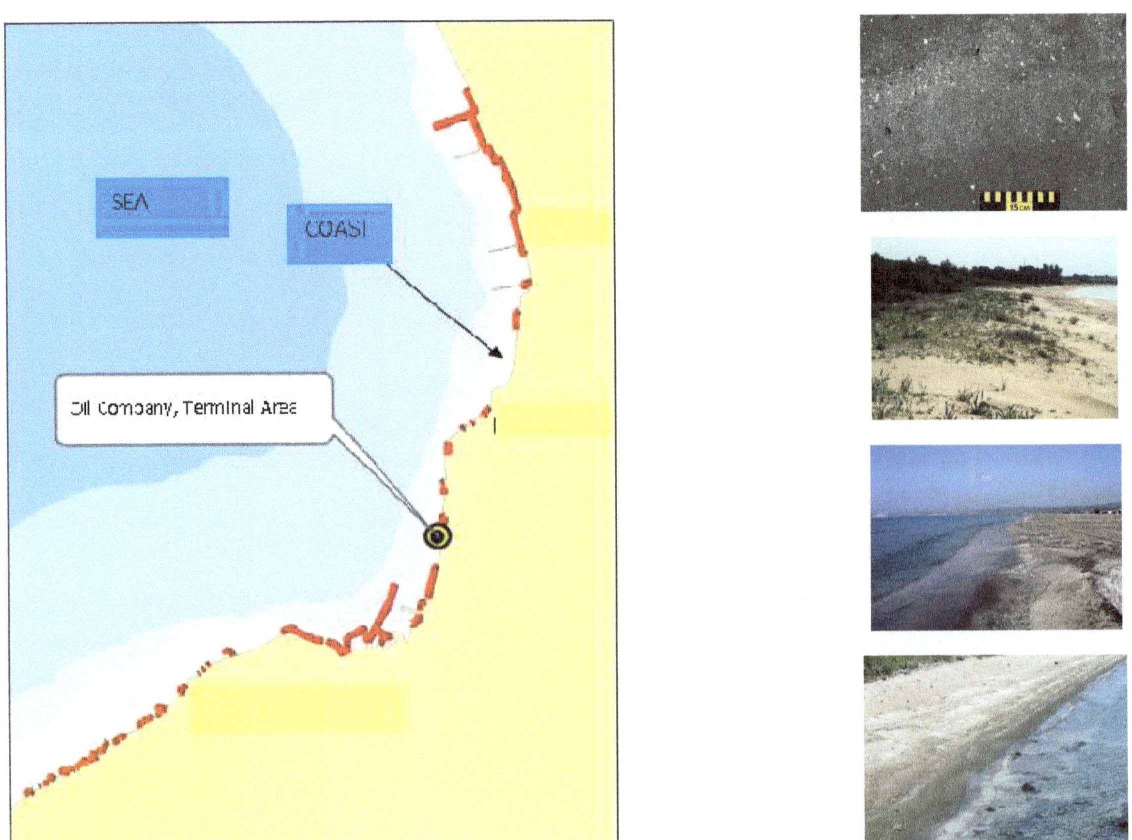


*ESI 3.* *Shore with sand of mid-fine thickness.*

A few amount of grit or sea shell might have been mixed with sand. Stranded sea algae and plastic sediments can be seen. In certain zones, ovulation and nutrition areas of birds and of turtles are present. Shore fauna being rare, crabs are observed on some of the shores. Generally, plant cover is common on sand hills, since the shore is densely sandy, vehicles except ones with 4 tires should not be used.

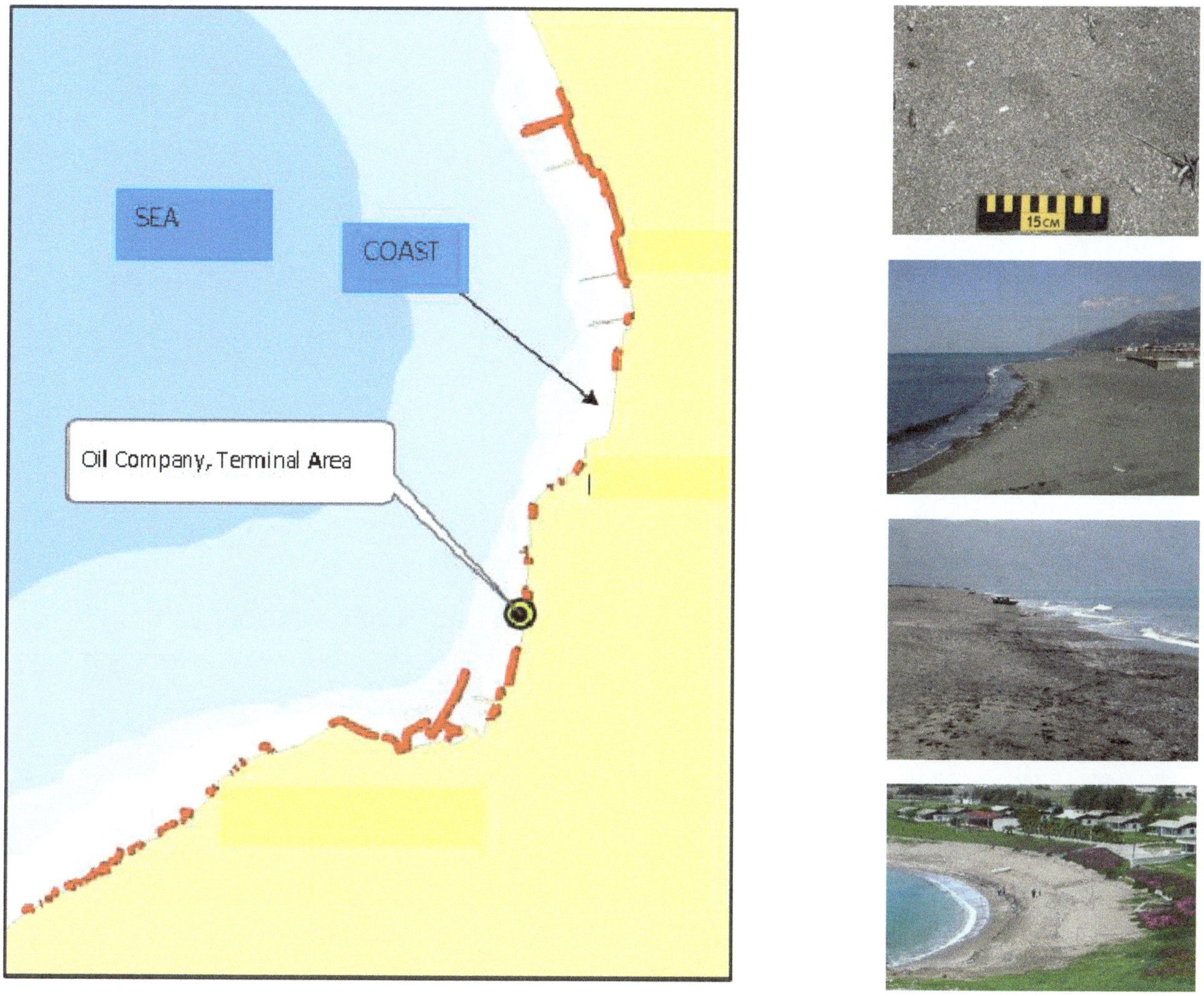


***ESI 4.*** *Shore with sand of thick thickness.*

These shores are lightly inclined, having soft layers and they do not enable the vehicles to operate. It also includes uncommon organisms such as crabs and scallops.

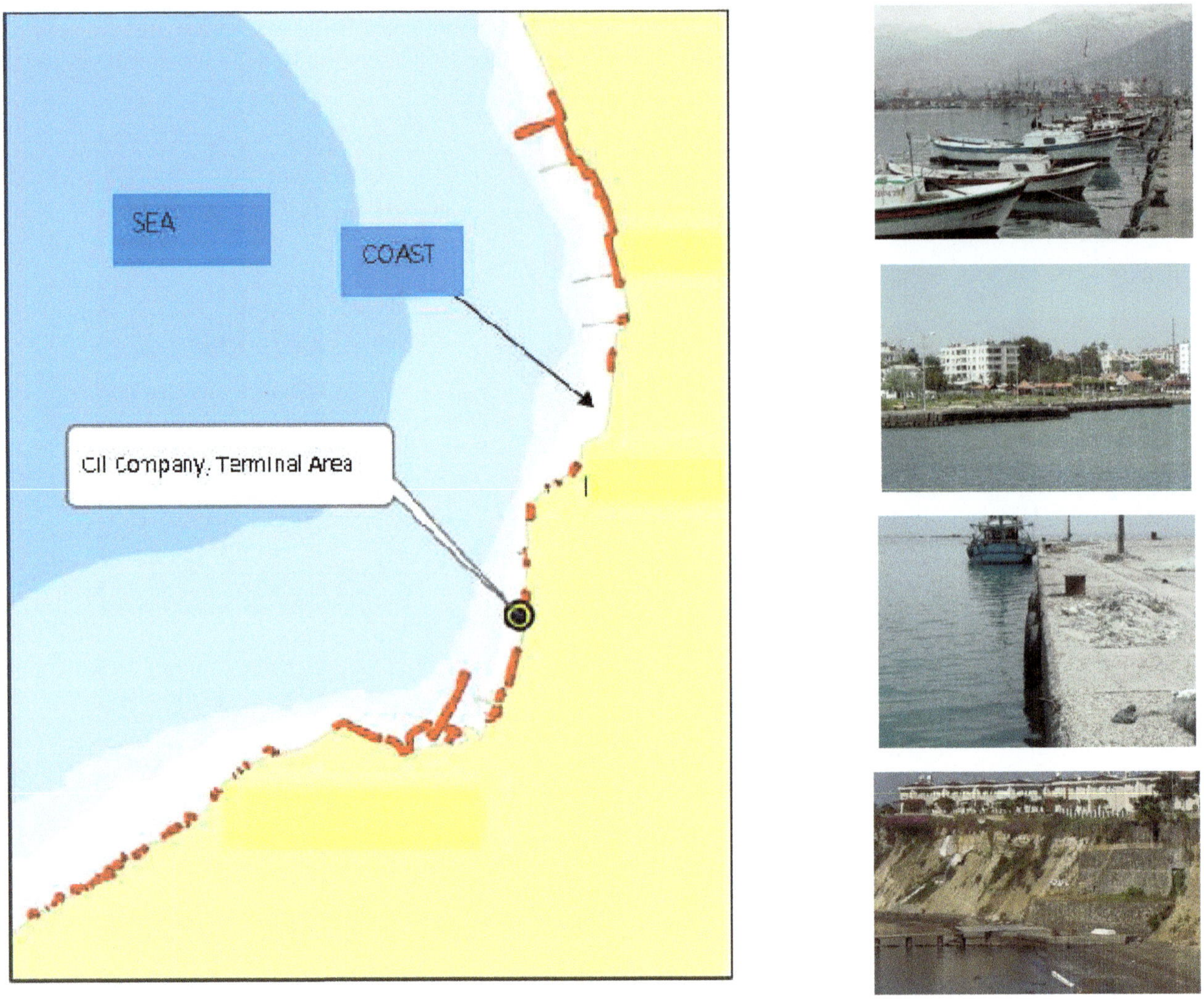


***ESI 5.*** *Art buildings.*

Constitutions such as breakwater and sheathing are generally built of stone and concrete to protect the dock and shoreline. On the sides of the breakwater facing the sea; sand, sand with grit and grit are observed. Crustaceans, whelks, lichens and sea algae can be seen on the breakwaters in tide areas, In case of an oil spill, it is necessary to take cautions as these areas are socio-economically important.

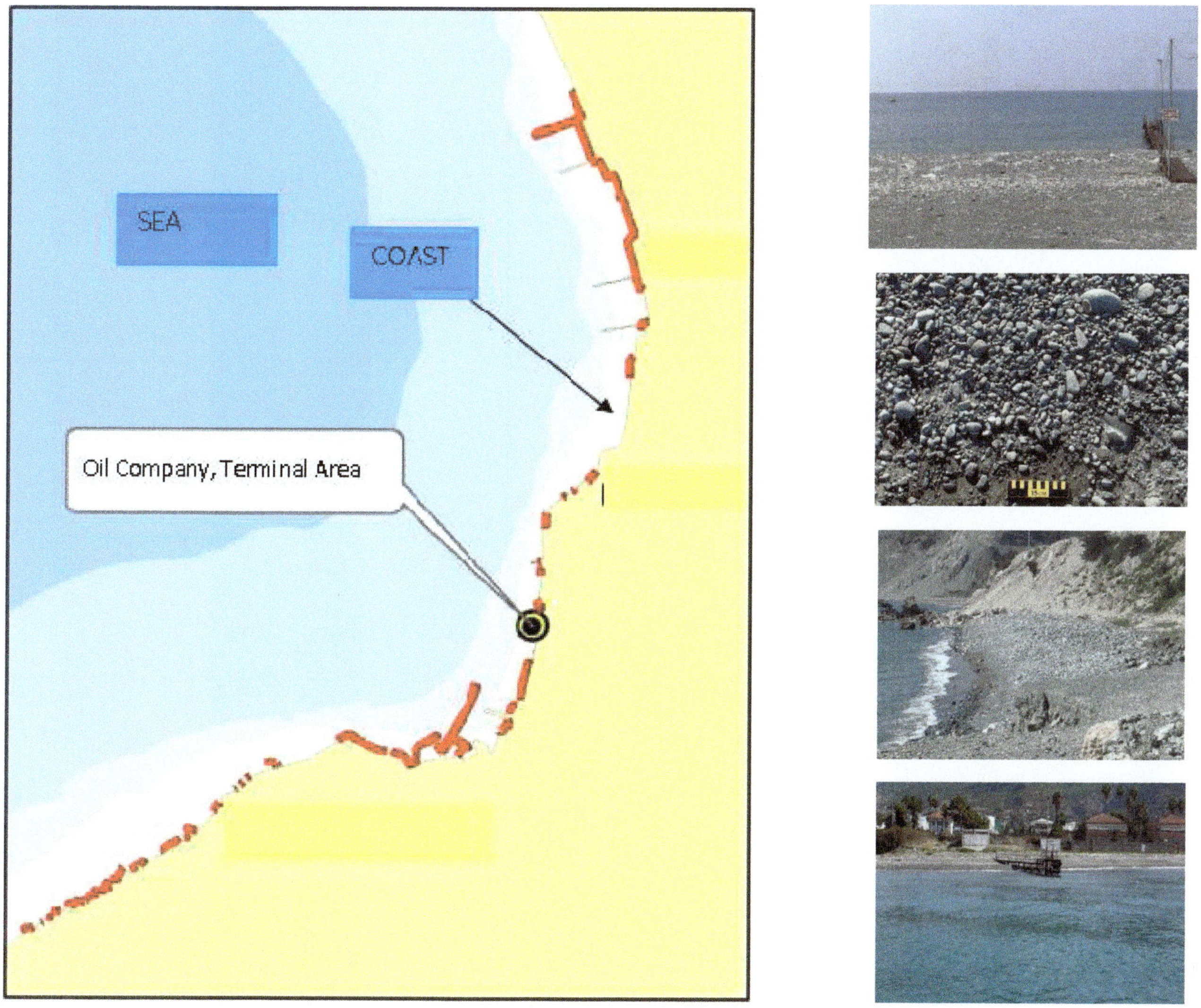


*ESI 6. Shore with a mixture of sand and grit.*

Shores of this type are partly inclined. Tide zone is formed of the mixture of sand and grit. Depending on sediment growth sand, grit or curbstones are present especially in upper tide zone. Small numbers of vegetables and of animals are encountered.

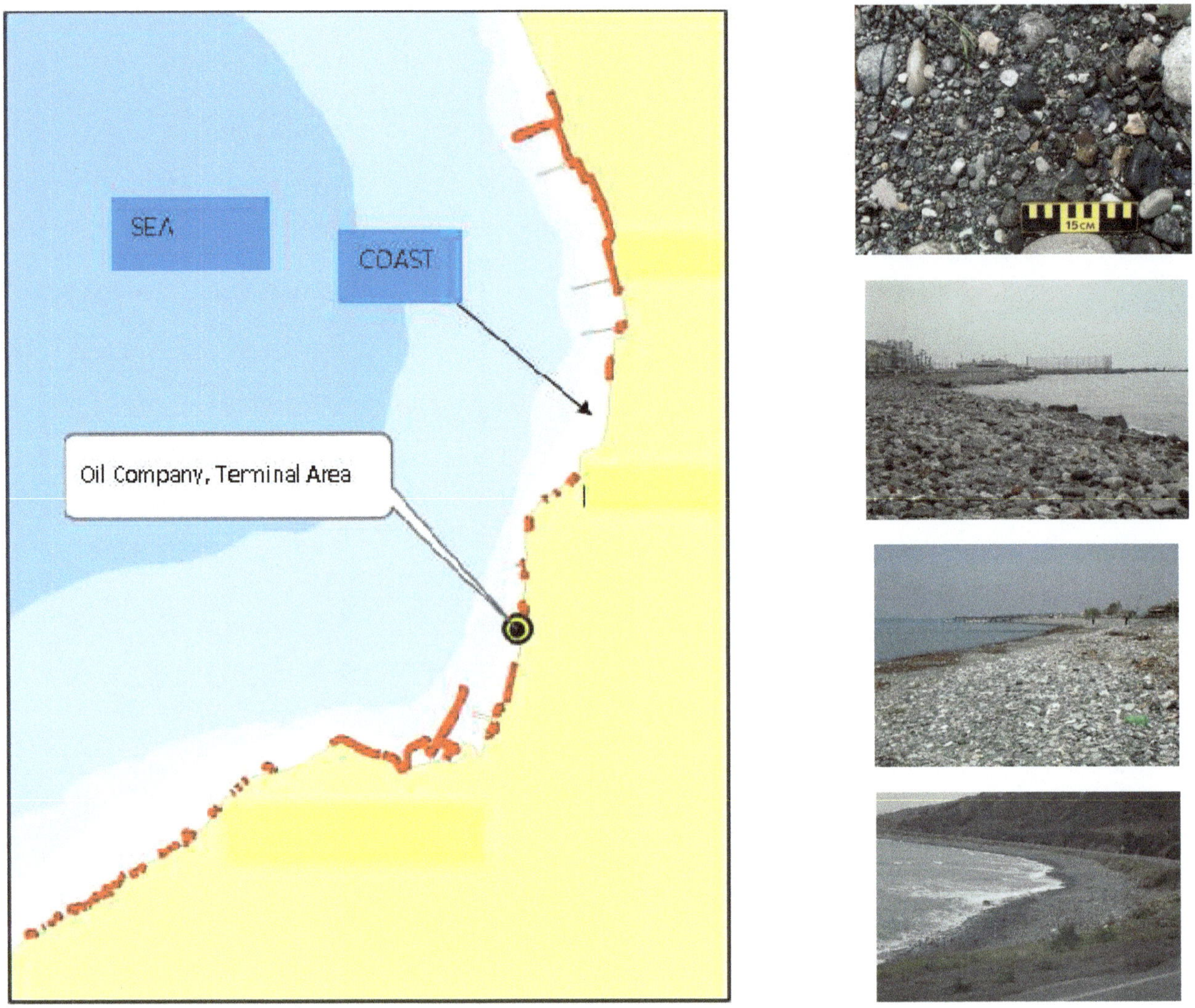


***ESI 7.*** *Shore with a mixture of grit and stone.*

These are the sediments formed of a mixture of grit, stone and unattached. They could be found as a range of mountains close together. Organisms such as snails and mussels as well as algae, which are held at lower areas of the shore where the sediment is decisive, are encountered.

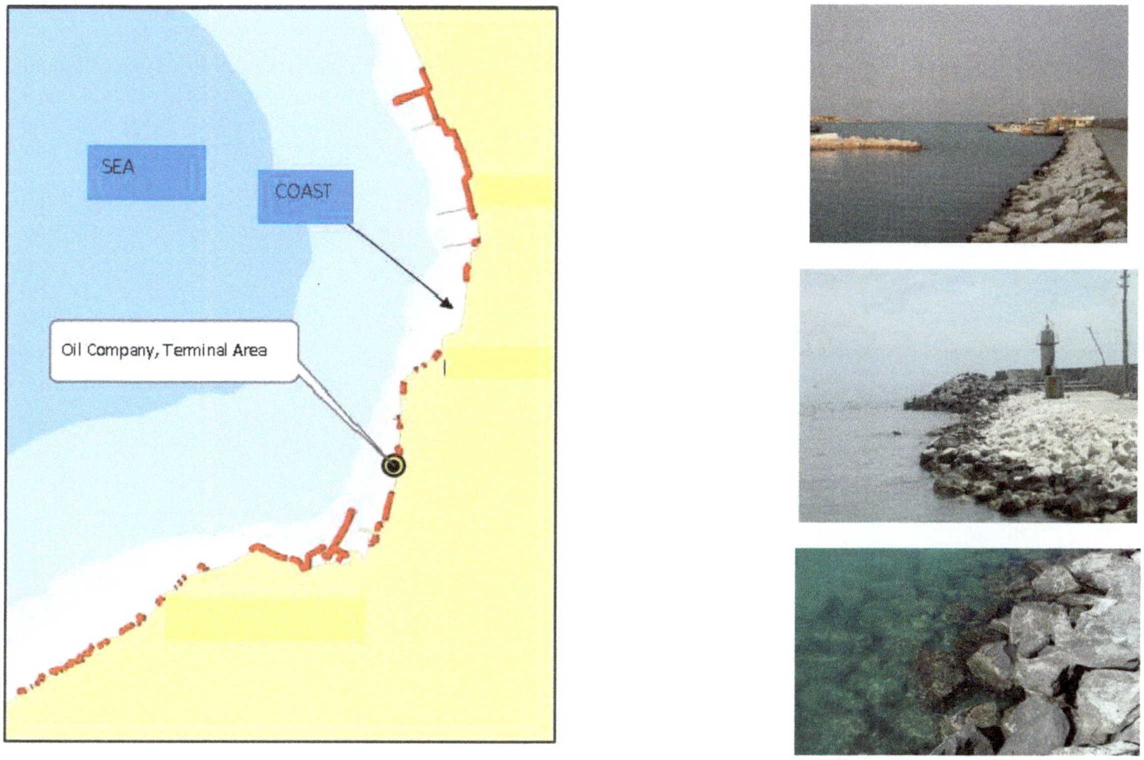


***ESI 8.*** *Shore with big stones.*

Shores with big stones are formed of massive blocks made of concrete or limestone and in general, are used for shore, dock protection and to consolidate the docks. As these areas are important from a socio-economical view, it is necessary to take good care of its cleaning in case of an oil spill. As a consequence of pulling back of the sea after tide, sea algae, mussels and other crustaceans are observed.

## References

[1] Aristotelis B. Alexopoulos, Managing and assessing oil spills through systematic methods, J. of Middle East Fr. 8, 111-123, 2008.

[2] Aristotelis B. Alexopoulos, Problems encountered when Tankers lie at Ports, The Case of Pylos and the Establishment of Marine Protection Zones, Honorary volume for Professor Lekkas edited by University of the Aegean, 79-90, 01/2011.

[3] J. R. van Dorp and J. R.W. Merrick, On a risk management analysis of oil spill risk using maritime transportation system simulation, Annals of Operations Research, published online before print December 12, 2010, DOI: 10, 1007/s10479-009-0678-1, (http://www.springerlink.com/content/w168111018754401/), 2009.

[4] National Research Council, Review of Prince William Sound Risk Assessment, Review of the Prince William Sound, Alaska, Risk Assessment Study Committee on Risk Assessment and Management of Marine Systems, Marine Board, ISBN: 0-309-55835-2, 78 pages, 1998.

[5] J. R. W. Merrick. J. R. van Dorp, T, Mazzuchi, J, Harrald, J, Spahn and M, Grabowski, The Prince William Sound Risk Assessment, Interfaces, and Vol. 32 (6): pp. 25-40, 2002.

[6] J. R. van Dorp. J. R. W. Merrick. J. R. Harrald. M. Gabowksi, Assessment of Oil Spill Risk due to Potential Increased Vessel Traffic at Cherry Point, Washington, Final Report (http://www.seas.gwu.edu/~dorpjr/tab4/publications_technical .html), August 2008.

[7] Kjell Skognes and Øistein Johansen, Statmap - a 3-dimensional model for oil spill risk assessment, Environmental Modelling & Software, 19, 727–737, 2004.

[8] Lisa Woolgar, Assessing the Increasing Risk of Marine Oil Pollution Spills In China, International Oil Spill Conference Proceedings, Vol. 2008, No. 1, pp. 711-715, May 2008.

[9] Oyvind Endresen, Oyvind Breivik, Odd Willy Brude, Ingrid H Ellingsen, Kjell Røang, Jarle Hauge and Per Olaf Brett, Prevention of oil spill from shipping by modelling of dynamic risk, Marine Pollution Bulletin, 54(10).1619-33, 11/2007.

[10] Magnus S. Eide, Øyvind Endresen, Øyvind Breivik, Odd Willy Brude, Ingrid H, Ellingsen, Kjell Røang, Jarle Hauge and Per Olaf Brett, Prevention of oil spill from shipping by modelling of dynamic risk, Marine Pollution Bulletin, 54, 1619–1633, 2007.

[11] Assessment of Marine Oil Spill Risk and Environmental Vulnerability for the State of Alaska, Department of Commerce National Oceanic and Atmospheric Administration Final Report September, 2014.

[12] Cristiana Dumitran and Ion Onutu, Environmental Risk Analysis for Crude Oil Soil Pollution Carpathian, Journal of Earth and Environmental Sciences, Vol. 5, No, 1, p, 83 - 92, April 2010.

[13] P. A. Kassomenos, Risk Analysis for Environmental Hazards: The Case Of Oil Spills in Crete, Global Nest: the Int. J. Vol. 6, No 1, pp 39-51, 2004.

[14] J. C. Udoh and E. M. Ekanem, GIS Based Risk Assessment of Oil Spill In The Coastal, African Journal of Environmental Science and Technology Vol. 5(3), pp. 205-211, March 2011.

[15] The Paris Protocol to the RAMSAR Convention on Wetlands, 1982.

[16] NOAA, Environmental Sensitivity Index Guidelines, version 3.0, NOAA Technical Memorandum, NOS, OR&R 11, Seattle: Hazardous Response and Assessment Division, National Oceanic and Atmospheric Administration, 129 p. 2002.

18

# Metal Pollution Assessment of Sediment and Water in Al-Ghadir River: Role of Continuously Released Organic Matter and Carbonate and Their Purification Capacity

**Amale Mcheik[1, 2*], Mohamad Fakih[2], Hussein Trabulsi[3], Joumana Toufaily[2], Taysir Hamieh[2], Evelyne Garnier-Zarli[1], Noureddine Bousserrhine[1]**

[1]Laboratory of Biogeochemistry and Ecology of continental regions (IBIOS – BIOEMCO), Department of Biology and Environmental Sciences, Paris-Est University, Creteil, France

[2]Laboratory of Materials, Catalysis, Environment and Analytical Methods (MCEMA), Department of Chemistry, Lebanese University, Hadath, Beirut, Lebanon

[3]Faculty of Economic Sciences and Business Administration, Lebanese University, Beirut, Lebanon

## Email addresses:

Amale.am@hotmail.com (A. Mcheik), fakih.mohamad@gmail.com (M. Fakih), husseintrabulsi@yahoo.fr (H. Trabulsi), joumana.toufaily@ul.edu.lb (J. Toufaily), tayssir.hamieh@ul.edu.lb (H. Taysir), garnier@u-pec.fr (E. Garnier-Zarli), bousserrhine@u-pec.fr (N. Bousserrhine)

---

**Abstract:** During the discharge of metals in the aquatic environment, metals are partitioned between the sediment and the water column phases. Further partitioning of metals occurs within the sediment chemical fractions. The present study focuses on one of the middle eat rivers, Al-Ghadir which is the smallest and the most polluted and found in the most populated region in Lebanon. The aim of this study was to integrate hydrochemical and sediment data- in order to evaluate the distribution pattern of the pollution of Al-Ghadir River. Bed sediments and water were collected from five locations in one dry season (Octobre 2010) and analyzed for more than 40 parameters, including (i) those of the chemical and the physico-chemical nature and (ii) those reporting the pollution caused by the heavy metals partitioned between the sediment and the water column phases by applying a sequential chemical fractionation scheme to the <75 μm sieved sediment fraction. The quality of waters for different uses has been tested and the state of sediments pollution was evaluated by comparison with general quality standards. Data showed that the highest percentages of total metal content in sediment are for: Fe and Mn in the residual and in the Fe/Mn oxides fractions, Cu in the oxidizable fraction, Cd and Zn in the carbonate and in the Fe/Mn oxides fractions, Pb and Cr in the Fe/Mn oxides fraction. Based on the geoaccumulation indices ($I_{geo}$), the river sediments are considered to be moderately polluted with the measured metals.

**Keywords:** Mobilization, Heavy Metals, Bed Sediments, Metal Pollution

---

## 1. Introduction

Rivers in Lebanon have always been the most important freshwater resources and most developmental activities are still dependent on river water. However, rivers have also been used for cleaning and disposal purposes. Huge loads of waste from industries, domestic sewage and agricultural activities find their way into rivers resulting in the deposition of contaminants, large scale deterioration of water quality and elevation of sediment concentrations that have the potential to cause toxicity to aquatic biota ([1], [2], [3]).

The optimal management of Lebanon's rivers is limited by a scarcity of both hydrological and water quality data. Both monitoring and research more or less ceased during the long civil war. [4] have discussed in detail the planning problems posed by the present lack of hydrometric data in relation to operational forecasting of water shortages and flood events. Both solid and liquid wastes are everywhere disposed to watercourses without treatment especially in inland villages and communities. It is essential therefore to conduct studies relating to: (a) surface and subsurface mapping to identify sources and types of contaminants, (b) assessing water quality for multipurpose utilization and (c) continuous monitoring to determine factors relating to self-purification.

This study was done during the dry season of 2010 on one of the smallest rivers in Lebanon, Al-Ghadir River, which has a total length of 15 km. This river originates from Alley (Mount Lebanon) and flows through Shweyfet, then passes through the Tyro in the industrial zone of Shweyfet till reaching the Mediterranean Sea at Ouzeieh, south of Beirut. The climate of the region is dry during the summer and cold wet during winter. The average annual rainfall is about 1000 mm annually, most part of which occurs between November and April while June, July and August are totally dry [4] and the average annual discharge of Al-Ghadir river in the year 2010 was 13.52 $m^3/s$ and the average flow rate was 0.47 $m^3/s$ according to the Litani River Authority.

Sixty years ago, this river was a source for irrigation, agriculture and industrial issues but overpopulation, agricultural activities, industrial and urban development situated on its banks has caused major reductions in river water quality. The main sources of pollution in the river include municipal and agricultural wastes, industrial effluents of pulp, paper, dyes, tans, batteries, ceramics, distilleries, car oils, stones and marbles and serum products and wastes from the farms of sheep and poultry released along the river and despite these inputs, the pollution state of this river has never been monitored, evaluated or documented.

During dry seasons, the flow of the river became slow and dry from its origin but the industrial effluents and the municipal wastewater generate the flow of the water in the river. Under these low flow river conditions, the system might approximate to that of a closed system, as justified by a correlation of the metal content between sediment and water column and the aqueous chemistry of the river became closer to an ideal condition. Therefore, pollutants enter the river under these conditions and partition between and within the various solid and aqueous phases at a state nearly similar to a closed system [5]. In this context, the main objective of this study has been to obtain information about both general chemical and physico-chemical parameters and pollution caused by the heavy metals partitioned between the sediment and the water column phases.

# 2. Materials and Methods

## 2.1. Sampling Methodology

Five sampling sites were chosen (Figure 1) along a 15 km stretch of the river Al-Ghadir upstream of its confluence with the Mediterranean Sea. The selection of these sites was based upon the practicability of collecting bed load sediments, the observed potential contamination from domestic waste or industrial discharge. Bed load sediments and water were collected simultaneously during the dry season (Octobre 2010). The collection of bed load sediments was in accordance with the methods used for shallow rivers ([6], [7]). The method involves wading in water and scooping sediments from a depth of 10 to 20 cm using an aluminium trowel. At each collecting site, two sub-samples were taken and bulked in polyethylene bag. Water samples were collected using two one-liter polyethylene bottles. All the sample bottles and other containers were soaked overnight in 10% (v/v) nitric acid and rinsed with ultrapure water several times prior to use. The bottles were rinsed twice with the river water and then filled with water running in the direction of river flow. The bottles were labeled with the corresponding sampling site, time and date of collection. The method of analyzing water and wastewater followed those reported by [8].

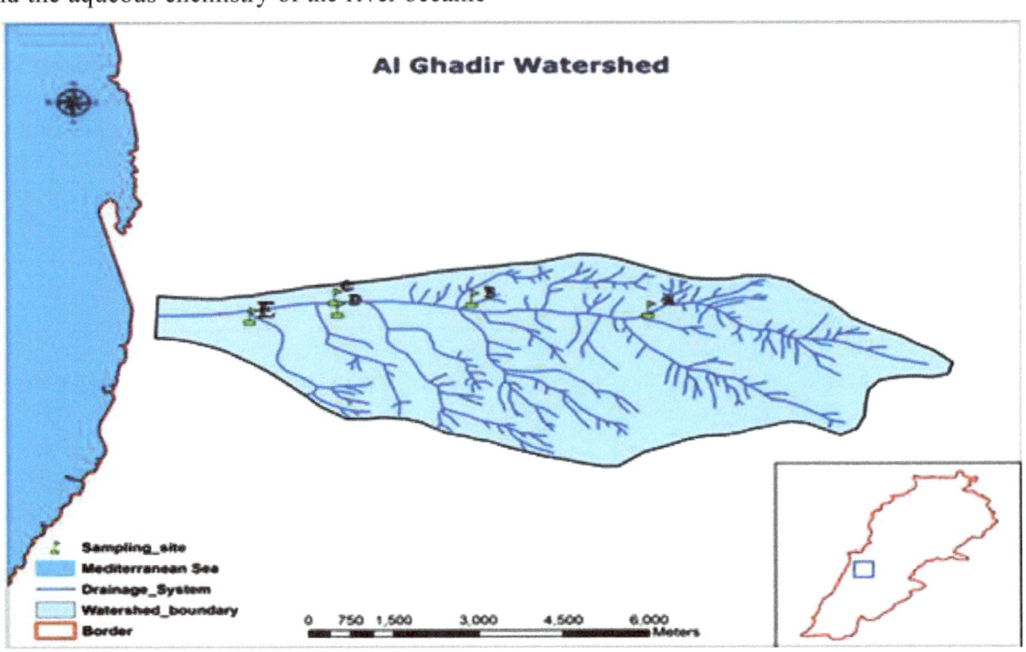


*Figure 1. A map of the study area showing the sampling site.*

### 2.2. Sediment Treatment and Analysis

Sediment samples were characterized for both general parameters (pH, organic matter granulometric and mineralogical analyses) and metal contents (Pb, Cu, Zn Cd, Cr, Fe and Mn). Analyses were performed using standard methods [9].

Analysis of pH in the sediment samples was performed by adding 10 g of the air dried sample into 50 ml distilled water and agitating for 5 min. The solution was left undisturbed for 1h with occasional shaking before measuring the pH. The granulometric analysis was performed from the whole untreated air-dried sample and was realized according to the normes AFNOR NF X31-107. The mineralogical analysis was applied by X-ray diffraction (Bruker, D8 Focus). The porosity of the sediment samples was measured according to the method adapted by [10]. The organic matter in the sediment was determined by wet oxidation-redox titration method using an acid dichromate solution. The reported data are the sum of soluble and insoluble carbon. For metal analysis, sediments were dried at room temperature and sieved while the fraction <75 µm sediment size was retained (clay/silt fraction). Analysis of this fraction is normally carried out in sediment studies because clay and silt particles generally contain the highest concentrations of pollutants and are most readily transported in suspension in natural waters [11]. Samples were digested in closed vessel microwave-assisted digested system for total metal analysis. In addition, one gram of each sediment sample was subjected to a sequential extraction procedure [12] to yield the following "operationally defined" chemical pools. The final volume of the prepared solutions was maintained 100 ml in each case.

*Exchangeable fraction ($F_1$)*: 1.0 g sediment sample was extracted at room temperature for 1 h with 10 ml of 1.0 M $MgCl_2$ (pH 7.0) with continuous agitation.

*Bound to carbonates ($F_2$)*: The residue obtained from step 1 was leached at room temperature with 10 ml of 1.0 M sodium acetate adjusted to pH 5.0 with acetic acid. The mixture was agitated for 5 h.

*Bound to Fe-Mn oxides ($F_3$)*: The residue obtained from step 2 was refluxed at 100°C with 20 ml of 0.04 M $NH_2OH.HCl$ in 25% (v/v) acetic acid for 6 h.

*Bound to organic matter ($F_4$)*: To the residue obtained from step 3, 3 ml of 0.02 M $HNO_3$ and 5 ml of 30% $H_2O_2$ were added and pH adjusted to 2.0 with $HNO_3$. The mixture was refluxed at 100°C for 2 h. A second 3 ml aliquot of 30% $H_2O_2$ (pH 2 with $HNO_3$) was then added and the sample was again refluxed for 3 h. After cooling, 5 ml of 3.2 M ammonium acetate in 20% (v/v) $HNO_3$ was added. The sample was diluted to 100 ml and agitated continuously for 30 min. The addition of ammonium acetate is designed to prevent adsorption of extracted metals onto the oxidized sediment.

*Residual ($F_5$)*: The residue obtained from step 4 was digested with a HF-$HClO_4$ mixture as per procedure followed for the analysis of trace elements.

All recovered supernatants were kept at 4°C before analysis. Metals in solution were subsequently determined by flame atomic absorption spectrometry (Ray Leigh Model WFX-210 AAS) using air-acetylene flame. 7 major elements were analyzed: Cu, Cd, Zn, Pb, Fe, Cr and Mn. Quantification of metals was based upon calibration curves of standard solutions of respective metals. Standard solutions were prepared from Merck. All wares used for metal analysis were cleaned with detergent, thoroughly rinsed with tap water, soaked in 10% $HNO_3$ solution overnight and finally rinsed with ultrapure water several times prior to use.

### 2.3. Water Treatment and Analysis

At each collection site, bottles were rinsed three times with the river water before filled. Each sample was analyzed for 20 parameters. The determination of pH, water temperature, electrical conductivity ($EC_w$), total dissolved solids was performed in situ using a Hach Model 44600 Conductivity/TDS meter. After arriving at the laboratory, water samples were filtered through Whatman 0.45µm pore diameter membrane filters and divided between two one-litre polyethylene bottles. Analytical parameters were determined with reference to official methods ([13], [14]). One bottle was immediately stored at 4℃, for the analysis of chloride, nitrate, nitrite, sulphate and phosphate and were determined using a spectrophotometer from Shimazdu (CTO-20A prominence column oven). Potassium and sodium were measured using the flame photometer model Sherwood, CL-420.The other polyethylene bottle was acidified with 2% $HNO_3$ and stored at 4℃ for trace metal analysis (Fe, Zn, Cd, Cr, Pb, Cu, Mn). Trace metals were determined by flame atomic absorption spectrometry (Ray Leigh Model WFX-210 AAS) using air-acetylene flame. All chemicals used in the study were obtained from Merck, and were of analytical grade. Working standards were prepared by dilution of stock solution (1mg metal/ml in 2% $HNO_3$) with ultrapure water and ultrapure water blanks were subjected to the same operations (storage, filtration, analysis) as real samples to control pollution along the sample manipulation stage. All the determinations were in triplicate and average values and their standard deviations were reported.

## 3. Results and Discussion

### 3.1. General Characteristics of River Water

All the conductivity values (Table 1) are over the limit for potable water production (1.000µs/cm) [15]. This indicated that we are dealing with waters with a high mineralization. The high conductivity at the different sites could be justified by the calcareous characteristics of this area and the calcareous wastes released. The pH of the river water is highly alkaline (7.2-8.9) (Table 1). These maximum values are related to the gypsum material released from the different types of industries and sewage and due to the agricultural run-off that is usually alkaline in nature due to the presence of ammoniacal compounds. The high concentration in TDS at the different sites (Table 1) is due to the release and the mobility of the particulate matter. For sodium, the water samples at the site D and E (945 and 1085mg/l, respectively) were above the level for human consumption [16]. The percentage of sodium ions is often taken as an important parameter in deciding the suitability of water for irrigation. The concentrations are over 100mg/l in 90% of the samples (Table 1) and it would be a potential sodium hazard for the crops if this water was used for irrigation. The major anions are sulphate and bicarbonate with the vast majority of samples has sulphate concentrations above 100mg/l (Table 1). The concentrations of bicarbonate expressed the alkalinity of the water which serves as a pH reservoir for inorganic carbon and is taken as an index of productivity potential of the water [17] and high value of alkalinity indicates a great ability of the river to support algal growth and other aquatic life. Chloride concentrations over 100mg/l were found in more than 80% of samples (Table 1). An excess of chloride in river water is usually taken as an index of pollution. Sewage water and industrial effluents are rich in chloride and hence the discharge of these waters results in high chloride levels in fresh water [18]. The nitrate in river water promotes high primary productivity and an excess of this ion in surface waters can be taken as a warning for algal blooms. The highest values are found at the site A (13.58 mg/l) and seems to be due to local run-off from the adjacent crop fields in this area where the farmers used nitrogen fertilizer.

In contrast, there is no indication of nitrate pollution at the different sites (2.34-13.58 mg/l) (Table 1). The nitrite ions are indicators of organic pollution on very spot and/or recent sources. The nitrite concentrations were over the level of tolerance of 0.5mg/l in all the samples.

Phosphate is present in natural waters as soluble phosphate and organic phosphate species. The concentration of the available phosphate is high at the different sites with the maximum value found at the site B (165mg/l) (Table 1) probably due to waste water containing detergents and agricultural activities. These concentrations are high enough to cause eutrophication, thus the organic pollution is very serious in the river. The reason for this level of pollution is the waste water input released all along the river. Table 1 lists the average values of heavy metals and physico-chemical parameters for the five sampling sites. A one-way analysis of variance ANOVA (Table 1) revealed no statistically significant differences in the order of each parameter and each metal among the study sites.

***Table 1.*** *Characteristics of Al-Ghadir River water at the different sites.*

| Parameter | A | B | C | D | E | Sig |
|---|---|---|---|---|---|---|
| pH | 7.20 | 6.40 | 7.10 | 7.80 | 8.90 | ** |
| TDS (mg/l) | 620 | 2000 | 1095 | 2000 | 2000 | ** |
| Conductivity (µs/cm)) | 1227 | 3999 | 2158 | 3999 | 3999 | ** |
| T (°C) | 18.50 | 24.80 | 24.30 | 25.30 | 25.0 | ** |
| Chloride (mg/l) | 63.37 | 3508.50 | 160.50 | 817.75 | 1193.5 | ** |
| Phosphate (mg/l) | 10.63 | 165 | 24.20 | 5.84 | 5.80 | ** |
| Nitrite (mg/l) | 2.14 | 2.14 | 0.28 | 1.17 | 1.21 | ** |
| Nitrate (mg/l) | 13.58 | 3.47 | 2.34 | 3.45 | 3.74 | ** |
| Sulphate (mg/l) | 103.14 | 3317.10 | 75.09 | 146.60 | 216.75 | ** |
| Potassium (mg/l) | 7.30 | 10.0 | 18.50 | 35 | 35.0 | ** |
| Sodium (mg/l) | 53.30 | 100 | 115.30 | 945 | 1085 | ** |
| Bicarbonate (mg/l) | 427.47 | 752.30 | 620.82 | 945.13 | 1025.72 | ** |
| Cd ($\mu g.l^{-1}$) | 3.04 | 3.65 | 2.83 | 3.48 | 4.11 | ** |
| Cr ($\mu g.l^{-1}$) | 62.80 | 41.80 | 25.30 | 53.30 | 53.70 | ** |
| Cu ($\mu g.l^{-1}$) | 23.74 | 73.85 | 29.18 | 63.35 | 74.12 | ** |
| Fe ($\mu g.l^{-1}$) | 506.30 | 613.20 | 609.70 | 516.60 | 559.70 | Ns |
| Mn ($\mu g.l^{-1}$) | 197.70 | 167.20 | 109.60 | 423.50 | 458.40 | ** |
| Pb ($\mu g.l^{-1}$) | 15.12 | 19.50 | 13.90 | 42.57 | 51.10 | ** |
| Zn ($\mu g.l^{-1}$) | 63.30 | 136.20 | 195.70 | 266 | 279.70 | ** |

** P<0.01; ns: not significant

### *3.1.1. Characteristics of the River Sediments*

The river sediments are made up of fine sand, silt and clay. The triangle of texture [19] whose three sides correspond to percentages of sand, silt and clay in a given soil is used to classify the studied sediments in the 5 sites under the classification of sediments with a clay texture for the sites A and D, clay-loamy texture for the sites B and E and sandy-clay-loamy texture for the site C. The fine silt and clay (<50µm fraction) content in sediment samples ranges from 15.12-23.12 and 30.15-51.25%, respectively (Table 2).

The content of limestone is, as expected, very high as in all alkaline soils. Active limestone represents the fractions of calcium carbonate present in the silts and clays [20]. The content of active limestone is high due to the high contents of silts and clays.

The pH of the river sediments varies from 7.52 to 8.25 indicating slightly alkaline nature of the sediments. The pH of the sediments and the soil being alkaline, heavy metals will tend to insolubilise hence reducing their toxicity toward microorganisms and delaying their toxicity to underground water. The pH measured in the sediments was shown higher than that in the water and this may be partly contributing to the increased concentration of heavy metal ion in the sediments due to precipitation of dissolved metals. The organic content of the sediment vary from 10.54 to 14.72% (Table 2). The maximum value of organic content is observed at site E (14.72%, respectively) which may be due to direct discharge of large amount of sewage and municipal solid wastes and the microflora susceptible to develop in these sediments is characteristic of organic sediments. This tendancy is confirmed by the enumeration of the total microflora on nutrient agar. The results obtained (Table 2) are characteristics of the sediments rich in microflora.

***Table 2.*** *Characteristics of Al-Ghadir River sediments at different sites.*

| Parameter | A | B | C | D | E | Sig |
|---|---|---|---|---|---|---|
| pH | 7.52 | 7.85 | 7.81 | 8.2 | 8.25 | * |
| Sand (%) | 25.6 | 41.12 | 50.62 | 40.12 | 44.87 | ** |
| Silt (%) | 23.1 | 20.62 | 19.20 | 15.12 | 16.37 | ** |
| Clay (%) | 51.25 | 38.25 | 30.15 | 44.75 | 38.75 | ** |
| Total limestone (g/kg) | 157 | 92 | 45 | 74 | 122 | ** |
| Active limestone (g/kg) | 16 | 10 | 7 | 13 | 15 | ** |
| Bulk density (g/cc) | 1.19 | 1.20 | 0.99 | 1.01 | 1.05 | Ns |
| Particle density (g/cc) | 2.48 | 2.24 | 1.94 | 1.90 | 1.87 | * |
| Porosity (%) | 52.04 | 46.26 | 49.00 | 46.85 | 43.78 | Ns |
| Organic matter (%) | 14.10 | 10.98 | 10.54 | 12.81 | 14.72 | ** |
| CEC (meq/kg) | 70 | 52 | 41 | 57 | 50 | ** |
| Cd (μg.g$^{-1}$ dry sediments) | 7.42 | 10.37 | 9.78 | 13.10 | 25.40 | ** |
| Cr (μg.g$^{-1}$ dry sediments) | 73.52 | 63.76 | 59.07 | 59.17 | 61.45 | Ns |
| Cu (μg.g$^{-1}$ dry sediments) | 57.42 | 107.40 | 44.02 | 131.05 | 157.90 | ** |
| Fe (μg.g$^{-1}$ dry sediments) | 2215.57 | 3227.07 | 3202.04 | 5266.12 | 5207.3 | ** |
| Mn (μg.g$^{-1}$ dry sediments) | 277.30 | 356.13 | 187.20 | 640 | 747.22 | ** |
| Pb (μg.g$^{-1}$ dry sediments) | 27.18 | 48.17 | 45.42 | 73.22 | 78.10 | ** |
| Zn (μg.g$^{-1}$ dry sediments) | 93.50 | 234.50 | 244.74 | 2.85 | 353.71 | ** |
| Bacterial count (CFU.g$^{-1}$ dry sediments) | $10^7$ | $10^7$ | $10^6$ | $10^7$ | $10^7$ | ** |

** P<0.01; * 0.05< P<0.01; ns p>0.05 (not significant)

Cation exchange capacity (CEC) allows to evaluate the amount of cations retained by the soil and this amount is related to the presence of negatively charged clay and organic soils colloïds [21]. The values found at the different sites are relatively high due to the high organic matter and clay contents (Table 2). This high CEC greatly increases the probability of adsorption of metallic cations on the clay-humus complex. It also corresponds to the values found in the soil with clay texture.

The mineralogical analysis of the sediments collected from the 5 sites showed the presence of large amounts of quartz and calcite in addition to the presence of different minerals and many contaminants. Most of the heavy metals showed maximum concentration at site E due to influx of large volume of domestic sewage and industrial discharges. Heavy metal adsorption was also found high at locations with high organic matter content. However, the distribution of heavy metals in the sediment is not uniform in the entire course of the river.

The variation in the total concentration of heavy metals may be due to differences in the sources of the heavy metals and prevailing physico-chemical conditions and complex reactions such as adsorption, precipitation and redox conditions that may be taking place in the sediments. In general, the concentration of organic bound fractions of heavy metals in the sediments has been high where there has been high organic matter.

Therefore, the organic fraction released in the oxidisable step is not considered very mobile or available because it may be associated with stable, high molecular weight humic substances that release small amount of heavy metals slowly [22]. Table 2 lists the average values of physico-chemical parameters and heavy metals for the 5 sampling sites in river sediments. A one-way analysis of variance ANOVA (Table 2) revealed no statistically significant differences in the order of each parameter and each metal among the study sites.

### *3.1.2. Heavy Metals in the River Sediments*

Figure 2 represents the variation of the total metal content in sediment and water along the sampling sites. The sediments existing at the bottom of the water column play a major role in the pollution scheme of the river systems [23]. They reflect the current quality of the water system and can be used to detect the presence of contaminants that do not remain soluble after discharge into surface waters. Moreover, sediments act as carriers and possible sources of pollution because pollutants are not permanently fixed by them and can be released back to the water column due to changes in environmental conditions [23]. Therefore, the analysis of pollutants in sediments permits to detect pollution that could escape to water system and also provides information about the critical sites of the water system under consideration [24]. For the spatial distribution of the heavy metals, results obtained in figure 2 showed that the trend of the heavy metals at the different sites of the river occurred to be the same as that of the metal concentrations in the associated water column where the maximum concentrations of these metals occurred downstream site of the river due to water impoundment for longer periods. These results let us to focus our next studies on downstream sites where the results obtained at these sites can be representative of the pollution present at the different sites of the river. These same results showed that the concentration of heavy metals in sediments are higher than those obtained in river water but a high positive correlation coefficient ($r^2$) was exhibited between the total metal concentrations in sediments and water among the different sampling sites for Pb, Zn, Cd and Mn and to a lesser extent for Fe, Cr and Cu suggesting that some kind of water-sediment interaction is taking place especially in the dry season when the river discharge is very low and where an interaction or reaction occurs between bed load sediments and water, thus an equilibrium between the bed load and water may be achieved ([6], [25], [26]). In addition, due to the very low discharge of the river water, it would be plausible to assume that the metal pollutants enter the river where the conditions approximately are that of the closed system, and then interact with sediment and with the existing water complexes and differentiate into different chemical species ([5], [27]). This positive correlation of metals in sediments and water suggests that their content in sediment is due to human activities (Pb, Zn, Mn). The occurrence of heavy metals in river water and sediments is due to discharge of industrial effluents from various sources including untreated sewage, municipal waste and agrochemical runoff from nearby industries directly into the river. Zinc was found to have its highest sediment content at site E in the industrial zone with content higher than the standard shale value (Table 3).

The other sites showed a lower content for Zn. Lead and Cd content in sediments were, at all sites, higher than the standard shale value, the highest appearing at sites D and E in the industrial zones. The higher values of Fe in this river may be attributed to geological source coupled with anthropogenic inputs from the catchment area.

***Table 3.*** *Comparison between metal contents of average shale (after Martin and Meybeck 1979) and results obtained.*

| Metal | Standard shale | Sites values |
|---|---|---|
| Zn (µg/g) | 95 | 234 (site B), 244 (site C)and 353 (site E) |
| Pb (µg/g) | 20 | 27(site A), 48 (site B), 45(site C),73(site D) and 78 (site E) |
| Cd (µg/g) | 0.3 | 7.42 (site A), 10.37 (site B), 9.78(site C), 13.10 (site D)and 25.4(site E) |

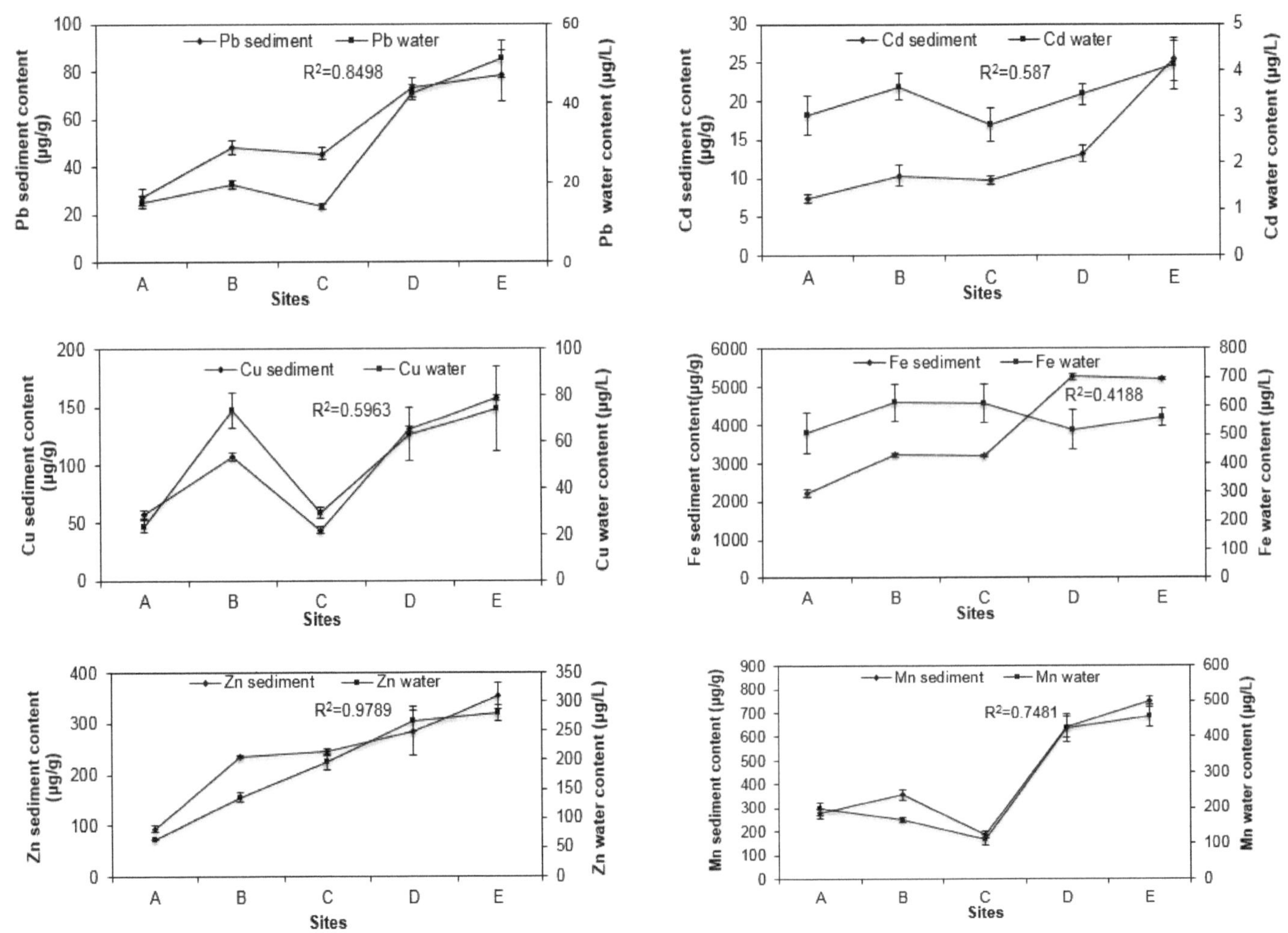

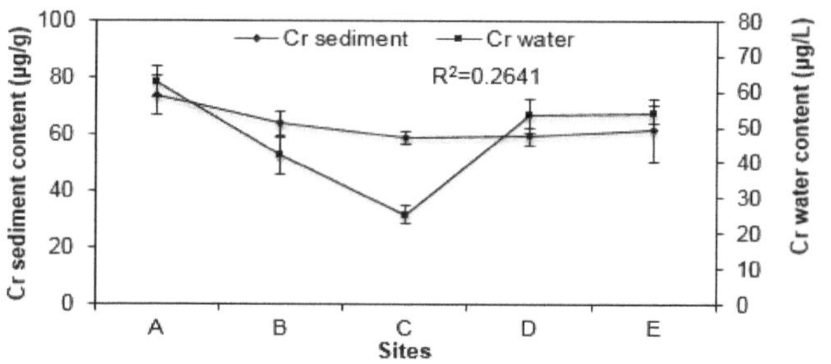


*Figure 2. Variation of total metal content in sediment and water along sampling sites ($R^2$ is correlation coefficient). Data columns and error bars represent mean (n=3) and S.D., respectively.*

### 3.2. Metal Fractionation

In this study, the sequential extraction procedure proposed by [12] has been used to obtain the five fractions. By studying the distribution of the metals between the different phases, their bioavailability and toxicity can be ascertained. The fractions introduced by man's activity include the adsorptive and exchangeable and bound to carbonates which are considered to be weakly bound and may equilibrate with aqueous phase thus becoming more rapidly bioavailable [28]. The Fe-Mn oxide and the organic matter have a scavenging effect and may provide a sink for heavy metals. The fractionation pattern of different metals in the bed sediments of the river at the different sites are shown in figure 3.

Iron was found the most abundant metal in all sediments because it is one of the most common elements in the earth's crust. Fractionation profile of iron in bed sediments of the river indicated that a high portion (>50%) is associated with the residual fraction (Figure 3). This is consistent with a natural source for Fe in the river sediments. The chemistry of Fe can explain the speciation of Fe in the reducible and organic matter sediment fractions and its small concentration in the carbonate fraction. Iron occurs in the oxide and hydroxide forms under alkaline oxidizing conditions, while the formation of Fe carbonates needs reducing conditions [29]. The samples represented the oxidized surface sediments. The assumption of an oxic sediment surface is based on the natural setting of the river basin during the dry season and the depth at which sediments were collected [30].

Manganese, which is also abundant in nature, behaves in a way similar to that of iron. The fractionation profile of manganese indicates that it is mostly bound to residual followed by reducible fractions, the rest being associated with organic matter, carbonate and exchangeable fractions. The high percentage of Mn total content in the residual sediment fraction (Figure 3) are ascribed to the natural source of Mn in the sediment (similar to speciation of Fe pattern). Fractionation profile of copper indicated that a high portion is bound to the organic matter fraction. As reported in earlier studies on polluted sediments ([30], [31], [32], [33]) extractable copper is mainly associated with oxidizable phase, where it is likely to occur as organically complexed metal species. This behavior can be explained by the well-known high affinity of copper to humic substances, which are a fraction of natural organic matter and chemically very active in complexing such metals [34]. Under an oxic water column, a significant fraction of the copper reaching the sediment surface may be returned to the overlying water column by mineralization of the host organic material at the surface and in the oxic upper layers of the sediment. A high portion of cupper was found associated with the exchangeable fraction indicating the dominance of anthropogenic sources through atmospheric deposition and municipal discharges.

The concentration of total cadmium was found to vary from 7.42 to 25.4 µg.g$^{-1}$ at the different stations. The fractionation profile of cadmium indicated that a high portion of cadmium is associated with carbonate fraction and a lower association with the reducible fraction (Figure 3). This shows an important role for sediment carbonates in the deposition of Cd irrespective of a higher content in other sediment fractions. The strong association of Cd with the carbonate fraction is most probably owing to the similarity of the ionic radii of Cd (0.97 A°) and Ca (0.99 A°), since Cd will co-precipitate with carbonates and enter the calcite crystal as a camouflaged element ([35], [36], [37]).

The concentration of lead was found to vary from 27.18 to 78.10 µg.g$^{-1}$ in the sediments of the river. The high association of Pb in the reducible fraction is due to the higher stability constant of Pb-oxides than Cd-oxides [38]. Furthermore, the variation in speciation of Cd and Pb in the carbonate and in the reducible fractions parallel that reported by [39] and [40] in alkaline media. Cadmium associates mainly with the carbonate fraction whereas Pb associates with the carbonate fraction for lower Fe oxides. [41] reported that the percentage of the total Pb in the carbonate fraction ranged (30–40%) when the iron oxide was less abundant in sediment. Others ([42], [43]) have reported that the reducible fraction is dominant for Pb speciation when the carbonates constituted a minor portion in sediment.

This important role for the carbonates would not have been observed if the sediments had been unpolluted. Studies of unpolluted sediments [44] in carbonate basins have reported low metal speciation (Pb and Cd) in the carbonate sediment fraction, the speciation of Pb and Cd was mainly residual and this speciation was a reflection of catchment geology. Al-Ghadir river sediments were found highly contaminated with Pb and Cd metals and the percentages of both metals in the residual sediment fraction are lowest.

The speciation pattern of Zn in the sediments in some respects paralleled Cd speciation pattern. The highest (~30–40% total Zn sediment content) was in the carbonate fraction, followed by the reducible (~20– 30%) fraction. At the sites A and D, the total Zn sediment contents were lower than the average standard shale value, whereas total Zn sediment contents at sites B, C and E were higher than the average standard shale value (near the industrial zone) (Table 3). These sites have a low concentration of Zn total content in the residual fraction and high concentration in the carbonate fraction.

This observation supports a role for carbonates in scavenging Zn introduced into the river through pollution.

The nearly equal percentages of the total Zn content in the carbonate and the reducible fractions (Figure 3) are a consequence of the limestone-dominated river basin ([5], [45], [46]). Therefore, a water chemistry dominated by carbonate species implies that carbonates precipitated on to sediment particles have a major role in adsorbing Zn. In this aspect, Zn showed the parallel speciation pattern in sediments with those of Pb and Cd speciation pattern. The carbonate sediment fraction of the river which is produced mostly from the industrial effluents is affected less by environmental changes, such as microbial sediment activity. Whereas, metals bound to reducible (Fe oxides and hydroxides) and oxidized (organic) sediment fractions have a potential for remobilization due to microbial activity [43].

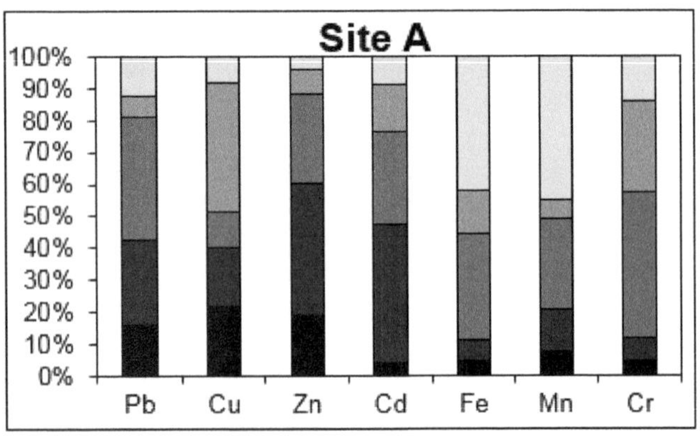


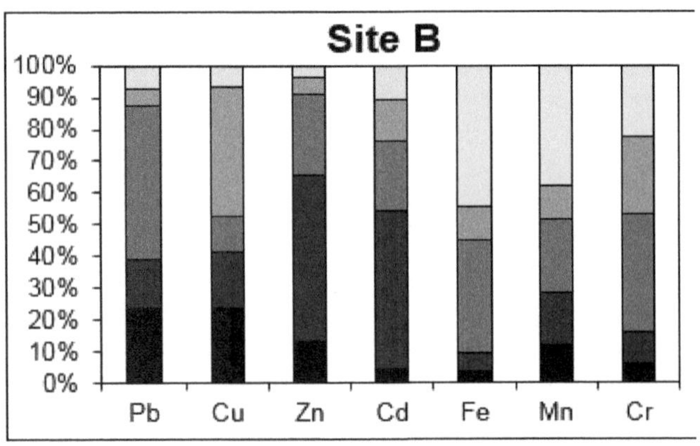


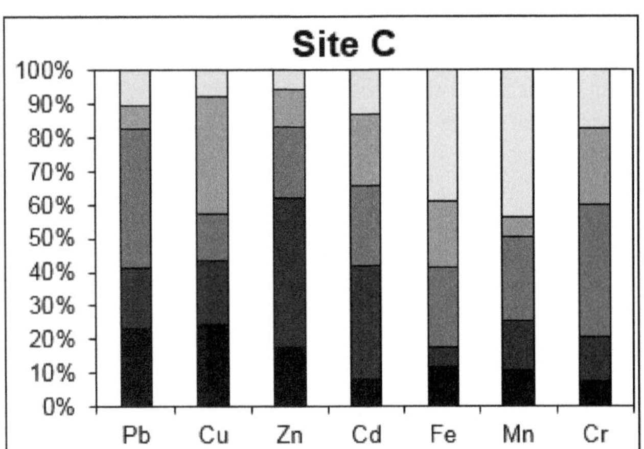


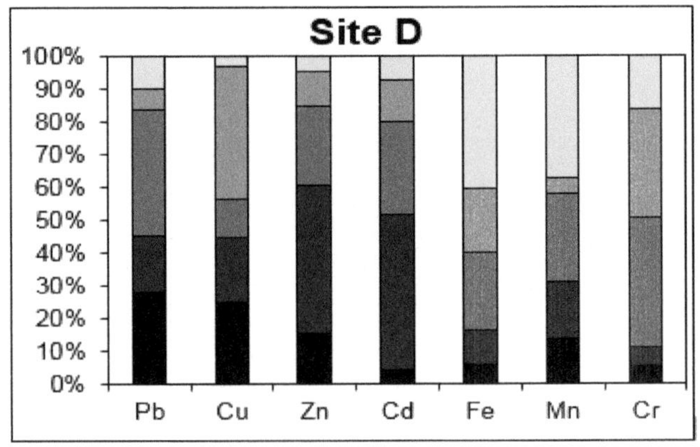


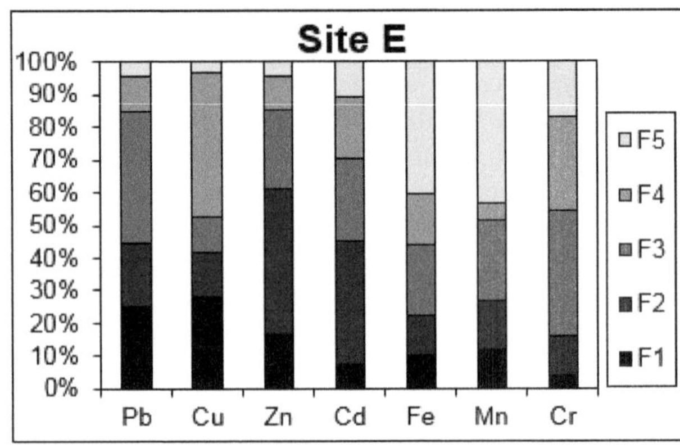


***Figure 3.*** *Heavy metal fractionation in bed sediments of Al-Ghadir river ($F_1$-Exchangeable; $F_2$-Bound to carbonates; $F_3$- Bound to Fe-Mn oxides; $F_4$-Bound to organic matter; $F_5$-Residual).*

In the sediment samples of the river, the concentration of total chromium was found to vary from 59.07 to 73.52 $\mu g.g^{-1}$. The fractionation profile of chromium indicated that more than 50 % of chromium is associated with the reducible and the organic matter fractions at most of the stations.

### *3.2.1. Environmental Impact*

Normally, reducing microorganisms in sediment are able to reduce insoluble Fe (III) oxides to soluble Fe (II) oxides and thereby release metals associated with oxide phase. Whereas, oxidizing microorganisms in sediments might oxidize the organic sediment constituent and expel the metals associated with the organic fraction [5].

Results of fractionation of heavy metals showed that 5 to 10 % of some metals (Pb, Cd and Cr) have been mobilized from the reducible phase.

In our case, wastewaters which contain large amounts of organic matter are continuously released at the different sites and may result in a continuous readsorption of heavy metals. What happens is that when wastewaters are released, for example between the source (point A) and point B, we remark that the sequence of reduction in the sediments is similar to that reported by [47]. At point B, when a new quantity of organic matter is released, this release may affect the sequence of reduction [47] and can poise the soil Eh for several days where autochthonous bacteria consume the newly released organic matter as a source of carbon [48] and release another type of organic matters (organic acids). The increase in the specific area of the degraded organic matter may explain the raise of absorption or adsorption of the heavy metals to the newly released organic matter.

Between the points B and C, a new typical reduction sequence will takes place. At the point C, when wastewater is newly released this release can poise the soil Eh for several days during which reduction will be stopped and the released metals will be adsorbed to the newly released organic matters and the cycle continuous at each site. This alternation is of great importance for the immobilization of heavy metals in and can be planned and used to reduce the toxic effect of mobilized metals at sites of high contamination.

#### 3.2.2. Geoaccumulation Index ($I_{geo}$)

The geoaccumulation index ($I_{geo}$) developed by [49] is used to determine the quantitative extent of metal pollution in aquatic sediment using the following expression

$$I_{geo} = \log_2 C_n / 1.5 B_n$$

Where $C_n$ is the concentration of the examined metals in the sediment; $B_n$ is the geochemical background value of a given metal in shale (average shale) [50] and 1.5 is the background matrix correction factor. The factor 1.5 is used for possible variation in the background data due to lithogenic effects. The

$I_{geo}$ consists of seven grades (0 to 6), indicating various degrees of enrichment above the background values ranging from unpolluted to very highly polluted sediment quality. Table 4 presents the geoaccumulation index for the quantification of heavy metal accumulation in the river sediments.

The following description classification is given for the index of geoaccumulation by [51] as follows:

<0 Practically unpolluted
0-1 Unpolluted to moderately polluted
1-2 Moderately polluted
2-3 Moderately to strongly polluted
3-4 Strongly polluted
4-5 Strong to very strongly polluted
> 5 Very strongly polluted

The $I_{geo}$ for the river sediments in this river varies from metal to metal and place to place. Chromium, manganese and iron remain in class 0 (Table 4). The $I_{geo}$ for lead and zinc is found to be 0 at the sites A, B and C and 1 at the sites D and E. The $I_{geo}$ for copper is found to be 0 at the first 4 sites and increases to 1 at the last site.

*Table 4. Geoaccumulation index of Al-Ghadir sediments.*

| Sites | A | B | C | D | E |
|---|---|---|---|---|---|
| Cd | 4 | 4 | 4 | 4 | 5 |
| Cr | 0 | 0 | 0 | 0 | 0 |
| Cu | 0 | 0 | 0 | 0 | 1 |
| Fe | 0 | 0 | 0 | 0 | 0 |
| Mn | 0 | 0 | 0 | 0 | 0 |
| Pb | 0 | 0 | 0 | 1 | 1 |
| Zn | 0 | 0 | 0 | 1 | 1 |

Throughout the mainstream, Cd remains in class 4 except at the site E where it falls in class 5. Based on the geoaccumulation indices (Table 4), the river sediments from the site A to E are considered to be unpolluted with respect to chromium, copper, iron, manganese, lead and zinc. In case of cadmium, it varies from moderately to highly polluted which may be attributed to the atmospheric deposition of vehicular and industrial pollutants.

## 4. Conclusion

This study showed that during the dry season, Al-Ghadir River whose drainage basin is characterized by a high aqueous pH and alkalinity, possess a self purification capacity for metal pollution primarily through an association of the metal with the carbonate species and to a certain extent with the organic matter species. This was demonstrated by the speciation of the metals in the sediments where Fe and Mn exhibited a high percentage association with the residual fraction followed by the Fe/Mn oxide fraction, Cu showed its high percentage association with the oxidizable fraction, Cd and Zn exhibited a high percentage occurrences as carbonate species which assumedly reduces the toxic effect of the free hydrated ion, and finally Pb and Cr showed their high percentage association with the oxidizable fraction. The association of Pb and Cr and the other metals but to a lesser extent with the Fe/Mn oxide fraction may play an important role in limiting the toxic metal content. Furthermore, work is needed to investigate the important role of the continuously released organic materials in forming adsorbed organic metal species, which might reduce the toxic effects of pollutant metals and their mobilisation toward underground water.

## References

[1] Jurdi, M. (1992). 'A National Study on the Quality of Potable Water in Lebanon', in *Proceedings of the National Workshop of the Status of Water in Lebanon*, Beirut, Lebanon: pp. 145–173(in Arabic).

[2] Jurdi, M. (1998). Follow up National Environmental Surveillance, 1996, Unicef Publications, Beirut (in Arabic).

[3] Khair, K.; Aker, N.; Haddad , F. ; Jurdi, M. and Hachach, A. (1994). The environmental impacts of humans on ground water in Lebanon', *Water, Air Soil Pollut.* 78: 37–49.

[4] Sene, K..J.; Marsh, T.J. and Hachache, A. (1999). An assessment of the difficulties in quantifying the surface water resources of Lebanon. Hydrological Sci. J. 44 (1):79–96.

[5] Korfali, S.I. and Davies, B.E. (2000). Total and extractable trace elements in Lebanese river sediments: dry season data. Environ. Geochem. Health 22: 265-273.

[6] Chapman, D. (1992). Water Quality Assessments. University Press, Cambridge, UK.

[7] Lewis, D.W. and McConchie, D.M. (1994). Analytical Sedimentology. Chapman and Hall, New York, USA.

[8] APHA, AWWA, WEF. (1995). Standard Methods for the Examination of Water and Wastewater, 19th ed. American Public Health Association/America Water Works Association/ Water Environment Federation. Washington, DC.

[9] Chapman ,H.D. (1991). Métodos de análisis de suelos, plantas y aguas. Mexico DC: Trillas Editor.

[10] Danielson, R.E. & Sutherland, P.L. (1986). Porosity. In: A. Klute (Eds) *Methods of soil analysis-Part1: Physical and mineralogical methods*. Second Edition. SSSA Inc.z : 443-462.

[11] Fytianos, K. & Lourantou, A. (2004). Speciation of elements in sediment samples collected at lakes Volvi and Koronia, N. Greece. Environment International, 3: 11-17.

[12] Tessier, A. ; Campell, P.G.C. & Bisson, M. (1979). Sequential extraction procedure for the speciation of partition of particulate trace metals. Analytical Chemistry, 51: 844–851.

[13] American Public Health Association (APHA). (1992). *Métodos normalizados para el análisis de aguas potables y residuales*, 17th edn. Madrid, Spain: Díaz de Santos Chapman, D., 1992.Water Quality Assessments. University Press, Cambridge, UK.

[14] Rodier, J. (1981). *Análisis de auas*. Barcelona, Spain: Omega Printer

[15] EEC. (1975). Directive 75/440/EEC. Require quality of superficial waters for production of drinking water. Off. J.Europ. Econ. Comm., No., L, 193 (16.06.75).

[16] EEC. (1980). Directive 80/778/EEC. Quality of Waters destined for human consumption. Off. J. Europ. Econ. Comm., No.L, 229 (30.08.80).

[17] Manahan, S.E. (1994). *Environmenntal chemistry*. USA: Lewis Public, CRC Rodier, J. (1981). *Análisis de aguas*. Barcelona, Spain: Omega Printer.

[18] Hasalan, S.M. (1991). *River pollution –An ecological perspective*. Great Britain: Belhaven Lewis, D.W., McConchie, D.M., 1994.Analytical Sedimentology. Chapman and Hall, New York, USA.

[19] Alexander, M. (1977). Introduction to soil microbiology. New-York: J. Wiley & Sons: 467.

[20] Duchaufour, Ph. (1995). Pédologie – sol, végétation, environnement. Abrégés, 4ème édition. Masson: 324.

[21] Cherier, G. (1991). Infiltration-percolation des eaux résiduaires. Thèse de doctorat, Sciences de l'eau, Université Paris VI: 129.

[22] Singh, S.P.; Tack, F.M. & Verloo, M.G. (1998). Heavy metal fractionation and extractability in dredged sediment derived surface soils. Water, Air and soil Pollution, 102: 313-328.

[23] Förstner, U. (1985). Chemical forms and reativities of metals in sediments. In Leschber, R., Davis, R. D. & L'Hermie, P. (Eds.), Chemical methods for assessing bioavailable metals in sludges and soils. London: Elsevier: 1-30.

[24] Förstner, U. & Salomons, W. (1980). Trace metal analysis on polluted sediments. Part I: Assessment of sources and intensities. *Environmental Technology Letters*, 1: 494-505.

[25] Salomons, W. and Förstner, U. (1984). Metals in the Hydrocycle. Springer-Verlag, Berlin.

[26] Piron, M.; Pineau, A. and Mabele, R.M. (1990). Sediment parameters and distribution of metals in fine sediments of the Loire esturary. Water, Air, Soil Poll. 50: 267-277.

[27] Korfali, S.I. (1999). Metal concentrations in The Nahr-Ibrahim River, Lebanon, PhD. Thesis, University of Bradford, UK.

[28] Gibbs, R.J. (1977). Transport phases of transition metals in the Amazon and Yukon rivers. *Geological Society of America Bulletin*, 88: 829-843.

[29] Guo, T.Z.; Delaune, R.D.; Patrick, W.V. (1997). The effect of sediment redox chemistry on solubility—chemically active forms of selected metals in bottom sediments receiving produced water discharge. Spill Sci. Technol. Bull. 4: 165–175.

[30] Gambrell, R.P. (1994). Trace and toxic metals in wetlands–a review. J. Environ. Qual. 23: 883–891.

[31] Samanidou, V. & Fytianos, K. (1987). Partitioning of heavy metals into selective chemical fractions in sediments from rivers in northern Greece. *The Science of the Total Environment*, 67: 279-285.

[32] Pardo, R.; Barrado, E. and Castrillejo, Y. *(*1993). Study of the contents and speciation of heavy metals in river sediments by factor analysis. Analytical Letters, 26: 1719-1739.

[33] Marin, B. ; Valladon, M. ; Polve, M. *et al.* (1997). A reproducibility testing of a sequential extraction scheme for the determination of trace metal speciation in a marine reference sediment by inductively coupled plasma mass spectrometry. Analytic Chimica Acta, 342: 91-112.

[34] Pempkowiase, J. ; Sikora, A. & Biemacka, E. (1999). Speciation of heavy metals in marine sediments vs. their bioaccumulation by Mussels. Chemosphere, 39(2): 313-321.

[35] Reeder, R.J. (1996). Interaction of divalent cobalt, zinc, cadmium, and barium with the calcite surface during layer growth. Geochim. Cosmochim. Acta 60: 1543–1552.

[36] Nilsson, O. and Sternbeck, J. (1999). A mechanistic model for calcite growth using surface speciation. Geochim. Cosmochim. Acta 63: 217–255.

[37] Ianni, C.; Magi, E. ; Rivaro, P. and Ruggieri, N. (2000). Trace metals in Adriatic coastal sediments: distribution and speciation pattern. Toxicol. Environ. Chem. 78: 73–92.

[38] Ramos, L.; Hernandez, L.M. and Gonzalez, M.J. (1994). Sequential fractionation of copper, lead, cadmium and zinc in soils from near Donana National Park. J. Environ. Qual. 23: 50–57.

[39] O'Day, P.A.; Carrol, S.A. and Waychunas, G.A. (1998). Rock–water interactions, controlling zinc, cadmium and lead concentrations in surface waters and sediments, UST ri-state mining district. 1. Molecular identification using X-ray absorption spectroscopy. Environ. Sci. Technol. 32: 943–955.

[40] Carrol, S.A.; O'Day, P.A. and Piechowski, M.P. (1998). Rock–water interactions, controlling zinc, cadmium and lead concentrations in surface waters and sediments. Environ. Sci. Technol. 32: 956–965.

[41] Stone, M. and Droppo, I.G. ( 1996). Distribution of lead, copper and zinc in size-fractionated river bed sediments in two agricultural catchments of southern Ontario, Canada. Environ. Pollut. 93: 353–362.

[42] Baruah, N.K.; Kotoky, P.; Bhattcharyya, K.G. and Borah, G.C. (1996). Metal speciation in Jhanji river sediments. Sci. Total Environ. 193: 1–12.

[43] Jones, B. and Turki, A. (1997). Distribution and speciation of heavy metals in surficial sediments from the Tees Estuary, Northeast England. Mar. Pollut. Bull. 34: 768–779.

[44] Kļaviņš, M.; Briede, A.; Rodi, V.; Kokorite, I.; Parele, E. and Kļaviņš, I. (2000). Heavy metals in rivers of Latvia. Sci.Total Environ. 262:175–183.

[45] Gonzalez, M.J. ; Ramos, L. ; Hernandez, L.M. (1994). Distribution of trace metals in sediments and relationship with their accumulation in earthworms. *Int. J. Environ. Anal. Chem.* 57:135–150.

[46] Surija, B. and Branica, M. (1995). Distribution of Cd, Pb, Cu, and Zn in carbonate sediments from Krka river estuary obtained by sequential extraction. Sci. Total Environ. 170: 101–11.

[47] Ramesh, R. and Ronald, D. (2008). Biogeochemistry of Wetlands: Science and Applications: 442-467.

[48] Mcheik, A.; Fakih, M.; Bousserrhine, N; Toufaily, J.; Garnier-Zarli, E. and Hamieh, T. (2013). Biomobilization of heavy metals from the sediments affect the bacterial population of Al-Ghadir River (Lebanon). *Agriculture, Forestry and Fisheries.*Vol. 2, No. 3:116-125.

[49] Muller G. (1979). Schwermetalle in den sedimenten des rheins-Veranderungen seit. Umschau, 79: 778-783.

[50] Turekian, K.K. and Wedepohl, K.. (1961). Distribution of the elements in some major units of the earth's crust. Bulletin of geological society of America. 72: 175-92.

[51] Förstner, U.; Ahalf, W.; Calmano, W. and Kersten M. (1990). Sediment Criteria Development –Contributions from Environmental Geochemistry to Water Quality Management', in: D. Heling, P. Rothe, U. Forstner and P. Stoffers (eds), *Sediments and Environmental Geochemistry: Selected Aspects and Case Histories*, Springer-Verlag, Berlin Hidelberg: 311-338.

# Modeling Cheetah *Acinonyx jubatus* Fundamental Niche in Kenya

**Bernard Mungoma Kuloba[1], Hein Van Gils[2], Iris Van Duren[2], Shadrack Muvui Muya[3], Shadrack Mumo Ngene[4, *]**

[1]Kenya Wildlife Service, Biodiversity Research and Monitoring, Nairobi, Kenya
[2]Faculty of Geo-Information Science and Earth Observation (ITC), University of Twente, Enschede, the Netherlands
[3]Department of Zoology, Jomo Kenyatta University of Agriculture and Technology, Nairobi, Kenya
[4]Kenya Wildlife Service, Parks and Reserves, Nairobi, Kenya

**Email address:**
sngene@kws.go.ke (S. M. Ngene)

**Abstract:** The cheetah is currently recognized by IUCN as a threatened species. Kenya is one of country with significant cheetah population in the world though it fundamental niche is not well known. Wildlife can live in an area only if basic resources such as food, water, and cover are present and if the species is adapted in ways that allow them to cope with the climatic extremes, selection involves several levels of discrimination and spatial scales and a number of potentially interacting factors. This study was to model cheetah fundamental niche using suitable environmental predictors and evaluates suitability of current protected area coverage in its conservation. Two types of model input data used were cheetah occurrence locations and a suite of environmental variables thought to have a direct physiological role in limiting the ability of the species to survive. The species occurrence records and environmental variables were entered into a MaxEnt model which uses maximum entropy algorithm to identify environmental conditions that are associated with species occurrence. Ideal fundamental niche for cheetah were found to be localities within an elevation range of 1600-2100 meters above sea level, receiving mean annual precipitation of 800 mm, with the warmest quarter of the year receiving 250 mm, the wettest month receiving 120 mm, precipitation of coldest quarter 10 mm and temperature seasonality ranges of 150°. Cheetah fundamental niche run across protected areas and for effective conservation, Results could be used to direct conservation effort go beyond parks and reserves by encouraging community conservancies and development of ecological corridors. Protected area planning could benefit too from these results.

**Keywords:** Cheetah, Fundamental Niche, Protected Areas, Community Conservancies

## 1. Introduction

The fundamental niche of a species is the set of environmental conditions within which a species can survive and persist. The fundamental niche may be thought of as an 'n-dimensional hyper-volume', every point in which corresponds to a state of the environment that would permit the species to exist indefinitely (Hutchinson, 1957). The fundamental niche describes conditions suitable for survival of the species, and is thus of great importance for conservation. It is also referred to as potential or physiological niche, which denotes the full range of conditions under which the species actually does best hence potentially could be found. It can also be used to estimate the species' ecological niche, for example by removing areas where the species is known to be absent because of con-specifics (Butler 1980), inter-specific competitors (Werner and Hall 1979), predators (Werner et al. 1983), human settlement or any other habitat destruction. Although a species' realized distribution may exhibit some spatial correlation, the potential distribution does not, so considering spatial correlation is not necessarily desirable during species distribution modeling.

An animal's habitat is, in the most general sense, the place where it lives. Wildlife can live in an area only if basic resources such as food, water, and cover are present and if the animals have adapted in ways that allow them to cope with the climatic extremes and the competitors and predators they encounter (Morison 2006). Habitat selection involves several levels of discrimination and spatial scales and a number of potentially interacting factors.

Ecological niche is realized or actual observed abundance distribution of a species. Usually will not include the full range of conditions under which the species could potentially be found and, in particular, may not include the physical conditions in which the species actually does best. The observed distribution of a species may differ from its fundamental niche if the species is excluded from the conditions in which it does best by competition, predation or human activities. Modeling of these factors with species presence data will result in a wealth of information about why and where we find a species. This is referred to as species distribution modeling (Guisan & Zimmermann 2000).

Species distribution models are empirical models relating field observations to environmental predictor variables based on statistically or theoretically derived response surfaces (Guisan & Zimmermann 2000). They estimate the relationship between species records at sites and the environmental and/or spatial characteristics of those sites (Franklin, 2009). Actual or potential geographic distribution of a species is estimated best by characterizing the environmental conditions that are suitable for the species, and to then identify where suitable environments are distributed in space. The environmental conditions that are suitable for a species may be characterized using either a mechanistic or a correlative approach. Mechanistic models aim to incorporate physiologically limiting mechanisms in a species' tolerance to environmental conditions. It require detailed understanding of the physiological response of species to environmental factors and are therefore difficult to develop for all but the most well understood species.

Correlative models aim to estimate the environmental conditions that are suitable for a species by associating known species' occurrence records with suites of environmental variables that can reasonably be expected to affect the species' physiology and probability of persistence. The central premise of this approach is that the observed distribution of a species provides useful information as to the environmental requirements of that species. Since spatially explicit occurrence records are available for a large number of species, the vast majority of species' distribution models are correlative, hence its use this research. Two types of model input data are needed: 1) known species' occurrence records; and 2) a suite of environmental variables thought to have a direct physiological role in limiting the ability of the species to survive.

The model can be used to predict species' occurrence in areas where the distribution is unknown. Thus, a set of environmental variables for the area of interest is input into the model and the suitability of conditions at a given locality is predicted. In many cases the model is used to 'fill the gaps' around known occurrences (e.g., Anderson et al., 2002a; Ferrier et al., 2002). In other cases, the model may be used to predict species' distributions in new regions (e.g. to study invasion potential, for review see Peterson 2003) or for a different time period (e.g. to estimate the potential impacts of future climate change, Pearson and Dawson, 2003).

Several recent studies comparing up to 16 of these approaches indicated that maximum entropy (MaxEnt) modeling performed as well or better than the other approaches (Elith, 2006, Hernandez, 2006 and Phillips 2006). As such, MaxEnt model was chosen to model and understand the suitable habitats for cheetah in Kenya.

Cheetah is a threatened species and to save it, one first needs to know where the species prefers to live, and what its requirements are for survival, i.e., its fundamental niche (Hutchinson, 1957). Studies in Serengeti, Tanzania shows cheetah inhabiting large, overlapping ranges averaging 833 km²(Caro, 1994). This large area requirement frequently brings them into contact with humans as they may not be contained in national parks and reserves (Woodroffe & Ginsberg 2000). In Kenyan rangelands, predators including cheetah, may kill livestock and are therefore themselves killed by local pastoralists (Woodroffe, 2006). Gros, 1997 estimated total number of 793 cheetahs in protected areas of Kenya. Based on prey availability Masailand (Kajiado and Narok districts) and the dry northern Districts appear to offer the best prospects for cheetah conservation in Kenya based on relative prey availability (Gros,1997). Contrary to the prediction that prey biomass can be used as a predictor of cheetah numbers, a Serengeti study found that cheetahs frequently use areas with low prey density (Durant 1998). By selecting a variety of prey species the cheetah can survive in a range of conditions and can adapt to changes in the environment by changing hunting style (Frame and Frame 1977, Caro 1994). We aim to design a predictive cheetah distribution model, identify the set of most suitable environmental predictors of cheetah occurrences, analyze suitable areas for cheetah conservation in Kenya, and evaluate suitability of current protected area coverage in cheetah conservation.

## 2. Materials and Method

### *2.1. Study Area*

The study was undertaken in Kenya, located in East Africa, which has a total area of 582,650 km$^2$. Kenya borders the Indian Ocean to the east, with a coastline of 536 kilometers, Somalia to the northeast, Ethiopia to the north, Southern Sudan to the northwest, Uganda to the west, and Tanzania to the south. It is located between latitudes 5°40' N and 4°4' S and between longitudes 33°50' and 41°45' E (Figure 1).

The country has a tropical climate, hot and humid at the coast, temperate inland and very dry in the north and northeast parts of the country. The long rains occur from April to June and short rains from October to December. The hottest period is from February to March and coldest in July to August. About 10% of Kenya area is set been set aside as either national or reserves for wildlife conservation, first national Nairobi national park having been established in 1946.

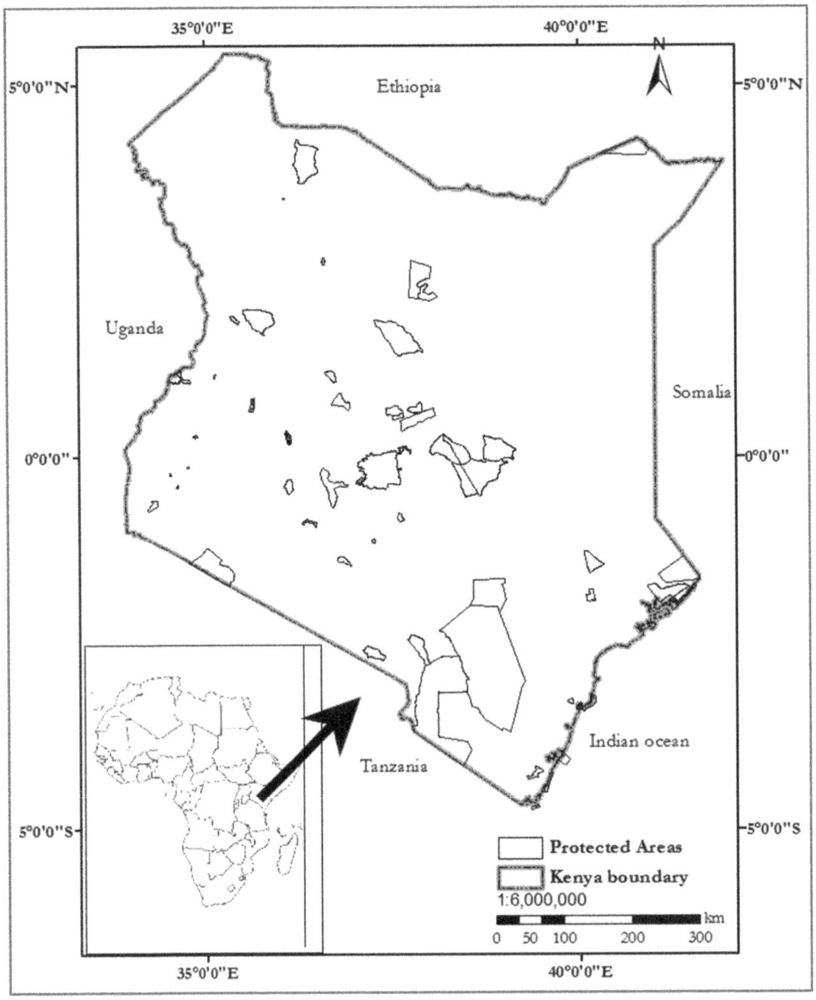


***Figure 1.*** *A map showing the study area.*

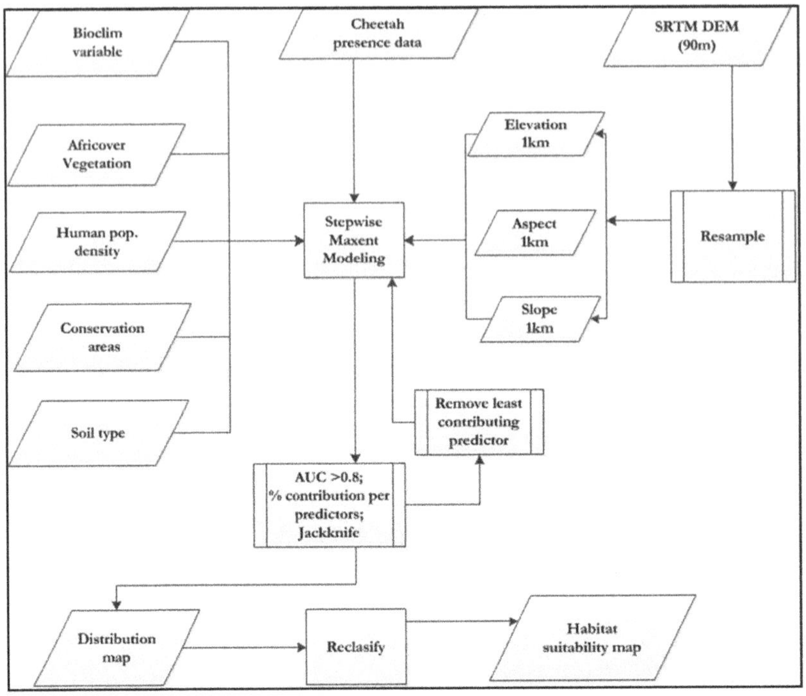


***Figure 2.*** *Flowchart of the modeling process using MaxEnt Model.*

### *2.2. Research Work Flow*

MaxEnt model require two types of data input: biological data species geo-referenced locations, describing the species' distribution, and environmental data, a suite of environmental variables of Vegetation cover types, Soil type, Bioclimatic data, slope, elevation, Conservation areas and human population density, describing the landscape in which the species is found (Figure 2).

### *2.3. The Target Species*

Cheetah (*Acinonyx jubatus*) is a medium-large cat, with females weighing an average of 35.9 kg and males 41.4 kg (Caro 1994). Genetic analysis suggests that cheetah cats have diversified in three main lineages, with the cheetah (*Acinonyx jubatus*) occupying the pantherine lineage together with lions, tigers and lynxes (Caro 1994). With the exception of a small population in Iran (Jackson 1998) there is scant evidence of cheetahs outside Africa. The main remaining populations exist in southern and eastern Africa (Caro 1994).

Cheetahs are diurnal and hunt during the day (Caro 1994). Various strategies are employed including stalking, approaching prey in full view and flushing hidden prey from long vegetation (Caro 1994), all utilizing the cheetah's exceptionally high running speed. Preferred food of cheetahs varies with location. For example, on the Serengeti plains, Thomson's gazelles are the preferred prey, whilst impala are favoured in South Africa and puku in Zambia (Caro 1994) although most preferred species weigh less than 40 kg (Schaller 1972a).

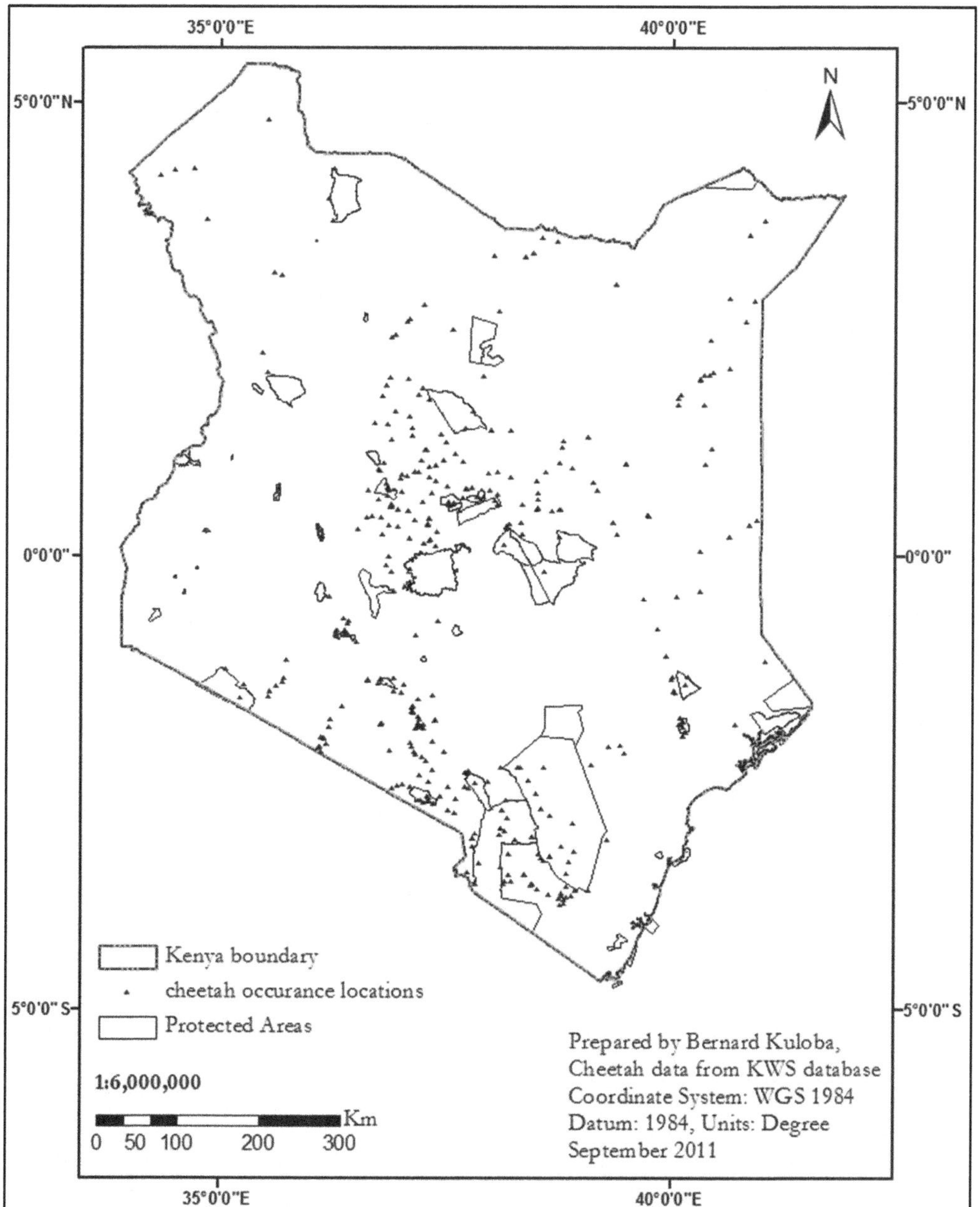


***Figure 3.*** *Map showing spatial data on cheetah species known occurrence in Kenya (2004-2007).*

Cheetah is currently recognized by IUCN as a threatened species (IUCN 2010). The current conservation status of the cheetah (*Acinonyx jubatus*) in Kenya is a controversial issue. Because cheetahs are elusive and typically roam over large home ranges, it is difficult to discern whether the species is actually rare or merely rarely seen, let alone if it has recently become scarce, (Gros, 1997). Over the past 30 years, the fear that cheetah numbers could be declining in the country instigated several surveys of the cheetah's distribution and abundance, (Graham and Parker, 1965; Myers, 1975; Hamilton, 1986).

### *2.4. Species Observation Data*

309 geo referenced cheetah presence data points in the entire cheetah range in Kenya were used in this research. The data was collected during a 2004- 2007 joint national survey by Action for Cheetah in Kenya, Cheetah Conservation Fund, East African Wildlife society and Kenya Wildlife Service. A field interview survey was used, where data were collected principally by means of detailed interviews. This was done by driving to villages, farms, ranger's outpost and protected areas throughout the country to make direct contacts with informants.

### *2.5. Environmental Predictors*

Table 1 below provides a summary of the environmental variables used in the survey including the source of the environmental variable data.

***Table 1.*** *Suite of potential environmental variables used in modeling.*

| Environmental Variables | Units | Data type | Spatial | Source |
|---|---|---|---|---|
| Soil type | types | Categorical | Raster | www.africover.org |
| Vegetation cover | types | Categorical | Raster | www.africover.org |
| Bioclimatic data | °C and mm | Continuous | Raster | www.worldclim.org |
| Slope | Degree | Continuous | Raster | www.worldclim.org |
| Aspect | Degree | Continuous | Raster | www.worldclim.org |
| Elevation | M.a.s.l | Continuous | Raster | www.worldclim.org |
| Conservation areas | 2 units | Categorical | Vector | www.kws.org |
| Human Pop.density | 3 units | Categorical | Vector | www.wri.org |

### *2.6. The Maxent Model*

*Modeling approach:* Two types of data used were known cheetah species geo referenced occurrence locations and a suite of environmental variables of Vegetation cover types, Soil type, Bioclimatic data, slope, elevation Conservation areas and human population density. The occurrence points and environmental variables were entered into maximum entropy algorithm that aims to identify environmental conditions that are associated with species occurrence using MaxEnt software Version 3.3.3e, (Phillips et al. 2005). Fifty individual MaxEnt models were run in batch mode with the following settings: Auto features (feature types are automatically selected depending on the training sample size), perform jackknife tests, logistic output format, random test percentage = 30, regularization multiplier = 1, maximum iterations = 1000, convergence threshold = 0.0001, and maximum number of background points =10,000. This default setting was used because it has been validated in studies over a wide range of species and environmental conditions, and found to be robust to unknown prevalence, easier to interpret (Phillips, 2008).

*Determining the most suitable environmental variable:* While the model was being trained, track of environmental variable with highest gain when used in isolation to the model were noted. Each step of the MaxEnt algorithm increases the gain of the model by modifying the coefficient for a single feature; the program assigns the increase in the gain to the environmental variable(s) that the feature depends on. Jackknife test was used to assess the importance of each variable. Following this approach, the least important variable from the full model was determined and dropped from subsequent models. Then a new model was constructed with all predictor variables except the previously dropped variable. The weakest of the remaining variables was again determined using the jackknife test and was dropped from subsequent models. This procedure was continued until only minimum variables remained out of the initial 21 with at least AUC equal or more than 0.8. AUC ranges from 0.5 for models that are no better than random to 1.0 for models with perfect predictive ability. Phillips et al. (2006) demonstrated how the test can be applied using randomly selected 'pseudo-absence' records in lieu of observed absences. In this case, AUC tests whether the model classifies presence more accurately.

*Model evaluation:* The accuracy of this model was tested to determine its relevance, through receiver operating characteristic (ROC) plots, ROC plot is a plot of sensitivity and 1–specificity, with sensitivity representing how well the data correctly predicts presence, whereas specificity provides a measure of correctly predicted absences. Available data is partitioned into calibration and test datasets. The relative proportions of data included in each data set are somewhat arbitrary, and dependent on the total number of locality points available 70% for calibration and 30% for testing is common, (Huberty, 1994).

*Mapping probability of presence:* The point-wise mean image of the 50 output grids at 95 % confidence interval was generated from the MaxEnt model with probability of presence ranging from 0 to 1. The map represents probability of finding cheetah in any given location in the country, where 0 probability indicates absence while one 1 presence is certain given the prevailing environmental predictors. The image uses colours to indicate variation in predicted probability of presence. The map range from high probability of suitable conditions for the species, to typical conditions of those where the species is found and to shades indicating low predicted probability of suitable conditions.

*Suitability of protected areas:* The predicted conditions

within protected areas were determined. The point wise image was reclassified into three classes, highly suitable where probability of presence is greater than 0.5, suitable where probability of presence is greater than 0.2 but less than 0.5 and not suitable habitat where probability of presence is less than 0.2. The country protected area coverage boundary was overlaid on the predicted probability map. The three classes within protected areas were determined. The effectiveness of protected area coverage was determined by estimating proportion of suitable areas within protected areas and how much of each class within protected areas. The suitable areas outside the protected areas were determined by subtracting areas within protected areas from total area generated by the model.

# 3. Results

### *3.1. The Full Model*

Figure 4 below summarizes results of the full model. The jackknife test variable of importance, the environmental variable with highest gain when used in isolation was soils, which therefore appears to have the most useful information by itself. The environmental variable that decreased the gain most when it is omitted is precipitation of warmest quarter of the year (bio_180), which therefore appears to have the most information that isn't present in the other variables. Values shown are averages over 50 replicate runs. Although soil predictive power was the best it was not used in the subsequent modeling because its resolution was low at 3km² compared to the rest of other environmental variables of 1 km². Re-sampling other variables to 3km would have lead to loss of details.

### *3.2. Model Selection*

The environmental variable with highest gain when used in isolation was annual precipitation (bio_12), which therefore appears to have the most useful information by itself (Figure 5). The environmental variable that decreases the gain the most when it is omitted is elevation, which therefore appears to have the most information that isn't present in the other variables in this model.

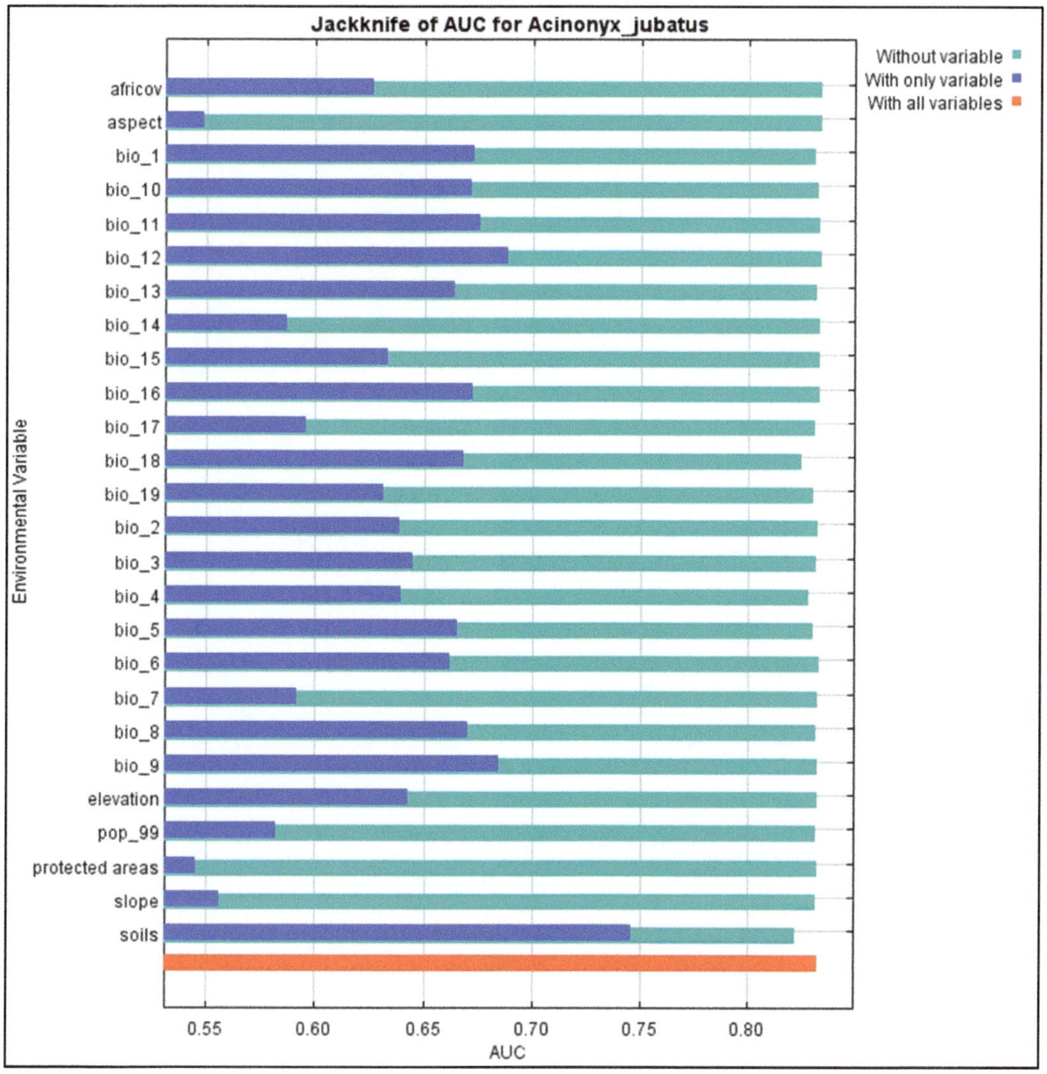


***Figure 4.*** *Initial full model with all 26 potential environmental variables.*

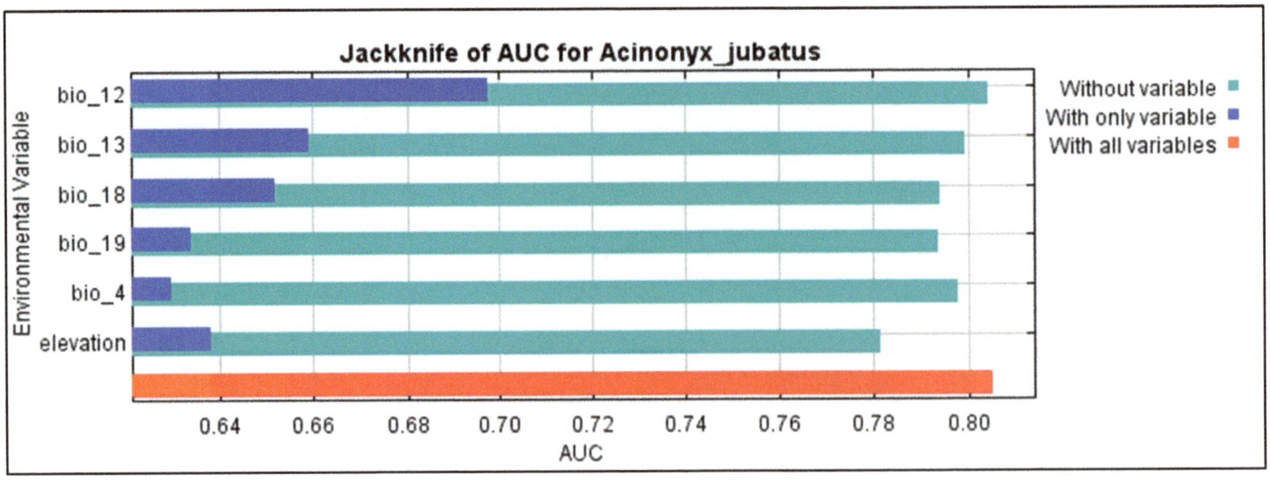


***Figure 5.*** *Cheetah distribution predictive model using the most suitable environmental variables.*

Values shown are averages over 50 replicate runs. Other variable in the model were precipitation of the wettest month (Bio13), temperature seasonality (bio4) and precipitation of warmest month (bio18). Temperature seasonality appears to contribute little but when left out the model AUC fell under 0.8 indicating prediction was no better than random.

### *3.3. Analysis of Environmental Variable Contributions*

Table 2 below gives estimate of relative contributions of each environmental variable to the model. Values shown are averages of 50 replicate runs. 70% of influence of the model was from precipitation, with precipitation during the warmest quarter of the year contributing about 28%. Elevations contributed 17% whereas temperature seasonality contributed below 10%. Relationship between precipitation and temperature were very important as seen from precipitation of warmest and coldest quarter contributing 50% of the total. In terms of permutation importance precipitation of the coldest quarter was the highest with about 40% followed by elevation. Precipitation of the wettest month of the year was the least.

The six final environmental variable used are shown in figure 6 below with their ranges, indicating Kenya to have varied environmental conditions. Annual precipitation range from 172-2625mm, precipitation of the wettest month of the year 42-697, precipitation of warmest quarter 23-1011, precipitation of coldest quarter 0-617,elevation -1-4725masl and temperature seasonality 48-172°.

***Table 2.*** *Contribution and permutation importance of each variable in the model.*

| Environmental variable | Percent contribution % | Permutation importance % |
|---|---|---|
| Precipitation of warmest quarter(bio_18) | 26.8 | 14.4 |
| Precipitation of coldest quarter (bio_19) | 23.7 | 39.3 |
| Elevation | 17.4 | 21 |
| Annual rainfall (bio_12) | 12.3 | 2.9 |
| Temperature seasonality (bio_4) | 10.4 | 11.9 |
| Precipitation of wettest month (bio_13) | 9.4 | 10.6 |

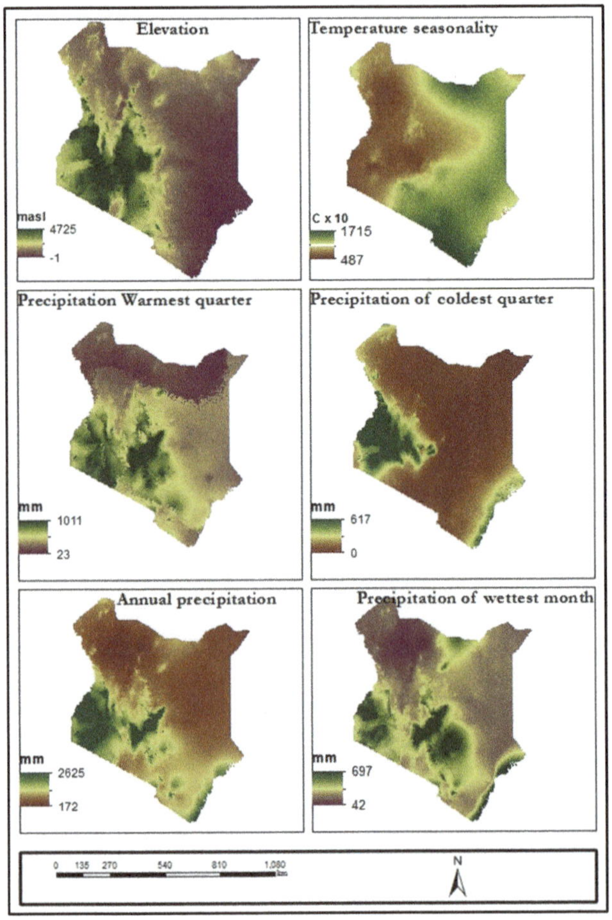


***Figure 6.*** *Environmental variables with highest influence on cheetah distribution.*

### *3.4. Model Evaluation*

Test omission rate and predicted area as a function of the cumulative threshold, averaged over the replicate 50 runs. The omission rate is close to the predicted omission, because of the definition of the cumulative threshold. The omission rate is very close to the predicted omission hence the model is good (Figure 7).

Figure 8 below shows receiver operating characteristic (ROC) curve of the model for the same data, averaged over the 50 replicate runs. The specificity is defined using predicted area, rather than true commission. The model was also evaluated using the receiver operating characteristic (ROC) curve, averaged over the 50 replicate runs. Specificity was defined using predicted area, rather than true commission. The average test AUC for the 50 replicate runs is 0.803, and the standard deviation is 0.015, which was above 0.8. AUC range is 0.5-1.0, with values between 0.5-0.7 indicating low discriminate ability, values between 0.7-0.9 indicating moderate discriminate ability, and values > 0.9 indicating high discriminate ability (Pearce and Ferrier 2000, Manel *et al.*, 2001). The AUC value of 0.8 is an indication that the model is good.

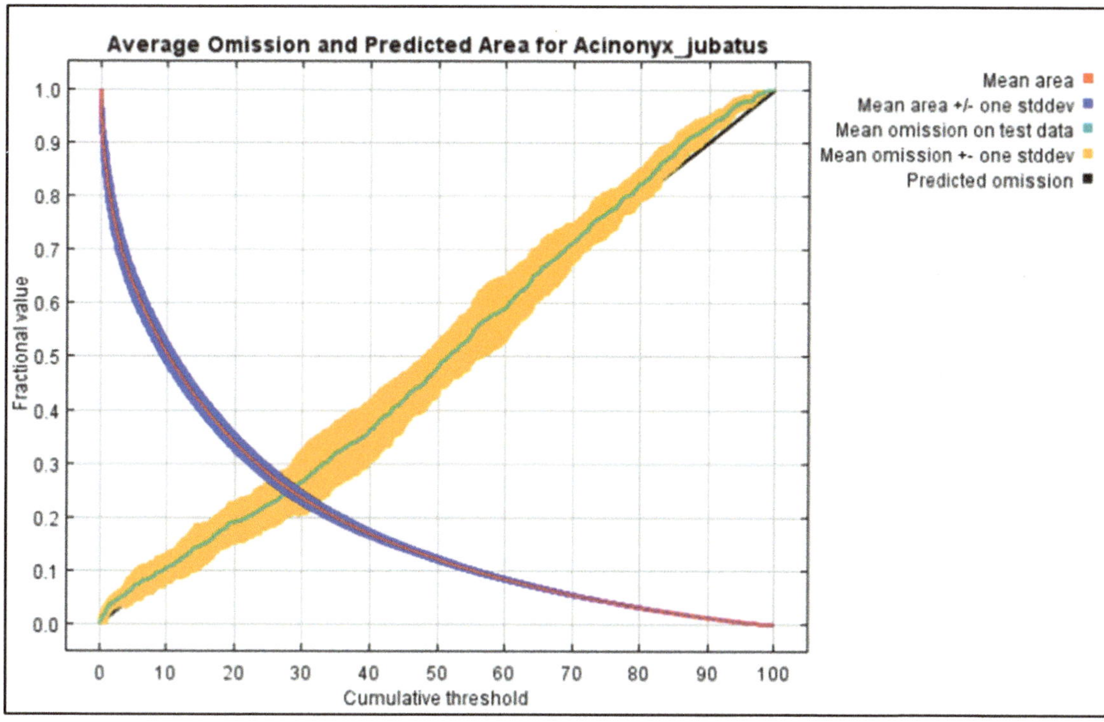


***Figure 7.*** *Test omission rate and predicted area as a function of the cumulative threshold.*

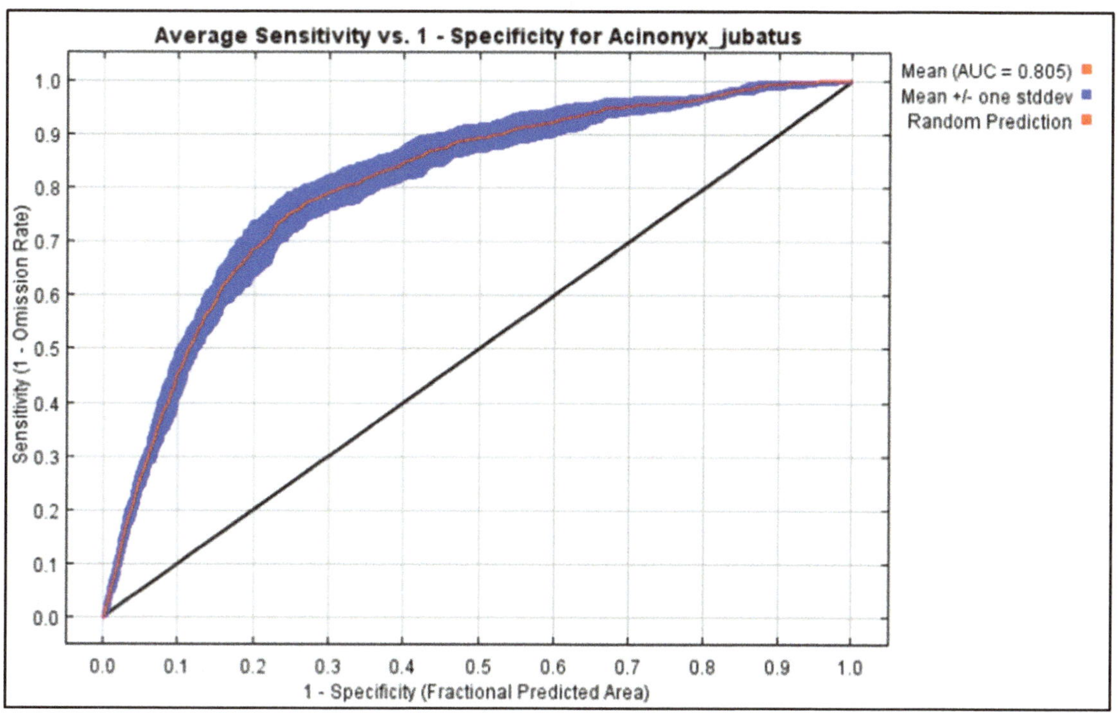


***Figure 8.*** *The receiver operating characteristics (ROC) of the model.*

### 3.5. Suitable Environmental Conditions

***Table 3.*** *Optimum performance level of each variable in the model.*

| Variable | Kenya | Fundamental niche |
|---|---|---|
| Precipitation of Coldest Quarter (Bio_19) | 0 – 600mm | 10- 110 mm |
| Precipitation of the Warmest Quarter (Bio_18) | 20 – 700mm | 160- 320mm |
| Precipitation of Wettest Month (Bio_13) | 40 - 500mm | 120mm |
| Annual Precipitation (Bio_12) | 200 – 2600mm | 650-800mm |
| Temperature Seasonality (Bio_4) | 50 - 170 | 70-90, 130-150° |
| Elevation | 0 – 4000m.a.s.l | 0-2100m.a.s.l |

Cheetah distribution is predicted to occurs at localities of an elevation ranging from 0-2100m above sea level, with climatic condition of mean annual precipitation of 650-800mm, where precipitation of the warmest quarter is between 160-320mm, precipitation of the wettest month is 150mm, precipitation of coldest quarter is 10-110mm and temperature seasonality ranges of 70°-90° or 130°-150° (Table 3).

### 3.6. Suitable Areas for Cheetah Observations

Figure 9 below shows areas where the probability of sighting cheetah in Kenya is high. The open savannah ecosystems are the main locations where the probability of sighting cheetah is high.

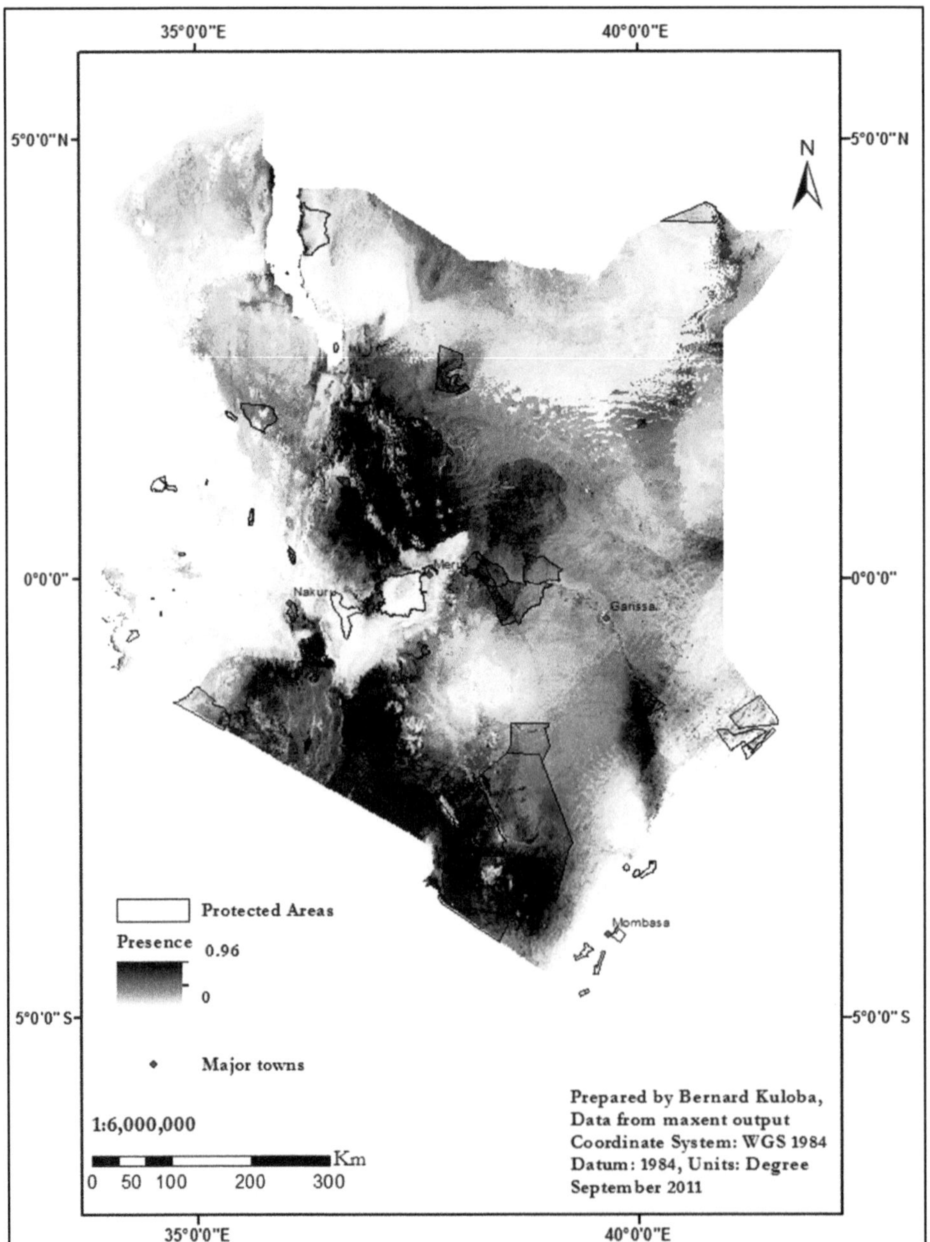


***Figure 9.*** *Spatial distribution of areas where there is high probability of sighting cheetahs in Kenya.*

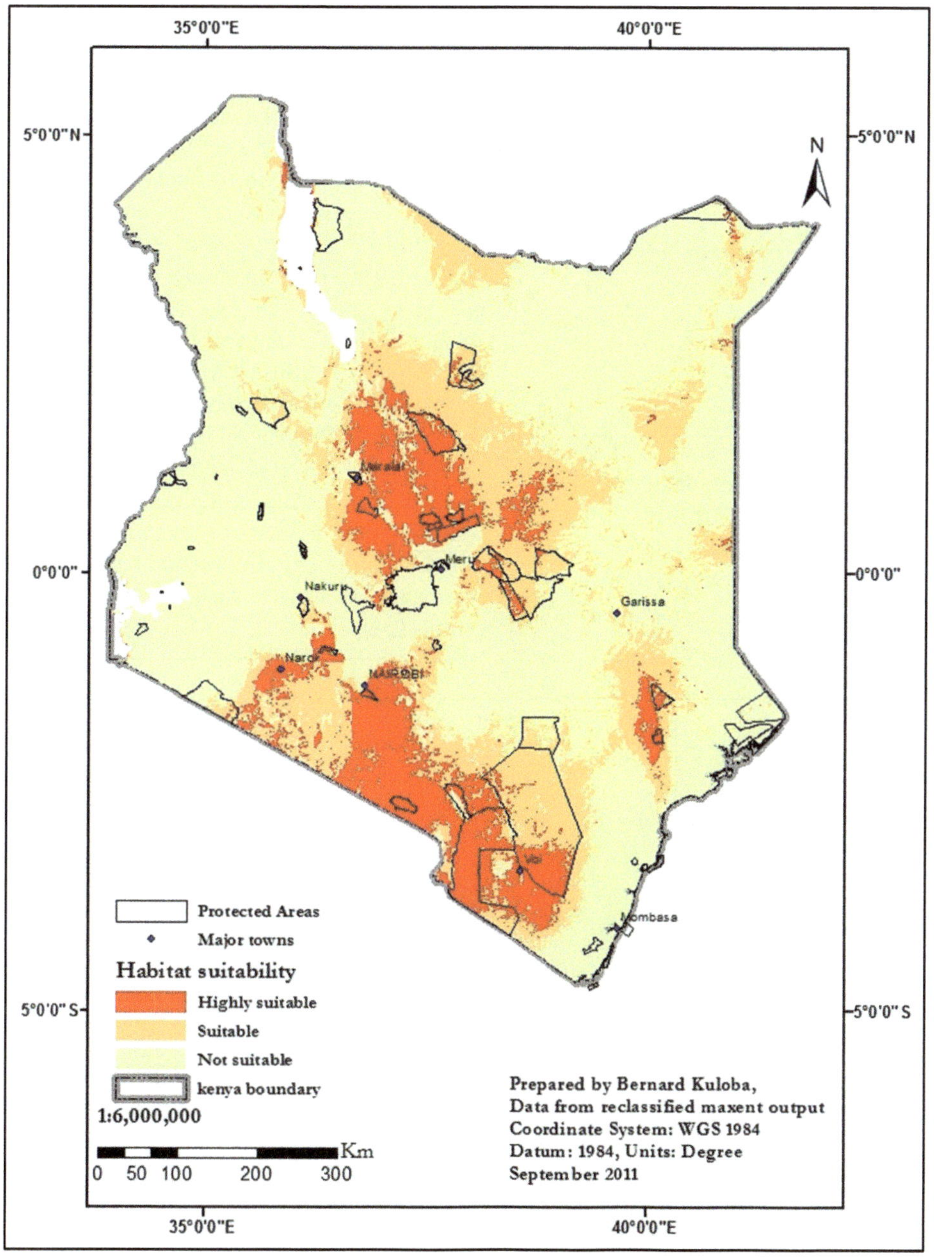


***Figure 10.*** *Suitable habitats for cheetah in Kenya.*

The suitable areas for cheetah are located in two areas in the country, central part around Laikipia-Samburu-Isiolo and in southern part along Kenya-Tanzania border, which include the Masai Mara ecosystem, Amboseli ecosystem, and the Tsavo Conservation areas (Figure 10).

Three areas are identified as key from the distribution map, Masai Mara-Amboseli-Machakos-Tsavo, Samburu-laikipia-Isiolo and TanaRiver. Much of the highly suitable habitats are outside protected area coverage though in close proximity. Amboseli and Tsavo west are predicted to have best environment for cheetah conservation.

### *3.7. Suitability of Cheetah Conservationin Current Protected and None Protected Areas in Kenya*

11.6% of Kenya is highly suitable for cheetah conservation, while 22.8% is categorized as suitable (Table 4). In total 40% of Kenya has suitable environment for cheetah conservation, in contrast about 10% of the country has been set aside as parks or reserves. For area under parks and reserves 31.4% is considered highly suitable whereas 43% is suitable. A quarter of protected areas are not suitable for cheetah conservation (Table 4; Table 5).

***Table 4.*** *Suitable cheetah habitat distribution in Kenya.*

| Habitat category | Country (%) | Protected areas (%) | None protected areas (%) |
|---|---|---|---|
| Not suitable | 65.48 | 25.60 | 69.64 |
| Suitable | 22.88 | 43.00 | 21.12 |
| Highly suitable | 11.64 | 31.40 | 9.24 |
| Area % | 100.00 | 100.00 | 100.00 |

*__Table 5.__ Proportion of land in Kenya showing the amount which is under parks and reserves.*

| Habitat category | Country (%) | Protected areas (%) | None protected areas (%) |
|---|---|---|---|
| Not suitable | 380,142.55 | 11,843.88 | 367,719.46 |
| Suitable | 132,845.04 | 19,853.86 | 111,527.61 |
| Highly suitable | 67,548.61 | 14,510.37 | 48,762.72 |
| Area size | 580,536.21 | 46,208.11 | 528,009.79 |

# 4. Discussion

### *4.1. Cheetah Distribution*

Cheetah distribution could be classified as found in Mara conservation, Meru, Samburu-Isiolo-Marsabit and Tsavo conservation areas (Figure 11). The distribution spread across the protected areas. The study used both intrinsic or natural ecological variables such as climate and landform and extrinsic variables related to human like population densities and land conservation status. Extrinsic variables like population and conservation status did not explain the model well. Protected areas are found in different parts of the country at different elevation hence some were not suitable either because of the amount of precipitation, temperature seasonality or elevation. Important environmental features that are expected to influence the presence of cheetahs include vegetation cover; the presence of predators and the availability of water and prey (Caro, 1994; Bissett & Bernard, 2007), in this model presence of other predators and prey base were not use but used primary environmental variables. In Narok district the highly suitable areas were found outside Masai Mara national reserve. Gros 1997 explained high cheetah density on community ranches compared to the Masai Mara National Reserve in terms of predation of cheetah cubs by large predators within the reserve. The density in the reserve was found to be half that occurring on adjacent rangelands, result from this study could explain in environmental suitability unlike the disparity in terms of lower large predator numbers control by pastoralists.

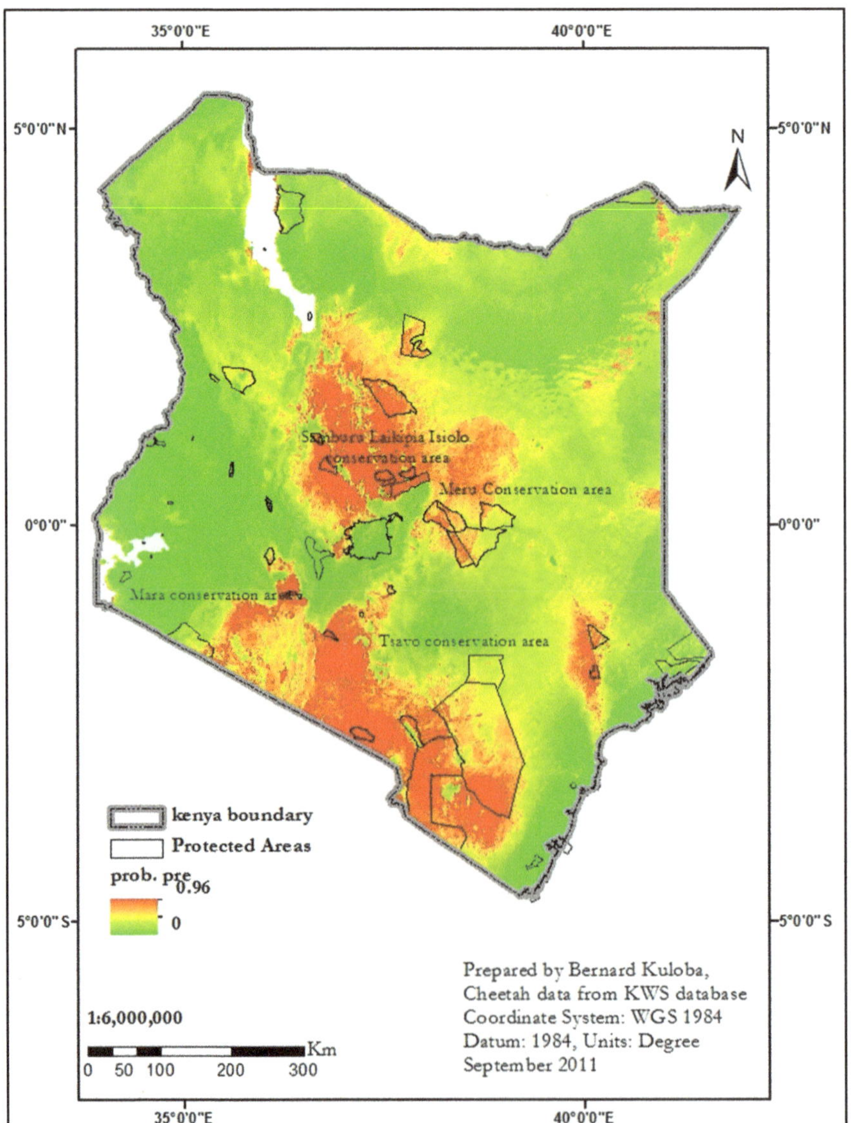


*__Figure 11.__ The distribution of cheetah in Kenya.*

### 4.2. Predictive Model

The predicted model was statistically robust given the AUC of 0.805. AUC range is 0.5-1.0, with values between 0.5-0.7 indicating low discriminate ability, values between 0.7-0.9 indicating moderate discriminate ability, and values > 0.9 indicating high discriminate ability (Pearce and Ferrier 2000, Manel *et al.*, 2001). The distribution of cheetah could not be detected on basis of whether areas is protected or not protected as the difference was not significant. Species distribution model works well when presence data cover geographic and ecological extent of the species. The extent of data used here compares well with past surveys which covered the same districts of the country (Gros 1997). MaxEnt uses background environmental data for the entire study area. It focuses on how the environment where the species is known to occur relates to the environment across the rest of the study area (background). One important factor with MaxEnt is the occurrence localities are also included as part of the background.

The model will be better if depending on degree to which the species is at 'equilibrium' with current environmental conditions. A species is at equilibrium with the physical environment if it occurs in all suitable areas, while being absent from all unsuitable areas. The degree of equilibrium depends both on biotic interactions, competitive exclusion from an area) and dispersal ability, cheetah with higher dispersal ability are expected to be closer to equilibrium than animals with lower dispersal ability (Araújo and Pearson, 2005).

The extent to which observed occurrence records provide a sample of the environmental space occupied by the species. In cases where very few occurrence records are available, perhaps due to limited survey effort (Anderson and Martinez-Meyer, 2004) or low probability of detection (Pearson et al., 2007), the available records are unlikely to provide a sufficient sample to enable the full range of environmental conditions occupied by the species to be identified. The study used 309 points compared to the minimum of 20 point for MaxEnt to work effectively (Philips, 2008).

### 4.3. Environment Condition Suitable for Cheetah

Cheetah distribution was predicted to occurs at localities of an elevation ranging from 0-2100m above sea level, with climatic condition of mean annual precipitation of 650-800mm, where precipitation of the warmest quarter is between 160-320mm, precipitation of the wettest month is 150mm, precipitation of coldest quarter is 10-110mm and temperature seasonality ranges of 70°-90° or 130°-150°. Snyman (2001) documented precipitation, temperature and elevation as primary factors influencing African vegetation or phytomass, with soil composition and precipitation accounting for 66% of the variation in phytomass. Gastone *et al.* (2009) while studying lion demography in different protected area across Africa showed that climatic parameters explained 62% of overall variance, whereas landscape features explained 32%; climatic parameters were fairly balanced between the effects of temperature (34%) and rainfall (28%).

This model relied more on intrinsic or natural ecological factors such as climate and landform. However, extrinsic factors related to human interference [trophy hunting, poaching (Whitman et al., 2007; Frank *et al.*, 2006), agricultural and urban development (Bourn & Blench, 1999; Herrmann, 2004), habitat conversion, civil conflicts (Siefert, 2005) severely impact on cheetah.

### 4.4. Management Implications

*Communication Tool:* The distribution maps can be used as a tool to equip conservationists with a powerful communication tool in their negotiations with land and livestock owners. Suitable areas range will be related well with farms hence local people will appreciate why they need to tolerate and conserve cheetah on their farms. This might be important to the long-term survival of cheetahs if they are to occupy much of their fundamental niche, which is outside the protected areas. Resources for wildlife survey is very limiting, using this distribution map, available resources can be directed to high potential. This will focus the surveys and might yield a better result. Contrary to the prediction that prey biomass can be used as a predictor of cheetah numbers, a Serengeti study found that cheetahs frequently use areas with low prey density (Durant 1998). By selecting a variety of prey species the cheetah can survive in a range of conditions and can adapt to changes in the environment by changing hunting style (Frame and Frame 1977, Caro 1994).

*Protected Area planning framework*: Conservation of wildlife in Kenya is tourism based, this has allowed infrastructure and tourism facilities development within parks and reserves. To ensure balance between conservation and tourism activities cheetah distribution map should be one of a guiding tool in location of these facilities. Parks could be zoned, where developed is prohibited in highly suitable areas. Over 80% of the predicted suitable areas are found outside protected areas, this means that conservation activities outside protected areas are absolutely critical to ensure that these habitats. This result could be used to guide in setting up of community conservancies. This could be done by mobilizing communities whose land is suitable for conservation. KWS cheetah conservation strategy recognizes the need to establish carnivore conservation zones outside protected areas (KWS, 2007).

*Ecological corridors:* Three core areas were identified in this study for cheetah conservation, Narok-Kajiado areas, Tsavos and Laikipia- Samburu-Isiolo. These areas could be connected through ecological corridors. Such initiative will allow cheetah movements. Studies in Serengeti have shown, cheetah inhabiting large, overlapping ranges of an average 833km² (Caro 1994). The distribution maps could be used by relevant authorities to support identification and prioritization of corridor and dispersal areas for improved connectivity of cheetah movement.

# 5. Conclusion

Suitable areas for cheetah conservation were established by means of predictive habitat modeling using MaxEnt. Ideal fundamental niche for cheetah were localities within an elevation range of 1600-2100 meters above sea level, receiving mean annual precipitation of 800 mm, with the warmest quarter of the year receiving 250 mm, the wettest month receiving 120 mm, precipitation of coldest quarter 10 mm and temperature seasonality ranges of 150°. Comparison of protected areas versus non protected areas could not reveal any difference in cheetah distribution because cheetah niche cut across the protected areas boundaries. Cheetah fundamental niche run across protected areas and for effective conservation, results could be used to direct conservation effort go beyond parks and reserves by encouraging community conservancies and development of ecological corridors. Protected area planning could benefit too from these results.

# Acknowledgements

Firstly, I would like to acknowledge and thank my two supervisors Dr. Hein van Gils who closely guided me develop a conceptual research focus on use of MaxEnt model and critical review of my initial research concept. Dr. Iris van Duren for guiding me through research report writing by critical review of my work and giving very constructive advice. Dr. A.G. Toxopeus for the dramatic way he introduced Species distribution modeling especially MaxEnt model that captured my imagination and interest into the modeling world. To Director Kenya Wildlife Service for granting study leave to Bernard Kuloba to undertake this work and permission to use data from the organization for my academic work. We thank all KWS scientists (Bernard Ngoru, Lekishon Kenana, Fredrick Lala, Daniel Muteti, Mohamed Omar, Winston Ouna and Alice Bett) for providing point data on the distribution of cheetah in Kenya. We thank the Netherland Government, through NUFFIC for funding this study and University Twente, ITC for providing Bernard Kuloba, the lead author, with opportunity to study at the institution.

# References

[1] Araujo, M.B. & Williams, P.H. (2000). Selecting areas for species persistence using occurrence data. Biol. Conserv. 96, 331–345.

[2] Baldwin, R. A. (2009), Use of Maximum Entropy Modeling in Wildlife Research, Entropy, 11, 854-866; doi: 10.3390/e11040854 www.mdpi.com/journal/entropy.

[3] Bissett, C. & Bernard, R.T.F. (2007). Habitat selection and feeding ecology of the cheetah in thicket vegetation: is the cheetah a savanna specialist? J. Zool. (Lond.) 271, 310–317.

[4] Cardillo, M., Purvis, A., Sechrest, W., Gittleman, J. L., Bielby, J. & Mace, G.M. (2004). Human populationdensit-y and extinction risk in the World's carnivores. PLoS Biol. 2, 909–914.

[5] Caro, T.M. (1994). Cheetahs of the Serengeti plains: group living in an asocial species. Chicago, IL: The University of Chicago Press.

[6] Caro, T.M. & Collins, D.A. (1987).Ecological characteristics of territories of male cheetahs. J. Zool. (Lond.) 211, 89–105.

[7] Caro, T. M. & Collins, D. A. 1987a Ecological characteristics of territories of male cheetahs (Acinonyxjubatus). J. Zool., Lond. 211, 89-105.

[8] Chefaoui, R. M., Hortal, J. & Lobo, J. M. (2005) Potential distribution modelling, niche characterization and conservation status assessment using GIS tools: A case study of Iberian Copris species. Biological Conservation, 122, 327-338.

[9] Durant, S. M. 1998. Competition refuges and coexistence: an example from Serengeti carnivores. Journal of Animal Ecology 67: 370-386.

[10] Elith J, Steven J. Phillips, Trevor Hastie, Miroslav Dudi´k, Yung En Chee and Colin J. Yates, (2011) A statistical explanation of MaxEnt for ecologists, Diversity and Distributions, 17, 43–57.

[11] Guisan, A. & Zimmermann, N. E. (2000) Predictive habitat distribution models in ecology. Ecological Modelling, 135, 147-186.

[12] Guisan A, Thuiller W (2005) Predicting species distribution: offering more than simple habitat models. Ecol Lett 8: 993–1009.

[13] Gros P, 1997 Status of the cheetah Acinonyxjubatus in Kenya: a field-interview assessment. Biological Conservation 85 (1998) 137-149.

[14] Hamilton, P. H. (1986) Status of the cheetah in Kenya, with reference to sub-saharan Africa. Cats of the World: Biology, Conservation and Management (eds S. D. Miller & D. D. Everett). National Wildlife Federation, Washington, D.C..

[15] Hamilton, P. 1981. The Leopard Pantherapardus and the cheetah Acynonixjubatus in Kenya. US Fish and Wildlife Service, African Leadership Foundaton and the Government of Kenya, Nairobi.

[16] Hirzel, A. H., Hausser, J., Chessel, D. & Perrin, N. (2002) Ecological-niche factor analysis: How to compute habitat suitability maps without absence data? Ecology, 83, 2027 - 2036.

[17] Jared A. Stabach, Nadine Laporte and William Olupot. (2009) Modeling habitat suitability for Grey Crowned-cranes (Balearicaregulorumgibbericeps) throughout Uganda International Journal of Biodiversity and Conservation Vol. 1(5) pp. 177-186 Available online http://www.academicjournals.org/ijbc.

[18] Pettorelli, N. A. Hilborn, F. Broekhuis & S. M. Durant (2009). Exploring habitat use by cheetahs using ecological niche factor analysis Journal of Zoology 277 141–148.

[19] Pearce, J., and S. Ferrier. 2000. Evaluating the predictive performance of habitat models developed using logistic regression. Ecological Modelling 133: 225-245.

[20] Stamps, J.A. & Swaisgood, R.R. (2007). Some place like home: experience, habitat selection and conservation biology. Appl. Anim. Behav. Sci. 102, 392–409.

[21] Urbina-Cardona, J. N. and Loyola R.D. 2008. Applying niche-based models to predict endangered-hylid potential distributions: are neotropical protected areas effective enough? Tropical Conservation Science Vol.1 (4):417-445. Available online: Hwww.tropicalconservationscience.orgH.

[22] KWS. 2010. Kenya National Strategy for the Conservation of Cheetahs and Wild Dogs.in Research, editor. Kenya Wildlife Service, Nairobi.

[23] Lack, D. (1933) Habitat selection in birds. With special reference to the effects of afforestation on the Breckland avifauna. The Journal of Animal Ecology, 2, 239-262.

[24] Laurenson, M. K. (1995) Implications of high offspring mortality for cheetah population dynamics.

[25] Serengeti II: Dynamics, management and conservation of an ecosystem (eds A. R. E. Sinclair & P. Arcese). The University of Chicago Press, Chicago.

[26] Myers, N. 1975.The Cheetah Acinonyxjubatus in Africa. IUCN Monograph, IUCN Morges Switzerland no. 4.

[27] M. S. Wisz, R. J. Hijmans , J. Li, A. T. Peterson, C. H. Graham, A. Guisan, (2008) Effects of sample size on the performance of species distribution models Diversity and Distributions, (Diversity Distrib.) 14, 763–773.

[28] Morrison, M. L., Marcot, B. G. & Mannan, R. W. (2006) Wildlife-habitat relationships: concepts and applications, 3rd edn. Island Press, Washington, D.C.

[29] Phillips, S. J. et al. (2005) Maxent software for species distribution modeling. http://www.cs.princeton.edu/_schapire/maxent/_.

[30] Richard G. Pearson, Christopher J. Raxworthy, Miguel Nakamura and A. Townsend Peterson (2007) Predicting species distributions from small numbers of occurrence records: a test case using cryptic geckos in Madagascar. Journal of Biogeography (J. Biogeogr.) 34, 102–117.

[31] Rotenberry, J. T., Preston, K. L. & Knick, S. T. (2006) GIS-based niche modeling for mapping species' habitat. Ecology, 87, 1458-1464.

[32] Sunil Kumar and Thomas J. Stohlgren. (2009) Maxent modeling for predicting suitable habitat for threatened and endangered tree Canacomyricamonticola in New Caledonia. Journal of Ecology and Natural Environment Vol. 1(4), pp. 094-098, Available online at http://www.academicjournals.org/JENE.

[33] Tomoko Doko, Hiromichi Fukui, Andre Kooiman, A.G. Toxopeus, Tomohiro Ichinose, Wenbo Chene, A.K. Skidmore. (2011) Identifying habitat patches and potential ecological corridors for remnant Asiatic black bear (Ursusthibetanusjaponicus) populations in Japan. Ecological Modelling 222 748–761.

[34] Wilting A, Cord A, Hearn AJ, Hesse D, Mohamed A, et al. (2010) Modelling the Species Distribution of Flat-Headed Cats (Prionailurusplaniceps), an Endangered South-East Asian Small Felid. PLoS ONE 5(3): e9612. doi:10.1371/journal.pone.0009612.

[35] Woodroffe, R. & Ginsberg, J.R. (1998). Edge effects and the extinction of populations inside protected areas. Science 280, 2126–2128 WRI. 2011. World Resources Institute. http://www.wri.org/.

# Heavy Metal Pollution Investigation of Left Bank Outfall Drain of Coastal District Badin, Sindh, Pakistan by using Arc GIS

**Mohammad Amin Qureshi[1, *],Ghulam Murtaza Mastoi[1], Zulfiqar Ali Laghari[2], Khalid Hussain Lashari[3], Fouzia Panhwar[4], Zameer Ali Palh[3]**

[1]Centre for Environmental Sciences, University of Sindh, Jamshoro, Sindh, Pakistan
[2]Department of Physiology, University of Sindh, Jamshoro, Pakistan
[3]Department of Fresh Water Biology and Fisheries, University of Sindh, Jamshoro, Pakistan
[4]Department of Pharmacology, University of Sindh, Jamshoro, Pakistan

**Email address:**
m-ameen-q@hotmail.com(M. A.Qureshi)

**Abstract:**Coastal district Badin of Sindh Pakistan is known as hub of pollution because of surface drain canal network, which continuously carries industrial effluents, municipal wastes, pumped saline water and excess agriculture water of three districts. Agriculture lands, ground water, lakes and Lagoons of the district have been contaminated with toxic heavy metals due to improper infrastructure of Left Bank Outfall Drain. The object of this study was to study the pollution contributed by drains of coastal areas of Badin. Nineteen samples of water were collected from different locations of drains on quarterly basis. Arc GIS was applied for vector mapping. Average results of parameters were EC 5.975 ± 4.956 mS/cm , TDS 3254.17 ± 2810.342 mg/L, Hg 0.0234 ±0.0118 µg/L , Ni 0.0076 ±0.0106mg/L, Cd 0.0245±0.0179mg/L , Zn 0.0393±0.0254mg/L , Cu 0.0890 ± 0.0911 mg/L, Fe 0.0546 ±0.0883mg/L.The results indicated that all parameters were within National Environmental Quality Standards (NEQS) limits of Pakistan for Industrial effluents. High concentration of TDS, and E.C, near thecoastal sampling areas, confirmed that, sea water has been facilitated, with backward flow by same drains due to sea tides. Heavy metal results were also within permissible limits of Food Agriculture Organization (FAO) for agriculture use. Drain water may be used for saline agriculture purpose, if properly checked under Environmental Protection Act of Pakistan 1997.

**Keywords:** LBOD (Left Bank Outfall Drain),Coastal Area of Badin, GIS (Geographic Information System), Industrial Effluent, KPOD (Kadhan Pateji Outfall Drain), DPOD (Dhoro Puran Outfall Drain)

## 1. Introduction

The river Indus of Pakistan irrigates 35.7 million acres of land. The irrigation network system has also given rise to water logging and salinity threats to agriculture, with passage of time. The coastal areas have been adversely affected .The Left bank outfall drain (LBOD) project was implemented from 1987 to 1997. The project was to mitigate water logging and salinity of 1.27 million acres of three districts, of Sindh province. The LBOD initially was aimed to drain out pumped saline water and agricultural runoff but municipal and industrial wastewas also thrown in this drain.The coastal district Badin became the final path way of this contaminated water to Arabian Sea. LBOD has a total discharge capacity of 4400 cusec and flows throughout the year [1, 2, 3].

LBOD is bifurcated into two main surface drains named KPOD (KadhanPateji Outfall Drain) and DPOD (DhoroPuran Outfall Drain). DPOD discharges into Shakoor Lake which is a joint Lakeof Pakistan and India (Fig.1).KPOD, flows with damaged infra structurepath of Tidal Link and Cholri Weir, reaches at Shah Samdo Creek and finally joins ArabianSea. It contaminates all wetlands of the coastal areas.Cholri weir and Tidal link, the important components of LBOD were constructed to maintain wet lands and high tides of sea. Both infrastructures became completely damaged after cyclone of 1999[1, 2, 3, 4, 5, 6].Consequently sea tidal water got quick access to wards inland areas which was not possible before. Hence the ecology of the coastal areas, mostly of wetlands, has been completely changed since two decades[7, 8].The World Bank was a major contributor of finance and technical assistance of the mega project, LBOD. The World Bank's Panel in 2005-6 reported that the failure of LBOD project was due to faulty design, underestimating the extreme meteorological risks and improper follow up of the recommended policies and procedures[9, 10, 3].The Panel also concededthat the little attention had been paid to the impacts on the environmentally important and Ramsar recognized coastal lakes and lagoons [10].This study was conducted to investigate the heavy metal pollution, carried by LBOD and other link surface drains near the coastal lines of Badin. At the same time to find out pollution contributed by sea tidal water from Arabian Sea.The results were compared with permissible limits of NEQS of Pakistan for municipal and industrial effluent. The results were interpreted by using Arc GIS application.

## 2. Experimental

### *2.1. History Background of Area of Study*

Badin is positioned between 24°-5` to 25°-25` north latitude and 68 21' to 69 20' east longitude [5].The district joins in south with the Arabian Sea and Rann of Kutch. The agriculture and agro based industry are the sources of economy. Coastal district of Badin is a flat, flood prone land, where required fresh canal water does not reach.The land and groundwater are saline. Climate of this area is unreliableand has proved to be very fatal to natives.Badin is situated in Zone II with respect to cyclonic activity which suggests that the district is exposed to tropical cyclones[2]. Stormy cyclones, monsoon rains, earthquakes and droughtshave aggravated already deteriorated conditions of the coastal areas.

Ecological degradation of Badin coastal area is result of neglecting attitude, considering this area, only a hub and pathway for disposable liquid waste towards Arabian Sea.Environmental, socio economical issues of these areas have never been taken properly and seriously. Deteriorated condition of this region is due to, two main surface drain system introduced in this area in 1960s and 1980s. Kotri Barrage drain system fromT.M.Khan,comprised of KaroGhungro(1960) and FuleliGuni(1959) Outfall drains, both drains, enter from west side of district,polluting coastal wet lands.FulleliGuni outfalls drain ends into MehroLake. KaroGhungro surface drainage System ends in Sanhro Lake. Both lakes are part of Ramsar recognized Jabholagoon (Fig. 1).Both drains carrying sugar mill effluents along with agriculture runoff, reducing bio life of wetlands[1, 2, 3, 6, 7].

The contaminated water of Kotri drain and LBOD, meet at saline Cholri shallow lake. It is zone of mixed polluted water (Fig. 1).A hub of pollution sink has been developed mostly after the destruction of Cholri weir and disappearance of tidal link after cyclone 1999[7].Costal tidal water strikes near village Golo Mandhro boundary wall at present time (Fig1).

Area of study

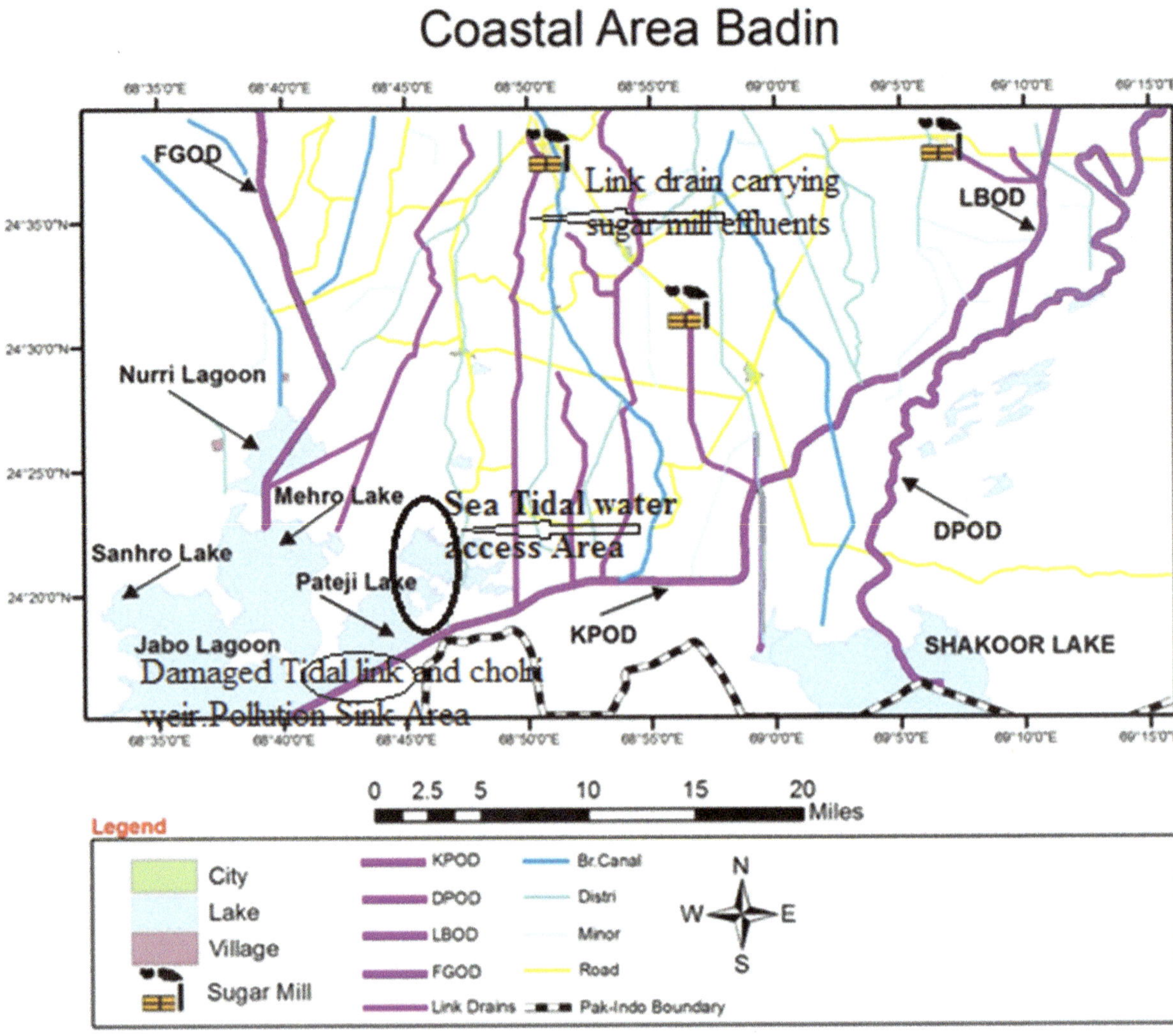


*Fig. 1. Base map of area study.*

### 2.2. Sample Collection

Nineteen surface drain water samples, with three replicates, were collected from the coastal belt of district Badin, after every three months (March, June, September and December) throughout the year 2013. Samples were collected from 9 am to 4 pm.

### 2.3. Method

Samples were collected in previously clean 1.5 L plastic bottles, temperature of air and water, electric conductivity, total dissolved salts, readings weretaken on the spot. Latitude and longitude were recorded by using GPSeTrex Legend Garmin. EC, TDS, of water samples was measured using pre calibrated Orion 115 conductivity meter [11]. Iron, Copper, Zinc, Cadmium, Nickel, Mercurry, concentrations were determined using Flame Atomic Absorption Spectrophoto-meter and Mercury by cold vapor method by Flame Atomic Absorption Spectrophoto-meter (FAAS) [11].

GIS is apowerful tool for collecting, storing, organizing, transforming and displaying data, for specific aims from the real world [12, 13, 14].GIS is increasingly used in Environmental pollution studies because of its ability in spatial analysis and understanding thematic maps [15, 16].Arc GIS 9.2 software was used for mapping and data interpretation. Arc GIS software was applied in this study as an important analyzing tool. Based on the sampling locations, with the utilization of coordinates of GPS, a point feature map showing the position of sampling location was prepared. By using Arc map of arc GIS software, Monitored water quality data was stored as attribute table.Graduated symbols and colors were selected using symbology option.

## 3. Result

Average results ( Table 3) of surface drains of coastal area of Badin indicated the obtained % concentration range of parameters studied were TDS(mg/L) 92.9 % , Hg (µg/L)0.234 % , Ni (mg/L)0.76 % , Cd (mg/L) 24.5 % , Zn(mg/L) 0.786 % , Cu (mg/L) 8.9 % , Fe(mg/L)0.68 % in comparison to defined limits of NEQS for Industrial effluents of Pakistan.

# 4. Discussions

The resultsof Electric Conductivity concentration indicated in Fig. 2, represented by graduated circles and colors.Red color circle symbol was selected to high light obtained results in four classes. Electric Conductivityof sampling areas waswithin range 1.298- 21.775 mS/cm (table1&2). E.C was alarmingly high at S-I, S-2, S-9, S-18, S-19, locations.The E.C was found to be high near coastal sea water. An increasing trend is seen as KPOD drain flow reaches near damaged cholri weir and tidal link area.This confirmed that coastal area was not only polluted by drainsbut back flow of sea water through tidal waves was also a major contributor (Fig. 1). Our results of EC (1.298-21.775 mS/cm) were found very high as compared toWorld Bankreport indicatingEC range from 624- 685 mS/cm of KPOD [3].

***Table 1.*** *Showing heavy metal results of drains of coastal area of Badin.*

| S.No | Sampling Stations | | | | | | | | | | | |
|---|---|---|---|---|---|---|---|---|---|---|---|---|
| 1 | Sample codes | S-1 | S-2 | S-3 | S-4 | S-5 | S-6 | S-7 | S-8 | S-9 | S-10 | NEQS for Municipal &Industerial effluents of Pakistan . |
| 2 | Latitude | N 24° 37.665 | N 24°37.660' | N 24° 37.698 | N 24 33.406 | N 24 33.325 | N 24° 30.612 | 24 21.490 | N 24°24.888 | N 24° 33.313 | N 24° 23.871 | |
| 3 | Longitude | E069° 13.115 | E069° 12.656 | E069° 10.876 | E069 09.822 | E069 09.873 | E069 °09.3 09 | E069 04.090 | E069° 00.473 | E068°54 .429 | E068° 54.191 | |
| | | Physicochemical parameters | | | | | | | | | | |
| 7 | EC (mS/cm) | 12.70 | 12.06 | 4.52 | 4.68 | 4.7 | 4.80 | 4.60 | 4.83 | 1.29 | 1.78 | * NGV |
| 8 | TDS (mg/L) | 7245 | 6995 | 2462 | 2488 | 2515 | 2485 | 2415.5 | 2557 | 696 | 933 | 3500 mg/L |
| | | Heavy metals | | | | | | | | | | |
| 19 | Hg(µg/L) | 0.025 | 0.0225 | 0.0213 | 0.0206 | 0.0185 | 0.0205 | 0.0155 | 0.0200 | 0.0155 | 0.018 | (10 µg/L) |
| 20 | Ni (mg/L) | 0.0377 | 0.03585 | 0.00665 | 0.0062 | 0.00685 | 0.0062 | 0.00675 | 0.0068 | 0 | 0 | 1 mg/L |
| 21 | Cd(mg/L) | 0.0698 | 0.0638 | 0.0272 | 0.0223 | 0.0221 | 0.0350 | 0.0368 | 0.0202 | 0.0014 | 0.011275 | 0.1 mg/L |
| 22 | Zn (mg/L) | 0.0582 | 0.0507 | 0.0478 | 0.056 | 0.0525 | 0.0524 | 0.0509 | 0.1227 | 0.0285 | 0.0303 | 5 mg/L |
| 23 | Cu(mg/L) | 0.1764 | 0.2674 | 0.1627 | 0.1647 | 0.1725 | 0.1508 | 0.1649 | 0.2444 | 0.0021 | 0.0227 | 1 mg/L |
| 24 | Fe (mg/L) | 0.177 | 0.1616 | 0.1429 | 0.1622 | 0.1461 | 0.0987 | 0.1140 | 0.1092 | 0.0295 | 0.1108 | 8 mg/L |

***Table 2.*** *Showing heavy metal results of drains of coastal area of Badin.*

| S.No | Sampling stations | | | | | | | | | | |
|---|---|---|---|---|---|---|---|---|---|---|---|
| 1 | Sample codes | S-11 | S-12 | S-13 | S-14 | S-15 | S-16 | S-17 | S-18 | S-19 | NEQS for Municipal & Industrial effluents of Pakistan. |
| 2 | Latitude | N 24° 24.122 | N 24°29.519 | N 24°30.143 | N 24 °24.204 | N 24° 18.528 | N 24°17.990 | N 24°17.490 | N 24°17.000 | N 24°16.652 | |
| 3 | Longitude | E068° 49.473 | E068° 41.643 | E068°45.4 47 | E068° 39.332 | E068° 46.514 | E068° 45.600 | E068°44.6 85 | E068°43.7 91 | E068°43.04 1 | |
| Physicochemical parameters | | | | | | | | | | | |
| 7 | EC (mS/cm) | 1.76 | 3.38 | 1.81 | 3.04 | 5.51 | 5.47 | 5.63 | 9.12 | 21.77 | * NGV |
| 8 | TDS (mg/L) | 889.5 | 1833.75 | 943.75 | 1557.5 | 2856.5 | 2887.5 | 3007.5 | 5000.5 | 12060 | 3500 mg/L |
| | | Heavy metals (mg/L) | | | | | | | | | |
| 19 | Hg (µg/L) | 0.0147 | 0.023 | 0.021 | 0.0262 | 0.0553 | 0.0526 | 0.0308 | 0.0149 | 0.0086 | (10 µg/L) |
| 20 | Ni (mg/L) | 0.0020 | 0.0012 | 0.0021 | 0.0016 | 0.0059 | 0.0086 | 0.0049 | 0.0061 | 0 | 1 mg/L |
| 21 | Cd(mg/L) | 0.0170 | 0.0142 | 0.0179 | 0.0202 | 0.0349 | 0.0226 | 0.0207 | 0.0092 | 0 | 0.1 mg/L |
| 22 | Zn (mg/L) | 0.0252 | 0.0285 | 0.0285 | 0.0315 | 0.0239 | 0.0193 | 0.0156 | 0.0145 | 0.0093 | 5 mg/L |
| 23 | Cu (mg/L) | 0.0102 | 0.0171 | 0.0161 | 0.0357 | 0.0297 | 0.0216 | 0.0152 | 0.0126 | 0.0042 | 1 mg/L |
| 24 | Fe (mg/L) | 0.0317 | 0.0234 | 0.0276 | 0.1194 | 0.0715 | 0.0574 | 0.0378 | 0.0459 | 0.0112 | 8 mg/L |

* NGV (No health-based guideline value recommended by NEQS of Pakistan).

***Table 3.*** *Showing average and St.Dev with min. max.*

| | E.C(mS/cm) | TDS(mg/L) | Hg( µg/L) | Ni(mg/L) | Cd(mg/L) | Zn(mg/L) | Cu(mg/L) | Fe(mg/L) |
|---|---|---|---|---|---|---|---|---|
| Average | 5.975 | 3254.17 | 0.0234 | 0.0076 | 0.0245 | 0.0393 | 0.0890 | 0.0546 |
| St.Dev. | 4.956 | 2810.342 | 0.0118 | 0.0106 | 0.0179 | 0.0254 | 0.0911 | 0.0883 |
| Min | 1.298 | 696 | 0.0086 | 0.0012 | 0.00148 | 0.0093 | 0.00215 | 0.011225 |
| Max | 21.775 | 12060 | 0.05535 | 0.037775 | 0.069825 | 0.12275 | 0.267475 | 0.177 |

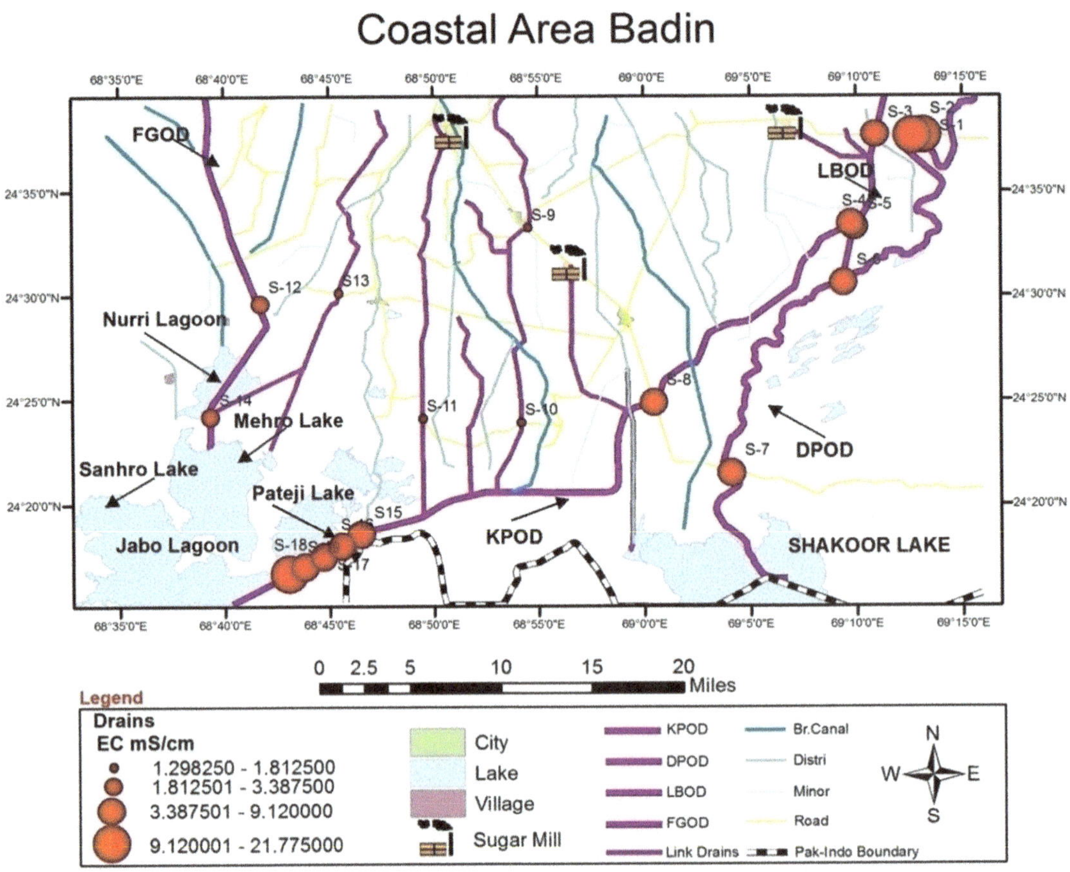


*Fig. 2.* *Map showing detection of E.C.*

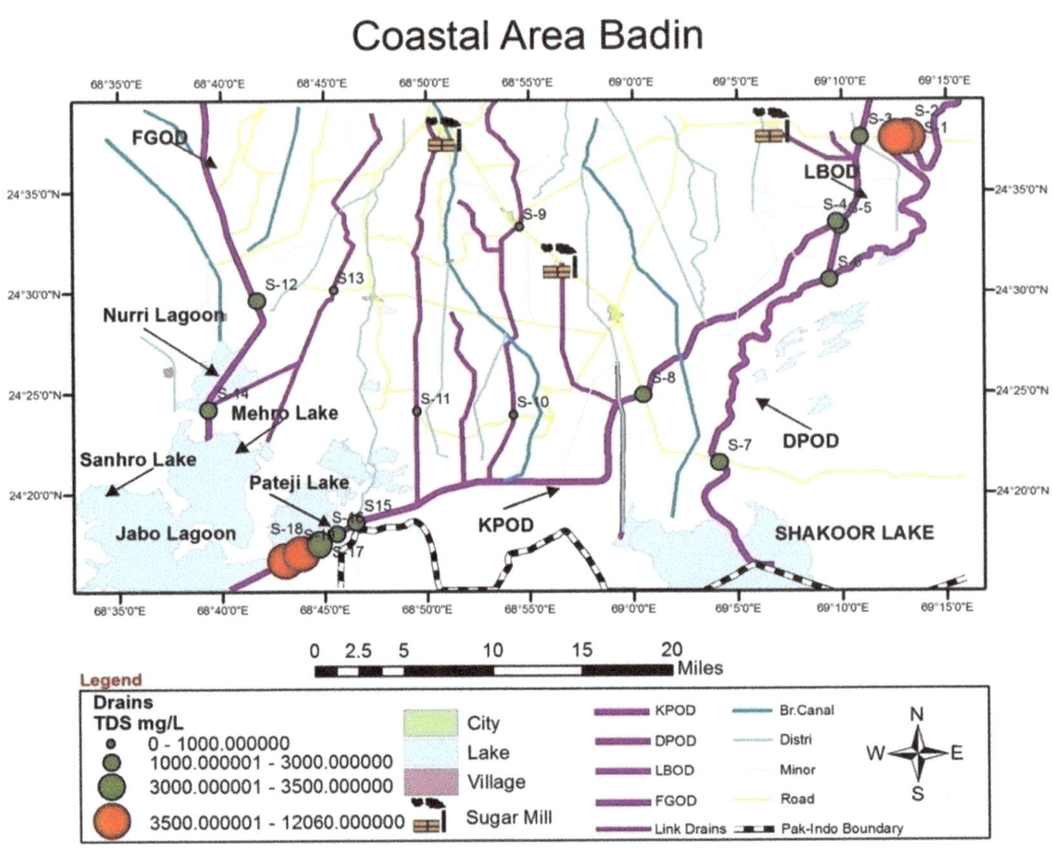


*Fig. 3.* *Map showing TDS detection.*

The results of TDS are shown in Fig.3, showing different ranges of concentrations from 696 - 12060 mg/L. Results were divided into four classes with graduated circles, represented by green and red colors. Green color was selected for representing permissible limit values (3500mg/L) forNEQS of Pakistan for industrialeffluents.The concentration above NEQS limit was indicated by color red. Sampling areas of S-1, S-2, S-18, S-19(table1 & 2), had high concentrations of TDS than the NEQS limits for industrial and municipal effluents. Samples S-18 and S-19 were closer to damaged tidal link area of LBOD, which was linked to Arabian Sea by shah SamdoCreek (Fig1).Increasing trend of TDS was seen close to coastal areas. The same trends continue,as was reported for, Amir Shah Drain of Badin coast,which had TDS concentration range within 4221 mg/L [2].

The result of concentration of Mercury indicated in fig 4, with graduated circles. The concentration of Hg range was found to be between 0.0086 - 0.05535µg/L (table 3). Mercury, represented with green color, was in all sampling locations within defined limit of NEQS (10 µg/L) for industrial effluents. The concentration of mercury reduced on moving closer to coastal sea water. The reduced concentration of mercury may be due to huge quantity of water, where mercury concentration gets diluted.

The concentration of Nickel as shown in Fig. 5, highlighted by graduated circles, with green color, by using symbology from arc map. Nickel concentration results (0.0012 - 0.0377 mg/L),were within describedpermissible limits (1 mg/L) of NEQS.It was observed that near damaged tidal link area, where huge coastal tidal water intrudes towards wetlands (Fig. 01), nickel concentration was found to be reduced. Theresults are in agreement withstudy conducted bySaif et al. [17].

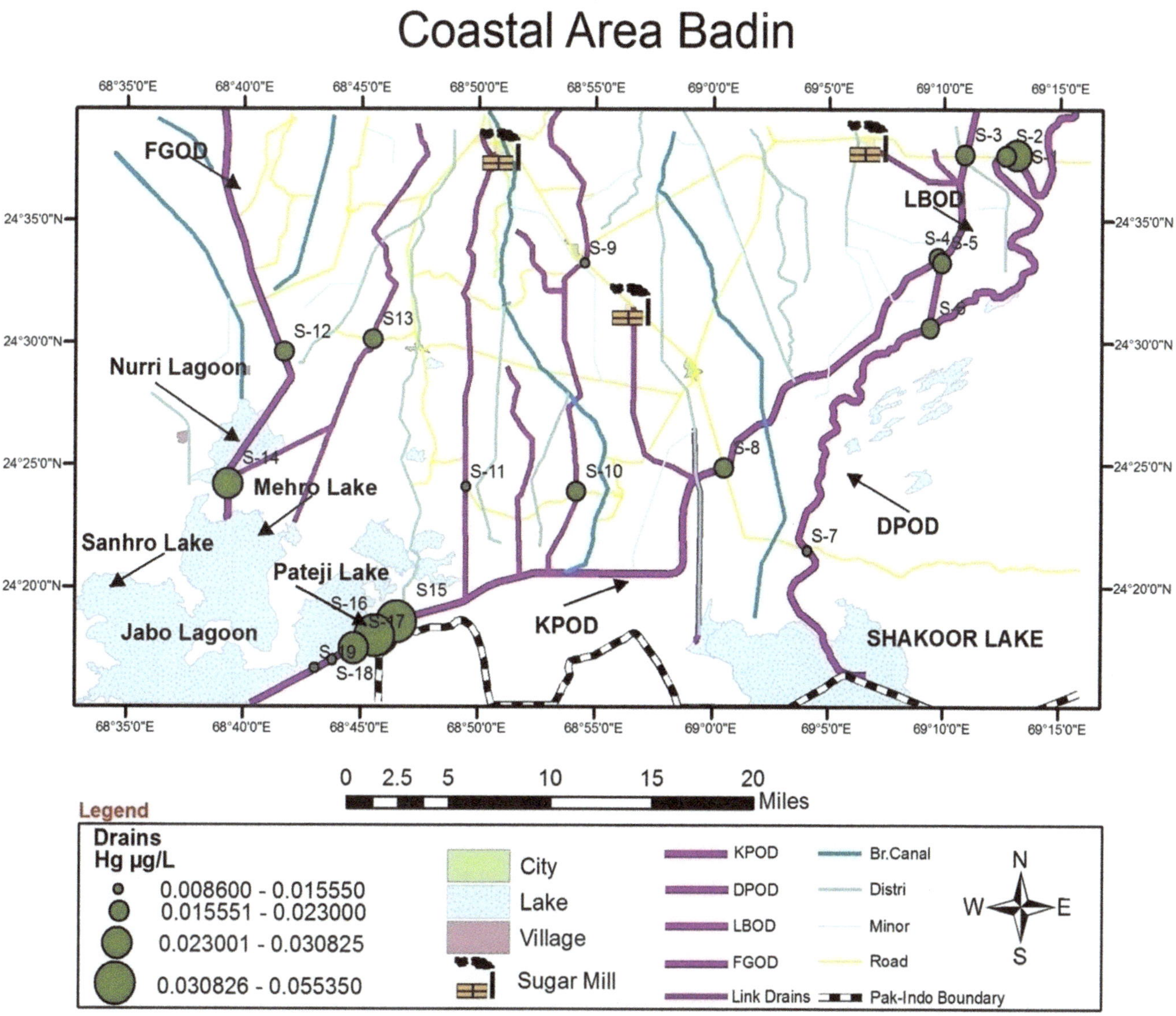


*Fig. 4. Map showing detection of mercury.*

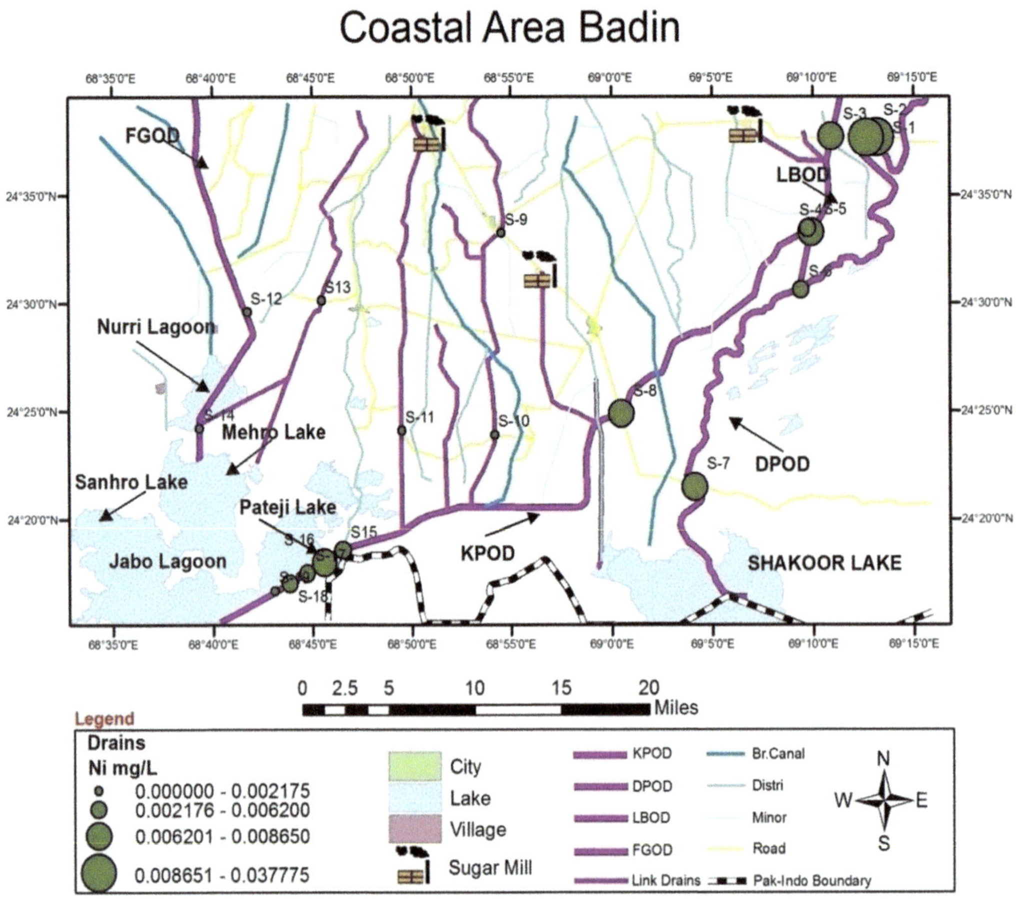


*Fig. 5. Map showing areas for Ni detection.*

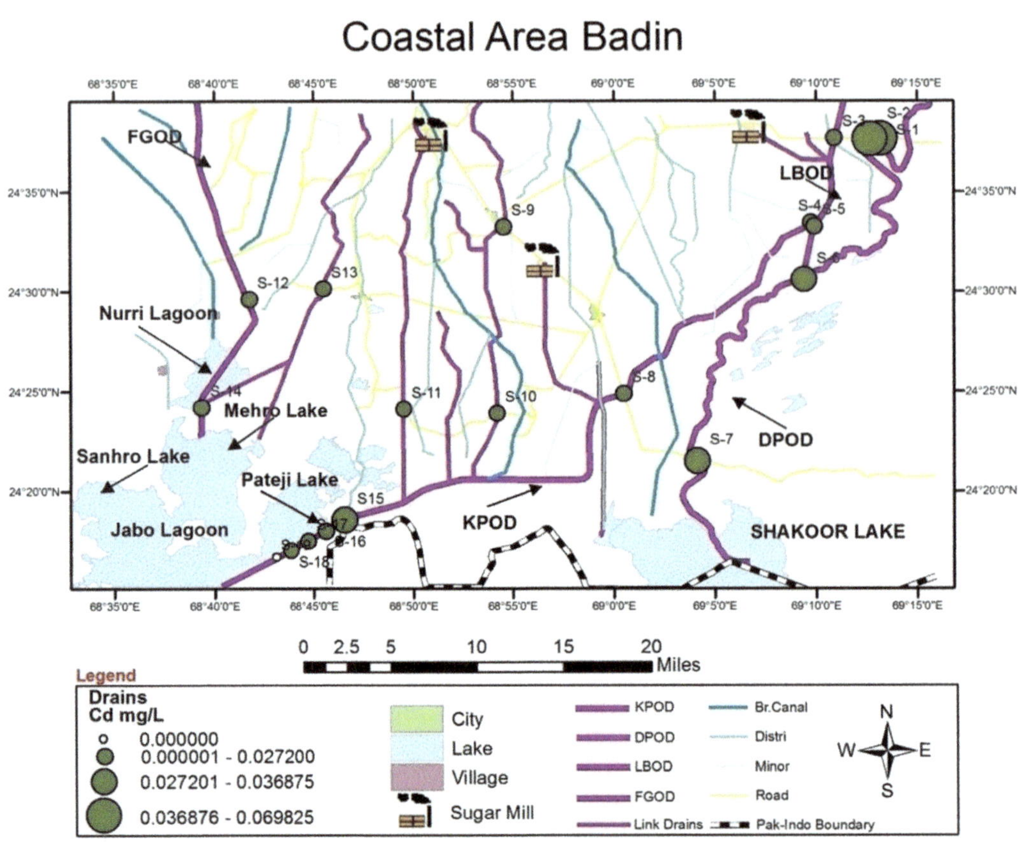


*Fig. 6. Map showing results of Cadmium detection.*

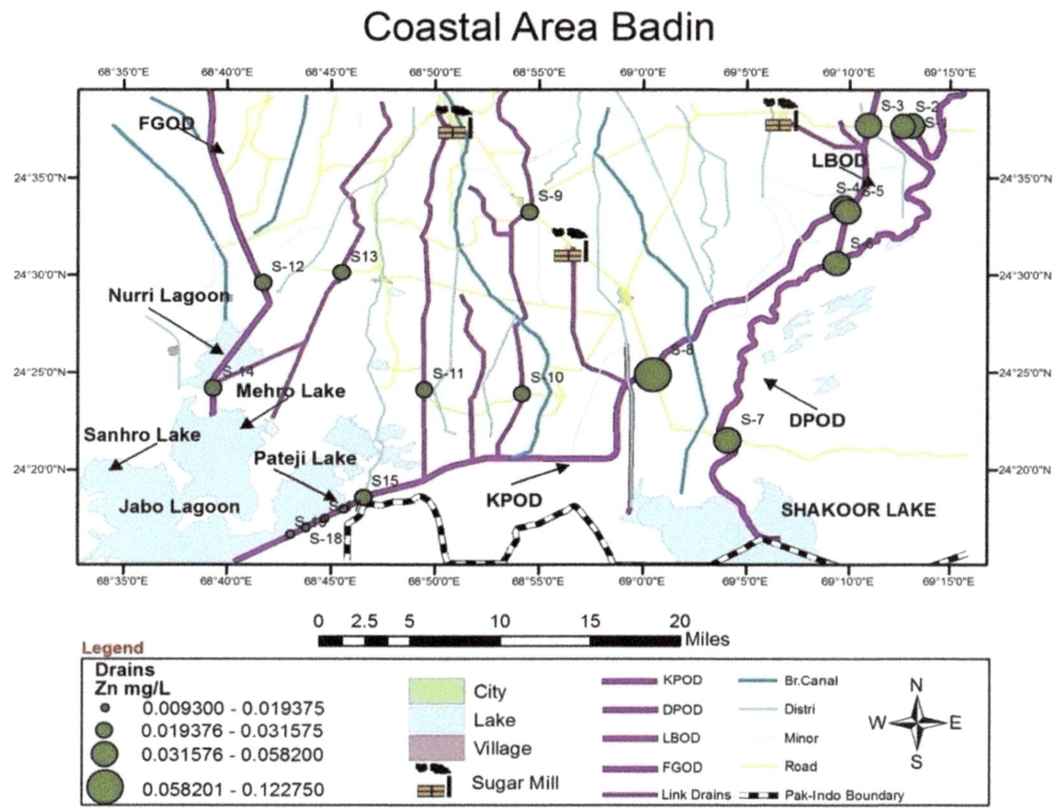


*Fig. 7. Map showing detected Zinc.*

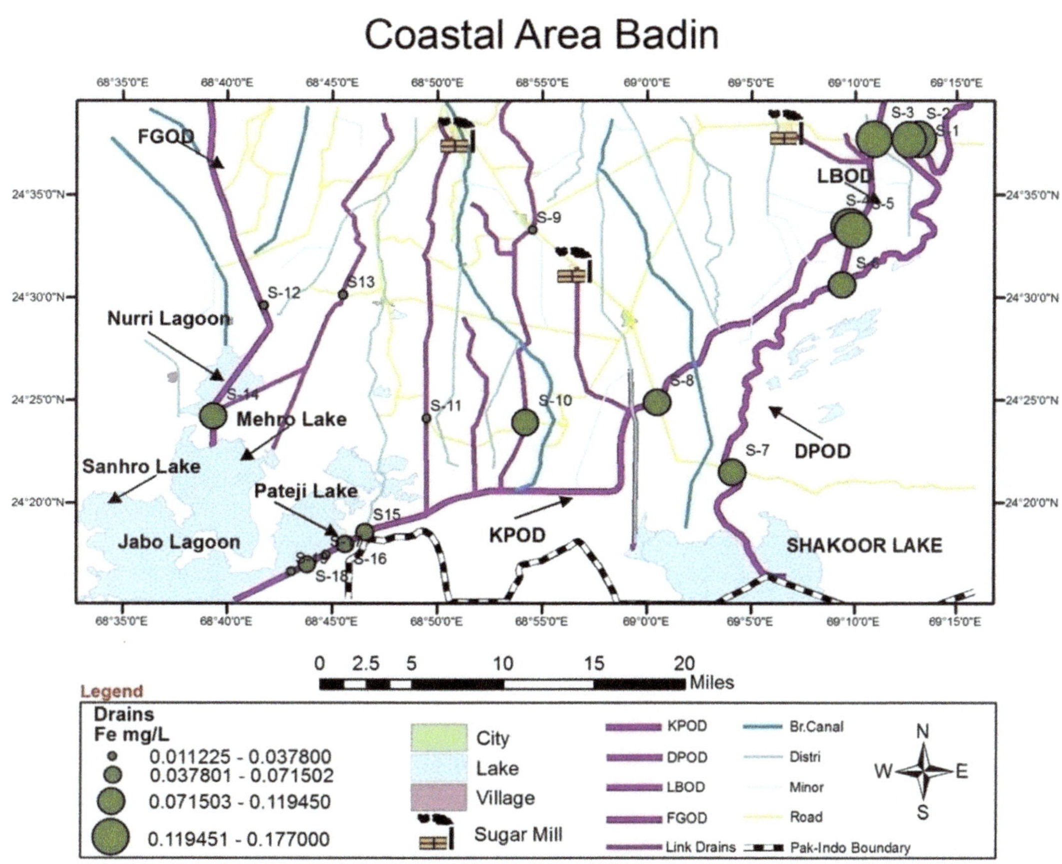


*Fig. 8. Showing detection of Iron.*

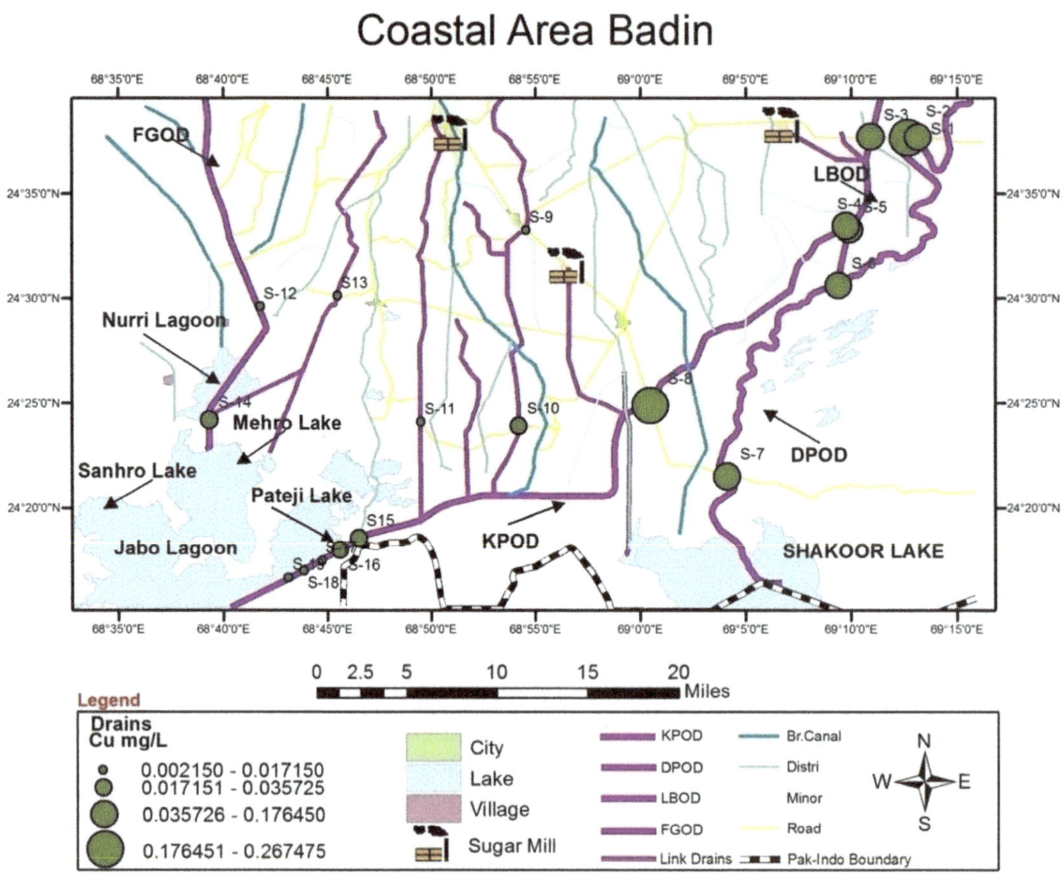


***Fig. 9.*** *Map showing copper detection.*

The Results of concentration of cadmium are showing Fig6, ranged between 0.00148 -0.0698 mg/L. The cadmium concentration in all sampling areas was found to be within limits (0.1 mg/L) set by NEQS. High concentration of Cd was observed at sampling area S-1, S-2.Concentration of cadmium was found to be diluted where KPOD enters near disappeared Cholri weir area (Fig.1).The cadmium concentration in all sampling areas was also found to be within permissible limits of FAO (0.1 mg/L)for agriculture use [18].According a study conducted for industrial effluent near Karachi, reported concentration of Cd was 0.004-2.4 mg/L, by Saif et al. [17]. Similar results were also reported by Tariq et al.[19] andAli et al. [24] for assessment of industrial and sewage effluents.According Iqbal and Edyvean,Khuhawar and Majidano , Devnani and Satsangee , andIqbal et al.[20,21,22 and 23] the major sources of Cd contamination are batteries, fertilizers,electroplating, smelting, paint pigments, mining and alloy industries.Though only sugar mill industries can be traced in district Badin(Fig 1),to find out other industries contributing cadmium pollutantsis very dificult task for present study, as LBOD caries industrial effluent of three distrits which are hundreds of km away from present study of areas.

The results of concentration of Zinc are shown in Fig 7, with graduated green circles, ranged from 0.0093 - 0.1227 mg/L, divided within four classes. Zinc was within permissible limits (5 mg/L) of NEQS in all the sampling areas. Same reducing pattern of Zinc concentration was found as water quantity increased near coastal mixing zone (Fig 1).Zinc concentration in all the sampling areas was also within permissible limits of FAO (2 mg/L) for agriculture use[18].Similar results for zinc concentration were also found by [17, 19, 24].It has been reported that exceeded concentration of zinc in water bodies, causes fever, depression, malaise, cough, vomiting and headache [25,26].

The results of the concentration of Iron as shown in Fig8, presented in measured green color circles, ranging concentration from 0.0112 -0.177 mg/L. The result values obtained in all sampling stations are within NEQS (8 mg/L) limits for municipal and industrial effluents. It is obvious that the concentration of iron was higher near entrance of drain and it reduced on reaching the coastal sea water mixing zone (Fig 8). It may be due to high quantity of water near coast, causing heavy metal concentration get diluted.Iron concentration in all the sampling areas was also within permissible limits of FAO (5 mg/L) for agriculture use [18].The results of the concentration of Fe are in agreement with Saif et al. [17].Similar findings were also reported by Tariq et al. [19] and Ali et al. [24].

The result of the concentration of copper indicated in figure 9. Copper was found to be within range (1 mg/L) ofNEQS for industrial effluents, of sampling areas.The result of copper concentration was ranged from0.002-0.267 mg/L .It is obvious in figure that the concentration of copper was higher near entrance path of mega drains and it reduced on reaching the coastal sea water. The concentration of copper in all the sampling areas except S-2 and S-8 was found within permissible limits of FAO (0.2 mg/L) for agriculture use [18].According Elenhorn and Barceloux [27] mostly water contamination of copper occurs due to untreated industrial wastewater pollution.

**5. Conclusion**

Pollution caused by LBOD,and link drains of the coastal areas of Badin have caused great problems. A positive approach for understanding the situation was developed by using GIS software. GIS maps revealed the extent of concentration of contaminatingpollutantsof drains and link drains of the coastal areas.Present study of drainsrevealed that main LBOD, KPOD, DPOD as well as link drains are carriers of pollutants including heavy metals.They bring industrial effluents,along with, pumped saline water, agriculture run off, and municipal waste water. Present study indicated that heavy metals are present within NEQS limits for municipal and industrial effluents of Pakistan.Low detection of heavy metal concentration in KPOD and DPOD is because of huge flow of water quantity exceeding 2000 cusecs, most of times. Therefore heavy metal concentration gets diluted.It was observed that at near coastal wetlands, the heavy metal pollutants were found in low concentration due to presence of excess tidal sea water. The Phenomenon for TDS and E.C was very opposite as compared with those of heavy metals.All drains indicated that they are very saline. High concentration of TDS, and E.C, near the coastal sampling areas confirms that sea water has been facilitated with backward flow by same drains due to sea tides.Poor infrastructure of LBOD has polluted the coastal zone, mostly due breaches in KPOD near zero point and completely disappearance of tidal link and cholri weir infrastructures. The results were found to be within limits of NEQS and FAO for industrial and agriculture purposes respectively.Hence water of LBOD can be used as a saline agriculture if other contaminants checked properly according to NEQS of Pakistan.

## References


[1] SIDA,"Sindh Water Sector Improvement Project-I.Regional master plan of Indus delta and coastal zone. Revival of Natural Dhoras"Sindh Water Sector Improvement Plan for Left Bank of Indus, Delta and Coastal Zone, 2013.

[2] EIA,"Environmental Impact Assessment for Exploration Activities in Badin Concession" – Protected Area Final report, 2012.

[3] [3]LBOD,"Review of the performance of the Left Bank Outfall Drain Stage IKPOD, DPOD, Tidal Link and Cholri Weir" Report of the World BankInternational Panel Experts, POE Report,2005.

[4] M. Memon, M. S.Soomro , M. S. Akhtar,K.S. Memon"Drinking water quality assessment in Southern Sindh (Pakistan)" Journal of Environmental Monitoring Assessment,2010.doi: 10.1007/s10661-010-1616-z.

[5] IUCN,"District Vision Badin A Framework for Sustainable Development"2006.

[6] S.M. Leghari,S.I.H. Jafri,A .Mahar,K. H. Lashari ,S.S. Ali,T.M. Jahangir, and M.Y. Khuahawar"Limonological study of Sanhro,Mehro,Pateji and Cholri lakes of district Badin ,Sindh, Pakistan" Pakistan Journal BiologyScience,vol.3 pp1904-1909, 2000

[7] M.A. Qureshi and G.M.Mastoi"The physiochemistry of sugar mill effluent pollution of coastlines in Pakistan" Ecological Engineering,vol. 75pp 137–144, 2015

[8] [8]SIDA,"Sindh Water Sector Improvement Project-I. Regional master plan of Indus delta and coastal zone (WSIP-I), Preparation of regional plan for the left bank of Indus, Delta and Coastal Zone Main report final - Phase – II" Sindh Irrigation and Drainage Authority 2012.

[9] DDRMP,"District Badin: District Disaster Risk Management Plan Badin, Sindh Province Second Draft District Badin, Sindh" UNDP for DDRMP District Badin ,2007.

[10] NDP,"Management report and recommendation in response to the inspection panel investigation report Pakistan: National Drainage Program" Project International Bank for Reconstruction and Development, International Development Association 2006.

[11] APHA,"Standard Methods for the Examination of Water and Wastewater" 17th ed. American Public Health Association, Washington D C , 1989.

[12] P.A. Burrough,"Principles of Geographic Information System for land Resources Assessment Monograph on Soils and Resources Survey No. 2" Clarendon Press, Oxford, England 1986.

[13] H.M.Baalousha,"Mapping groundwater contamination risk using GIS and groundwater modeling: A case study from the Gaza Strip, Palestine" Arabian Journal of Geosciences,vol.4 pp 483 – 494, 2011.

[14] M.O. Arnous& A.E. El-Rayes,"An integrated GIS and hydro chemical approach to assess groundwater contamination in West Ismailia area, Egypt" Arabian Journal of Geosciencesvol. 6:pp2829- 2842, 2013.

[15] A. Facchinelli,E. Sacchi, L. Mallen,"Multivariate statistical and GIS-based approach to identify heavy metal sources in soils". J. Environ. Pollut. Vol. 114,pp 313–324, 2001.

[16] F. Wang,"Quantitative Methods and Applications in GIS" Taylor & Francis, New York,2006.

[17] M. S. Saif, Midrar-ul-Haq , K.S.Memon,"Heavy Metals Contamination through Industrial Effluent to Irrigation Water and Soil in Korangi Area of Karachi Pakistan"Internationa Journal of Agrriculture andBiology, vol. 4pp 646–648,2005.

[18] A.A. Othman, S.A.Rabeh,M. Fayez,M, Monib,N.A. Hegazi,"El-Salam canal is a potential project reusing the Nile Delta drainage water for Sinai desert agriculture: Microbial and chemical water quality"Journal of Advance Research pp99-108,2012.

[19] M. Tariq, M. Ali, Z. Shah,"Characteristics of industrial effluents and their possible impactson quality of underground water"Soil & Environment,vol. 25pp 64-69,2006.

[20] M.Iqbal, R.G.J. Edyvean, "Loofah sponge immobilized fungal biosorbent a robust system for cadmium and other dissolved metal removal from aqueous solution" Chemosphere, vol. 61, pp 510–518,2005.

[21] M. Y. Khuhawar and S. A. Majidano,"An Investigation of Quality of Groundwater of TalukaNawabshah" Pakakistan Journalof Chemistery vol. 1(2)pp 65-71,2011.

[22] H. Devnani, S.P.Satsangee,"Green gold nano particle modified anthocyanin-based carbon paste electrode for voltammetric determination of heavy metals"International Journal of Environmental Science and Technology. 2014. DOI10.1007/s13762-014-0497-z

[23] M. Iqbal, A. Saeed, S.I. Zafar,"Hybrid biosorbent: an innovative matrix to enhance the biosorption of Cd(II) from aqueous solution" Journal of Hazard Materials.vol.148pp 47–55, 2007.

[24] H.M. Ali, E.M. EL-Mahrouk, F.A. Hassan, M.A.EL-Tarawy,"Usage of sewage effluent in irrigation of some woody tree seedlings. Part 3: Swieteniamahagoni (L.) Jacq" Saudi Journal of Biology and Science,vol.18,pp 201–207. 2011.

[25] M. Shakirullah , I. Ahmed , K.Mehmood , A .Khan , H. Rehman , S. AlamA. A. Shah,"Physicochemical Study of Drinking Water from Dir Districts" Journal of Chemical Society of Pakistan vol. 27(4), pp 374-387,2005.

[26] S.Farid,M.K. Baloch and S.A. Ahmad ,"Water pollution: Major issue in urban areas" International Journal of Water Resources and Environmental Engineering vol. 4,pp55-65,2012.

[27] M. J. Ellenhorn,D.G.Barceloux,"Medical toxicology: diagnosis and treatment of human poisoning", New York, Elsevier Science publishing company 1988.

# Evaluation of Radioactivity Concentration in Drinking Water Collected from Local Wells and Boreholes of Dutse Town, North West, Nigeria

**Shittu Abdullahi[1, *], Chifu E. Ndikilar[2], Abdussalam B. Suleiman[2], Hafeez Y. Hafeez[2]**

[1]Department of Physics, Gombe State University, Gombe, Nigeria
[2]Department of Physics, Federal University Dutse, Dutse, Nigeria

**Email address:**
ashittu@gsu.edu.ng (S. Abdullahi)

**Abstract:** Water is necessity to human and other life, human uses water for drinking, agricultural activities, domestic activities and so on. Therefore, it become necessary to study the level of radiation in our drinking water, because it poses many biological risks to human and other life when exceeded the maximum contamination level. This study examined the level of radioactivity of drinking water collected from local hand-dug wells and boreholes in Dutse town of Jigawa State. A total of 24 samples were collected and analyzed using MPC-2000, a Low Background alpha and beta counter. The equipment is a gasless proportional counter with ultra-thin window. The results showed that the alpha activity ranges from 0.006 to 0.485 Bq/l with a geometric mean of 0.060 Bq/l for borehole water samples and 0.003 to 1.08 Bq/l with a geometric mean of 0.094 Bq/l for well water samples and that of beta activity ranges from 0.009 to 84.68 Bq/l with a geometric mean of 5.288 Bq/l for borehole water samples and 6.830 to 6284.50 Bq/l with a geometric mean of 29.70 Bq/l for well water samples. Most of the sampling areas were above the maximum recommended level set by World Health Organization (WHO) of 0.5 Bq/l for alpha and 1.0 Bq/l for beta and United State Environmental Protection Agency (USEPA) of 0.55 Bq/l for alpha and 1.85 Bq/l for beta activity. An attempt has been done to examine the correlation between the alpha and beta activity and a very poor correlation was obtained. It was further observed that well water samples recorded higher activity compared to borehole water samples. The lower activity in the borehole water samples may be due to the absorption of the activity in the pipes unlike the well water samples that was drawn directly from it source. The overall results showed that many of the sampling areas are not suitable for drinking, agriculture and other domestic activities. It is therefore, recommended that more studies should be carried out to ensure the safety of the general public in these locations.

**Keywords:** Gross Alpha, Gross Beta, Radioactivity, Boreholes, Wells, Dutse Town

## 1. Introduction

Radioactivity in water is relatively easily measured, given the appropriate equipment and radio-analytical expertise, the interpretation of the significance of the measured radioactivity to the domestic water user is beset with uncertainties and imponderables, especially when it comes to evaluating the actual risk to the consumer. Water is a necessity to man and his environment; it has existed throughout the history of the earth crust even before the existence of man. Man uses water for the following activities: irrigation, power generation and domestic activity. There are two sources of water: rain and ground waters. It is found in rivers, wells, lakes and streams. The activities of human and natural phenomena constantly pollute the sources of water and affect water quality. Water pollution arises as a result of waste and sewage disposal into the environment and rivers by industries, hospitals and use of materials such as fertilizers by farmers. These disposed materials often contain radio nuclides [1]. In developing countries like Nigeria, lack of good drinking water is one of the serious threats to the human health as a result of that, rivers, streams, well and borehole waters are often used as supplement for the scarce pipe-borne water for drinking and domestic activities without any treatment [2].

Water bodies are also polluted from Naturally Occurring Radioactive Materials (NORM) that emits alpha, beta and gamma radiation. These usually have elements in the uranium and thorium series whose radioactive cause an appreciable airborne particulate activity and contribute to the radioactivity of rain and ground waters. It also affects drinking water from deep wells which are expected to have a higher concentration of radioactive elements compared to surface water. Furthermore, man-made alpha emitters, such as plutonium and americium could be transported into springs or wells there upon extending the activity level of the water [1].

Radioactivity in drinking water is an important mode of transfer of radio nuclides from the environment to man. The most important natural radio nuclides in drinking water are tritium, potassium-40, radium and their decay products, which are in essence beta and gamma emitters [1].

There is evidence of widespread contamination of water resources in many areas of our country. The Environmental Protection Agency's National Water Quality Inventory of 1994 has identified agriculture, urban runoff/storm water, and municipal point sources as the largest pollutant sources for rivers, lakes, and estuaries. Contaminants from these sources include pesticides, metals, nitrates, solvents, and other wastes. More than 200 chemical constituents have been documented in groundwater alone. But it should be noted that not all contamination events pose a threat to our health. The National Water Quality Inventory 1994 Report to Congress states that 40% of flowing river and stream miles can be used for drinking water after conventional water treatment, and 37% of lake and reservoir acres meet the designated use criteria for drinking water. Pollutant concentrations become diluted when they enter water sources and are further reduced by biological degradation, filtration, and adsorption to soil. Some chemicals, such as the man-made chlorinated hydrocarbons, are very stable in the environment. Some of these compounds accumulate in living organisms and are not readily metabolized and excreted.

The impacts of contamination events to lakes and reservoirs are more severe and persistent than streams and rivers because there is no natural flushing process as is characterized by the flow in streams and rivers. Contamination is even more persistent in groundwater due to lack of biological degradation. The most biologically active bacteria live within the soil above groundwater supplies. Inflow of waste-water from industries and rainwater flowing over farmland into the rivers and other water bodies constitute a source of pollution to drinking water in a particular area. The pollution may constitute heavy elements as well as radioactive elements whose existence could prove to be hazardous to health. Data on radioactivity in drinking water is therefore important in determining the quality and health impact on the populace. In Dutse town there is yet no established data on the radioactivity level in drinking water sources. Most of the rural inhabitants of the area depend on water from boreholes and local dug wells for drinking, household activities and irrigation. It may be possible that the water they used contain excess of naturally occurring radioactive elements and Heavy metals which may be harmful to human and other life. The aim and objectives of this study is to evaluate the gross alpha and beta activity in the sample locations and to establish the distribution pattern of the radioactivity measured in the locations in order to identify areas with elevated activity if any.

# 2. Materials and Methods

## *2.1. Counting Equipment*

The gross alpha and beta counting equipment used in this work was MPC-2000, a Low Background alpha and beta counter. The equipment is a gasless proportional counter with ultra-thin window. The sample was placed in a planchet and later placed in a sample carrier and the carrier was then placed on the sample drawer and closed for the purpose of gross alpha and beta counting, the counting was done automatically according to the selected count mode when the appropriate sample parameters were entered [3].

## *2.2. Sampling*

a) Sampling Frame.

The area under study is the Dutse metropolis and this investigation is limited to underground water sources in local boreholes and wells used by people for drinking, domestic activities, irrigation and animal husbandry. The method adopted for the sampling is stratified random sampling [2].

b) Sampling Procedure.

The sampling procedure involves the following:

- The sample container was rinsed three times with the water being collected, to minimize contamination from the original content of the sample container.
- An air space of about 1% in the container of water collected was left for each sample to give room for thermal expansion. There was marked on each container indicating two liters of volume corresponding to the air gap. 20 ml of diluted nitric acid was added to the sample immediately after collection to reduce the pH and to minimize precipitation and absorption on container walls.
- The samples were tightly covered with container cover and masking tape and kept in the laboratory for analyses
- Two samples were collected per sampling point which gives the total of 48 samples from 24 different sampling points.

## *2.3. Sample Preparation*

The specified volume of each sample was measured into a beaker and evaporated on a hot plate to 50 ml volume and then transferred to a clean dry planchet. The samples were evaporated to dryness on hot plate. The residues were transferred onto to a clean, dry and previously weighed planchet and the difference between the mass of empty planchet and that of the empty planchet plus residue gives the mass of the residue. The residues were uniformly spread on the planchet by dropping a few drops of ethanol. The residues were allowed to dry and then covered with Mylar film ready for counting.

$$\text{Sample efficiency} = \frac{(w_2-w_1)-(w_3-w_1)}{w_2-w_1} \times 100\% \quad (1)$$

where $w_1$= weight of empty planchet, $w_2$ = weight of planchet plus residues obtained after evaporation and $w_3$ = weight of empty planchet – weight of residue

### 2.4. Counting and Analysis

The counting equipment is automated. The protocol involves entering preset time, counting voltage and number of counting cycles. Also, the counter characteristics (efficiency and background) volume of sample used and sampling efficiency specified. Results are displayed as raw count, (count/min) count rate, activity, and standard deviation [4]. The calculation formulae for count rate activity and other parameters for a given sample are shown below:

a) Count Rate

$$\text{Rate } (\alpha,\beta) = \frac{Raw\ (\alpha,\beta) Count}{Count\ Time} \quad (2)$$

In all modes except mode alpha then beta:

b) Activity

$$\text{Activity } (\alpha,\beta) = \frac{Rate\ (\alpha,\beta) - Bgd\ (\alpha,\beta)}{Sample\ Efficiency \times Channel\ Efficiency \times Volume} \times \frac{1}{60} \quad (3)$$

The statistical precision is calculated for each channel, on each measurement and it depends only on the preset count whose value is declared indirectly. Assume N measurements are made during a time T, the average is given by $x = \frac{\Sigma xi}{N}$ and the standard deviation is given by

$$\sigma = \sqrt{\left[\frac{R_s}{t_s} + \frac{R_b}{t_b}\right]} \quad (4)$$

$R_s$ = Sample counting rate
$R_b$ = Background counting rate
$t_s$ = Sample counting time
$t_b$ = Background counting time

$$CR = 1.96\sigma \quad (5)$$

Where CR = Counting Error

c) Statistical Analysis

In order to analyze the activity measured, statistical analysis employed are estimation of the central tendencies and deviations, correlation analysis and histograms. The correlation coefficient (r) is given by the expression below [5]:

$$r = \frac{n\sum xy - \sum x \sum y}{\sqrt{[n\sum x^2 - (\sum x)^2][n\sum y^2 - (\sum y)^2]}} \quad (6)$$

and the geometric mean is given by as:

$$\text{Geometric Mean} = \sqrt[n]{(x_1, x_2 \ldots x_n)} \quad (7)$$

Where μ is the population mean and $\sigma$ is the standard deviation. All statistical analyses were made using the Microsoft Excel.

## 3. Result and Discussion

### 3.1. Measured Activity for Boreholes and Wells Samples

In this section we shall discuss the measured activity for the samples collected from boreholes and local dug wells. It was observed that the average alpha activity measured from boreholes and wells were 0.09 and 0.33 Bq/l respectively and the average beta activity measured for boreholes and wells samples were 18.6 and 508 Bq/l respectively. In addition, the geometric mean for both alpha and beta activity was also computed, for the samples collected from local dug wells and boreholes: the geometric mean of alpha activity were 0.09 and 0.06 Bq/l respectively and 29.70 and 5.29 Bq/l respectively for beta activity.

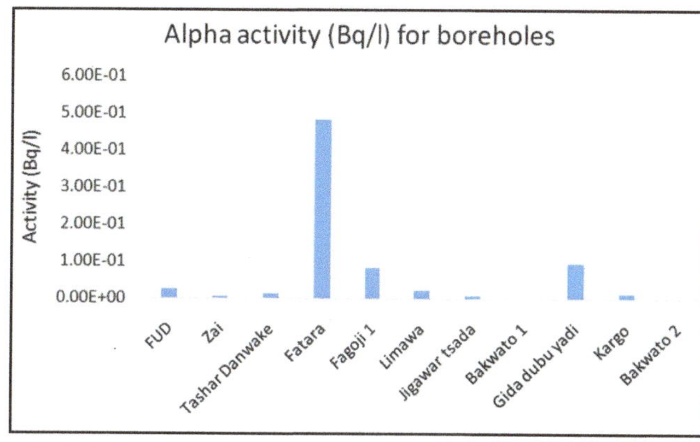


*Fig. 1. Plot of Alpha Activity for Boreholes Water.*

Figure 1 shows an alpha activity for boreholes water samples collected from Dutse town and the chart behavior fluctuates with some elevated activity and low activity at some other areas. The maximum and minimum recorded activity at fatara and Zai with corresponding values of 0.485 and 0.0062 Bq/l respectively. It could be seen that Bakwato 1 and 2 were recorded as 0 Bq/l in the plot above but they were below detection limit (BDL) of the system.

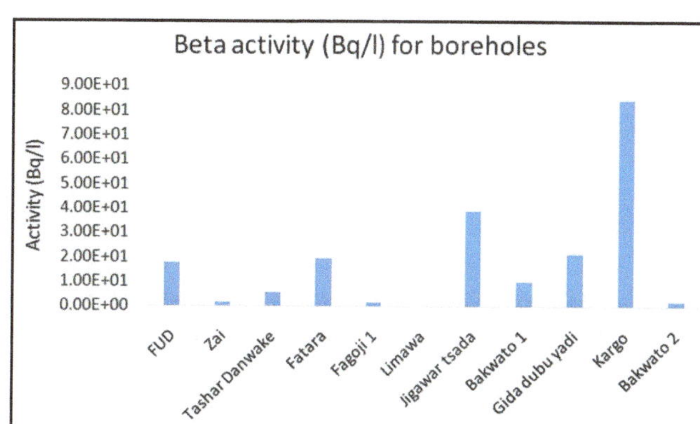


*Fig. 2. Plot of Beta Activity for Boreholes water.*

Figure 2 shows the plot of beta activity for boreholes water samples collected from Dutse town. The chart indicated a fluctuating behavior with some elevated activity and low activity in some other areas. The maximum and minimum activity were recorded at Kargo and Limawa with corresponding values of 84.68 and 0.0089 Bq/l respectively. Further observation was made almost all the sampling area exceeded the maximum permissible limit set WHO and USEPA except for Fagoji 1 and Zai which were below the maximum contamination level set by USEPA of 1.85 Bq/l and only Limawa comply with the recommended level set WHO of 1.0 Bq/l.

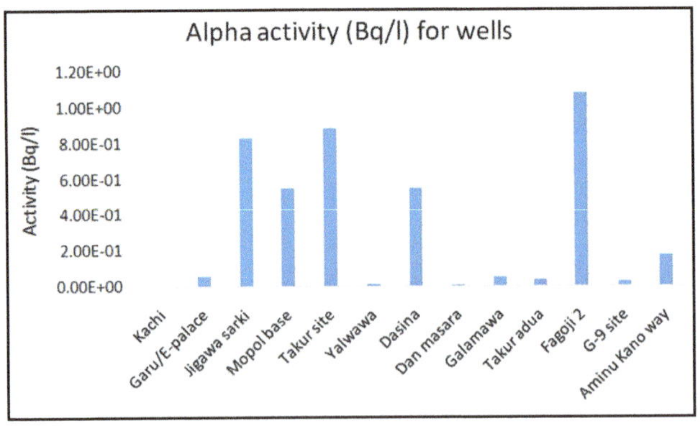


***Fig. 3.*** *Plot of Alpha Activity for Wells water samples.*

Figure 3 shows the bar chart of alpha activity for wells water samples collected from Dutse town. The chart indicated a fluctuating behavior with some elevated activity and low activity at some other areas. The maximum and minimum activity were also recorded at Fagoji 2 and Kachi with corresponding values of 1.08 and 0.0031 Bq/l respectively.

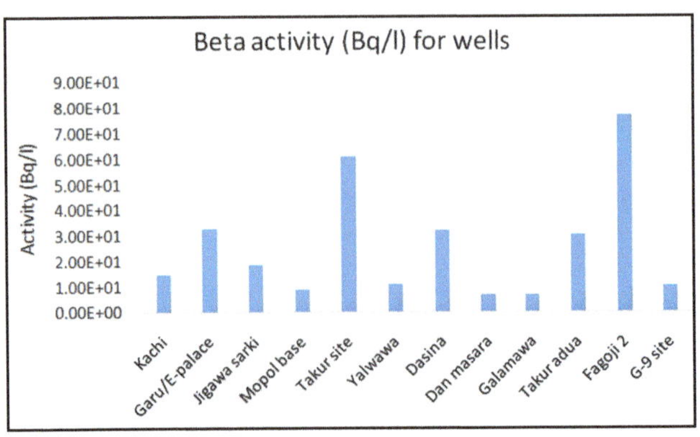


***Fig. 4.*** *Plot of Beta Activity for Wells water samples.*

Figure 4 shows the beta activity for wells water samples collected from Dutse town. The chart indicated that the activity varies across the sampling areas. However, all sampling areas exceeded the maximum recommended limit set by WHO and USEPA for safe drinking of values 1.0 and 1.85 Bq/l respectively. The maximum and minimum activity were recorded at Aminu Kano way and Galamawa with corresponding values of 6284.5 and 6.83 Bq/l respectively.

### *3.2. Alpha and Beta Activity for the Combined Boreholes and Wells Water Samples*

Below illustrated the alpha and beta activity for Dutse town it could be seen that the maximum and minimum alpha activity observed at Fagoji 2 and Kachi with corresponding values of 1.08 and 0.0031 Bq/l respectively. Similar observation was made for beta activity at Aminu Kano way and Limawa with corresponding values of 6284.50 and 0.0083 Bq/l respectively. It would be observed that Aminu Kano way dose not captured in the plot for beta activity due its magnitude which would make it difficult to observe the other activities as shown in the figures 5 to 6.

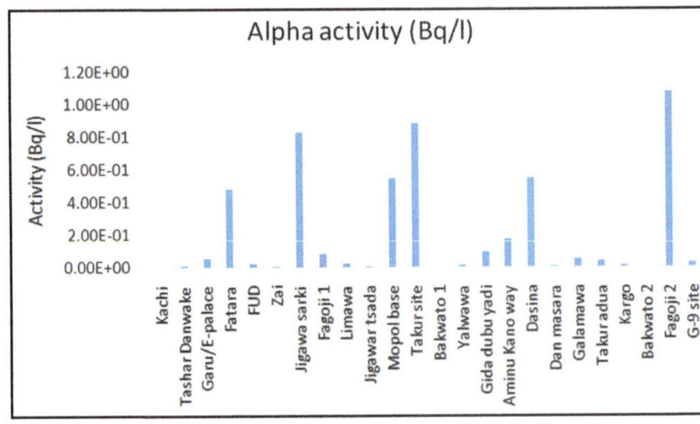


***Fig. 5.*** *Plot of Alpha activity (Bq/l) against Sampling Points.*

Figure 5 shows the plot of alpha activity against sampling points for Dutse town. The results indicated that Jigawa sarki, Mopol base, Takur site, Dasina and Fagoji 2 showed elevated activity which is above the maximum recommended level set by World Health Organization (WHO) of 0.5 Bq/l. Apart from Jigawa sarki, Takur site and Fagoji 2 which displayed very high activity, all other sampling areas comply with maximum recommended level set by United State Environmental Protection Agency (USEPA) of 0.55 Bq/l. the average alpha activity computed was 0.229 Bq/l which was below the maximum permissible limit set by WHO and USEPA for safe drinking water.

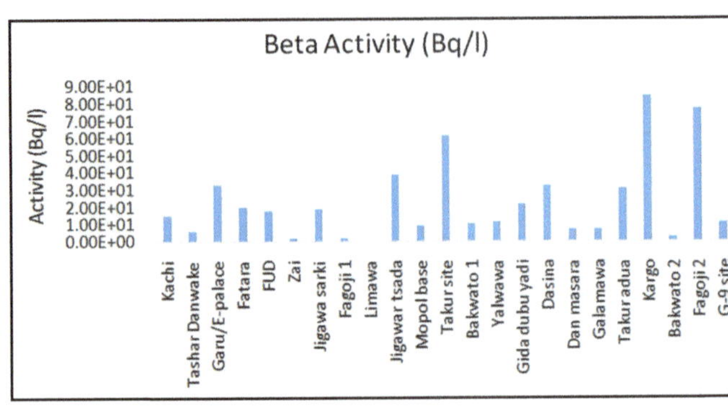


***Fig. 6.*** *Plot of Beta activity (Bq/l) against Sampling Points.*

Figure 6 shows the plot of beta activity against sampling points for Dutse town. The results indicated that almost all the sampling sites showed elevated activity except for Fagoji 1 and Zai which are below maximum recommended level set by USEPA which is 1.85 Bq/l and only Limawa comply with the recommended level set by WHO which is 1.0 Bq/l. The mean value computed was 283.00 Bq/l which exceeded the maximum permissible limit set by WHO and USEPA for safe drinking water.

### 3.3. Comparison Study Between Alpha and Beta Activity

Below is the correlation between the measured alpha and beta activity in Dutse town of Jigawa State, Nigeria.

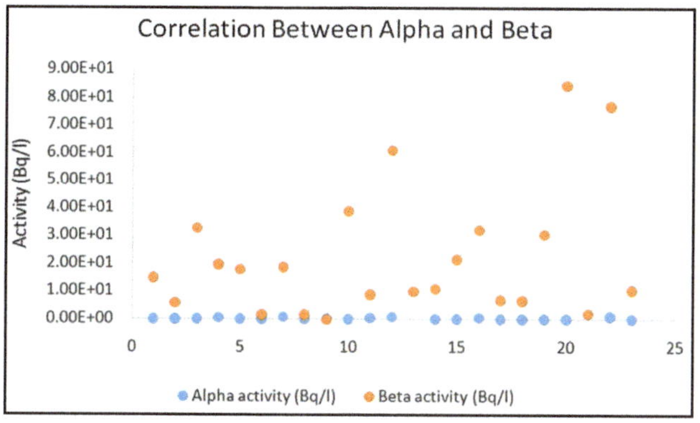


*Fig. 7. Plot of Correlation between Measured Alpha and Beta Activity.*

Figure 7 shows the correlation study between measured alpha and beta activity for Dutse town. The correlation between alpha and beta gave the coefficient of determination ($R^2$) value of 0.0046 and a Pearson regression coefficient (r) of 0.068 showing that there is no significance statistical relationship between alpha and beta activity as indicated from the correlation, that only 6.8% of the alpha activity is correlated with beta activity.

### 3.4. Comparison of Measured Activity in Dutse with Other Locations

The results of measured gross alpha and beta activity in Dutse were compared with other locations within and outside Nigeria. The maximum and minimum recorded in China were compared with the results measured in Dutse town and the following were observed: the maximum alpha activity measured in China was 88% less than that of Dutse town and the minimum measured was three times that of Dutse town. Similar comparison was conducted for beta activity, the maximum beta activity measured in China was negligible compared to the one measured in Dutse and the minimum activity measured in China was almost eight times that of Dutse town. Other comparisons were described in the table 1 below.

*Table 1. Comparison of Measured Activity in Dutse with Other Locations.*

| Location | Measured Activity (Bq/l) | | | | | | | | Source |
|---|---|---|---|---|---|---|---|---|---|
| | Alpha | | | | Beta | | | | |
| | Min | Max | Gm | Mean | Min | Max | Gm | Mean | |
| Gombe | 0.001 | 0.063 | 0.01 | - | 0.132 | 3.828 | 0.526 | - | [2] |
| KUST, Wudil | 0.0005 | 0.022 | - | 0.006 | 0.008 | 0.345 | - | 0.048 | [6] |
| Sokoto | 0.01 | 6.00 | 0.26 | - | - | - | - | - | [7] |
| Bayelsa | - | - | - | 4.02 | - | - | - | 54.23 | [8] |
| Turkey | 0.08 | 0.38 | - | 0.192 | 0.12 | 3.47 | - | 0.579 | [9] |
| China | 0.009 | 0.20 | - | - | 0.067 | 0.32 | - | - | [10] |
| Zaria | 0.035 | 0.01 | - | 0.015 | 0.06 | 0.91 | - | 0.33 | [11] |
| Dutse | 0.003 | 1.08 | 0.006 | 0.229 | 0.008 | 6284.5 | 13.48 | 283.0 | This work |

Table 1 above shows the comparison between the measured alpha and beta activity in Dutse town with other places including Gombe, Wudil, Sokoto, Bayelsa, Turkey, China and Zaria. It was observed that the alpha activity measured in Gombe was 0.4% greater than that of Dutse town and that of beta activity was extremely less than that of beta activity measured in Dutse town. For Wudil; it was 22.3% less than that of Dutse for alpha and extremely less than that of beta activity measured in Dutse. For Sokoto; it was 25.4% greater than that of Dutse for alpha activity. For Bayelsa, it was extremely greater than that of Dutse for alpha and extremely less than that of Dutse for beta activity. For Turkey; it was 3.7% less than that of Dutse for alpha and extremely less than that of Dutse for beta activity. For Zaria, it was 21.4% less than that of Dutse for alpha and extremely less than that of Dutse for beta activity.

It could be seen that arithmetic and geometric mean computed indicated that the alpha activity measured was less than maximum contamination level set by WHO of 0.5 Bq/l and for beta activity, it was above the maximum contamination level set by WHO of 1.0 Bq/l. The elevated activity recorded for beta may be due to the nature of the sample locations and geological activities in these areas.

Water samples were collected from both local dug wells and boreholes in Dutse town. The samples were studied and analyzed for gross alpha and beta activity.

The first study was to examine alpha and beta activity for the samples collected from boreholes and wells respectively. In the study the maximum and minimum alpha activity recorded for boreholes water samples were 0.485 and 0.006 respectively and the corresponding beta activity were 84.68 and 0.009 Bq/l respectively. Similar results were obtained for wells water samples of values 1.08 and 0.003 Bq/l respectively for alpha activity and 6284.5 and 6.83 Bq/l respectively for beta activity. Among the study areas, only two sampling points namely: Bakwato 1 and 2 were below detection limit of the system.

Study for the combined boreholes and wells water samples has been done and the following observations were made as shown in the table 2 below:

***Table 2.*** *Arithmetic and Geometric Mean of Alpha and Beta Activity.*

| Observations | Alpha Activity (Bq/l) | Beta Activity (Bq/l) |
|---|---|---|
| Overall Maximum | 1.08 | $6.28 \times 10^{3}$ |
| Overall Minimum | $3.09 \times 10^{-3}$ | $8.29 \times 10^{-3}$ |
| Arithmetic mean | 0.23 | 283.00 |
| Geometric mean | $5.84 \times 10^{-3}$ | 13.48 |

Finally, a correlation between alpha and beta activity was also determined and a Pearson regression coefficient of value 6.8% was obtained which implied that there is no significant relation between alpha and beta activity.

# 4. Conclusion

The level of gross alpha and beta radioactivity in drinking water samples drawn from local hand-dug wells and boreholes in Dutse town were examined using MPC-2000, a Low Background alpha and beta counter. The results showed that most of the sampling areas showed elevated activity. The overall maximum and minimum activity concentrations were 1.08 and 0.0031 Bq/l respectively with mean value of 0.23 Bq/l for alpha activity and 6284.5 and 0.0083 Bq/l respectively with a mean value of 283.0 Bq/l for beta activity. The result obtained was compared with other locations within and outside Nigeria. A study was made to determine the correlation between the alpha and beta activity and a Pearson correlation coefficient of 6.8% was obtained which implied very poor correlation. It was also observed that the study area was covered with rocks mostly metamorphic type which was believed to have affected the level radioactivity in the sample locations especially the well water samples. The reason of having lower activity in the borehole water samples was due to the absorption of some part of the radiation by the wall of the pipes. Therefore, it is recommended that further studies to investigate the level of radioactivity and radionuclides concentration in these locations need to be carried out.

# Acknowledgment

The authors acknowledged the financial support of Nigerian Tertiary Education Trust Fund (TETFUND) through the Federal University Dutse, Jigawa State, Nigeria.

# References

[1] R. A. Onoja, "Survey of Gross Alpha and Beta Radioactivity in well water from Zaria area," ABU Zaria, Zaria, 2004.

[2] A. Shittu, M. L. Madugu and S. T. Lawan, "Measurement of Gross Alpha and Beta Radioactivity in selected Local Water Boreholes in Northern Nigeria, Gombe State," *Nigerian Association of Mathematical Physics,* vol. 25(2), pp. 243-248, 2013.

[3] E. W. Mangset, L. D. Christopher, A. O. Solomon, E. E. Ike, S. P. Mallam and R. A. Onoja, "Gross Alpha and Beta Radio Activity Concentrations and Estimated Committed Effective dose to the General Public Due to Intake of Groundwater in Mining Areas of Plateau State, North Central Nigeria," *Advances in Physics Theories and Applications,* vol. 38, 2014.

[4] T. C. Akpa, S. P. Mallam, I. G. Ibeanu and R. A. Onoja, "Characteristics of Gross Alpha/Beta Proportional Counter," *Nigerian of Journal of Physics,* vol. 16(1), pp. 13-18, 2004.

[5] P. D. Nicholas, Elementary Statistics, 2006.

[6] E. Yusuf, G. Iliyasu, U. A. Danbatta and A. Y. Maitama, "Effect of Gross Alpha and Beta in Groundwater Intake and Estimation of Groundwater Table in Kano University of Science and Technology, Wudil," *American Scientific Research Journal for Engineering, Technology, and Sciences (ASRJETS),* vol. 14(1), pp. 46-54, 2015.

[7] A. Saidu and E. E. Ike, "Survey of Gross Alpha Radioactivity in Bore Hole and Well Water in Sokoto City North-Western Nigeria," *Nigerian Journal of Basic and Applied Science,* vol. 20(3), pp. 20-26, 2013.

[8] R. K. Meindinyo and E. O. Agbalagba, "Radioactivity concentration and heavy metal assessment of soil and water, in and around Imirigin oil field, Bayelsa state, Nigeria," *Journal of Environmental Chemistry and Ecotoxicology,* vol. 4(2), pp. 29-34, 2012.

[9] S. Turhan, E. Ozc¸ıtak, H. Taskın and A. Varinlioglu, "Determination of natural radioactivity by gross alpha and beta measurements in ground water samples," *Elsevier,* vol. 47, pp. 3 1 0 3 -3 1 0 8, 2013.

[10] X.-X. Miao, Y.-Q. Ji, X.-Z. Shao, H. Wang, Q.-F. Sun and X. Su, "Radioactivity of Drinking-Water in the Vicinity of Nuclear Power Plants in China Based on a Large-Scale Monitoring Study," *International Journal of Environmental Research and Public Health,* vol. 10, pp. 6863-6872, 2013.

[11] M. L. Garba, A. S. Arabi and D. J. Adeyemo, "Assessment of Gross Alpha and Beta Radioactivity in Groundwater by Liquid Scintillation," *Journal of Applied Environmental and Biological Sciences,* vol. 3(7), pp. 1-5, 2013.

# PERMISSIONS

All chapters in this book were first published in IJEMA, by Science Publishing Group; hereby published with permission under the Creative Commons Attribution License or equivalent. Every chapter published in this book has been scrutinized by our experts. Their significance has been extensively debated. The topics covered herein carry significant findings which will fuel the growth of the discipline. They may even be implemented as practical applications or may be referred to as a beginning point for another development.

The contributors of this book come from diverse backgrounds, making this book a truly international effort. This book will bring forth new frontiers with its revolutionizing research information and detailed analysis of the nascent developments around the world.

We would like to thank all the contributing authors for lending their expertise to make the book truly unique. They have played a crucial role in the development of this book. Without their invaluable contributions this book wouldn't have been possible. They have made vital efforts to compile up to date information on the varied aspects of this subject to make this book a valuable addition to the collection of many professionals and students.

This book was conceptualized with the vision of imparting up-to-date information and advanced data in this field. To ensure the same, a matchless editorial board was set up. Every individual on the board went through rigorous rounds of assessment to prove their worth. After which they invested a large part of their time researching and compiling the most relevant data for our readers.

The editorial board has been involved in producing this book since its inception. They have spent rigorous hours researching and exploring the diverse topics which have resulted in the successful publishing of this book. They have passed on their knowledge of decades through this book. To expedite this challenging task, the publisher supported the team at every step. A small team of assistant editors was also appointed to further simplify the editing procedure and attain best results for the readers.

Apart from the editorial board, the designing team has also invested a significant amount of their time in understanding the subject and creating the most relevant covers. They scrutinized every image to scout for the most suitable representation of the subject and create an appropriate cover for the book.

The publishing team has been an ardent support to the editorial, designing and production team. Their endless efforts to recruit the best for this project, has resulted in the accomplishment of this book. They are a veteran in the field of academics and their pool of knowledge is as vast as their experience in printing. Their expertise and guidance has proved useful at every step. Their uncompromising quality standards have made this book an exceptional effort. Their encouragement from time to time has been an inspiration for everyone.

The publisher and the editorial board hope that this book will prove to be a valuable piece of knowledge for researchers, students, practitioners and scholars across the globe.

# LIST OF CONTRIBUTORS

**Adhena Ayaliew Werkneh and Angaw Kelemework Abay**
Department of Chemistry, College of Natural Science, Jigjiga University, PO. Box: 1020, Jigjiga, Ethiopia

**Anbisa Muleta Senbeta**
Department of Food Science and Nutrition, College of Dryland Agriculture, Jigjiga University, PO. Box: 1020, Jigjiga, Ethiopia

**Arthur Mngoma Omari**
College of engineering and Technology, Mbeya University of Science and Technology, Mbeya Tanzania

**Geoffrey Reuben John nad Peter Lucas Mtui**
College of engineering and Technology, University of Dar es Salaam, Dar es Salaam, Tanzania

**Karoli Nicholas Njau**
School of Materials, Energy, Water and Environmental Sciences, Nelson Mandela African Institution of Science and Technology, Arusha Tanzania

**Tubonimi Joseph Kio Ideriah and Udonna Joseph Ikoro**
Institute of Pollution studies, Rivers State University of Science and Technology Nkpolu Oroworukwo Port Harcourt, Rivers State, Nigeria

**Alie Wube Dametew**
Department of Mechanical Engineering, and Research and technology Transfer Head, KIOT Wollo University, Dessie, Ethiopia

**Godson Rowland Ana, Olusola Olabisi Ogunseye**
Department of Environmental Health Sciences, Faculty of Public Health, College of Medicine, University of Ibadan, Ibadan, Nigeria

**Hossam Ismael**
Geography and GIS Department, Faculty of Arts., Assiut University, New Valley Branch, Egypt

**Eboua Narcisse Wandan**
Environmental Management Research Laboratory, Department of Water, Forestry and Environment, INP-HB, Yamoussoukro, Côte d'Ivoire

**Gnammytchet Barthelemy Koffi**
Civil Engineering Research Laboratory, Geosciences and Geographical Sciences, Department of Earth Sciences and Mineral Resources, INP-HB, Yamoussoukro, Côte d'Ivoire

**Kouakou Bob Kouadio**
Sustainable Forest Management Research Laboratory, Department of Water, Forestry and Environment, INP-HB, Yamoussoukro, Côte d'Ivoire

**Zoly Desiré Gomba**
Departmental Directorate of Water and Forests, Bouna, Côte d'Ivoire

**Emeka Udokporo and Kevin Ejike Chukwu**
1Department of Geography and Meteorology, Enugu State University of Science and Technology, Enugu, Nigeria

**Martin Atu Ngozika Anikwe**
Department of Agronomy and Ecological Management, Enugu State University of Science and Technology, Enugu, Nigeria

**Allian Majhi and Subhra Keshari Biswal**
Department of Chemistry, Indira Gandhi Institute of Technology, Denkanal, Odisha, India

**Kedar Mohapatra**
Department of Chemistry, Gandhi Institute of Technological Advancement, Bhubaneswar, Odisha, India

**Britta Jahnke**
Institute of Coastal Research / Molecular Recognition and Separation, Helmholtz-Zentrum Geesthacht, Geesthacht, Germany

**Bernd Niemeyer**
Institute of Thermodynamics, Helmut-Schmidt-University / University of the Federal Armed Forces Hamburg, Hamburg, Germany

**Amale Mcheik**
Laboratory of Biogeochemistry and Ecology of continental regions (IBIOS - BIOEMCO), Department of Biology and Environmental Sciences, Paris-Est University, Creteil, France
Laboratory of Materials, Catalysis, Environment and Analytical Methods (MCEMA), Department of Chemistry, Lebanese University, Beirut, Lebanon

**Mohamad Fakih, Hiba Noureddine, Joumana Toufaily and Taysir Hamieh**
Laboratory of Materials, Catalysis, Environment and Analytical Methods (MCEMA), Department of Chemistry, Lebanese University, Beirut, Lebanon

**Hussein Trabulsi**
Faculty of Economic Sciences and Business Administration, Lebanese University, Beirut, Lebanon

**Evelyne Garnier-Zarli and Noureddine Bousserrhine**
Laboratory of Biogeochemistry and Ecology of continental regions (IBIOS - BIOEMCO), Department of Biology and Environmental Sciences, Paris-Est University, Creteil, France

**Abdul Ali Juma Mohamed, Sara Abdalla Khamis, Haji Mwevura, Miza Ali Kombo and Said Suleiman Bakari**
Department of Natural Sciences, State University of Zanzibar (SUZA), Zanzibar, Tanzania

**Yoshiko Muto, Eiichi Kurashima, Kiyohito Yamamoto and Koji Harashina**
Faculty of Agriculture, Iwate University, Morioka, Japan

**Muhammad Shamsuddeen Abdullahi and Halil Zafer Alibaba**
Department of Architecture, Eastern Mediterranean University (E.M.U.), Gizamagusa, North Cyprus via Mersin 10, Turkey

**Yang Hui, Sun Yanfei and Wang Tiedong**
New Star Application Technology Institute, Hefei, China

**Zhao Xuesong**
Key Lab. of Environmental Optics & Technology, CAS, Hefei, China

**Hanaa S. M. Abd El-Rahman, Abeer M. N. H. El-Dakak and Zein H.**
Special Food and Nutrition Dept., Food Technology Research Institute, Agricultural Research Center, Giza, Egypt

**Huseyin Murat Cekirge**
Department of Mechanical Engineering, Prince Mohamed Bin Fahd University, Al Khobar, KSA

**Amale Mcheik**
Laboratory of Biogeochemistry and Ecology of continental regions (IBIOS - BIOEMCO), Department of Biology and Environmental Sciences, Paris-Est University, Creteil, France

Laboratory of Materials, Catalysis, Environment and Analytical Methods (MCEMA), Department of Chemistry, Lebanese University, Hadath, Beirut, Lebanon

**Mohamad Fakih, Joumana Toufaily and Taysir Hamieh**
Laboratory of Materials, Catalysis, Environment and Analytical Methods (MCEMA), Department of Chemistry, Lebanese University, Hadath, Beirut, Lebanon

**Hussein Trabulsi**
Faculty of Economic Sciences and Business Administration, Lebanese University, Beirut, Lebanon

**Evelyne Garnier-Zarli and Noureddine Bousserrhine**
Laboratory of Biogeochemistry and Ecology of continental regions (IBIOS - BIOEMCO), Department of Biology and Environmental Sciences, Paris-Est University, Creteil, France

**Bernard Mungoma Kuloba**
Kenya Wildlife Service, Biodiversity Research and Monitoring, Nairobi, Kenya

**Hein Van Gils and Iris Van Duren**
Faculty of Geo-Information Science and Earth Observation (ITC), University of Twente, Enschede, the Netherlands

**Shadrack Muvui Muya**
Department of Zoology, Jomo Kenyatta University of Agriculture and Technology, Nairobi, Kenya

**Shadrack Mumo Ngene**
Kenya Wildlife Service, Parks and Reserves, Nairobi, Kenya

**Mohammad Amin Qureshi and Ghulam Murtaza Mastoi**
Centre for Environmental Sciences, University of Sindh, Jamshoro, Sindh, Pakistan

**Zulfiqar Ali Laghari**
Department of Physiology, University of Sindh, Jamshoro, Pakistan

**Khalid Hussain Lashari and Zameer Ali Palh3**
Department of Fresh Water Biology and Fisheries, University of Sindh, Jamshoro, Pakistan

**Fouzia Panhwar**
Department of Pharmacology, University of Sindh, Jamshoro, Pakistan

**Shittu Abdullahi**
Department of Physics, Gombe State University, Gombe, Nigeria

**Chifu E. Ndikilar, Abdussalam B. Suleiman and Hafeez Y. Hafeez**
Department of Physics, Federal University Dutse, Dutse, Nigeria

# Index